Series Title: Frontiers in Clinical Drug Research – Alzheimer Disorders
(Volume 1)

Editor

Atta-ur-Rahman, *FRS*

Kings College
University of Cambridge
Cambridge
UK

Bentham Science Publishers
Executive Suite Y - 2
PO Box 7917, Saif Zone
Sharjah, U.A.E.
subscriptions@benthamscience.org

Bentham Science Publishers
P.O. Box 446
Oak Park, IL 60301-0446
USA
subscriptions@benthamscience.org

Bentham Science Publishers
P.O. Box 294
1400 AG Bussum
THE NETHERLANDS
subscriptions@benthamscience.org

CONTENTS

PREFACE

Alzheimer's disease (AD) is a disease that causes dementia, or loss of brain function. It affects the parts of the brain that deal with memory, thought, and language. The brain of a person with Alzheimer's contains abnormal clumps of cellular debris and protein (plaques) and collapsed microtubules (support structures of the cell).

The chapters in this volume are written by leading authors in their respective fields. There are several AD related topics in this introductory volume covering immunotherapy, AD diagnosis, symptomatic AD pathogenesis along with neurodegenerative effects, AD drug development models and therapy, and age related effects leading to AD.

In the first chapter of the eBook, Katharina Schindowski Zimmerman reviews the immunological basis for the mechanism of action of passive Aβ immunotherapy, anti-Aβ mAbs and scaffolds. In the next chapter, Wright and Harding review the approach of employing multi-target-directed ligands (MTDL) in the development of drugs to treat AD. The authors of this review demonstrate that there is a paradigm shift in drug screening, away from single lead-target models, and it could be potentially useful in treating AD. Chapter 3, written by Cavallaro and Fuso, summarizes current drugs used in AD therapy and new developments in therapeutic interventions in clinical phases. In the subsequent chapter Yohsiyama reviews the progress in tau-directed therapies. This work is significant from the standpoint of current clinical trials of therapies focusing on restricting beta – amyloid (Aβ) production in order to reduce Aβ pathogenesis and neurodegeneration in patients suffering from symptomatic AD. Asija provides a comprehensive insight into the different, currently available, approaches in the treatment of AD in chapter 5. The review presents findings of ongoing clinical trials and drug research studies in advanced stages of development and their application in the prevention and treatment of AD.

Physical activity and exercise is thought to play a positive role in improving the quality of life among patients suffering from brain disorders. Chapter 6 by Eef Hogervorst *et al.* takes on this perspective. The authors discuss factors that may influence the relationship between physical activity and cognitive functioning with the implications in theory theory and practice. Zheng *et al.* elaborate on some examples of neurogenic molecules which can reduce amyloid beta (Aβ) production, prevent oxidative stress and tau hyperphosphorylation, and restore metal dyshomeostasis while demonstrating neuroprotective/neurorestorative effects. These include ladostigil and rasiline.

The subsequent chapter addresses sleep apnea in the elderly and its role in hypoxia, hypoxemia, cerebral hypoperfusion and compromised blood supply, brain hypometabolism in aging and AD. Epidemiological data is employed in this study and the review points to an ever increasing need for researchers to address sleep disorders as the cost of healthcare for AD patients is rising. Chapter 9 of the eBook deals with various natural and synthetic drug therapies and also presents rational guidelines for the prevention of AD. Jellinger presents a timely review on the diagnosis of Alzheimer subtypes – dementias. Morphological dementia is incurable at present, but Jelinger proposes a revision of current diagnostic protocols and frameworks as a basis for early diagnosis and efficient new therapy options against dementias.

In the last chapter of this eBook, the role of various trace metals as cofactors of a poor cognitive function and of an increased risk of dementia leading to AD is reviewed. In particular, this study is a critical analysis of the role played by copper, iron, zinc and aluminum in neurodegeneration.

I hope that readers will find value in these reviews and draw inspiration for conducting further research in the quest for a cure for Alzheimer's disease and related disorders.

Atta-ur-Rahman, FRS
Kings College
University of Cambridge
Cambridge
United Kingdom

List of Contributors

Chrystelle Mavoungou

Institute of Applied Biotechnology, Faculty for Biotechnology, Biberach University of Applied Science, Karlstrasse 11, D-88400 Biberach/Riss, Germany

Katharina S. Zimmerman

Institute of Applied Biotechnology, Faculty for Biotechnology, Biberach University of Applied Science, Karlstrasse 11, D-88400 Biberach/Riss, Germany

John W. Wright

Departments of Psychology, Veterinary and Comparative Anatomy, Pharmacology, and Physiology, and Programs in Neuroscience and Biotechnology, Washington State University, Pullman, Washington, USA

Joseph W. Harding

Departments of Psychology, Veterinary and Comparative Anatomy, Pharmacology, and Physiology, and Programs in Neuroscience and Biotechnology, Washington State University, Pullman, Washington, USA

Rosaria A. Cavallaro

Department of Surgery "P. Valdoni" , Sapienza University of Rome, Italy

Andrea Fuso

Department of Psychology - Section of Neuroscience, Sapienza University of Rome, Italy

Yasumasa Yoshiyama

Clinical Research Center, Department of Neurology, Chiba East National Hospital, Chiba, Chiba 260-8712, Japan

Zaciragic Asija

Department of Physiology, Medical Faculty University of Sarajevo, Bosnia and Herzegovina

Angela Clifford

SSEHS Loughborough University and Center for Aging Studies Universitas Indonesia, Depok, Indonesia

Jennifer Stock

SSEHS Loughborough University and Center for Aging Studies Universitas Indonesia, Depok, Indonesia

Stephan Bandelow

SSEHS Loughborough University and Center for Aging Studies Universitas Indonesia, Depok, Indonesia

Tri B. Rahardjo

SSEHS Loughborough University and Center for Aging Studies Universitas Indonesia, Depok, Indonesia

Eef Hogervorst

SSEHS Loughborough University and Center for Aging Studies Universitas Indonesia, Depok, Indonesia

Hailin Zheng

Department of Medicinal Chemistry, Intra-cellular Therapies Inc. 3960 Broadway, New York, NY 10032, USA

Mati Fridkin

Department of Organic Chemistry, The Weizmann Institute of Science, Rehovot 76100, Israel

Moussa B.H. Youdim

Eve Topf and USA National Parkinson Foundation Centers of Excellence for Neurodegenerative Diseases and Department of Pharmacology, Technion-Rappaport Family Faculty of Medicine Haifa, 31096, Israel, and Department of Biology, Yonsei World Central University, Seoul, South Korea

Mak A. Daulatzai

Sleep Disorders Group, EEE/MSE, The University of Melbourne, Parkville, Victoria 3010, Australia

Chuang Wang

Ningbo University School of Medicine, 818 Fenghua Road, Ningbo, Zhejiang 315211, PR China

Rui Wang

Program of Pharmacology, Weill Graduate School of Medical Science, Cornell University, New York, NY 10021,USA

James M. O'Donnell

Department of Behavioral Medicine and Psychiatry, West Virginia University, Morgantown, WV 26508, USA

Ying Xu

Department of Behavioral Medicine and Psychiatry, West Virginia University, Morgantown, WV 26508, USA

Kurt A. Jellinger

Institute of Clinical Neurobiology, Vienna, Austria

Guido Crisponi

Department of Chemistry and Geology, University Campus, 09042 Montserrat Cagliari, Italy

Valeria M. Nurchi

Department of Chemistry and Geology, University Campus, 09042 Montserrat Cagliari, Italy

Daniela Fanni

Department of Surgical Sciences, Sec. Pathology, Via Ospedale 46, 09124 Cagliari, Italy

Clara Gerosa

Department of Surgical Sciences, Sec. Pathology, Via Ospedale 46, 09124 Cagliari, Italy

Sonia Nemolato

Department of Surgical Sciences, Sec. Pathology, Via Ospedale 46, 09124 Cagliari, Italy

Miriam Crespo-Alonso

Department of Chemistry and Geology, University Campus, 09042 Montserrat Cagliari, Italy

Leonardo Toso

Department of Chemistry and Geology, University Campus, 09042 Montserrat Cagliari, Italy

Joanna I. Lachowicz

Department of Chemistry and Geology, University Campus, 09042 Montserrat Cagliari, Italy

Gavino Faa

Department of Surgical Sciences, Sec. Pathology, Via Ospedale 46, 09124 Cagliari, Italy

Send Orders of Reprints at reprints@benthamscience.net
Frontiers in Clinical Drug Research – Alzheimer Disorders, Vol. 1, **2013, 3-85**

CHAPTER 1

Immunotherapy with Anti-Aβ Monoclonal Antibodies in Alzheimer's Disease: A Critical Review on the Molecules in the Pipelines with Regulatory Considerations

Chrystelle Mavoungou[1] and Katharina Schindowski Zimmerman[2,*]

[1]Professor of Pharmaceutical Quality and Regulations, Institute of Applied Biotechnology, Faculty for Biotechnology, Biberach University of Applied Science, Karlstrasse 11, D-88400 Biberach/Riss, Germany; [2]Professor of Molecular Pharmacology, Institute of Applied Biotechnology, Faculty for Biotechnology, Biberach University of Applied Science, Karlstrasse 11, D-88400 Biberach/Riss, Germany

Abstract: Alzheimer's disease (AD) is the most common dementia in the industrialized world with prevalence rates far over 30% in the over 80 years old population. The dementia causes enormous cost to the social health care systems besides the personal tragedies for the patients, families and caregivers.

One of the pathological protein aggregations that occur in AD is the Amyloid-beta (Aβ) aggregation in extracellular plaques, accompanied by Tau hyperphosphorylation, chronic neuroinflammation and oxidative stress, leading to severe neurodegeneration of brain areas involved in learning and memory. Plaques, thus Aβ, appeared to be the more druggable and promising target for disease-modifying therapeutic strategies like passive immunotherapy with monoclonal antibodies (mAbs) against Aβ, though today it is clear that Aβ is a dreadful target. Meanwhile, the first-in-class mAb Bapineuzumab and the fast-follower mAb Solanezumab failed in Phase III whereas several other candidates – some of them modified 2nd generation mAbs – now entered Phase I (PF-05236812, BAN2401, SAR228810 and BIIB037) and Phase II (Gantenerumab and Crenezumab) respectively. Others are known to be in preclinical stages. On the first view, the above-cited mAbs cleared or improved amyloid burden and validated the proposed Aβ read-out biomarkers, but have yet not shown relevant improvement in the major aim in AD therapy: cognition. Also, currently under Phase III investigation, are human IgG from healthy donors. The latest, so-called IVIG, are nowadays interestingly reported to stabilize cognition in AD patients.

In this review we discuss the immunological basis for the mechanism of action of passive Aβ immunotherapy, anti-Aβ mAbs and scaffolds in the pipelines and patents, their preclinical and clinical outcome and strategies for 2nd generation biobetters.

***Address correspondence to Katharina Schindowski Zimmermann**: Institute of Applied Biotechnology, Faculty for Biotechnology, Biberach University of Applied Science, Karlstrasse 11, D-88400 Biberach/Riss, Germany, Tel: +49 7351 582498; E-mail: zimmermann@hochschule-bc.de

Keywords: Passive immunization, dementia, therapeutic antibodies, effector function, Aβ-oligomers, ADDLs, protofibrils, regulatory strategy, market considerations, biobetters

"I fear I am not in my perfect mind.

Me thinks I should know you and know this man;

Yet, I am doubtful; for I am mainly ignorant

What place this is; and all the skill I have

Remembers not these garments; nor I know not

Where I did lodge last night. Do not laugh at me."

(William Shakespeare (1605) King Lear, Act IV, Scene 7.

Referred as first description of Alzheimer's disease)

ALZHEIMER'S DISEASE DEMENTIA, A GENERAL OVERVIEW

Epidemiology

Alzheimer's disease (AD) is the most common form of dementia. It accounts for 60-70% of all cases [1]. Prevalence and incidence projections indicate that the number of people with dementia will continue to grow, particularly among the oldest old, and countries in demographic transition will experience the greatest growth. According to the latest data from the world health organization, in 2010 the worldwide total number of people with dementia was estimated to be at 35.6 million. This number is projected to almost double every 20 years, to 65.7 million in 2030 and 115.4 million in 2050. Each year, the worldwide total number of new cases of dementia is estimated to almost 7.7 million, implying one new case every four seconds. In Western Europe, the prevalence for the 60 - 64 years old is 1.6, which is increasing up to 43.1 in the over 90 years old [2]. Furthermore, dementia shortens the lives of the demented patients: the median survival with AD is at 7.1 years [3]. Dementia is not only overwhelming for the people who are affected including their caregivers and families; it is one of the major causes of disability and dependency among older people worldwide. Therefore dementia should be considered a part of the public health agenda. The total estimated worldwide costs of dementia were US$ 604 billion in 2010. In high-income countries, informal care (45%) and formal social care (40%) account for the majority of costs, while the proportionate contribution of direct medical costs (15%) is much lower [2].

Research identifying modifiable risk factors of dementia is in its infancy. In the meantime, primary prevention should focus on targets suggested by current evidence. These include countering risk factors for vascular disease, including diabetes, midlife hypertension, midlife obesity, smoking, and physical inactivity [2].

Clinical Pathology

AD is a neurodegenerative disorder that is characterized by global cognitive decline including a progressive loss of memory, orientation, and reasoning. At the beginning of the last century, the neurologist and psychiatrist Alois Alzheimer described extensively a dementia syndrome of his patient Auguste Deter whom he had treated in Frankfurt am Main/Germany [4]. He recorded a rapidly progressing memory loss of the 52-year old women. After her death, he examined her brain and found histological changes consisting of misfolded proteins: extracellulary aggregated Amyloid-β (Aβ) in senile plaques and intracellularly hyperphosphorylated Tau in neurofibrillary tangles. Moreover, aggregated Aβ is also found in vascular plaques in cerebral blood vessels (also termed cerebral amyloid angiopathy, CAA).

Neurofibrillary Tangles

Neurofibrillary tangles consist of hyperphosphorylated Tau that aggregates intracellulary in neurites and finally in the soma. Tau belongs to the microtubule-associated protein family, predominantly found in neurons. Non-neuronal cells usually display trace amounts of Tau but in some diseases Tau accumulation can be detected in glial cells [5]. In a low phosphorylated state, Tau binds to microtubules through the microtubule-binding domains and stabilizes their polymerization and assembly. However, microtubule assembly depends somewhat upon the phosphorylation state. Phosphorylated Tau is less effective than non-phosphorylated Tau on microtubule polymerization. Phosphorylation inside and outside the microtubule-binding domains can strongly influence tubulin assembly by modifying the affinity between Tau and microtubules. However, properly assembled microtubules are essential to maintain axonal transport processes and hyper- or AD-specific abnormal phosphorylation of Tau causes an impairment of axonal transport [6, 7] and neuronal death [8].

Kinases involved in Tau phosphorylation include Glycogen synthase kinase-3β (GSK-3β), mitogen activated protein kinase, Tau-tubulin kinase, cyclin-dependent kinases and stress-activated protein kinases. Some phosphorylation- and conformation-dependent epitopes are specific for abnormal AD specific Tau-phosphorylation. Thus, antibodies against such epitopes could be of interest for an anti-Tau immunotherapy approach. There is a direct relationship between hyperphosphorylation, abnormal phosphorylation and Tau aggregation. During normal ageing Tau hyperphosphorylation occurs in the transentorhinal cortex and spreads from here *via* the entorhinal cortex to the hippocampus [9, 10]. Tau, however, is not only limited to AD, it can be found in several other dementia, the so-called tauopathies (for review see [11]).

Amyloid Plaques

The other prominent hallmark of AD is the accumulation and deposition of Aβ - a peptide of predominantly 39 to 43 residues; but also smaller species can occur. The aggregation of Aβ species is thought to play a pivotal role in neuronal loss or dysfunction through a cascade of events that include the generation of free radicals, mitochondrial oxidative damage and inflammatory processes. The primary event that results in the abnormal accumulation of Aβ is the dysregulated proteolytic processing of its parent molecule, the amyloid precursor protein (APP) [12]. APP is a transmembrane glycoprotein that is proteolytically processed by two competing pathways, the non-amyloidogenic and amyloidogenic (Aβ forming) pathway. Three major secretases are postulated to be involved in the proteolytic cleavage of APP. These include α-secretase, which is the metalloprotease ADAM 10, (but ADAM17/TACE is also capable of processing APP; [13, 14]), beta APP cleaving enzyme (BACE, formally known as β-secretase) and the γ-secretase complex. The α-secretase cleaves within the Aβ domain of APP thus preventing the formation Aβ and generating the membrane-bound intracellular C-terminal fragment C83 and sAPPα. In the non-amyloidogenic pathway C83 is broken down from the γ-secretase to the APP intracellular domain (AICD, a transcription factor) and the extracellular p3 peptide. In the amyloidogenic pathway, BACE cleaves near the N-terminus of the Aβ domain on the APP molecule, liberating another soluble form of APP, sAPPβ and a membrane bound C-terminal fragment (C99) containing the whole Aβ domain. The last step in the amyloidogenic pathway is the intramembranous cleavage of the C99 fragment by γ-secretase, to liberate a number of Aβ isoforms of 39-43 amino acids and AICD [15]. The most common isoforms are $A\beta_{40}$ and $A\beta_{42}$; the shorter form is typically produced by cleavage that occurs in the endoplasmic reticulum, while the longer form is produced by cleavage in the trans-Golgi network. The $A\beta_{40}$ is the more common of the two, but $A\beta_{42}$ is - due to its more hydrophobic nature - the more fibrillogenic.

Even though extensive data support a central pathogenic role for Aβ in AD, the amyloid hypothesis assuming that increased production and accumulation of Aβ leads to all other observed neuropathologies of AD [16-18] still remains controversial. Relative weak correlations between fibrillar plaque density and severity of dementia are found in Alzheimer's diseased brains, whereas correlations between soluble Aβ levels and the extent of synaptic loss and cognitive impairment are somewhat stronger. Aβ oligomers appeared to be the missing link in Aβ toxicity, being more toxic than the mature fibrils [19]. Those oligomers were identified to impair cognition and disrupt synaptic plasticity [20]. Immunodepletion with antibodies against the free C-terminus (2G3 and 21F12) could completely abrogate this effect. In addition, pre-treatment of the medium with insulin-degrading enzyme (IDE), which is believed to degrade Aβ monomers but not oligomers, did not prevent the inhibition of long term potentiation (LTP, needed for learning and memory) by Aβ oligomers. $A\beta_{42}$ aggregation into oligomers takes place within endosomal vesicles and along microtubules [21]. Subsequently, anterograde axonal transport delivers Aβ to plaques [22, 23].

Autosomal-dominant mutations in APP cause hereditary early-onset AD, likely as a result of altered proteolytic processing. Increase in either total Aβ levels or the relative concentration of both $A\beta_{40}$ and $A\beta_{42}$ has been implicated in the pathogenesis of both familial and sporadic AD [24].

The presence of CAA is often found in AD brains, but it is hard to establish whether its appearance is related to the initial manifestations of the disorder or is a process that occurs later during the development of the disease.

Neuroinflammation

Very early in the pathogenesis, the extracellular deposits and the neuronal cell death in AD brains trigger an activation of inflammatory processes predominantly of the innate immune system (for review see [25]). Interestingly, polymorphisms in cytokines and other immune-related molecules is linked to AD, and epidemiological studies indicated that use of nonsteroidal anti-inflammatory drugs (NSAIDs) reduces the risk for developing AD. The severity of the neuroinflammatory process is dependent from different levels of Aβ, Tau and ubiquitin and the presence or absence of the *apolipoprotein E4* (*ApoE4*) allele [26]. The brain is said to be an immune-privileged region, *i.e.* immunity is strongly reduced. In healthy brains the expression and presentation of major histocompatibility complex (MHC) class II in glial cells is minimized. In contrast, in AD this receptor complex is up-regulated in microglia surrounding amyloid plaques [27, 28] and expression of cytokines, chemokines, complement proteins and other inflammatory proteins is elevated [29, 30].

Epidemiological analysis revealed that persons with arthritis using anti-inflammatory NSAIDs have a lower incidence for AD [31]. Unfortunately, the results from several studies using NSAIDs to treat AD are puzzling and discouraging [32]. However, probably here again it was a matter of trying to treat pathology at late stages that means when the cognitive decline is already difficult to halt. PPAR-γ that is activated by NSAIDs is believed to mediate the positive effects of long-term NSAID-treatment and many studies currently focus on pharmaceutical manipulation of PPAR-γ.

Since neuroinflammation appear to trigger cognitive impairment, anti-inflammatory treatment is potentially beneficial on the cognitive outcome. Unfortunately, a recent study using Interferon-1β (Rebif) in patients with mild to moderate AD could not confirm a benefit for treated patients [33]. By contrast, intraspinal treatment with an Fc-fusion of the tumor necrosis factor (TNF)-α-receptor (Etanercept) to capture pro-inflammatory cerebral TNF-α revealed possible promising results in a pilot study [34], which have to be confirmed by ongoing trials.

AD Pathogenesis and Cognitive Impairment

The molecular mechanisms causing neuronal loss in AD are still not fully elucidated, but well-known is the specific neurodegeneration of cholinergic neurons leading to a loss of cholinergic transmission. The loss of cholinergic neurons seems to be specifically associated with typical clinical symptoms, like memory deficits, impaired attention, cognitive decline, and reduced learning abilities [35, 36]. Symptomatic treatment with acetylcholinesterases inhibitors like Donezepil support cholinergic transmission and can delay the progression of AD by 1 –2 years. More recently, cholinergic effects have been proposed as a potential causative agent for the formation of plaques and tangles [37].

According to data from *in vitro* and animal studies, Aβ oligomers induce Tau-phosphorylation and inactivate the insulin receptor substrate (IRS-1) linking diabetes mellitus type 2 to AD (see diabetes type 3; [38]). This triggers a complex pathological cascade believed to lead to neurodegeneration and a loss of memory and cognition [39, 40]. In addition, impaired axonal transport is considered to be a major cause for the symptoms and pathology observed in AD and other neurodegenerative dementias [41, 42]. Oligomeric $A\beta_{42}$ in microtubule-associated endosomal vesicles impairs axonal transport [21, 43, 44] by activating the casein kinase 2 [45]. In addition, hyperphosphorylated and aggregated Tau impairs axonal transport in mouse models [7, 46-48], invertebrate models [49-51] and cellular models [52-54]. So, oligomers and hyperphosphorylated Tau are playing the same fatal game: disrupting axonal transport [55]. To understand why this is so critical one should consider the role of axons and neuronal projections within the brain. Protein synthesis takes place in the cell bodies and the cytoskeleton components like microtubules function as rail tracks to transport proteins but also mitochondrias along the axons far away to the nerve terminals far away [56]. Neuronal transmission and synaptic plasticity in the nerve terminals are highly energy consuming events. Hence, our brains have the highest metabolic rate relative to their weight [57]. A loss of mitochondria and thereby a loss of energy at the synapses is therefore a pivotal and dramatic scenario for cognition processes. In addition, some transported proteins are more critical for neuronal function and survival than others, *e.g.* neurotrophic factors like nerve growth factor (NGF), brain-derived neurotrophic factor (BDNF), neurotrophin-3 and -4

(NT3, NT4) and its receptors (*e.g.* TrkA and TrkB). Cholinergic neurons are dependent on the supply of neurotrophic factors since they do not produce them by their own [58, 59], thus the early cholinergic loss in AD is presumably a cause of deficits in axonal transport (for review see [42]). Though, cholinergic transmission plays an important role in cognitive deterioration in AD, many other neurotransmitter systems like glutamate are affected.

Genetic Predisposition and Lifestyle Risk Factors for AD

Though most AD cases are sporadic, there are some genetic determinants that cover more than 99% of all sporadic late onsets AD cases. These include polymorphism of ApoE (see below), the receptor for advanced glycation end-products (RAGE, [60]), the nicotinic acteylcholine receptor, nAChR; [61], BDNF [62], NGF [63], Interleukin 1α and 6 (IL1α, IL6; [64, 65]), α2-Macroglobulin [66], and interestingly for Aβ itself [67-69].

ApoE

Well recognized and doubtless is the *ApoE* polymorphism that is also a risk for cardiovascular diseases [70]. ApoE is a member of the class of apolipoproteins found on chylomicrons and binds the LDL receptor-related protein (LRP or scavenger receptor). The three major isoforms, ApoE2, ApoE3, ApoE4 have different risks for developing AD: Homozygous *ApoE4* carrier have for example a 10 - 30 times higher risk for developing AD by 75 years compared to non-*ApoE4* carriers. *ApoE3* is considered to have a 'normal' risk since it is the most common genotype. *ApoE2* in contrast is considered to be rather protective. Interestingly, ApoE is also thought to play a prominent role in the deposition of Aβ peptides in cerebral vessels [71]. *ApoE* may confer some of its risk for the disease by modulating immune responses and thereby neuroinflammation. It is involved in the presentation of lipid antigens *via* MHC class I–like CD1 molecules [72] and appears to play a role in extracellular Aβ clearance [73]. For reviewing the genetic polymorphism in AD see [74].

AD as Diabetes Type 3

Insulin, insulin-like growth factor type 1 and type 2 (IGF-1 and IGF2), and their corresponding receptors are abundantly expressed in neuronal and glial cells throughout the brain [75]. Insulin and IGFs modulate cerebral blood flow and energy metabolism. They regulate learning and memory by influencing neuronal growth, survival, differentiation, migration, metabolism, gene expression, protein synthesis, cytoskeletal assembly, synapse formation and plasticity [76]; for review see [77].

Patients with type 2 diabetes mellitus have an increased risk for cognitive dysfunction and AD [78]. Although the exact pathogenesis is not yet understood, it is becoming more and more evident that there must be several mechanisms, including hyperglycemia, insulin resistance, microvascular disease, glucose toxicity, oxidative stress, inflammation and atherosclerosis, through which type 2 diabetes may modulate brain function. Any disturbance in insulin signaling appears to have negative effects on memory [79]. Moreover, insulin increases the extracellular concentration of Aβ by inhibiting IDE responsible for the extracellular degradation of Aβ [80]. Critically, Aβ inhibits insulin degradation in a dose-dependent manner [81] causing a vicious circle.

GSK3 is a constitutively active kinase that plays a role in glycogen metabolism, gene transcription and, interestingly, also memory and cognition [82]. GSK3 plays a central role in the pathogenesis of AD and over-activity of GSK3 accounts for memory impairment, Tau hyperphosphorylation, increased Aβ production, reduced acetylcholine synthesis and neuroinflammation [83]. Epidemiological and genetic studies suggest that GSK3 is dysregulated in AD for example through differences in the insulin signalling pathway, implying a direct association of AD with diabetes and insulin resistance [78].

Peripheral insulin resistance appears to affect CNS insulin levels, cognition and Aβ levels [84]. Peripheral insulin resistance downregulates functional insulin receptors at the blood brain barrier that transport insulin to the brain. Abnormalities in insulin, IGF1 and IGF2 signaling mechanisms have been reported in brains of AD patients [85]. Moreover, sedentary is a high risk factor for AD [86] and exercise decreases the chance for developing a dementia [87]. Therefore, AD is also termed type 3 diabetes that shares molecular and biochemical features with both type 1 diabetes mellitus and type 2 diabetes mellitus [88].

Clinical Biomarkers

Well-documented clinical biomarkers for AD in cerebrospinal fluid (CSF) include alterations in $A\beta_{42}$, total-Tau or phospho-Tau or an alteration of these three, detectable in early dementia stages as well as in individuals with mild cognitive impairment (MCI) who are at high risk of conversion to AD [89]. These CSF biomarkers generally allow the diagnosis of AD and even the prediction of the conversion from MCI to AD with a specificity and sensitivity of about 85%.

In view of clarifying the question if the use of CSF biomarkers (among others) may contribute to the diagnosis of AD or enhance the odds to find a appropriate diagnosis, the Scientific Advice Working Party (SAWP)/ CHMP (The Committee for Medicinal Products for Human Use) of the EMA came to the following conclusion: The position of the CHMP would depend on the benefit/risk profile of a given medication as regards the indication statement at the time of marketing authorization procedure. Nevertheless, it is believed that CSF biomarkers based on a low $A\beta_{42}$ and high total-Tau are qualified and therefore appropriate to accurately design AD clinical trial. In addition, they can be used to identify patients with clinical diagnosis of mild to moderate AD who are at greater risk to have an underlying neuropathology of AD. It is more likely that patients with the above-mentioned pathologic CSF signature have underlying neuropathology of AD. They were supposed to benefit from amyloid-modulating therapies. Finally, the CHMP does not see any basis for qualification of the said CSF biomarkers as diagnostic tool unless there are used for accurately design AD clinical trials [90]. This conclusion is based on a data assessment from 2001 until 2011 used for qualification of the above-mentioned biomarkers, on request of the applicant Bristol-Myers Squibb International Corporation.

Nowadays, neuroimaging is used for functional imaging with positron emission tomography (PET), determination of structural/volume changes with magnetic resonance imaging (MRI) or computed tomography (CT) and the assessment of amyloid burden with the radiotracer Pittsburgh compound B (PIB) during a PET scan in AD patients *in vivo* [91].

Given the long pathogenesis process of AD, any AD therapeutics will be hypothetically more effective in the treatment of preclinical patients at risk for AD but with minimal or no cognitive or functional deficits than in established AD patients. Therefore, biomarker strategies are more and more entering clinical AD trials and have an important impact on the outcome of the studies although, no biomarker has yet been proven to be a successful predictor for AD therapy.

Therapeutic Treatment with Aβ Immunotherapy

Although AD has been known about for over a century, there is still only symptomatic treatment available on the market. Symptomatic therapy strategies aim on increasing synaptic acetylcholine levels (like Donezepil) or decreasing glutamate toxicity (like Memantine), while disease modifying strategies (like possibly Aβ reducing therapies) should delay or stop the progression of the disease *e.g.* the loss of cognition.

More than 10 years ago active immunotherapy using aggregated, full-length $A\beta_{1-42}$ was shown to be effective in AD animal models lowering cerebral Aβ levels and improving cognitive deficits especially if given early in the disease process [92]. These observations initiated a clinical trial (AN-1792; NCT00021723) in which patients with mild to moderate AD were vaccinated with $A\beta_{1-42}$ and the adjuvant QS 21. However, aging is linked to a reduced immune response [93-95]. Thus, active immunization induces less protective immunity in elderly than in young individuals [96]. A large percentage of elderly humans did not generate a robust anti-Aβ antibody response following active Aβ immunization comparable to what was known before from mice studies [97]. In addition, active vaccination in aging individuals risks an increase in autoimmune responses despite the lower immune response to the foreign antigen [98].

The Phase IIa trial of active $A\beta_{42}$ immunization was halted after 6% of the patients developed meningoencephalitis [99]. The exact cause of these adverse events had been speculated to be an autoimmune response to $A\beta_{42}$, which appear to have enhanced the strong Th1, cytotoxic lymphocyte-inducing adjuvant QS-21 [100] used in this study. A small number of T cell infiltrates were observed in the *post mortem* brain tissues of all autopsied patients showing that in the patients without clinical symptoms, the T cells were restricted to the leptomeninges. One study described a vaccinated patient with perivascular distributed $CD4^+$ T cells [101]. While neither the function nor the specificity of the infiltrating T cells was determined, it could be possible that the age-associated tendency to generate autoimmune reactions led these patients to generate autoreactive $CD4^+$ T cells that entered the brain and contributed to encephalitis [96]. So, apparently it was not clear before going into a clinical trial, that a Th2 (humoral antibody-mediated) rather than a Th1 reaction (cellular cytotoxic) would be needed to reduce Aβ pathology.

However, *post mortem* analysis of patients that died during the course of the study demonstrated a dramatic reduction in amyloid load and other AD-associated histopathology, encouraging the scientific community to continue with Aβ immunotherapy. Consequently, passive immunotherapy was considered as safer and more controllable than active immunization. The clear advantage of passive immunization is that therapy can be discounted if severe adverse effects occur while an active vaccination cannot be revoked in a single patient. Passive immunization with anti-Aβ monoclonal antibodies (mAbs) was demonstrated to be effective in clearing Aβ plaques in AD animal models [102]. Moreover, passive immunization also reversed cognitive deficits and memory loss in APP transgenic mice [103].

For the passive immunotherapy strategy the Irish company Elan humanized a well-known murine research antibody named 3D6 and developed this molecule as the frontrunner for passive anti-Aβ approaches [104, 105]. 3D6 or its humanized form Bapineuzumab recognize the N-terminus of Aβ and binds all species of Aβ (*e.g.* soluble, oligomers, fibrils, aggregates and plaques.)

Meanwhile, Eli Lilly & Company developed a fast follower also from a murine research antibody named m266 by humanizing it (Solanezumab; [106]). m266 recognizes an epitope in the center of Aβ and therefore does not bind any aggregated species but the soluble Aβ. This was a courageous decision since the mode of action of Aβ immunotherapy was and appears still to be a matter of debates (see mechanism of action).

Actually, there are many more antibodies and non-antibody scaffolds against Aβ in the pipelines of the pharmaceutical companies. So, before going into the details of passive immunotherapy, let's first have a look on the target or antigen itself being recognized by the monoclonal anti-Aβ antibodies: the peptide Aβ.

Aβ Species and Epitopes as Targets - The Good, the Bad, and the Ugly

Aβ exists not only as monomer or as full length peptide: there is a bundle of different Aβ species that have not only differential aggregating activities, but also form completely different kind of aggregates and posttranslational modification. So evidently, it is not only important which epitope to choose in monomeric Aβ, it is also important to make the right choice for an Aβ species or a conformational (*i.e.* related to the three-dimensional structure) epitope since the different aggregates have different conformational structures.

It should be kept in mind, that although there is a strong suspicious it is still not clear which species cause AD, if at all amyloid is the entity that causes AD.

EPITOPES IN MONOMERIC AB

It is hypothesized that the immunogen used in AN-1742, full-length $A\beta_{1\text{-}42}$, has led to an autoimmune response, mainly triggered by autoreactive T cells against the C-terminus of the peptide. It is believed that immunization with the extracellular N-terminus $A\beta_{1\text{-}15}$ targets Aβ-specific B cell epitopes while $A\beta_{16\text{-}42}$ appears to produce Aβ-specific T cell epitopes [107]. Thus, choosing different parts of the peptide for active immunization induces differential kinds of immune responses. But even for passive immunization, the choice of the epitope is crucial: the N-terminus is freely accessible even in aggregated forms of Aβ while the central region is only available in monomeric not aggregated Aβ. Passive immunization with anti C-terminal Aβ mAb is reported with variable results. Treatment with 16C11 against $A\beta_{33\text{–}42}$ failed to lower plaque burden and did not improve cognitive deficits [108]. Treatment with 2286, an IgG_1 anti C-terminal Aβ mAb against $A\beta_{28\text{–}40}$ showed an improvement in cognition, a reduction in amyloid deposits and microglial activation [109].

N-TERMINAL SPECIES AND PYRO-GLU

N-terminally truncated Aβ peptides are much more abundant than the full-size secretase generated Aβ. Those species are present in early and diffuse plaques in AD and their amount seems to be directly correlated to the severity and the progression of the disease [110]. Significant quantities of $A\beta_{11\text{-}42}$ were found in

Down Syndrome's brains, a pathology with high APP expression and early onset of AD. In an AD mouse model similar observations were made concluding a crucial role of N-terminal fragmented Aβ species [111].

N-terminally truncated and pyro-Glu modified $A\beta_{3\text{-}x}$ (pE3-Aβ) is formed upon removal of the first 2 N-terminal residues of Aβ followed by cyclization by glutaminyl cyclase to convert the third residue, glutamic acid, to pyro-Glu [112, 113]. pE3-Aβ is more hydrophobic than Aβ, which accelerates its aggregation. In addition, pE3-Aβ resists degradation, favoring the formation of stable, neurotoxic aggregates [114]. pE3-$A\beta_{40/42}$ represent the prominent form among the N-truncated species, and may account for more than 50% of Aβ accumulated in plaques. Comparing the toxic properties, fibrillogenic capabilities and *in vitro* degradation profile of $A\beta_{1\text{–}40}$, $A\beta_{1\text{–}42}$, pE3-$A\beta_{40}$ and pE3-$A\beta_{42}$ it was shown that fiber morphology of Aβ peptides is greatly influenced by the C-terminus while toxicity, interaction with cell membranes and degradation are influenced by the N-terminus. pE3-$A\beta_{40}$ induced significantly more cell loss than the other species both in neuronal and glial cell cultures. While pE11-$A\beta_{42}$-modified Aβ species exist predominantly in mature plaques in AD, both unmodified free $A\beta_{11\text{-}40}$ and pyro-Glu modified $A\beta_{11\text{-}40}$ are detected in vascular amyloid deposits [115]. Mouse models that produce pE3-$A\beta_{42}$ show neuronal loss, age-dependent behavioral deficits and pE3-$A\beta_{42}$ accumulation [116]. Interestingly, there is a reported correlation between pyro-Glu modified Aβ forms and severity of AD, implying a pivotal role in AD pathogenesis and making these Aβ species a target for immunotherapy.

Very recently, a mechanism was reported by which pE-Aβ may trigger AD pathogenesis. pE3-$A\beta_{42}$ co-oligomerizes with excess $A\beta_{1\text{–}42}$ to form metastable low-n oligomers that are structurally distinct and far more cytotoxic to cultured neurons than comparable low-n oligomers made from $A\beta_{1\text{–}42}$ alone. Tau is required for cytotoxicity and low-n oligomers comprising of 5% pE3-$A\beta_{42}$ plus 95% $A\beta_{1\text{–}42}$ seed in the absence of additional pE3-$A\beta_{42}$ new cytotoxic low-n oligomers through multiple serial dilutions into $A\beta_{1\text{–}42}$ monomers. Low-n oligomers isolated from human AD brain contained pE3-$A\beta_{42}$ and enhanced its formation in young mice. In addition, neuronal loss and gliosis was triggered, but only in the presence of Tau, concluding that pE3-$A\beta_{42}$ confers Tau-dependent neuronal death and causes template-induced misfolding of $A\beta_{1-42}$ into structurally distinct low-n oligomers that propagate by a prion-like mechanism [117].

Interestingly, the level of IgM antibodies against pE3-Aβ was significantly decreased in AD patients as compared to healthy controls. A significant positive correlation between pE3-Aβ-IgM and cognitive decline was observed in a group of MCI patients [118].

Innogenetics and Sanofi share a patent [110] for the diagnostic and therapeutic use of mAbs against N-terminal species and Probiodrug holds a diagnostic patent for antibodies against pyro-Glu modified Aβ species [119].

OLIGOMERS

A large and confusing body of literature describes many types of assembly forms of synthetic Aβ [120], and not too much is known, which of these also occur *in vivo*. In general, soluble oligomers are defined as Aβ assemblies that are not pelleted from physiological fluids by high-speed centrifugation,

According to recent Solid-State NMR (Nuclear Magnetic Resonance Spectroscopy) studies of Aβ oligomerization, the Aβ residues 12–24 and 30–40 adopt β-strand conformations and form parallel or anti-parallel β-sheets through intermolecular hydrogen bonding [67, 121]. This process is followed by oligomerization of the Aβ peptides. After fibrillization, the C-terminal part of the Aβ peptide is masked in the most conformations.

Natural oligomers, principally dimers, are sufficient to induce Tau hyperphosphorylation at AD-relevant epitopes and then disrupt the microtubule cytoskeleton and cause neuritic dystrophy. Soluble dimers from the human brain appear to have a conformation that is highly potent in inducing these neuronal changes, as concentrations in the subnanomolar range caused cytoskeletal collapse, whereas levels of pure synthetic dimers at least two orders of magnitude higher were needed to produce otherwise indistinguishable effects in the same experiments [122]. In addition, soluble Aβ oligomers, mainly trimers and dimers, interfere with memory and cognition in animal models without inducing permanent neurological deficits [123-126]. In some studies natural oligomers isolated directly from the cortex of typical, late-onset AD patients were used [127]. Significantly higher levels of Aβ dimer (but also monomer) were found in the blood of AD subjects and their levels correlated with clinical markers of AD including brain amyloid burden, cognitive impairment and brain atrophy [128]. Dimers alone are capable to impair cognition in animal studies [125]. Thus, preventing or reversing the formation of aggregated amyloid would appear to be a promising strategy for AD treatment [129].

Abbott's antibody A-887755 binds oligomeric Abeta only, without recognizing monomeric or fibrillar species [39, 130-134]. Chronic treatment of an APP transgenic mouse model starting at the age when cognitive deficits occur but before plaque pathology starts (which is – to be honest - not comparable to human AD) revealed improved synaptic plasticity and memory.

PROTOFIBRILS

Protofibrils are intermediates that were observed in the course of studying the fibrillization of synthetic Aβ [120]. Synthetic Aβ protofibrils have been shown to contain substantial β-sheet structure. Unfortunately, many reports do not distinguish between higher oligomers and protofibrils, *i.e.* the term protofibrils is used whenever Aβ oligomerization occurred *in vitro*. Some authors refer to protofibrils if more than 4 Aβ peptides are fibrillated others for soluble high molecular weight oligomers. Therefore, mAbs that were reported to bind protofibrils are often also reported to bind oligomers.

Mechanism of Action of Anti-Aβ mAbs or Scaffolds

The mechanistic of monoclonal antibodies cannot be explained without having a closer look on the immune system since antibody therapeutics have a dual mode of action (Fig. **1**).

The first mode of action is the capture of the target/antigen. This can be already a therapeutic effect like blocking TNF-α activity in rheumatoid arthritis with mAbs (Adalimumab or Infliximab) or with a TNF-α-binding domain fused to an Fc-domain (Etanercept). The therapeutic effect is achieved by blocking/antagonistic or agonistic interaction with the pharmaceutical target. A direct interaction of the target with the immune system is not wanted or needed. The Fc-domain is not necessarily needed and can be omitted by using Fc-free fragments and scaffolds unless it is needed for increased half-life *e.g.* for chronic treatment. Typical indication could be for diagnostic purposes [135].

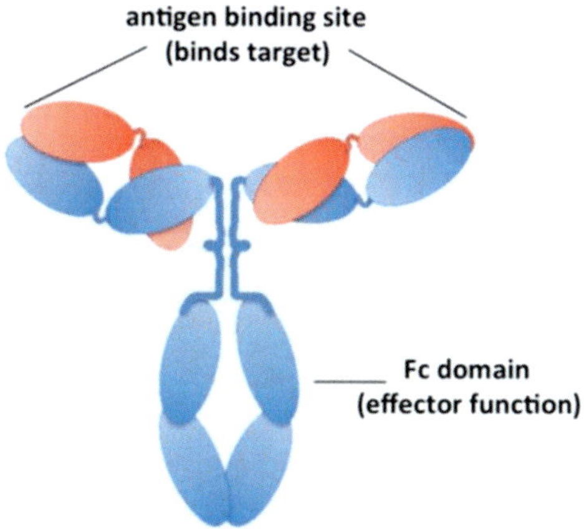

Figure 1: An antibody combines target binding with effector function. A regular IgG antibody has two antigen-binding sites and an Fc domain that mediates different effector functions (*e.g.* opsonisation, complement activation *etc.*) depending on the antibody's subtype.

A big advantage of antibodies is their high specifity and affinity towards their antigen. As antibodies have two binding sites, the combination of the two leads to avidity with a much higher binding of the antigen. In the case of Aβ, avidity with a monospecific mAb can only be achieved for oligomeric or fibrillar Aβ since there are no repetitive structures in the Aβ monomer. To achieve avidity for Aβ monomer a combination of two epitopes is needed like with a bispecific antibody or with an antibody fragment combining two epitopes [136]. Genentech developed recently a bispecific trojan antibody with one binding site to target the transferrin receptor for being transcytosed to the brain lumen and the other binding site against the catalytic site of BACE1 inhibiting Aβ processing [137].

The second action is the effector function linked to the Fc domain, that is widely recognized by several immune cells with Fc receptor (FcR) or the complement system that execute effector function once the antibody has bound its target. The most relevant effector functions for IgGs are opsination/phagocytosis, ADCC, CDC and feedback inhibition (see Box below). The therapeutic effect of the IgG's effector function is desired for most oncological targets (*e.g.* Transtuzumab or Rituximab) to destroy the cancer cells and to overcome the immune escape of the tumor. These function are mostly destructive killer functions and trigger inflammation. Only the feedback inhibition acts anti-inflammatory, however, the interaction with the responsible FcRIIB is rather weak for IgG_1 and it occurs only at high mAb concentrations or at a stronger polyclonal immune response.

Box: AD-relevant effector function of IgG antibodies (for details see [138] or [139]):

<u>Involving the antigen binding site</u>

- neutralizing/binding the target (all Ig classes).

→ this interaction can either be neutral, agonistic or antagonistic concerning the target's activity. Neutral binding can cause steric hindrance and thereby acting as antagonist.

<u>Involving the Fc domain</u>

- Opsonization for phagocytosis by macrophages/microglia and neutrophils

→ FcγRI (CD64) on macrophages, neutrophils, eosinophils and dendritic cells binds monomeric Fcγ-domain of IgG and leads through an Immunoreceptor tyrosine-based activation motif (ITAM) domain to the activation of phagocytosis and an inflammatory response. Crosslinking FcγRIIA (CD32) on macrophages, neutrophils, eosinophils and platelets by Fcγ/IgG leads to a less effective phagocytosis (IgG and IgM only; also see Fig. **2**).

- ADCC (antibody dependent cellular cytotoxicity) by NK cells

→ NK cells (and some other cell types) can bind antibody-coated cells *via* the FcγRIII (CD16) and destroy these cells if they are sensitive to perforin and granzyme. This interaction needs crosslinking of the Fc receptor and is well known from antibody therapeutics against cancer (IgG only).

- Complement-dependent cytotoxicity (CDC)

→ antigen-binding induces an allosteric shift in the antibody's structure that enables complement factors (C1q) to bind the antibody-antigen complex and to trigger a classical complement pathway activation. The final step of this pathway is the cell lysis of the antibody-binding structure through the membrane attack complex (MAC; see also Fig. **4**; IgG and IgM).

- **Feedback inhibition**

→ Crosslinking of FcγRIIB on B cells, mast cells, macrophages, neutrophils and eosinophils counterbalances activation by inhibiting cellular activity through an immunoreceptor tyrosine-based inhibition motif (ITIM) domain. For B cells this crosslink can occur with the B cell receptor and reduces antibody production [140] (IgG only).

- Mast cell degranulation (IgE only) and mucosal immunity (IgA only) are not referred here since they are not implicated in IgG effector functions. Up to now, all full immunoglobulins used for AD immunotherapy are IgGs.

Other Fc-function for IgG related to neonatal Fc receptor (FcRn) [141]

- **Fc-recycling**

FcRn protects Fc-bearing molecules from intracellular catabolism and causes a half-life of three weeks for IgG_1 and IgG_4.

- **FcRn mediated transport**

Outside the brain, crossing the placenta, uptake from the gut.

It should be noted that all interaction with Fc receptors depend on Fc glycolysation and production of mAbs in non-human non-glycoengineered cells produces molecules with altered interaction with Fc receptor.

But how can antibodies against Aβ interfere with AD pathology? Despite the rapid advance of this therapeutic strategy into clinical trials and the hundreds of research papers, that had been done, there remain some enigmatic aspects to Aβ immunotherapy [25]. Most importantly is the mechanism of action that is still not elucidated in mice, let alone in humans, although many hypotheses have been proposed including microglia-mediated phagocytosis, antibody-mediated alterations of Aβ aggregation and neutralization of Aβ toxicity, peripheral sink, and intracerebral sequestration of Aβ in a monomeric state [142]. Finally, it appears that more than only one of these processes takes place in passive Aβ vaccination.

MICROGLIA-MEDIATED PHAGOCYTOSIS/CENTRAL SINK

Peripherally administered anti-Aβ antibodies may cross the blood brain barrier, bind to Aβ in plaques and trigger FcR-mediated microglial phagocytosis (Fig. **2**; [108, 143]. Active vaccination with $A\beta_{42}$ in the AN-1742 trial [101, 144] showed evidence that the reduction of Aβ deposits was associated with microglial phagocytosis. The results of numerous other active and passive Aβ immunotherapy studies in animal models suggest that FcR-mediated activation of microglia may be a central mechanism of reducing Aβ load in the brain. But clearance of Aβ deposits was also observed after $A\beta_{42}$ active immunization in APP transgenic mice lacking the FcγR ($Fc\gamma R^{-/-}$; [145]) or in APP transgenic mice treated with a 3D6 $F(ab)_2$ (a 3D6 antibody without Fc domain [146, 147], thus FcR-mediated phagocytosis is apparently not the only mechanism involved in antibody mediated plaque removal.

Not to forget, that Aβ is not only present in cerberal plaques but also in vascular deposits as well as Aβ epitopes in APP-expressing cells like monocytes. Binding of Fc-bearing molecules to vascular amyloid plaques can trigger adverse effects like vasogenic edema (see adverse effects).

INHIBITION OF Aβ AGGREGATION AND NEUTRALIZATION OF AB TOXICITY

Most anti-Aβ antibodies prevent oligomerization and fibril formation of Aβ [148, 149] or dissolve Aβ aggregates (Fig. **2**; [150, 151]. Antibodies, that recognize the N-terminal part of Aβ may not inhibit oligomerization and fibrillization of the peptide. To block these plaque forming processes an antibody appears to be useful that interferes with the C-terminal part of the Aβ peptide [152-155]. Specific conformational anti-Aβ mAbs disassemble existing plaques in the brain [156]. These findings suggest that direct interaction of anti-Aβ antibodies with Aβ seems to affect Aβ aggregation. Moreover, Aβ antibodies that recognize Aβ oligomers have been shown to reverse or improve cognitive deficits [157-159], indicating that one possible mechanism of Aβ immunotherapy may be the blockade of neurotoxicity induced by Aβ oligomers.

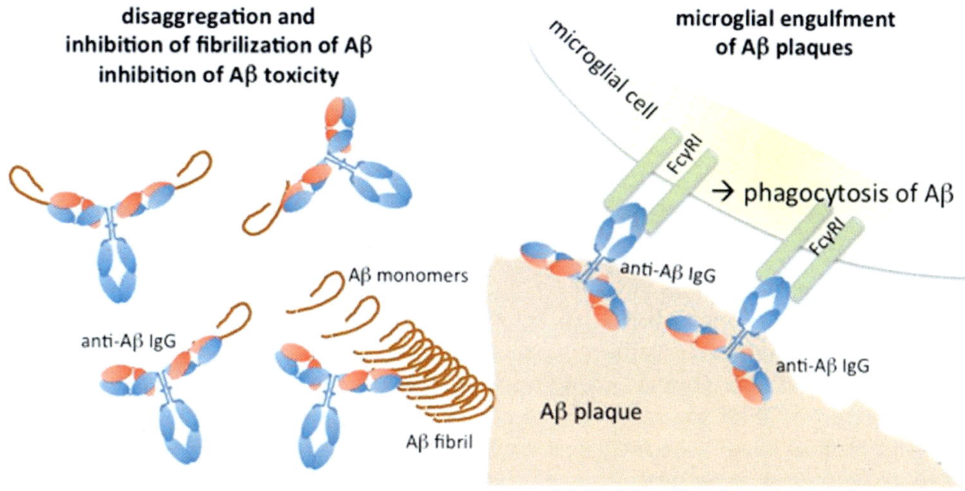

Figure 2: Putative central sink mechanism of passive anti-Aβ immunotherapy. Aggregated and soluble Aβ is in equilibrium. Whenever the anti-Aβ IgG binds to soluble Aβ, re-aggregation is inhibited. Disaggregation and sequestering of Aβ may lead to reduced amyloid burden. Anti-Aβ IgG bind on cerebral Aβ plaques and mark them thereby for opsonisation. Microglia recognizes the Fc domain of the IgG *via* its FcγRI that initiates phagocytosis of the plaques.

PERIPHERAL SINK

In contrast to the above mentioned mechanisms; even antibodies that bind monomeric Aβ only (like m266/Solanezumab) are effective in passive immunotherapy in animal models [160]. A peripheral action of anti-Aβ mAbs presumes that Aβ that typically is in steadiness between periphery and brain is captured in the periphery and can thereby not enter the brain anymore (Fig. **3**). Receptors involved in Aβ transport at the blood-brain barrier are LRP, RAGE and FcRn [161]. Evidences were found for an improved clearance of Aβ as well as a decreased Aβ influx from blood to the brain due to complex formation of immunoglobulins with free Aβ in the periphery.

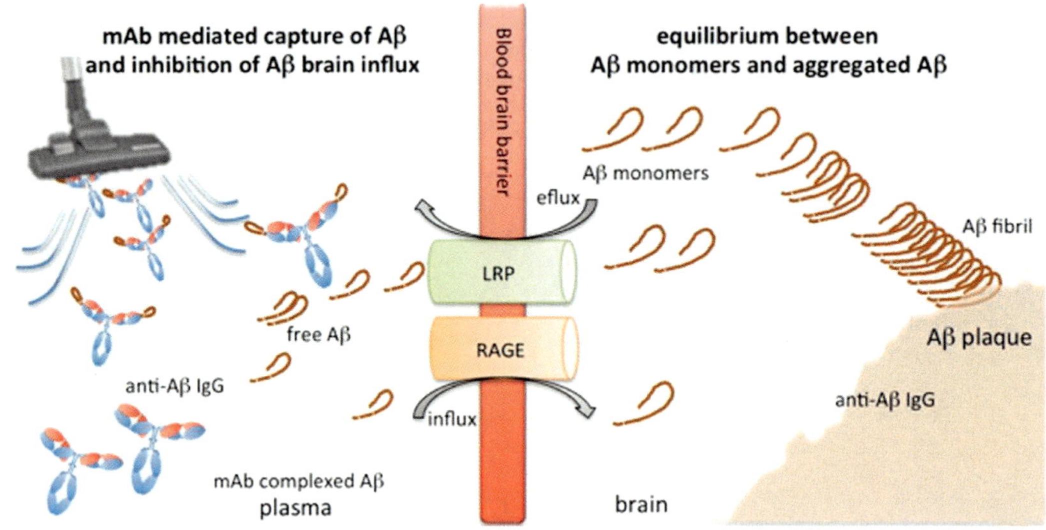

Figure 3: Putative peripheral sink mechanism of passive anti-Aβ immunotherapy. Aβ is continuously transported *via* the blood brain barrier in both directions. The presence of an anti-Aβ IgG that sequesters Aβ in the plasma compartment disturbs the equilibrium between brain and plasma Aβ is and Aβ is 'hovered' to the plasma (as Martin Citron named it). In addition, since soluble Aβ is permanently removed from the brain and thereby from the equilibrium, aggregated Aβ plaques disaggregate. Binding of Aβ to an IgG increases the half-life of Aβ from some minutes to roughly 20 days for IgG_1. Therefore, the accumulation of Aβ can be detected as plasma increase.

This mechanism was fortified by active [162-165] and passive [166-174] immunization studies in AD models. Passive Aβ immunization studies suggests that capturing Aβ in plasma reduces free Aβ levels and increases total Aβ levels through prolongation of Aβ half-life [136]. This in turn disrupts the brain-blood equilibrium of Aβ and drives the removal of soluble Aβ from the brain, thereby disturbing the equilibrium between soluble and aggregated Aβ in the brain.

The peripheral sink theory is in accordance with the very limited distribution of antibodies to brain through an intact blood brain barrier. In general, antibodies do not accumulate in a healthy brain to a higher percentage than 0.1%. Therefore, this theory assumes that the antibodies do not work in the brain but in the periphery. However, it should be mentioned that the blood-brain barrier appears to be leaky rather early in AD and other neurodegenerative diseases [175].

INTRACEREBRAL SEQUESTRATION OF MONOMERIC Aβ

Peripheral injection of anti-Aβ monoclonal antibody m266/Solanezumab followed by intracerebral microinjection of radiolabeled $A\beta_{40}$ slowed Aβ clearance from mouse brains. Moreover, peripheral administration of m266 antibody in young, pre-plaque APP transgenic mice led to increased levels of monomeric, antibody-bound Aβ in brain [176]. Their data suggest that m266 binds and sequesters monomeric Aβ in the brain parenchyma (Fig. **2**). A reduction in the level of antibody-free Aβ, may influence the ability of free Aβ to develop into oligomers and amyloid fibrils [149]. In addition, passive and active Aβ immunization rescued LTP deficits induced by intracerebroventricular infusion of naturally-occurring Aβ oligomers from APP-transfected cell culture media in rats [177].

Mechanism-Related Adverse Effects of Anti-Aβ mAbs or Scaffolds

Using monoclonal antibodies to various regions of Aβ, it was observed that antibodies generated against the N-terminal region, especially $A\beta_{1-17}$, promoted dose-dependently amyloidogenic processing of APP *via* β-secretase activation [178].

Most of the first mAbs against Aβ are full IgG_1s that strongly mediate pro-inflammatory effector functions. Thereby, the Fc-domain of the antibodies can mediate toxicity if Aβ is also deposited in cerebral vasculature forming vascular plaques. In particular ApoE4 carriers reveal vascular amyloid plaques. Anti-Aβ antibodies capable to induce the complement system can lead to the formation of membrane attack complexes and thereby microbleedings *i.e.* microhemorrhages, vasogenic edema or if diagnosed with magnetic resonance imaging ARIA (amyloid-related imaging abnormalities). Microhemorrhage (Fig. **4**) has been reported following passive immunization with N-terminal Aβ antibodies in APP transgenic mice [109, 179, 180]. In contrast, passive immunization with m266 did not increase microhemorrhage in mouse brains [180]. A C-terminal antibody 2886 led to a significant increase of CAA-associated microhemorrhage in immunized mice [109].

NAbs-Aβ, our Natural Insurance Against AD? The 'Physiological' Anti-Aβ Treatment with IVIG

Commercially available Ig preparations purified from human plasma (called Intravenous immunoglobulins, IVIG) are approved for the treatment of immunosuppression, autoimmunity and a variety of other neurological conditions.

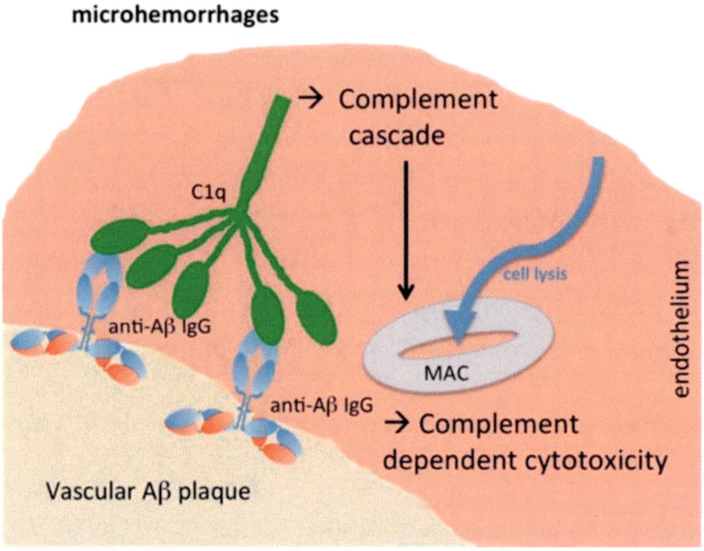

Figure 4: Putative mechanism of microhemorrhages during passive anti-Aβ immunotherapy. The complement cascade is initiated through C1q that binds to the Fc domain of antigen-bound IgGs. The cascade leads to an assembly of a pore, the membrane attack complex (MAC). In case an anti-Aβ IgG binds to vascular Aβ plaques - that occur more frequently in *ApoE4* carriers - cerebral microhemorrhages resulting in vasogenic edema and ARIA can be a consequence.

Regular IVIG treatment reduces the risk of developing AD by 40%. A decade ago, it was demonstrated that human serum and IVIG have a significant quantity of polyclonal naturally occurring autoantibodies against Aβ (NAbs–Aβ) and interestingly NAbs–Aβ are reduced in AD patients [152, 181, 182]. Naturally occurring autoantibodies possible make up a significant part of the human antibody pool. However, the underlying mechanisms are far from being completely understood. Those NAbs–Aβ as well as polyclonal human anti-Aβ antibody containing preparations like IVIG have been characterized in different experimental settings to inhibit the propensity of Aβ to fibrillize, thereby blocking its toxicity and to affect the clearance of Aβ [96]. NAbs–Aβ detectable in healthy individuals, interacted with Aβ, and promoted its degradation. NAbs–Aβ were able to interfere with Aβ peptide toxicity, but NAbs–Aβ did not readily clear senile plaques although early fleecy-like plaques were reduced. Administration of

NAbs–Aβ in transgenic mice improved cognition significantly [129]. In epitope mapping, NAbs–Aβ detected mainly the mid-/C-terminal epitope of Aβ, starting at the amino acid 28. Interestingly, they were not able to bind and clear plaques, but they were able to improve behavioral changes in a transgenic mice model. NAbs–Aβ preferentially capture dimers and trimers and interfere with their oligomerization [183]. This may be the mechanism why treatment with NAbs–Aβ in young animals led to a reduction in plaque number and size, suggesting that NAbs–Aβ prevented plaque formation. Naturally occurring autoantibodies do not bind to distinct protein sequences but rather recognize specific patterns of conservative molecules. Therefore, it is hypothesized that NAbs–Aβ may recognize a common conformational epitope among those proteins, a rather narrow epitopal recognition pattern [129].

Based on the concept of NAbs–Aβ, IVIG have been used in small pilot trials for the treatment of patients with AD [184], which showed promising effects on cognition, thus leading now to a Phase III trial [155]. In the active immunization studies, fibrillated $A\beta_{42}$ peptide was used as antigen, suggesting that N-terminal epitopes of the Aβ peptide were predominantly exposed and were available as binding sites. Accordingly, active immunization generated primarily antibodies that recognized the N-terminal epitope [185]. Likewise, Aβ peptide is deposited in a fibrillar form in the plaques, and the N-terminal part of the Aβ peptide is mostly available at the plaque surface. In contrast, NAbs–Aβ, as mainly directed to the mid-/C-terminal epitope of Aβ, are less capable of binding to the N terminus of Aβ, thus NAbs–Aβ do not efficiently clear plaques.

Efflux of central Aβ into the periphery (peripheral sink) appear to be the main mechanism for NAbs-Aβ or IVIG, however, a small amount of NAbs–Aβ/IVIG crossed the blood–brain barrier as was shown using radiolabeled NAbs–Aβ [154]. Earlier, those few autoantibodies available in the brain were suggested to be sufficient to induce uptake by opsonization in the brain [92, 108, 186]. Therefore, other mechanisms like central degradation of Aβ oligomers in the presence of NAbs–Aβ by microglial cells, may also play an important role.

Interestingly, antibodies, which bind to Aβ, also typically bind to the parent APP, suggesting that certain Aβ-targeting autoantibodies may bind to APP thereby altering its conformation and processing. NAbs–Aβ isolated from AD patients, but not from healthy controls, promoted β-secretase activity in cultured cells [178]. Moreover, some data indicate an autocatalytic function of NAbs–Aβ in hydrolyzing Aβ specifically [187]. The catalytic activity improved as a function of age and patients with AD produced catalytic antibodies at increased levels (more IgMs than IgGs). However, IVIG did not cleave Aβ in this study, possible as IgMs are rather low in IVIG.

The Protagonists: the Anti-Aβ mAbs and their Data from Preclinical and Clinical Studies

After the halt of AN-1792 many expectations focused on passive immunization with monoclonal antibodies. The adverse events in AN-1792 did not correspond to an antibody response and may have been initiated by activation of cytotoxic T cells and/or autoimmune reactions [99, 100]. Active Aβ vaccine trials are currently underway, sponsored by pharmaceutical companies such as ELAN/Wyeth, Novartis, Merck, Affiris, and United Biomedical [142]. In addition, passive anti-Aβ immunotherapy may be an easy and relatively safe method to provide Aβ antibodies without eliciting Th1-mediated autoimmunity.

Table 1: Passive immunotherapy for AD in clinical phases

Company	Antibody	Binds To	Backbone	Biomarker Effects	Summary of Clinical Trials
Developed by Elan, clinical development with Wyeth that was acquired by Pfizer, late clinical development with J&J/ Janssen	**Bapineuzumab**, humanized murine 3D6	N-terminus (aa 1-5)	humanized IgG_1	plasma Aβ increase, decrease of CSF phospho-tau	Phase II: no statistical significant effect on cognition nor activities of daily living, Reanalysis revealed some evidence for an improvement in study completers and ApoE4 non-carriers. Phase III: data not published yet. Trials were halted after completion of two trials demonstrated a failure to meet primary outcome measures of cognition and activities of daily living.
Eli Lilly	**Solanezumab**, humanized murine m266	central (aa 16-24), accessible only on soluble Aβ	humanized IgG_1	increase in CSF $A\beta_{42}$ and plasma Aβ	Phase II: correlation between plasma $A\beta_{1-42}$ and assessment of amyloid burden using PET. No significant cognitive benefit and no adverse effects. Phase III: data not published yet. Trials failed to meet their primary endpoints in cognition and activities of daily living. A subsequent analysis of mild AD patients pooled from both trials showed a significant effect on cognition. Clinical development continues, preventive trial in familial AD planned for 2013.
Roche and Morphosys	**Gantenerumab**, full human mAb	N-terminal (aa 3-12) and C-terminus (aa 18-27)	human IgG1	no increase in Aβ plasma and CSF	Phase I: ARIAs were observed in 2 *ApoE4* homozygous patients, cerebral amyloid reduction was observed. Phase III: ongoing in prodromal AD patients.
Developed by AC Immune, licenced by Genentech/Roche	**Crenezumab**, humanized murine mMABT	Immunized with $A\beta_{1-15}$ multiple conformational epitopes of Aβ, including oligomeric and protofibrillar forms, (aa 13-14	disulfide-stabilized IgG_4	Increase in total Aβ plasma levels	Phase I: no vasogenic edema was observed. Phase II: preventive trial in an extended family carrying a *presenilin-1* mutation, which causes early onset AD planned for 2013.

		appears relevant)			
Glaxo Smith Kline	**GSK933776**	N-terminus of Aβ	humanized IgG_1 with mutated Fc to reduce effector function	increase in total serum Aβ levels, decrease in free serum Aβ levels, increase in $Aβ_{1-38}$ CSF levels.	Phase I: two clinical trials for AD are completed and one for macular degeneration is ongoing. No drug related adverse events. Neuropsychological parameters were not improved. Further development for macular degeneration is in Phase II.
developed by Wyeth/Pfizer and Jansssen/Johnson&Johnson	**AAB-003**, Fc-engineered Bapineuzumab	N-terminal (aa 1-5)	mutations in the Fc domain to reduce effector function	No data published, similar action on biomarkers as Bapineuzumab can be expected	Phase I: ongoing. Lower toxicity compared to Bapineuzumab is expected.
generated from BioArtic Neuro-science and partly licensed by Eisai	**BAN2401**, humanized mAb158	binds large-size Aβ protofibrils (>100 kDa)	humanized IgG	No data published. As it does not bind monomers, no dramatic plasma increase is expected	Phase I: ongoing
developed by Rockefeller University, licensed by Sanofi-Aventis	**SAR228810**, humanized murine mAb 13C3	protofibrils, and low molecular weight Aβ	humanized IgG_4	No data published. As it does not bind monomers, no dramatic plasma increase is expected	Phase I: ongoing
developed by Neurimmune, licensed by Biogen Idec	**BIIB037/BART**, full human IgG_1	binds insoluble fibrillar human Aβ	fully human IgG_1	No data published. As it binds mono-mers with loe affinity, no dramatic plasma increase is expected	Phase I: ongoing
Baxter, Octapharma, Sutter Health, Grifols Biologicals	IVIG containing polyclonal NAbs-Aβ: **Gammagard**, **Octagam**, **New Gam**, **Flebogamma**	Most NAbs-Aβ bind central and C-terminus and pathogenic conformations of Aβ (focus on dimers)	95% IgGs 5% IgM 0.2% Nabs-Aβ	transiently increased plasma Aβ levels	Phase II (Gammagard): three-year stabilization of AD symptoms with IVIG, including no decline on measures of cognition, memory, daily functioning and mood.

(aa, amino acid; Aβ, Amyloid-β; ApoE4, ApolipoproteinE4; ARIA, amyloid-related imaging abnormalities; IVIG, Intravenous Immunoglobulin; J&J, Johnson&Johnson; NAbs-Aβ, natural occurring polyclonal Anti-Aβ antibodies).

BAPINEUZUMAB, THE FIRST PASSIVE N-TERMINAL SPECIFIC MAB

Bapineuzumab the humanized IgG_1 version of the mouse mAb 3D6 recognizes the epitope $A\beta_{1-5}$. It is specific for Aβ and does not cross-react with APP or its α-secretase product [188]. To share the financial risks of clinical development of Bapineuzumab, the trials were sponsored by a collaboration involving Pfizer, Johnson&Johnson/Janssen and Elan, with Pfizer maintaining a roughly 50% economic interest and both Johnson&Johnson, including the subsidiary Janssen, and Elan each maintaining an approximate 25% economic interest [189]. In Phase II, a total of 234 patients with a clinical diagnosis of probable AD with MMSE scores 16–26, received intravenous Bapineuzumab (0.5 mg/kg - 2.0 mg/kg) or vehicle over 18 months. Bapineuzumab was infused every 13 weeks due to its serum half-life of 24 days, *e.g.* 6 infusions all together [190]. After evidence of vasogenic edema (similar to microhaemorraghe preclinical data in mice) emerged in patients enrolled in the Phase I study, an additional dosing cohort of 0.15 mg/kg was added. Since the most frequent serious adverse event and reason to drop-out from the study, vasogenic edema, was more common in *ApoE4* carriers, there were fewer treated carriers than noncarriers in some of the analyses, possibly leading to decreased power of the analysis to detect an effect. In a Phase II study, approximately 10% (12 of 124) Bapineuzumab-treated patients developed vasogenic cerebral edema [190]. Bapineuzumab treatment was responsible for this adverse event, as it was observed in none of the placebo-treated patients, and it exhibited a clear dose-dependence, but also a dependency on *ApoE4* genotype. In a subsequent Phase II study, again 10% of Bapineuzumab-treated patients (2 out of 20, both *ApoE4* carriers) developed vasogenic edema [191]. *ApoE4* carriers develop more severe CAA as they have higher levels of Aβ in the walls of blood vessels [192]. Therefore, anti-Aβ immunotherapy may cause Fc-mediated effector function and inflammation on vascular Aβ plaques, leading to damage and leakiness of the blood-brain-barrier [193].

In the completed Phase II trial, the primary outcome measures were the Alzheimer's Disease Assessment Scale–Cognitive subscale (ADAS-Cog) and the Disability Assessment for Dementia (DAD). A not statistically significant trend towards improved scores on ADAS-Cog was observed. All Bapineuzumab-treated patients were pooled into one group, irrespective of dosing, in a *post hoc* analysis.

Nevertheless, these data were not statistically significant different of that from placebo treated patients unless all patients were excluded that did not complete the trial. In this rather arbitrary cohort a small however significant difference was apparent: roughly 6% less loss of scores on the ADAS-Cog or disability scale after 18 months of treatment. This observation justified moving to Phase III trials.

Also very interesting is the effect of *ApoE4* carrier status on outcomes [190]. In this Phase II study, patients bearing one or two alleles of *ApoE4* showed no treatment effect on any measure, *i.e.* they are less likely to experience clinical improvement, although they are also more likely to suffer a significant adverse event like vasogenic edema. Alternatively, since ApoE4 may decrease Aβ transport across the blood-brain barrier and considering peripheral sink as main mechanism, *ApoE4* carriers could be less sensitive to treatment with peripheral anti-Aβ antibodies. Therefore, the sponsor designed separate Phase III trials for *ApoE4* carriers (NCT00575055) and noncarriers (NCT00574132), suggesting that *ApoE* genotyping may become clinical routine test.

In the above mentioned study, Bapineuzumab treatment in a small cohort did not significantly influence CSF Aβ or phospho-Tau levels at week 52 compared to baseline [190], but recent data show a significant reduction of CSF Tau levels in treated patients [194]. *ApoE4* noncarriers revealed significantly less brain volume loss determined by MRI while in *ApoE4* carriers more ventricular enlargement but no change in brain volume were observed [190].

In a subsequent Phase II study of Bapineuzumab [191], the primary outcome was the extent of PiB signal on PET. PiB, a thioflavin analog that binds fibrillar Aβ, is used as an *in vivo* marker of cortical amyloid plaques [195]. 18 Bapineuzumab-treated patients showed reduced PiB retention at 18 months treatment compared to baseline, whereas the 7 placebo-treated patients showed more retention compared to baseline, suggesting that Bapineuzumab is effective in clearing cerebral Aβ. Moreover, these data were discussed as - at least - partial proof of the amyloid hypothesis. An autopsy reported for a patient in the Phase IIa Bapineuzumab trial confirms the PiB PET data [196]. The patient received four doses of 1 mg/kg of Bapineuzumab, before he withdrew from the trial due to progression of his dementia symptoms. He died 36 months after the PIB PET scan, and 11 months after the last dose of Bapineuzumab. At baseline, he exhibited extensive PiB binding. The *post mortem* histological findings were unusual for typical severe AD cases, and showed a mis-match between pre-treatment PiB PET and post-treatment pathology. This suggests that Bapineuzumab treatment may have cleared small non-neuritic Aβ plaques in areas previously identified with high PiB retention. Furthermore, because the neuropathological data were collected almost a year after treatment ended, these findings also suggest that Aβ clearance could persist for prolonged periods of time.

However, in general Bapineuzumab was well tolerated and safe, besides the vasogenic edema in *ApoE4* carriers. The Phase III studies are designed with a pharmacogenetical approach to observe the effects of Bapineuzumab separately for *ApoE4* carriers and non-carriers to avoid statitistical and artistical re-analysis of small cohorts as done in the Phase II trial.

In Summer 2012, the sponsors reported disappointing results from the Phase III trials in *ApoE4* carriers (Study 302) and non-carriers (Study 301; [197]). Some details of the studies were presented on the Clinical Trial in Alzheimer's Disease conference in October 2012 [198]. Both studies failed to show any significant benefit on their endpoints cognition and functional performance. In *ApoE4* carriers ADAS-Cog decreased roughly by 9 scores after 78 weeks in the placebo (n = 432) and the 0.5 mg/kg Bapineuzumab (n= 658) group without any statistical significance (p=0.8). However, the cognition score at 39 and 52 weeks appeared to reveal a very small benefit for the mAb-treated patients but the final analysis will hopefully shed more light on this. In parallel, in *ApoE4* non-carriers ADAS-Cog decreased roughly by 7 – 8 scores irrespective of Bapineuzumab (0.5 mg/kg (n = 314) and 1 mg/kg (n = 307)) or placebo treatment (n = 493) and without any statistical significance. Similarly, function was measured on the Disability Assessment for Dementia scale (DAD) after 78 weeks of treatment. No effect of Bapineuzumab was observed in *ApoE4* carriers (n = 658) *vs.* Placebo (n = 432; p = 0.34) nor in non-carriers (0.5 mg/kg, n = 314; 1 mg/kg, n = 307; placebo, n = 493; p = 0.07, p = 0.55, respectively). In ApoE4 carriers the placebo group decreases roughly 16 scores on the DAD scale over 78 weeks and the treatment group roughly 17. In non-carriers the placebo group decreased by 15.5 scores, the 0.5 mg/kg group roughly 12.5 and the 1 mg/kg group 14.5 scores.

However, if only mild non-carrier cases with an MMSE over 20 were investigated a statistical significant benefit on function (DAD) was observed in both treatment groups (0.5 mg/kg n = 204; 1 mg/kg n = 201) *vs.* placebo (n = 334; p = 0.04; p = 0.02, respectively). This effect was not clearly dose-dependent: the placebo group decreased in mean by 12 points over 78 weeks while the treatment groups decreased in the meantime by roughly 9 scores. In mild *ApoE4* carrier cases no benefit was observed and in the mild pooled population only the highest dose revealed significance (p = 0.04). No significant effects on cognition (ADAS-Cog) were observed with this re-analysis though the mild non-carrier displayed a positive tendency (roughly one score less over 78 weeks) for the treatment (p = 0.19).

At the same conference, data on safety and biomarker were presented [199]. ARIA (amyloid-related imaging abnormalities), asymptomatic ARIA (vasogenic edema) occurred in 15.3% in the treated *ApoE4* carrier while only in 0.2% of the placebo treated. Symptomatic ARIA was observed in 2.4% of the treated carriers while in no case of the placebo group. Intracranial hemorrhages were similar in Bapineuzumab and placebo treated carriers (1% *vs.* 1.6, respectively). Seizures were more abundant during the mAb treatment with 1% *vs.* 0.2% in the placebo group. The occurrence of adverse events was somewhat lower in the *ApoE4* non-carriers and asymptomatic ARIA (vasogenic edema) increased dose-dependently from 0.2% in the placebo, 4.2% in the 0.5 mg/kg and 9.4% in the 1 mg/kg group. In parallel, symptomatic ARIA was not observed in placebo, but in 1.5% of the patients in each treatment group. Seizures occurred in 1% of placebo treated and 0.3%, 1.8%, respectively in Bapineuzumab treated non-carriers. Intracranial hemorrhages occurred in similar percentages in placebo and mAb treatment. Furthermore, the estimated number of ARIA is expected to be even higher due to not detected ARIAs during the course of the study. The occurrence of ARIA is strongly related to *ApoE4* copy number and arose predominantly during the first three infusions.

Finally, the number of death was twice as high in the Bapineuzumab treated *ApoE4* carrier (2,2%) as in the placebo group (1.1%), especially due to cancer deaths. In non-carrier the higher dose group revealed a comparable percentage of deaths 2.1 % *vs.* 1.3% in placebo) like in the treated *ApoE4* carrier while the lower dose was not changed (1.2%). The increased rate of deaths in the 1 mg/kg dose were mainly due to cardiac events.

Amyloid burden measured by PiB in PET in a subgroup of 40-75 patients decreased significantly in treated ApoE4 carriers ($p = 0.004$), while in non-carriers due to large variations no significant benefit was observed. The authors speculate that a high number of 'amyloid negative' patients among the non-carriers could be responsible for the negative outcome, likewise 15-20% of 'AD cases' that do not meet the criteria for AD at autopsy. When pooling both studies and analysing the data for the mild cases, the benefit of Bapineuzumab treatment is even more pronounced. MRI brain volume was not significantly changed during treatment in both genotypes.

Phospho-tau in CSF decreased significantly ($p = 0.005$) in the treated *ApoE4* carriers [200] and decreased significantly ($p = 0.009$) in the high dose in the non-carriers [201].

This clinical outcome resulted in the discontinuation of all Bapineuzumab trials, except for one subcutaneous clinical Phase II trial (NCT01254773). AAB-003 (see below) is now the only anti-AD mAb in the sponsor's pipeline.

SOLANEZUMAB, A SPECIFIC ANTI-MONOMERIC Aβ mAb

Solanezumab (LY2062430) is a humanized IgG_1 of the murine antibody m266 against a central epitope of Aβ (residues 16–24) in a more hydrophobic region [202]. Humanization was done *via* targeted sequence replacements of the murine m266 antibody, including the disruption of an N-glycosylation site in the heavy-chain CDR2 [106]. The parent mAb was generated in mice using synthetic $A\beta_{13-28}$ (HHQKLVFFAEDVGSNGGC) conjugated to an anti-CD3 immunoglobulin [202, 203]. Therefore, Solanezumab can capture not only full-length Aβ but also N-terminal truncated or modified forms of Aβ [204]. In addition, Solanezumab provides little to no affinity to fibrillar Aβ and binds preferentially soluble species with an accessible central domain only. Thus, a peripheral sink mechanism has been considered as the mechanism of action for m266/Solanezumab [205]. Preclinical *in vitro* and *in vivo* studies with m266 demonstrated that m266 although unable to bind to Aβ aggregates, prevented completely $A\beta_{1-42}$

fibrillization [149]. Treatment with m266 rapidly increased CSF and plasma Aβ levels and reduced cerebral Aβ deposition without decorating (*i.e.* binding) the amyloid plaques [160, 205]. In rats, m266 antibodies have been shown to enter the CSF at concentrations that are approximately 0.1% of those found in plasma [160]. In addition, m266 markedly improved learning and memory under acute and subchronic conditions in APP transgenic mice, in parallel with the increase of Aβ in both plasma and CSF, but without any detectable alteration of the brain Aβ burden [103]. Conversely, the peripheral sink mechanism of m266/Solanezumab has recently been doubted [176], based on observations that intracerebral injection of radioisotope-labeled Aβ in mice. Without m266 injection, radioactive Aβ was rapidly eliminated from the brain, whereas m266 significantly retarded the elimination of radio-labeled Aβ, probably by entering the brain and sequestering Aβ, implying that this monoclonal may act by stabilizing Aβ monomers, thus preventing formation of neurotoxic Aβ oligomers.

Solanezumab entered as second anti-Aβ antibody the clinical stage. Very early in clinical development, Solanezumab had a complete different trial strategy than Bapineuzumab had: Eli Lilly focused from the beginning on on peripheral sink as mechanism and therefore plasma and CSF Aβ levels as biomarkers. An increase in Aβ levels is believed to go in hand with clearing Aβ and hopefully improving or halting cognition. In the Phase I trial, maximum concentrations of plasma Solanezumab increased in a linear fashion with increasing dose following a single administration of the drug. Half-life ranged from 26 - 30 days for the 1.5, 4 and 10 mg/kg doses [203]. In a subsequent Phase II study, plasma concentrations of Solanezumab fell to below 5% of the initial peak with monthly dosing [206].

In a Phase II trial (NCT00749216), pharmacokinetics, pharmacodynamics, safety and tolerability after a single dose (0.5, 1.5, 4.0, and 10.0 mg/kg) of Solanezumab were found to be comparable between Japanese and white patients with AD. A marked increase in plasma total $A\beta_{40}$ and $A\beta_{42}$ was observed; both the magnitude and time to reach maximum concentration tended to increase with increasing doses of Solanezumab. Indeed, similar to the results of previous studies in mice, a small (~0.1%) but significant fraction of the antibody crossed the blood-brain barrier and was found in the CSF [207]. Administration of Solanezumab was generally well-tolerated in both populations [208].

In a Phase II study of 52 patients with mild to moderate AD who received one of four doses of Solanezumab, plasma and cerebrospinal biomarker assays suggested a promising drug effect [203, 209]. Clinical laboratory values, CSF cell counts and magnetic resonance imaging scans were unchanged by treatment and no adverse events could be clearly related to antibody administration. Total (antibody-complexed and free) $A\beta_{40}$ and $A\beta_{42}$ in plasma increased in a dose-dependent manner. Antibody treatment similarly increased total $A\beta_{40}$ and $A\beta_{42}$ in CSF. For patients taking 400 mg weekly, antibody treatment decreased significantly free $A\beta_{40}$ in CSF, but increased free $A\beta_{42}$ in CSF in a dose-dependent manner. Studies examining biomarkers found that plasma pE3-Aβ as well as plasma and CSF N-terminally truncated Aβ, but not total Tau and phospho-Tau in CSF, exhibited significant changes in patients receiving Solanezumab, indicating the utility of these biomarkers for evaluating the pharmacodynamic effects of Solanezumab [207, 210-213]. In addition, there was a correlation between plasma $A\beta_{1-42}$ and assessment of amyloid burden using PET [214]. A preliminary study suggested that increased plasma levels of $A\beta_{42}$ might originate from brain plaques, indicating that Solanezumab may disrupt brain amyloid deposits. However, no significant cognitive benefit appeared on the ADAS-cog in this small sample over the 12-week treatment period, although the study was not powered to detect efficacy.

The Phase III trial, NCT01127633, explores ongoing safety data for Solanezumab. In preclinical studies, mice treated with m266/Solanezumab had a smaller incidence of developing cerebral microhemorrhage [180]. Trials of Solanezumab to date have demonstrated no signs of the proinflammatory or vasogenic edema reactions. No infusion reactions, meaningful laboratory anomalies or other drug-related adverse events have been reported for Solanezumab [206], implying a high safety profile of this monoclonal. However treatment related adverse events were reported equally in each dose group. Notably, there is no evidence to date that *ApoE4* genotype alters the efficacy or side effect profile of Solanezumab [204].

Two large multicenter randomized controlled Phase III trials (EXPEDITION, NCT00905372 and EXPEDITION2, NCT00904683) investigate the effects of Solanezumab *vs.* placebo on the progression of AD. First data has been released in August 2012 showing that Solanezumab did not meet its primary endpoints in improving or stabilizing cognition and function [215]. Adverse events with an incidence of at least 1% that occurred statistically significantly more in the Solanezumab group than in the placebo group were lethargy, rash and malaise (in EXPEDITION1) and angina (in EXPEDITION2).

An independent analysis of the raw data presented in October 2012 [216] caused 'cautious optimism' for Eli Lilly. A pre-specified secondary analysis of pooled data across both trials showed statistically significant slowing of cognitive decline in the overall study population of patients with mild-to-moderate AD after switching the cognition score. In addition, pre-specified secondary subgroup analyses of pooled data across both studies showed a statistically significant slowing of cognitive decline in patients with mild AD, but not in patients with moderate AD. Therefore, an on-going, open-label extension study, EXPEDITION-EXT, is fully enrolled and will continue as planned. In addition, secondary prevention trials are due to start in 2013 [217]. Those trials will focus on people who have biomarker evidence of AD but have yet to show any symptoms. Researchers are hopeful that targeting patients earlier may result in more effective treatments. Finally, no changes were noted in biomarkers such as tau, phospho-tau, hippocampal volume, whole brain volume, or amyloid accumulation.

GANTENERUMAB, THE *IN VITRO* GENERATED FULL HUMAN MAB

Gantenerumab (RO4909832 or R1450) is a fully human anti-Aβ IgG_1 generated from the HuCal phage display library of Morphosys [218] and engineered by *in vitro* maturation [219, 220]. It binds Aβ at sub-nanomolar affinity (KD 0.6 nM for Aβ-fibrills, 1.2 nM for Aβ oligomers and 17 nM for Aβ monomers) [221]. *Gantenerumab* recognizes two regions within Aβ: the N-terminal sequence EFR and in the central region around the sequence VFFAEDVGSN [222].

Gantenerumab cleared human Aβ plaques in *post mortem* brain slices cultured with primary human effector cells (macrophages and microglia isolated from brain tissue obtained from neurosurgery) after preincubation with different concentrations of Gantenerumab [221]. In APP transgenic mice, Gantenerumab decorated Aβ plaques thereby guiding microglia to the plaques for engulfement.

Noteably, cerebral plaque bound Gantenerumab peaked at two to three weeks and was detectable up to two months after bolus intravenous administration while plasma half-life of Gantenerumab was six to seven days. Chronic treatment of APP transgenic mice resulted in significant reduction of plaque load. In these preclinical studies, Gantenerumab was well tolerated and did not induce reactive inflammation following the degradation of plaques or exacerbate the manifestation of CAA or CAA-related microhemorrhage. Excitingly, treatment with Gantenerumab did not significantly increase plasma Aβ levels. Finally, inhibition of LTP by Aβ oligomers could be reduced by treating with Gantenerumab. However, it should be noted that due to the human origin of Gantenerumab, mice were also treated with anti-CD4 to avoid generation of ADAs against Gantenerumab by blocking humoral immune responses.

Clinical evidence from Gantenerumab has been delivered in a Phase I clinical trial (NCT00531804) with PiB PET and MRI imaging as biomarker readout [222]. Amyloid-related imaging abnormalities (ARIAs), which include microhemorrhages and vasogenic edema were observed in 2 *ApoE4* homozygous patients following treatment with Gantenerumab resolving spontaneously after discontinuation of the drug.

During this two to seven month treatment with Gantenerumab a dose-dependent cerebral amyloid reduction was observed, while the amyloid load in the placebo-treated patients continued to increase. AD patients in the low-dose group lost 15.9% of their plaques, while the ones in the high-dose group lost 35.7%. If this can be confirmed in subsequent studies, it would indicate faster amyloid removal than reported for Bapineuzumab [223]. Based on the imaging results, phagocytosis was suggested as the likeliest mechanism of amyloid reduction by treatment with Gantenerumab [222].

A Phase I study (NCT01636531) to evaluate lyophilized formulation *vs.* liquid is ongoing and a Phase II clinical trial (NCT01224106) is under way to investigate whether a clinical benefit can be achieved in Gantenerumab-treated patients with prodromal AD, *e.g.* patients are recruited who meet the criteria for CSF Aβ and cognitive impairment, but whose impairment falls short of a dementia diagnosis [223]. In addition, intravenous treatment was changed to subcutaneous administration. Recently, it was announced to expand the Phase II study to a potentially pivotal study as a favorable outcome to the trial could be used to support a marketing application for Gantenerumab [224].

GSK933776, AN N-TERMINAL SPECIFIC FC-ENGINEERED MAB

GSK933776 is a humanized IgG_1 monoclonal antibody directed against the N-terminus of Aβ. When administered to an AD mouse model, the parent molecule of GSK933776, GSK719556A, decreased the amyloid brain content. The Fc part of GSK933776 is inactivated by introduction of two mutations. This modification was done with the aim of improving the safety profile, in particular with regard to the risk of vasogenic oedema reported in association with other N-terminal anti-Aβ antibodies [225].

Very little is published about the preclinical performance of GSK719556A or GSK933776, however two clinical trials for AD are completed (NCT00459550 and NCT01424436) and one for macular degeneration is ongoing in Phase II (NCT01342926).

The *ApoE4* carrier frequency in the overall study population (64%) was not significantly different to the general AD population (64.8%), 20% were homozygous for *ApoE4*. The clinical safety was quite high as neither of the two serious adverse events that occurred after administration of GSK933776 were judged to be drug-related. No ARIA findings associated with drug administration were noted, in particular no cases of vasogenic oedema. One novel microbleed was recorded in a subject receiving placebo. Novel ADAs against GSK933776 were detected in two subjects, one of these reported as neutralizing GSK933776. The ADAs titres did not increase markedly during treatment and the sponsor cannot exclude whether the ADA assay was valid.

Neuropsychological testing was part of the safety assessment. None of the primary or secondary comparisons of interest on the neuropsychological tests attained statistical significance. Detection of signals of efficacy was unlikely due to the small sample size based on feasibility and short duration of treatment. The terminal phase half-life ranged from 10–15 days and clearance was higher in the lower dose group. This higher value appeared to be due to a higher estimated volume of distribution of the central compartment. GSK933776 CSF pharmacokinetics was measured at Day

78 and depending on the dose. 25 – 100% of the patients had detectable mAb concentrations in CSF. The mean of the CSF *vs.* plasma concentration ratios at Day 78 for the quantifiable concentrations was higher than 0.2%. In addition, a dose-dependent increase in total Aβ levels and a dose-dependent decrease in free Aβ levels was detected while placebo had no effect on free or total plasma Aβ levels. There was increase over baseline in $A\beta_{1\text{–}38}$ CSF levels at the highest dose and when pooling all dose groups also over placebo. No apparent changes from baseline on Tau or phospho-Tau could be determined.

However, GSK933776 will actually not be further developed for AD but for macular degeneration [226], a career that undoubtedly more anti-Aβ mAbs some day will share [227].

CRENEZUMAB, AN ANTI-PROTOFIBRILLAR IGG_4 TO REDUCE EFFECTOR FUNCTION

The disulfide-stabilized IgG_4 anti-Aβ mAb Crenezumab/MABT5102 (original name RG7412) is a humanized form of the mouse IgG_{2b} mMABT generated by immunizing mice with modifided $A\beta_{1\text{-}15}$ [228]. Several criteria were used to select candidate antibodies, including the ability to bind multiple Aβ species and to inhibit $A\beta_{1\text{–}42}$ assembly into small cytotoxic peptide aggregates. MABT5102A targets multiple conformational protofibrillar epitopes of Aβ, including oligomeric forms, while inhibiting aggregation and promoting disaggregation of Aβ. A recent study had demonstrated that the two adjacent histidines at position 13 and 14 of Aβ are necessary for cell membrane binding and uptake and thereby for Aβ-mediated cellular toxicity [229]. These histidine residues overlap with the epitope of Crenezumab, thus predicting the results of Aβ cellular binding studies in which robust inhibition of Aβ binding to neurons is observed. In addition, Aβ oligomers mediated toxicity is also inhibited by Crenezumab.

Crenezumab contains a human IgG_4 backbone. IgG_4 reveals reduced effector function besides the inhibitory ones when compared with IgG_1 [230, 231]. Thus, MABT5102A sequesters effectively Aβ, reduces $A\beta_{1\text{–}42}$-induced neuronal death and notably promotes microglial Aβ phagocytosis, but has a significantly reduced capacity to activate microglial FcγRs and thereby pro-inflammatory response compared to an IgG_1 subclass [232]. Unlike FcγR binding, FcRn binding is not substantially different between IgG_4 and IgG_1 implying no relevant consequences for the mAb's half-life. Although, it is notable that altering FcγR binding alone is sufficient to alter ability of microglial phagocytosis of Aβ, this does not rule out the possibility that anti-Aβ antibodies without effector function may be efficacious *in vivo via* a FcRn-mediated mechanism. Indeed, previous studies have shown that antibodies lacking effector function are efficacious *in vivo* [233]. Additionally, an FcRn-driven mechanism of anti-Aβ action has been directly investigated, suggesting that Aβ/anti-Aβ complexes are cleared across the blood–brain barrier *via* an FcRn-dependent pathway [171]. Thus, in addition to the IgG_4-dependent ability of MABT to enhance microglial engulfment of Aβ, it is also reasonable to propose that MABT5102A could use FcRn to enhance its activity *in vivo*.

MABT5102A was generated by AC Immune and was licensed for further development as Crenezumab by Genentech. In preclinical studies using an AD mouse model the parental mMABT with an IgG_{2b} backbone showed reduced plaque burden and improved memory performance. In *In vitro* cellular assays, MABT5102A prevented oligomer induced neurotoxicity [232].

A Phase I clinical trial (NCT00736775) was conducted with the primary objective to determine the safety and tolerability of intravenous MABT5102A in AD patients and to characterize pharmacokinetics and Aβ pharmacodynamics. Importantly, patients were genotyped and randomized to ensure that at least 40% of enrolled patients were *ApoE4* carriers. None of the patients receiving MABT5102A developed vasogenic edema. The serum concentration–time profiles for MABT5102A following both a single dose and four weekly doses increased in dose-responsive proportion and revealed a long half-life of 18–23 days. A dose-dependent elevation in total Aβ plasma levels was determined, that correlated well with serum MABT5102A concentrations. A second Phase I study was conducted in healthy controls (NCT00997919).

Clinical Phase II trials (NCT01343966 and NCT01397578) are underway to assess the efficacy of MABT5102A/Crenezumab. Very remarkably, efficacy of Crenezumab will be tested in a five year ‚prevention' trial in an extended Columbian family carrying a *presenilin-1* mutation which causes early onset AD [234]. The trial that is planned to start in 2013 will test the efficacy of

Crenezumab on 300 individuals who have that certain mutation, but do not yet have symptoms of AD.

PONEZUMAB, AN FC-ENGINEERED ANTI-C-TERMINAL mAb

Ponezumab/PF-04360365 is a humanized IgGΔ2A mAb with a deglycosylated Fc domain targeting the C-terminal end of Aβ. Its parental molecule was the murine 2H6. Ponezumab was initially developed and characterized as RN1219 by Rinat Neuroscience [235]. Pfizer acquired Rinat for a reported several hundred million dollars in 2006, in part to develop this therapeutic antibody.

As Ponezumab demonstrates reduced effector activity (see Table **2** and chapter: reducing effector function); plaque binding should not lead the massive phagocytosis. Therefore, a peripheral sink mechanism has been suggested like for Solanezumab [236]. However, unlike Solanezumab, Ponezumab targets the C-terminus of the $A\beta_{1-40}$ peptide; thus, it might be that the drug would cause encephalitis [237] like it occurred in AN-1792.

The pharmacokinetics and pharmacodynamics of Ponezumab were evaluated in two Phase I studies (NCT00722046 and NCT00945672) in patients with mild-to-moderate AD [238, 239]. The pharmacokinetics of Ponezumab were linear with a mean terminal half-life of 35-52 days. Ponezumab CSF concentrations at Day 29 post-dose were quantifiable in 25% of the patients with ~0.5% of plasma values. In both studies, plasma Aβ concentrations increased around 500 times in the 10 mg/kg group compared to the placebo group and mean CSF Aβ percentage increased significantly from baseline at Day 29 for $A\beta_{1-x}$, $A\beta_{1-40}$, and $A\beta_{1-42}$. There were no changes from baseline for CSF Tau and phospho-Tau concentrations. ADAs were not detected. The antibody appeared to be safe, with no evidence of encephalitis or vasogenic edema. Ponezumab is modified to reduce activation of the innate and cellular immune systems. These may be beneficial properties, since the antibody may be less likely to cause vascular damage that leads to microhemorrhage or vasogenic edema.

In 2011, Pfizer halted the development of Ponezumab for AD, based on recent data from ongoing studies but without giving more details [240, 241]. Instead of that, a new candidate popped up in the Pfizer pipeline: AAB-003 (PF-05236812).

AAB-003 (PF-05236812), AN FC-ENGINEERED BAPINEUZUMAB

AAB-003 is a humanized 3D6 with three mutations in the Fc domain (L234A/L235A/G237A) to significantly reduce FcγR and C1q binding (see also: reducing effector function) patented by Wyeth and Janssen [242]. Although, the three mutations in AAB-003 are closer to the hinge region rather than to the complement binding region, AAB-003 has reduced binding to both FcγR and to C1q, relative to Bapineuzumab (see Table **2**).

Treatment with AAB-003 revealed reduced plaque burden similar to Bapineuzumab in an APP transgenic model but reduced the incidence of microhaemorrhage and phagocytosis. Thus, the AAB-003 antibody has reduced capacity to induce both phagocytosis and the complement cascade. However, it should be noted that AAB-003 still binds murine FcγRs, though with a reduced affinity (see Table **2**). Therefore, it can be expected that the clinical safety of AAB-003may be more significantly reduced compared to Bapineuzumab than it can be extrapolated from animal studies.

Currently two clinical Phase I trials are ongoing to evaluate the safety (NCT01193608) and the tolerability (NCT01369225, open label extension study) of AAB-003.

AAB-003 appeared to serve as line extension for Bapineuzumab for the treatment of *ApoE4* carriers, thus roughly 65% of all AD patients, in order to prevent ARIA [242]. After the clinical failure of Bapineuzumab, AAB-003 is the only anti-Aβ backup in the Pfizer pipeline.

BAN2401 AGAINST Aβ PROTOFIBRILS DERIVED FROM THE ARCTIC MUTATION

BAN2401 is a humanized antibody against large-size Aβ protofibrils (>100 kDa) originally developed as mAb158 derived from mice immunized with protofibrils $A\beta_{42}$ or $A\beta_{42}$arc with an arctic mutation in the Aβ sequence. Arctic Aβ is much more prone to oligomerize than non-mutated Aβ [243]. BAN2401 was the first monoclonal antibody that binds selectively, neutralizes and eliminates soluble protofibrils. Vascular plaques of CAA are not recognized by the anti-protofibrils mAb since it does not bind fibrillar Aβ, therefore ARIA should be avoided using such mAbs. Moreover, it neither recognizes monomers nor full-length APP. In cell culture, BAN2401 lowered the binding of Aβ protofibrils to hippocampal neurons and neutralized their damaging effect on those cells [244]. This antibody was generated from BioArtic Neuroscience and Eisai obtained the global rights to study, develop, manufacture and market BAN2401 for the treatment of AD. Though, BioArctic Neuroscience retains the rights to market the antibody in Nordic countries. BAN2401 is currently in a Phase I study (NCT01230853) assessing for potential safety concerns, which include vasogenic edema, microhemorrhage, hypersensitivity reactions, hypotension, formation of ADAs or inflammation. By July 2012, Eisai is pursuing the development intends to submit regulatory applications for approval of BAN2401 in 2015 or 2016 [245].

SAR228810, A HUMANIZED IGG_4 AGAINST Aβ PROTOFIBRILS

SAR228810 is based on the murine mAb 13C3 developed and characterized at the Rockefeller University [246]. Synthetic $A\beta_{42}$ peptides were polymerized to form fibriliar protein structures and used to generate hybidomas by immunizing mice. The clones were screened to react with protofibrils but not with the low molecular weight (oligomeric) fractions of Aβ. The clone 13C3 was also used as diagnostic mAb to demonstrate a decline in the plasma levels of Aβ protofibrils during the onset of cognitive decline in the AD pathogenesis in a large prospective study [247]. 13C3 recognizes a conformational epitope of the protofibrillar Aβ peptide. Here, protofibrillar form means an oligomeric form of Aβ peptides, soluble *in vitro* and which can be isolated as an entity of molecular weight greater than 200 kDa, 300 kDa, 400 kDa or 500 kDa and which can fix agents such as thioflavin-S or Congo Red.

In 2007 a collaboration agreement was signed from Sanofi-Aventis with the Rockefeller University to give access to 13C3 [248], followed end of 2009 by an exclusive global licensing agreement to develop, manufacture and commercialize the mAb [249].

For humanization the variable chains were fused in an IgG_4 isotype [250] to reduce pro-inflammatory effector function compared to IgG_1 [231]. The humanized mAb displayed EC_{50} levels of 0.08 μg/ml for Aβ protofibrils and 14.7 μg/ml for low molecular weight Aβ [250]. In addition, the IgG binds to dense, mature senile plaques in AD brains and in an AD animal model while diffuse Aβ deposits and monomers are not recognized. Actually, a Phase I trial for safety and tolerability (NCT01485302) is ongoing [251].

BIIB037, A FULL HUMAN mAb DERIVED FROM AN ATYPICAL AD PATIENT

BIIB037/BART is a novel fully human IgG_1 and was generated using a Reverse Translational Medicine approach by Neurimmune screening anti-Aβ antibodies from an AD patient with an unusual stable clinical course. BIIB037 shows a high affinity for insoluble fibrillar human Aβ and a 100-fold less affinity for Aβ monomers. In addition, it decorates amyloid plaques in both AD brain and transgenic mice brain tissues, but does not cross-react with APP. BIIB037 enters dose-dependently the brain and binds to amyloid plaques after systemic administration in APP transgenic mice [252, 253]. To evaluate the chronic effects, a murine chimeric IgG_{2a} of BIIB037 was generated and administered to APP transgenic mice. BIIB037 chronic treatment in mice led to a significant reduction in amyloid burden but no Aβ plasma increase was observed. The Incidence of microhemorrhage and inflammation was not increased compared to vehicle treated group. Microglia in treated animals often encircled the plaques and were less dendritic [254], implying an important role of microglia in plaque clearance. Moreover, treatment with this mAb improved neurogenesis and neurogenesis related markers, angiogenesis and synaptic plasticity [255].

Neurimmune partnered with Biogen Idec for the worldwide development and commercialization of BIIB037 for AD treatment in November 2007 [256]. A Clinical Phase I trial (NCT01397539) is currently ongoing.

OTHER PASSIVE IMMUNOTHERAPY MOLECULES IN PRECLINICAL STAGES

A-887755, an Aβ Globulomer Specific mAb

$A\beta_{20\text{-}42}$ globulomers, a condensed and hydrophobic oligomer, was used to immunize mice, in order to raise antibodies that were now able to discriminate between Aβ oligomers and all other Aβ species, especially monomers and fibrils and soluble APP [130-133]. The antibody A-887755 developed at Abbott has a high affinity on immobilized $A\beta_{20-42}$ globulomers (K_D 1.56 nM) and detects natural Aβ species in AD and APP transgenic mice [257] but not in non-demented age-matched control patients nor in vascular Aβ deposits, limiting the risk for ARIA. In addition, it neutralizes Aβ oligomer-derived neuropathogenic effects in neuronal cell and slice cultures and restored synaptic transmission [258]. Chronic passive immunization with A-887755 at an age when cognitive deficits occur normalizes cognition and synaptic spine density in APP transgenic mice. In addition, A-887755 also reversed hyperactivity in APP transgenic mice [259].

Therefore, the authors conclude that the globulomer-specific mAb A-887755 does not interfere with other Aβ species than globulomers and is not consumed by binding to non-disease related Aβ species, in particular Aβ monomer. These characteristics synergistically provide globulomer specific antibody A-887755 with a highly favorable profile compared to Aβ unselective antibodies for AD immunotherapy. In 2010, Abbott announced that the company is currently pursuing immunotherapy strategies for AD, but the programs were in early-stage development [260].

A Non-mAb Anti-Aβ Scaffold

A single domain antibody is a 12-15 kDa antibody fragment consisting of the variable heavy chain of so called heavy chain antibodies (VHH) lacking the light chains and are found in cartilaginous fishes and camelids [261]. Due to an extended CDR3 loop VHH can bear similar affinities than Fab fragments or single chain variable fragments (scFv), although they only have three CDRs for antigen binding compared to six for a Fab/scFv.

Anti-Aβ VHHs against the N-terminus and a central epitope of Aβ were fused to achieve avidity by binding two epitopes in one Aβ molecule [136]. Monomeric and fibrillar Aβ is captured at low picomolar levels, pyro-Glu modified Aβ at low nanomolar levels and also amyloid plaques in APP transgenic mice. Moreover, Aβ aggregation is reduced and total plasma Aβ increases while unbound (bioactive [262]) Aβ decreases intensely in treated mice. Since VHHs are lacking an Fc-domain, no CDC, ADCC or other Fc-mediated pro-inflammatory reactions like ARIA are expected [263]. Thus, Fc-mediated proposed mechanism of action for Aβ immunotherapy with VHH can be excluded. Therefore, peripheral sink, anti-aggregation of Aβ or central sequestering remain as mechanisms. An alliance of Boehringer Ingelheim and Ablynx developed this polypeptide that is being prepared for Phase I [264, 265].

mAb Against Pyro-Glu Modified Aβ Oligomers

The pE3-Aβ oligomer-specific antibody 9D5 was generated by immunizing mice with pE3-$A\beta_{38}$ [266]. The 9D5 antibody detected only low-n oligomeric fractions (4–10×) of pE3-Aβ but not monomeric Aβ and inhibited pE3-Aβ aggregation *in vitro*. It stained AD brains specifically and mainly intracellulary. Occasionally, weak 9D5 blood vessel immunostaining was observed. Passive immunization of an AD mouse model with 9D5 demonstrated reduction of the intracellular pE3-Aβ oligomers. pE3-Aβ removal has also been recently proven by passive immunization with another highly specific mAb (mAb07/1) in a chronic treatment with APP transgenic mice [112]. Total plaque deposition was significantly reduced and gliosis appeared to be reduced while soluble Aβ levels were not affected. Microhemorrhage was not observed with anti-pE3-Aβ immunotherapy. mAb07/1 showed no crossreactivity with full-length Aβ or truncated, noncyclized $A\beta_{3-X}$. Moreover, pyro-Glu modified neuropeptides or hormones were not recognized by mAb07/1, suggesting a very high specificity of the molecule without potential of side effects. Further preclinical experiments are planned.

A mAb Against Oxidatively Modified Forms of Aβ

Oxidatively modified forms of the Aβ oligomers found in the AD brain [267], could serve as immunological targets for vaccine development [268]. Prana Biotechnology has identified the mAb 4E10, which targets a proprietary pathological Aβ target epitope but not the normal, endogenous Aβ. According to their annual report, technical hurdles have delayed the progress of the preclinical project, but the company aimed to re-initiate work on the mAb by 2012 [269].

IVIG, The Human-Derived Polyclonal Nabs-Aβ, Which Convince in the Clinic

Natural polyclonal Anti-Aβ antibodies (Nabs-Aβ) occur in healthy humans and therefore also in pooled preparations of intravenous immunoglobulin (IVIG) derived from healthy donors [166]. IVIG preparations are already authority-approved for the treatment of immunosuppression, autoimmunity and a variety of other neurological conditions.

Nabs-Aβ were found to interfere with the oligomerization and fibrillization of Aβ, to be neuroprotective and to promote the clearance of cerebral Aβ plauqes [270, 271], suggesting that IVIG treatment may be useful in humans as a type of passive immunotherapy to treat AD [142, 272]. Interestingly, IVIG treatment due to other indications reduced the risk of developing AD by more than 40% [273].

After the discovery of NAbs-Aβ in human plasma and thereby in IVIG, a pilot study for the use of IVIG in AD was initiated [152]. Five patients with AD were enrolled and received monthly IVIG (Octagam/Octapharma) over a 6 month at a dose of 0.4g/kg. Serum Aβ increased significantly subsequent to IVIG by more than 230% while Aβ CSF levels decreased significantly by 30% compared to baseline. Cognition improved on the ADAS-Cog scale by 3.7 points, but the scores in the mini-mental state score were not significantly altered following treatment (improved in four patients, stable in one patient). No serious adverse events associated with IVIG were noted during the clinical trial. There were neither clinically relevant changes in blood pressure, heart rate, and electrocardiogram findings nor clinically relevant differences in haematological or biochemical laboratory test values.

An open label dose-ranging study in eight mild AD patients, that received IVIG (Gammagard S/D Immune Globulin Intravenous Human, donated by Baxter Healthcare Corporation; 0.4 – 2 g/kg) for 6 months, discontinued, and then resumed for another 9 months was reported in 2007 [155]. IVIG was used as add-on to approved AD therapies. The amount of Nabs-Aβ in patient's serum increased dose-dependently with IVIG treatment. Interestingly, the half-life of the Nabs-Aβ was reduced compared to non-Aβ binding IgGs (anti-hepatitis antibodies). Plasma Aβ levels increased transiently after each infusion. The levels of CSF Aβ decreased significantly after 6 months IVIG treatment, returned to baseline after washout and decreased again after IVIG was re-administered for additional 9 months, suggesting that the antibody mixture induced efflux of Aβ from the brain to the periphery. Remarkably, cognition measured by mini-mental state scores increased an average of 2.5 points after 6 months, returned to baseline during washout and remained stable during subsequent IVIG treatment. Again, the IVIG infusions were generally well-tolerated.

In a subsequent small Phase II study sponsored again by Baxter, IVIG (Gammagard) exhibited a dose-dependent effect on brain atrophy [274]. Enlargement of the cerebral lateral ventricles is known to occur as a consequence of brain atrophy in AD. This increase in ventricular volume is correlated with cognitive decline and AD neuropathology [275]. Among 14 IVIG-treated patients that underwent volumetric MRI at baseline and after 18 months, the yearly increase in lateral ventricle volume measured with volumetric MRI was lowest in patients treated with the highest dose of IVIG over time. The volume of the lateral ventricles significantly increased 50% less per year during treatment with IVIG, compared with the annual rate of the six placebo-treated patients. The reduction in brain atrophy was significantly correlated with improvement in clinical outcomes at 18 months on the Clinical Global Impression of Change and the cognitive subscale of the Alzheimer's Disease Assessment Scale. Notably, patients' baseline characteristics were not correlated with volumetric MRI outcomes [276].

The determination of avidity constants of various NAbs-Aβ found in human plasma and IVIG [277] identified at least two pools of human NAbs-Aβ with distinguishing avidities: lower avidity antibodies with K_Ds in the micromolar range and higher avidity antibodies with K_Ds in the range of tens of nanomoles. Conversely, the reported benefits of IVIG in Alzheimer's patients followed an inverted dose response curve [155], with higher doses showing less apparent efficacy. Thus, the avidities of the IVIG NAbs-Aβ used should be specified to employ doses that better reflect the natural abundance and/or the levels of antibodies achieved in patients after IVIG infusions [278].

Anti-Aβ antibodies were detected in CSF of the patients subsequent to IVIG treatment and even earlier IVIG were found to be able to pass the blood brain barrier [279], indicating that IVIG antibodies may cross the blood-brain barrier and thereby also affect Aβ levels in the brain.

Though, the pilot studies only included a small number of patients, the preliminary clinical results, in accordance with preclinical data, justified a larger clinical trial to proof the effects of IVIG and human NAbs–Aβ in patients with AD [129]. Several Phase III trials are currently ongoing: a Phase III trial sponsored by Baxter and the NIH Alzheimer's Disease Cooperative Study (Gammagard Liquid; NCT00818662), a Phase II trial by Octapharma (Octagam10%; NCT00812565), a Phase II trial sponsored by Sutter Health (NewGam10%, NCT01300728) and a Phase II trial by Grifols Biologicals Inc evaluating the combination of IVIG (Flebogamma DIF5%) with human albumin (Albutein 20%) after plasmapheresis (NCT01561053).

Recently, preliminary data from the extension of a Gammagard Phase II study was presented at the Alzheimer's Association International Conference 2012 [280] and showed a three-year stabilization of AD symptoms with IVIG, including no decline on measures of cognition, memory, daily functioning and mood. The 24 participants in that study received six months of treatment followed by a 12-month open-label extension where all subjects received IVIG. Several doses were tested. To evaluate the long-term effects of IVIG, participants were offered additional IVIG treatment at a single standardized dose (0.4 g/kg every 2 weeks) for an additional 18 months. Sixteen of the originally enrolled subjects received treatment through month 36, including five originally given placebo and 11 treated with various doses of IVIG. It was observed that, study participants who were treated with IVIG 0.4 g/kg every two weeks for the full 36 months (n = 4) had the best outcome, with no cognitive decline. As a group, the 11 participants who received IVIG for the full 36 months had favorable outcomes in terms of their thinking abilities, behavior and daily function. And the five participants who were initially treated with a placebo and then switched to IVIG declined while on placebo but experienced less rapid decline while receiving a uniform dose of IVIG.

The above mentioned studies showed that IVIG treatment appeared to be efficacious in the treatment of AD [152, 155] and advantages to IVIG include that this blood product already has a long clinical track record, it is generally safe and well-tolerated, and it circumvents the high research and manufacturing costs associated with mAbs [281]. One potential disadvantage is the reliance on human blood donation; increasing demand for IVIG for other indications has already led to seasonal shortages. Modified purification processes for IVIG are discussed to increase the yield and thus the quantity [282], but this would also significantly affect the manufacturing costs.

Although, the enrichment of Nabs-Aβ from bodyfluids was patented quite early [283], there is no other commercial source than IVIG up to now, possibly due to the relatively low production costs of IVIG compared to the biopharmaceutical production of mAbs.

REGULATORY AND MARKET CONSIDERATIONS FOR ANTI-Aβ mAbs OR SCAFFOLDS

Market Potential for Passive Aβ-Immunotherapy

The 2009 sales of the four major AD drugs as reported by companies were US$ 6.65 billion with donepezil as the market leader. If symptomatic AD drugs can achieve blockbuster sales within a few years after approval, disease-modifying drugs, which can stop disease progress, could become blockbuster drug within 2 years after launch. A safe and effective drug that can stabilize, stop or reverse the disease and restore memory or cognitive functions could be a blockbuster straight away. Several companies and analysts estimate the market potential at US$ 8 billion for peak sales.

A business report on Alzheimer's disease market development estimates that in the seven major markets it will more than double over the next 10 years, from US$5.8 billion in 2011 to US$14.5 billion in 2020. The catalysts for this growth include an increasingly elderly population, earlier and improved diagnosis, and the introduction of immunotherapies [284]. Anti Aβ immunotherapies such as Bapineuzumab/AAB-003, Solanezumab, Gammagard or others have the potential to grow the market by US$7.6 billion per year. According to business analysts' forecasts around 600,000 patients with mild cognitive impairment or mild to moderate AD will receive immunotherapies annually by 2020 in the seven major markets.

Although, mAbs are expensive to manufacture, they have nevertheless generated top profits because of historically generous reimbursements [285]. Interestingly, mAbs are somehow immune to generic competition as the development of biosimilars is not as easy as for small molecule generics. In an optimistic view a successful disease-modifying anti-Aβ mAb would be administered chronically and extend the duration of AD that is at the same time expected to increase dramatically in incidence. In other words, the growth of the market would be multiplicative [281]. However, a more realistic view is that due to the current political and economic environment, the healthcare market will not sustain this level of expansion [285]. In case an active vaccination or an oral medication emerges with disease-modifying effects in AD, a mAb would likely have to demonstrate sizable advantages either in efficacy or safety in order to survive on market.

Nevertheless, Johnson and Johnson agreed to pay US$ 1.5 billion dollars to buy 18% stake in Elan and get majority share of its AD R&D portfolio including Bapineuzumab and AAB-003 [286, 287]. Several analysts have rated the move risky due to high failure rates of new drugs in AD. On top Johnson and Johnson has to share market sales and profits with Pfizer and pay a royalty to Elan.

Clinical Readouts to Achieve Marketing Authorization for Disease Modification

In spite of the current knowledge of the pathogenesis of AD – however, mostly based on nonclinical evidence – and possible therapeutic targets being nowadays identified, no new 'challenging' medicinal product has been granted for AD treatment by the European Medicinal Agency (EMA) since the approval of Axura and Ebixa (Memantine hydrochloride) in 2002. Whereas acetylcholinesterase inhibitors such as Donepezil (the reference marketing authorization is Aricept, first granted in the United Kingdom in 1997 to Eisai Limited), Rivastigmine (the reference medicinal product being Exelon, granted in 1998 by the EMA) and Galantamine (the reference medicinal product being Reminyl, first granted in Sweden to Janssen-Cilag AB in 2000) are all indicated for the symptomatic treatment of mild to moderately severe AD, only Memantine containing medicinal products (an NMDA receptor antagonist) are indicated for moderately severe to severe AD. However all the above-mentioned medicinal products, unfortunately, only provide a temporarily help since they act in a symptomatic way; their overall effect on the impairment as such is transitory. Hence, those approvals are based on symptomatic improvement of AD dementia; the treatment of AD and other dementias, therefore, still remaining an open research field [288].

Following the Committee for Medicinal Products for Human Use (CHMP) opinion, the treatment of AD should ideally improve the global condition of the patient in order to fulfill the conditions for marketing authorization in the EU by seeking to what follows:

- Improve or at least slow down the loss of memory and cognition. It is believed indeed that symptomatic improvement may consist in enhanced cognition, more autonomy and/or improvement in neuropsychiatric and behavioral dysfunction.

- Help maintain independent function and finally, control the behavioral symptoms associated with the disease [289].

In other words, criteria for assessment of the treatment efficacy with the aim to grant a marketing authorization are associated with two primary endpoints as well as a secondary endpoint, the so-called global endpoint. These criteria are as follows:

1) cognition, as measured by objective tests (cognitive endpoint);

2) activities of daily living (ADL) as functional endpoint;

3) an overall clinical response, as reflected by global assessment (global endpoint).

According to the current CHMP guideline on medicinal products for the treatment of AD and other dementias all the above-mentioned endpoints must be accompanied by strong supportive evidence from a biomarker program (surrogate endpoint).

One can expect the IVIG's disease-modifying effect to be confirmed in the ongoing Phase III IVIG 10% study is conducted in patients in early/mild AD stage with patients within 50 to 89 years of age. What is meant by disease modifying effect from a regulatory point of view is that the progression of the disease as measured by cognitive and functional assessment tools can be reduced or slowed down. Moreover, these results should be linked to an effect on the underlying disease process [288]. Following the aforesaid, IVIG products may thereby fulfill all the criteria for assessment of the treatment efficacy. The study results are in awaited February 2013 will demonstrate, if a marketing authorization approval for IVIG 10% can be expected in the next future.

Many of the proposed therapeutic strategies in connection with biopharmaceuticals are focused on the amyloid cascade hypothesis. Therefrom, apart from the development of biopharmaceutical seeking either the amyloid reduction, brain penetrant inhibitors of Aβ aggregation have nowadays also been identified with the aim to target the amyloid aggregation in the brain.

Given the fact that increased plaques accumulation in the brain can take place well before clinical symptoms are observed, moreover clearance of the N-terminal domain of Aβ has yet not provided sufficient satisfactory results as it has not shown so far disease-modifying effects with respect to the cognitive impairment [190, 191], one may believe that,

- At once, the reduction of Aβ plaques solely cannot be considered as an outcome measure in the effective treatment of AD,

- Second, there is reasonable ground to question the collective evidence suggesting that a reduction or a total clearance of plaques at any stage of the disease anyhow, may have the potential to intervene in the AD disease, leading to the recovery of cognitive function.

Knowing the fundamental limitations of the above-mentioned strategy, are the mAb experience targeted against amyloid plaques still relevant and crucial for a strategic AD treatment?

Passive and active immunization with various approaches is broadly discussed, tested and still being investigated in several on-going large Phase III clinical trials. Most of them can be still considered as key studies, where a total of 6000

AD patients were enrolled. Such studies concern Bapineuzumab and Solanezumab even though both studies were discontinued in 2012:

1) A double blind, placebo controlled, randomized, outpatient multiple dose study in male and female patients with mild to moderate AD, sponsored by JANSSEN Alzheimer Immunotherapy Research & Development, LLC in collaboration with Pfizer (ClinicalTrials.gov Identifier: NCT00574132). Patients with 50 to less than 89 of ages were randomized to receive either Bapineuzumab (AAB-001, ELN115727) or placebo in *ApoE4* non-carrier. The study was composed of three arms encompassing the two doses, 0.5 mg/kg and 1 mg/kg of Bapineuzumab and placebo.

2) Effect of LY2062430 (Solanezumab), an Anti-Amyloid Beta Monoclonal Antibody, on the Progression of Alzheimer's Disease as Compared With Placebo (EXPEDITION and EXPEDITON 2; respective ClinicalTrials.gov Identifier: NCT00905372 and NCT00904683), sponsored by Eli Lilly and Company. Patients aged 55 or older were randomized to receive either LY2062430 or placebo with the aim to targeting Aβ in AD. The study encompassed two arms with 400 mg of LY2062430 and Placebo, intravenously administrated.

3) Continued Efficacy and Safety Monitoring of Solanezumab, an Anti-Amyloid β Antibody in Patients With Alzheimer's Disease (EXPEDITION X; NCT01127633). As mentioned by the sponsor, this study is an open-label extension study in Alzheimer's patients who have completed participation in one of the above-mentioned Solanezumab trials.

In Phase II and Phase III trials regarding Bapineuzumab, unfortunately, no satisfactory results with respect to cognitive and functional endpoint could be provided. Beyond that, serious safety concerns for *ApoE4* carriers were claimed since they have developed cerebral vasogenic edema as the result of adverse event due to the intake of Bapineuzumab. However, following the outcome of the study in phase II, potential disease-modifying effects were expected for *ApoE4* non-carriers from, the CRO conducting the study, justified by the lack of statistical precision and doses variation. Unlike Bapineuzumab, no such safety concerns were claimed at first for Solanezumab but still, the ADAS-Cog and CSF Tau levels were unchanged in the mild to moderate AD patients treated, despite the clearance of toxic plaques. At the end, both the Bapineuzumab und Solanezumab clinical trials failed to meet their primary endpoints as well as the global endpoints as mentioned above.

Because of these unsatisfactory results, one can speculate on the fact that circulating antibodies to Aβ shift the equilibrium of the peptide from the CSF to the plasma; indirectly reducing the brain's Aβ burden and therefore, not the elimination of plaques may be critical for a cognitive improvement but the shift in equilibrium (peripheral sink) [280]. Nevertheless, currently investigated treatments may be rather crucial for the development of a potent AD medication for a patient subgroup like *ApoE4* non-carriers or those having a genetic predisposition at risk for early-onset AD on the other side, since the conditions for marketing authorization could be justified by delaying the AD progression somehow. Genentech/Roche has now taken the challenge to test such medicinal product with a preventing treatment in individuals at risk to develop AD early in life: The participants share a rare genetic mutation in *presenilin-1* that typically triggers Alzheimer's symptoms around an age of 45. Apart from the genetic mutation, the individuals have apparently the same rural background and similar activities, which could help to lower the potential for considerable variance in cognitive test scores influenced for example by level of education, time interval of testing *etc.* According to the Genentech, Crenezumab is thought to stimulate microglia just enough to clear Aβ, and therefore do not induce inflammatory responses. That means no big safety concerns are expected at high doses, when compared to other passive immunotherapies. 300 patients aged 50 to 80 will be enrolled to participate in a randomized, double-blind, placebo controlled, parallel-group, multicenter, Phase II Study to Evaluate the Efficacy and Safety of Crenezumab in patients with mild to moderate AD (Long term safety extension study NCT01723826). The study will consist of two arms encompassing respectively a repeating subcutaneous injection and a repeating intravenous infusion of Crenezumab, both tested against placebo.

What does preventive measure means in terms of therapeutic and regulatory strategy? In general, a strong connection with AD is made with elderly and thus, a suspicion for patient less than 60 of age might be excluded or possibly unpronounced by clinicians unless a genetic predisposition has been notified beforehand along with a preventative check-up. Moreover in the course of daily life, none may feel the necessity to undergo a treatment on spec unless AD signs and symptoms are assigned using the 'Disability Assessment in Dementia' or the 'Alzheimer Disease Cooperative Study ADL scale'. A possible justification for that lies in the nature of the subject matter as it has long been perceived and confirmed later on by *'The public health principles and neurological disorders'* published by the WHO [290] saying that only 2% of AD cases start before the age of 65 years. In particular, how big are the chances for clinicians to detect the time point of which the silent plaques formation commences without any visible signs of the disease? Basically, the odds to ascertain subsyndromal or prodromal AD in patients, thus to detect early AD process in individuals is still today rather unlikely and must be crucial point for consideration. Likewise, a growing demand from a high number of scientists has arisen nowadays to target a best possible time point at which patients would potentially be tested positive for AD but do not have clinical symptoms. Thanks to the use of some biomarkers, it is today rather clear that treating patients with AD symptoms is too late, since too much irreversible brain damage has meanwhile occurred making an effective treatment with long lasting results almost impossible. On the other hand, the use of PET, CSF and MRI biomarkers made it possible to test patients for a clinically probable AD. This can be highly beneficial to trigger preventive measure against the progression of the disease.

Also, other test methods for the prevention of AD have commenced very recently such as those studying the resting brain activity, using a new functional MRI (fMRI-SAM) procedure (NCT00412048). The procedure is being applied to 25 AD patients, 25 patients suffering from amnestic MCI - a clinical picture, which may be a prodromal form of AD - and 60 healthy controls. The first analysis of the data will search differences of brain activation profiles between the three groups. In the second step, the investigators will study the predictive value of fMRI-SAM to detect MCI patients, who will later convert to AD.

> *"At the current stage of knowledge, it is impossible to predict who will get AD. It can strike anyone irrespective of gender, caste, creed, culture or socio-economic status" [291].*

WHAT WILL THE BIOLOGICAL TREATMENT LANDSCAPE FOR AD COULD LOOK LIKE IN THE NEXT FUTURE?

According to the report of WHO South-East Asia Region [292], the intake of hormones such as the estrogens in menopausal women, NSAIDs, antioxidants such as vitamin E, vitamin B and lipid-lowering agents may lower the risk of getting AD. Also the CHMP encourage the initiation and maintenance of good health. Eliminating potential causes of disease in subsyndromal individuals or in those with likelihood risk factors for dementia such as individuals who have insulin resistance, hypertension, high cholesterol or those with *ApoE4* status may represent a measure to reduce the probability. As referred by the CHMP, apparently, length of follow-up, timing in relation to possible dementia onset, use of valid outcomes, which are sensitive to change, *etc.* were neither considered nor justified [288].

What should be given a lot of attention is the IVIG pilot study [152] as already mentioned above. Patients were enrolled, when the criteria for either 'clinically probable' or 'clinically possible' AD was met. Five patients took part to the study, from 56 to 64 of age. According to the published literature, the patient received IVIG (Octagam, Octapharma) at a total dose of 0.4 g/kg body weight on three consecutive days every 4 weeks over 6 months. Following the results of the study, the cognitive test ADAS-Cog could show an improvement in four patients and stabilization in one patient. But unfortunately, the results of this study could not be considered discriminating because of the small number of patients studied and the potential for considerable variance in cognitive test scores influenced by age, gender, level of education and time interval of testing, should the number of patients increase. Those results were confirmed in a placebo controlled Phase II study in mild to moderate AD in which, IVIG (Gammagard, Baxter) has been studied as an immunotherapy for AD. More participants (24) were enrolled in that study, where they received six months of treatment, followed by a 12-month open-label extension where all subjects received several doses of IVIG. To evaluate the long-term effects of IVIG, participants were offered additional IVIG treatment at a single standardized dose (0.4 mg/kg every 2 weeks) for additional 18 months. Sixteen of the originally enrolled subjects received treatment through month 36, including five originally given placebo and 11 treated with various doses of IVIG [280]. Just as in the pilot study encompassing five patients, the intake of IVIG 0.4 g/kg every two weeks for a time period of 36 months showed a very satisfactory outcome with respect to several standard measures of cognition, memory, daily functioning, mood as well as behavior.

As mentioned above, larger Phase III clinical IVIG trial in AD patients are ongoing to confirm those promising results. The encouraging clinical results of IVIG raise many regulatory questions: What if, a positive benefit/risk profile during the large Phase III study and a marketing authorization is granted afterward? How do the applicants of a future IVIG medication and the regulatory authorities will be phrasing the indication statement? Would the indication phrasing include age differentiation and/ or limitation taking into account the heterogeneity of patient groups? How does the ideal population may look like for prevention trials to delay the onset or slow the progression of cognitive decline?

The CHMP has expressed within a recent concept paper [293] the following statement:

> *"If a treatment claim for disease modification is strived for, it has to be shown that the treatment has an impact on the underlying pathophysiology of AD in addition to clinical improvement ... In addition the progress in qualification and validation of biomarkers might play an important role in future drug development for AD, and regulatory requirements in this process should be specified based on recent scientific advice procedures".*

Based on the above-mentioned results and current scientific findings, a revision of the guidelines should be strived to consider and to give the chance for subgroup treatments for AD to take place and furthermore define new criteria allowing the granting of new marketing authorization preventing AD in prodromal individuals. That marketing authorization may be for example granted to allow therapy strategies consisting of a first diagnosis *via* biomarkers such as PET imaging or biochemical CSF markers followed by a subsequent treatment with monoclonal antibodies, depending on the stage of the disease and/or the genetic history. One particularity of AD might be the finding of personalized treatments divided in geriatrics and those having a genetic predisposition due *e.g.* to a mutation of the *presenelin-1* gene, the *ApoE4* status *etc.*, when considering mild to moderate impairment. Also, based upon current scientific evidence, A β immunotherapy can be considered as proactive measure to prevent or reduce the high incidence of the disease.

In general, should any of the anti-A β mAbs listed above investigational medicinal products shows similar modest effects on cognition than Donezepil and Memantine, they might not be the first choice within the EU member states, unless a detection of AD plaques *via* MRI, CSF or PET biomarker has taken place prior to the treatment of the impairment. A differentiation could then be made, in case they are used to prevent a rapid progression of the AD, but costs benefits might remain an issue in the general public.

Should the running Phase III study for IVIG confirms the positive effects on the cognition to mild-to-moderate AD, an wide acceptance from the public based on well-established treatments in connection with other therapeutic areas is expected upon marketing authorization, particularly because it has been safely used in many of patients. IVIG treatment indeed, has been available for many years as it is approved for the US territory by the FDA and *via* a mutual recognition procedure in several EU member states for the treatment of several immune disorders.

In case the Phase III studies for Gammargard/ Octagam (IVIG) are completed, the timelines for time to market would be very advantageous in contrast to the other investigational biological medicinal products for the following reason: Adding the new AD indication to the already registered therapeutic area would cause a variation of the corresponding Summary of the Product characteristic, which take approximately 6 to 9 months. Other investigational medicinal products mentioned above will have to enter first a new application procedure for marketing authorization (MA) taking approximately 310 days following a six months scientific advice by the EMA (even longer at the FDA) before entering the market. The off-label used status of IVIG for AD treatment could be therefore abrogated.

SECOND GENERATION/BIOBETTER

Clinical results from Bapineuzumab and Solanezumab suggest that these mAbs may be beneficial in some patients. Though, use of Bapineuzumab appears to be associated with the occurrence of vasogenic edema, an urgent serious adverse event [294]. A number of Bapineuzumab's direct competitors were explicitly designed or some even re-designed to decrease the risk of inflammation and vasogenic edema and to increase efficacy in decreasing amyloid burden. The development of conformation-specific mAbs selectively directed against Aβ oligomers and protofibrils would hopefully lead to better clinical outcome as they leave the Aβ monomers intact, thus preserving its physiological functions.

Below two examples from Pfizer's pipeline and examples from Genentech and Sanofi are shown, how to potentially decrease the risk for inflammation and ARIA.

ELIMINATING OR REDUCING EFFECTOR FUNCTION

In case pro-inflammatory effector functions contribute to adverse events, an IgG can be re-designed by Fc engineering or by changing the IgG subclass.

Deglycosylation of the Fc-domain alters FcR binding of IgGs. Rinat developed a deglycosylated anti-Aβ C-terminal specific antibody (De-2H6, parent molecule for Ponezumab). Intracranial injection into the right frontal cortex and hippocampus of an APP transgenic mouse model demonstrated that the De-2H6 antibody had slightly lower affinity for several murine Fcγ receptors and human complement than intact 2H6 had. De-2H6 exhibited reduced binding to its effector proteins responsible for opsonization and the activation of microglia (FcγRI, FcγRIIb, FcγRIII and C1q; see Table **2**). Both antibodies reduced Aβ pathology in the mouse brain. In animals treated with the intact 2H6 antibody, a significant increase in FcγRII and FcγRIII immunoreactivity and for the microglial activation marker CD45 was observed compared to animals treated with a control IgG. No increase in FcγRII/III and CD45 was found with the De-2H6 antibody. Vascular amyloid and microhemorrhages were reduced in De-2H6 vaccinated mice [233]. Importantly, deglycosylation had no effect on the affinity of 2H6 for its antigen, $A\beta_{40}$.

Table 2: Affinities (K_D) of 2H6 and deglycosylated 2H6 (De-2H6) for antigen and effector proteins (taken from [233])

Antibody	Aβ 1–40 (nM)	mFcγRI (μM)	mFcγRIIb (μM)	mFcγRIII (μM)	hC1q (μM)
2H6	8	16	20	39	5
De-2H6	9	6,5	30	67	30

AAB-003 is a humanized 3D6 mutated in the Fc domain to decrease affinities to FcγR and C1q binding [242].

Table 3: Human FcγRs binding capacities of Bapineuzumab and AAB-003 (taken from [242])

	Relative Binding Capacity (%)		
3D6 derivative	**Human FcγRI**	**Human FcγRII**	**Human FcγRIII**
Bapineuzumab IgG1	100	100	100
AAB-003 IgG1 (humanized 3D6 3m)	0	8-12	0

Besides deglycosylation, antibody fragments without Fc-domains were shown to be effective in clearing Aβ plaques without mediating Fc effector function [147].

The recent advance of the IgG_4 mAbs Crenezumab and SAR228810 is due to the fact that IgG_4 mediates pro-inflammatory effector functions less than IgG_1. The table below demonstrates the FcγR affinities for IgG_1 and IgG_4 (it should be noted that murine IgG and human recombinant FcγRs were used in this study [231]).

Table 4: Differential binding affinities oft he IgG subclasses IgG1 and IgG4 on the ectodomains of human FcγR (adopted from [231])

	Mean K_A Binding Affinities to Polymorphic Variants of FcR Ectodomains (M^{-1})				
Subclass	FcγRI	FcγRIIA	FcγRIIB	FcγRIIIA	FcγRIIIB
IgG_1	6.5×10^7	4×10^6	2×10^5	1.5×10^6	2×10^5
IgG_4	3.4×10^7	2×10^5	2×10^5	2×10^5	not determined

Apart from FcγRI, binding affinities to pro-inflammatory FcγRIII and also C1q [295] are significantly reduced for IgG_4 while binding to the inhibitory FcγRIIB is not altered. This means that IgG_4 can mediate opsonization and phagocytosis but only weakly ADCC and CDC.

CONCLUSION

Aβ is a difficult target, although it has been intensely studied for more than two decades. There are too many different Aβ species, forming too many different conformations and last but not least, the physiological role of Aβ, its precursor and the other fragments generated during secretase cleavage is still a matter of debates.

The abundance of anti-Aβ mAbs and scaffolds in the pipelines of pharmaceutical companies may look wired at the first view. In truth, not all of these antibodies are created equally and target different species of amyloid in the brain [223]. But like Eric Siemers from Eli Lilly said: "What's not known at this point is which species of Aβ treatment should target?" [223].

If AD and its dramatic prevalence in our aging population wasn't behind Aβ, it is unlikely that pharmaceutical and biotechnological companies would put so much effort, work and finally money in to studies and trials even after the discouraging outcome of AN-1742, Bapineuzumab and Solanezumab. Would they invest that for a therapy of a rare disease?

Currently, there is a small chance of proving the clinical efficacy of anti-Aβ mAbs in 'ordinary' clinical trials, since early diagnosis based on CSF and imaging biomarkers for a successful treatment with disease modifying drugs has still not been achieved. One – probably last – chance would be preventive trials in familial AD cases that were currently started with Crenezumab. Remarkably, efficacy will be tested in a five-year prevention trial in an extended Columbian family carrying an AD-related mutation, which causes early onset AD [234]. Moreover, in 2013 the "Dominantly Inherited Alzheimer's Network", DIAN, will launch clinical trials with Solanezumab and Gantenerumab in large families with genetic mutations that make them susceptible to the disease [296].

In the competition of Bapineuzumab, Solanezumab, Gantenerumab, Crenezumab and AB-003 rushed the off-label use of IVIG for passive AD immunotherapy and could – if the Phase III data holds what the Phase II data promised - at least for a certain time win the race. However, to make IVIG treatment even more effective and obtainable, the use of recombinant biopharmaceutical engineered Nabs-Aβ possibly cloned from B cells of successfully aged non-demented humans appears very feasible.

In addition, a preventive treatment may probably commence when brain aging processes starts by the age of 40 [297] if not according to Braak already in the 30s [9]. A passive immunotherapy drug when used for AD prevention should be marketed for a reasonable prize affordable from health care systems of an aging society, thus reasonable cost of goods. If this will not be the case, preventive passive AD immunotherapy will be accessible only to a minority of prodromal 'patients' or – more likely – will not be able to compete with active Aβ immunization that very recently demonstrated to be safe in a Phase I study [298]. In that light, passive immunotherapy may be used for known and newly diagnosed AD cases and younger may receive an active safe vaccination against AD to grow old in a world without AD.

"I now begin the journey that will lead me into the sunset of my life"

Ronald Reagan, 1994

After being diagnosed with AD

ACKNOWLEDGEMENTS

The authors thank René Handrick and Jens Moreth for helpful discussions.

CONFLICT OF INTEREST

The author(s) confirm that this chapter content has no conflict of interest.

AUTHOR'S CONTRIBUTION

CM supported with insight from regulatory affairs and KS supported with insight on Aβ immunotherapy und neuroimmunology. All authors drafted the manuscript. All authors read and approved the final manuscript.

LIST OF ABBREVIATIONS

aa	=	amino acid
Aβ	=	Amyloid-beta
AD	=	Alzheimer's disease
ADL	=	Activities of daily living
ADAS-Cog	=	Alzheimer's Disease Assessment Scale–Cognitive subscale
ADA	=	anti-drug antibodies
ADCC	=	antibody dependent cellular cytotoxicity
APP	=	Amyloid Precursor Protein
ApoE4	=	ApolipoproteinE4
ARIA	=	amyloid-related imaging abnormalities
BACE	=	beta-amyloid converting enzyme
BDNF	=	brain-derived neurotrophic factor
C1q	=	complement factor 1q
CAA	=	cerebral amyloid angiopathy
CDC	=	complement-dependent cytotoxicity
CDR	=	complementary determining region
CHMP	=	The Committee for Medicinal Products for Human Use
CRO	=	contract research organisation
CSF	=	cerebrospinal fluid

CT	=	computed tomography
DAD	=	Disability Assessment for Dementia
DIAN	=	Dominantly Inherited Alzheimer's Network
EMA	=	European Medicinal Agency
Fab	=	antigen binding fragment
Fc	=	fragment crystallizable
FcγRI	=	Fcγ receptor I
FcRn	=	neonatal Fc receptor
FDA	=	Food and Drud Administration
GSK	=	Glycogen synthase kinase
IDE	=	insulin-degrading enzyme
Ig	=	immunoglobulins
IGF	=	insulin-like growth factor
IL	=	interleukin
ITAM	=	Immunoreceptor tyrosine-based activation motif
IRS-1	=	insulin receptor substrate
IVIG	=	intravenous immunoglobulins
J&J	=	Johnson & Johnson
LRP	=	LDL receptor-related protein
MAC	=	membrane attack complex

MCI = mild cognitive impairment

MRI = magnetic resonance imaging

nAChR = nicotinic acteylcholine receptor

NAbs–Aβ = naturally occurring autoantibodies against Aβ

NMR = nuclear Magnetic Resonance

NT = neurotrophin

NGF = nerve growth factor

PiB = Pittsburgh compound B

pE- Aβ = pyro-Glu modified Aβ

phospho-tau = hyperphosphorylated tau

VHH = variable domain of heavy chain antibody

RAGE = receptor for advanced glycation end-products

SAWP = Scientific Advice Working Party

scFv = single chain variable fragment

TNF = tumor necrosis factor

WHO = world health organisation.

REFERENCES

[1] Alzheimer's Disease International. World Alzheimer Report 2009. Available from: http://www.alz.co.uk/research/world-report [cited: 5th Jul 2012].

[2] World Health Organization and Alzheimer's Disease International. Dementia: a public health priority 2012. Available from: http://www.who.int/mental_health/publications/dementia_report_2012/en/index.html [cited: 5th Jul 2012].

[3] Fitzpatrick AL, Kuller LH, Lopez OL, Kawas CH, Jagust W. Survival following dementia onset: Alzheimer's disease and vascular dementia. J Neurol Sci 2005;229-230:43-9.

[4] Alzheimer A. About a peculiar disease of the cerebral cortex. By Alois Alzheimer, 1907 (Translated by L. Jarvik and H. Greenson). Alzheimer Dis Assoc Disord 1987;1(1):3-8.

[5] Bergeron C, Pollanen MS, Weyer L, Lang AE. Cortical degeneration in progressive supranuclear palsy. A comparison with cortical-basal ganglionic degeneration. J Neuropathol Exp Neurol 1997;56(6):726-34.

[6] Belarbi K, Burnouf S, Fernandez-Gomez FJ, *et al.* Loss of medial septum cholinergic neurons in THY-Tau22 mouse model: what links with tau pathology? Curr Alzheimer Res 2011;8(6):633-8.

[7] Belarbi K, Schindowski K, Burnouf S, *et al.* Early Tau pathology involving the septo-hippocampal pathway in a Tau transgenic model: relevance to Alzheimer's disease. Curr Alzheimer Res 2009;6(2):152-7.

[8] Schindowski K, Belarbi K, Bretteville A, Ando K, Buee L. Neurogenesis and cell cycle-reactivated neuronal death during pathogenic tau aggregation. Genes Brain Behav 2008;7 Suppl 1:92-100.

[9] Braak H, Braak E. Neuropathological stageing of Alzheimer-related changes. Acta Neuropathol. 1991;82(4):239-59.

[10] Delacourte A, Sergeant N, Champain D, *et al.* Nonoverlapping but synergetic tau and APP pathologies in sporadic Alzheimer's disease. Neurology 2002;59(3):398-407.

[11] Sergeant N, Buée L. Tau Pathology: Cytoskeleton of the Nervous System. In: Nixon RA, Yuan A, editors. Advances in Neurobiology. 3: Springer New York 2011; pp. 83-132.

[12] Selkoe DJ. Alzheimer's Disease: Genes, Proteins, and Therapy. Physiological Reviews. 2001;81(2):741-66.

[13] Kuhn PH, Wang H, Dislich B, *et al.* ADAM10 is the physiologically relevant, constitutive alpha-secretase of the amyloid precursor protein in primary neurons. EMBO J. 2010;29(17):3020-32.

[14] Tachida Y, Nakagawa K, Saito T, *et al.* Interleukin-1 beta up-regulates TACE to enhance alpha-cleavage of APP in neurons: resulting decrease in Abeta production. J Neurochem 2008;104(5):1387-93.

[15] Verdile G, Groth D, Mathews PM, *et al.* Baculoviruses expressing the human familial Alzheimer's disease presenilin 1 mutation lacking exon 9 increase levels of an amyloid beta-like protein in Sf9 cells. Mol Psychiatry 2004;9(6):594-602.

[16] Hardy J, Allsop D. Amyloid deposition as the central event in the aetiology of Alzheimer's disease. Trends Pharmacol Sci 1991;12(10):383-8.

[17] Hardy J, Selkoe DJ. The amyloid hypothesis of Alzheimer's disease: progress and problems on the road to therapeutics. Science 2002;297(5580):353-6.

[18] Verdile G, Fuller S, Atwood CS, *et al.* The role of beta amyloid in Alzheimer's disease: still a cause of everything or the only one who got caught? Pharmacol Res 2004;50(4):397-409.

[19] Kayed R, Head E, Thompson JL, *et al.* Common structure of soluble amyloid oligomers implies common mechanism of pathogenesis. Science 2003;300(5618):486-9.

[20] Walsh DM, Klyubin I, Fadeeva JV, *et al.* Naturally secreted oligomers of amyloid beta protein potently inhibit hippocampal long-term potentiation *in vivo*. Nature 2002;416(6880):535-9.

[21] Takahashi RH, Almeida CG, Kearney PF, *et al.* Oligomerization of Alzheimer's beta-amyloid within processes and synapses of cultured neurons and brain. J Neurosci 2004;24(14):3592-9.

[22] Lazarov O, Lee M, Peterson DA, Sisodia SS. Evidence that synaptically released beta-amyloid accumulates as extracellular deposits in the hippocampus of transgenic mice. J Neurosci 2002;22(22):9785-93.
[23] Stokin GB, Lillo C, Falzone TL, *et al.* Axonopathy and transport deficits early in the pathogenesis of Alzheimer's disease. Science 2005;307(5713):1282-8.
[24] Lue LF, Kuo YM, Roher AE, *et al.* Soluble amyloid beta peptide concentration as a predictor of synaptic change in Alzheimer's disease. Am J Pathol 1999;155(3):853-62.
[25] Wyss-Coray T. Inflammation in Alzheimer disease: driving force, bystander or beneficial response? Nat Med 2006;12(9):1005-15.
[26] Iqbal K, Flory M, Khatoon S, *et al.* Subgroups of Alzheimer's disease based on cerebrospinal fluid molecular markers. Ann Neurol 2005;58(5):748-57.
[27] Eikelenboom P, Stam FC. Immunoglobulins and complement factors in senile plaques. An immunoperoxidase study. Acta Neuropathol 1982;57(2-3):239-42.
[28] McGeer PL, Itagaki S, Boyes BE, McGeer EG. Reactive microglia are positive for HLA-DR in the substantia nigra of Parkinson's and Alzheimer's disease brains. Neurology 1988;38(8):1285-91.
[29] Katsel PL, Davis KL, Haroutunian V. Large-scale microarray studies of gene expression in multiple regions of the brain in schizophrenia and Alzheimer's disease. Int Rev Neurobiol 2005;63:41-82.
[30] Blalock EM, Chen KC, Stromberg AJ, *et al.* Harnessing the power of gene microarrays for the study of brain aging and Alzheimer's disease: statistical reliability and functional correlation. Ageing Res Rev 2005;4(4):481-512.
[31] McGeer PL, McGeer E, Rogers J, Sibley J. Anti-inflammatory drugs and Alzheimer disease. Lancet 1990;335(8696):1037.
[32] Hoozemans JJ, O'Banion MK. The role of COX-1 and COX-2 in Alzheimer's disease pathology and the therapeutic potentials of non-steroidal anti-inflammatory drugs. Curr Drug Targets CNS Neurol Disord 2005;4(3):307-15.
[33] ClinicalTrials.gov. A Pilot Trial of Interferon Beta-1a in Alzheimer's Disease (REAL) - Study Results 2012. Available from: http://www.clinicaltrials.gov/ct2/show/results/NCT01075763 [cited: 19th Jul 2012].
[34] Tobinick E, Gross H, Weinberger A, Cohen H. TNF-alpha modulation for treatment of Alzheimer's disease: a 6-month pilot study. Medscape General Medicine 2006;8(2):25.
[35] Kar S, Slowikowski SP, Westaway D, Mount HT. Interactions between beta-amyloid and central cholinergic neurons: implications for Alzheimer's disease. J Psychiatry Neurosci 2004;29(6):427-41.
[36] Hasselmo ME, Stern CE. Mechanisms underlying working memory for novel information. Trends Cogn Sci 2006;10(11):487-93.
[37] Shen ZX. Brain cholinesterases: III. Future perspectives of AD research and clinical practice. Med Hypotheses 2004;63(2):298-307.
[38] Ma QL, Yang F, Rosario ER, *et al.* Beta-amyloid oligomers induce phosphorylation of tau and inactivation of insulin receptor substrate *via* c-Jun N-terminal kinase signaling: suppression by omega-3 fatty acids and curcumin. J Neurosci 2009;29(28):9078-89.
[39] Barghorn S, Nimmrich V, Striebinger A, *et al.* Globular amyloid beta-peptide oligomer - a homogenous and stable neuropathological protein in Alzheimer's disease. J Neurochem 2005;95(3):834-47.

[40] Kokubo H, Kayed R, Glabe CG, *et al.* Oligomeric proteins ultrastructurally localize to cell processes, especially to axon terminals with higher density, but not to lipid rafts in Tg2576 mouse brain. Brain Res 2005;1045(1-2):224-8.

[41] Adalbert R, Gilley J, Coleman MP. Abeta, tau and ApoE4 in Alzheimer's disease: the axonal connection. Trends Mol Med 2007;13(4):135-42.

[42] Schindowski K, Belarbi K, Buee L. Neurotrophic factors in Alzheimer's disease: role of axonal transport. Genes Brain Behav 2008;7 Suppl 1:43-56.

[43] Hiruma H, Katakura T, Takahashi S, Ichikawa T, Kawakami T. Glutamate and amyloid beta-protein rapidly inhibit fast axonal transport in cultured rat hippocampal neurons by different mechanisms. J Neurosci 2003;23(26):8967-77.

[44] Maloney MT, Minamide LS, Kinley AW, Boyle JA, Bamburg JR. Beta-secretase-cleaved amyloid precursor protein accumulates at actin inclusions induced in neurons by stress or amyloid beta: a feedforward mechanism for Alzheimer's disease. J Neurosci 2005;25(49):11313-21.

[45] Pigino G, Morfini G, Atagi Y, *et al.* Disruption of fast axonal transport is a pathogenic mechanism for intraneuronal amyloid beta. Proc Natl Acad Sci U S A 2009;106(14):5907-12.

[46] Ishihara T, Hong M, Zhang B, *et al.* Age-dependent emergence and progression of a tauopathy in transgenic mice overexpressing the shortest human tau isoform. Neuron 1999;24(3):751-62.

[47] Lewis J, McGowan E, Rockwood J, *et al.* Neurofibrillary tangles, amyotrophy and progressive motor disturbance in mice expressing mutant (P301L) tau protein. Nat Genet. 2000;25(4):402-5.

[48] Probst A, Gotz J, Wiederhold KH, *et al.* Axonopathy and amyotrophy in mice transgenic for human four-repeat tau protein. Acta Neuropathol 2000;99(5):469-81.

[49] Kraemer BC, Zhang B, Leverenz JB, *et al.* Neurodegeneration and defective neurotransmission in a Caenorhabditis elegans model of tauopathy. Proc Natl Acad Sci U S A. 2003;100(17):9980-5.

[50] Mudher A, Shepherd D, Newman TA, *et al.* GSK-3beta inhibition reverses axonal transport defects and behavioural phenotypes in Drosophila. Mol Psychiatry 2004;9(5):522-30.

[51] Chee FC, Mudher A, Cuttle MF, *et al.* Over-expression of tau results in defective synaptic transmission in Drosophila neuromuscular junctions. Neurobiol Dis 2005;20(3):918-28.

[52] Seitz A, Kojima H, Oiwa K, *et al.* Single-molecule investigation of the interference between kinesin, tau and MAP2c. EMBO J 2002;21(18):4896-905.

[53] Stamer K, Vogel R, Thies E, Mandelkow E, Mandelkow EM. Tau blocks traffic of organelles, neurofilaments, and APP vesicles in neurons and enhances oxidative stress. J Cell Biol 2002;156(6):1051-63.

[54] Mandelkow EM, Thies E, Trinczek B, Biernat J, Mandelkow E. MARK/PAR1 kinase is a regulator of microtubule-dependent transport in axons. J Cell Biol 2004;167(1):99-110.

[55] Vossel KA, Zhang K, Brodbeck J, *et al.* Tau reduction prevents Abeta-induced defects in axonal transport. Science 2010;330(6001):198.

[56] Hollenbeck PJ, Saxton WM. The axonal transport of mitochondria. J Cell Sci 2005;118(Pt 23):5411-9.

[57] Attwell D, Laughlin SB. An energy budget for signaling in the grey matter of the brain. J Cereb Blood Flow Metab 2001;21(10):1133-45.

[58] Svendsen CN, Cooper JD, Sofroniew MV. Trophic factor effects on septal cholinergic neurons. Ann N Y Acad Sci 1991;640:91-4.

[59] Venero JL, Knusel B, Beck KD, Hefti F. Expression of neurotrophin and trk receptor genes in adult rats with fimbria transections: effect of intraventricular nerve growth factor and brain-derived neurotrophic factor administration. Neuroscience 1994;59(4):797-815.
[60] Daborg J, von Otter M, Sjolander A, *et al.* Association of the RAGE G82S polymorphism with Alzheimer's disease. J Neural Transm 2010;117(7):861-7.
[61] Vasto S, Candore G, Aquino A, *et al.* The nACHR4 594C/T polymorphism in Alzheimer disease. Rejuvenation Res 2006;9(1):107-10.
[62] Ventriglia M, Bocchio Chiavetto L, Benussi L, *et al.* Association between the BDNF 196 A/G polymorphism and sporadic Alzheimer's disease. Mol Psychiatry 2002;7(2):136-7.
[63] Nagata T, Shinagawa S, Nukariya K, *et al.* Association between nerve growth factor gene polymorphism and executive dysfunction in Japanese patients with early-stage Alzheimer's disease and amnestic mild cognitive impairment. Dement Geriatr Cogn Disord 2011;32(6):379-86.
[64] Du Y, Dodel RC, Eastwood BJ, *et al.* Association of an interleukin 1 alpha polymorphism with Alzheimer's disease. Neurology 2000;55(4):480-3.
[65] McGeer PL, McGeer EG. Polymorphisms in inflammatory genes and the risk of Alzheimer disease. Arch Neurol 2001;58(11):1790-2.
[66] Liao A, Nitsch RM, Greenberg SM, *et al.* Genetic association of an alpha2-macroglobulin (Val1000lle) polymorphism and Alzheimer's disease. Hum Mol Genet 1998;7(12):1953-6.
[67] Petkova AT, Leapman RD, Guo Z, *et al.* Self-propagating, molecular-level polymorphism in Alzheimer's beta-amyloid fibrils. Science 2005;307(5707):262-5.
[68] Paravastu AK, Leapman RD, Yau WM, Tycko R. Molecular structural basis for polymorphism in Alzheimer's beta-amyloid fibrils. Proc Natl Acad Sci U S A 2008;105(47):18349-54.
[69] Fandrich M, Meinhardt J, Grigorieff N. Structural polymorphism of Alzheimer Abeta and other amyloid fibrils. Prion 2009;3(2):89-93.
[70] Poirier J, Davignon J, Bouthillier D, *et al.* Apolipoprotein E polymorphism and Alzheimer's disease. Lancet 1993;342(8873):697-9.
[71] Kurian SM, Le-Niculescu H, Patel SD, *et al.* Identification of blood biomarkers for psychosis using convergent functional genomics. Mol Psychiatry 2011;16(1):37-58.
[72] van den Elzen P, Garg S, Leon L, *et al.* Apolipoprotein-mediated pathways of lipid antigen presentation. Nature 2005;437(7060):906-10.
[73] Kang DE, Pietrzik CU, Baum L, *et al.* Modulation of amyloid beta-protein clearance and Alzheimer's disease susceptibility by the LDL receptor-related protein pathway. J Clin Invest 2000;106(9):1159-66.
[74] Bertram L, Tanzi RE. Alzheimer's disease: one disorder, too many genes? Hum Mol Genet 2004;13 Spec No 1:R135-41.
[75] Frisardi V, Solfrizzi V, Capurso C, *et al.* Is insulin resistant brain state a central feature of the metabolic-cognitive syndrome? J Alzheimers Dis 2010;21(1):57-63.
[76] D'Ercole AJ, Ye P, Calikoglu AS, Gutierrez-Ospina G. The role of the insulin-like growth factors in the central nervous system. Mol Neurobiol 1996;13(3):227-55.
[77] Farooqui AA, Farooqui T, Panza F, Frisardi V. Metabolic syndrome as a risk factor for neurological disorders. Cell Mol Life Sci 2012;69(5):741-62.
[78] Biessels GJ, Kappelle LJ. Increased risk of Alzheimer's disease in Type II diabetes: insulin resistance of the brain or insulin-induced amyloid pathology? Biochem Soc Trans 2005;33(Pt 5):1041-4.

[79] Luchsinger JA, Tang MX, Shea S, Mayeux R. Hyperinsulinemia and risk of Alzheimer disease. Neurology 2004;63(7):1187-92.

[80] Gasparini L, Gouras GK, Wang R, *et al.* Stimulation of beta-amyloid precursor protein trafficking by insulin reduces intraneuronal beta-amyloid and requires mitogen-activated protein kinase signaling. J Neurosci 2001;21(8):2561-70.

[81] Perez A, Morelli L, Cresto JC, Castano EM. Degradation of soluble amyloid beta-peptides 1-40, 1-42, and the Dutch variant 1-40Q by insulin degrading enzyme from Alzheimer disease and control brains. Neurochem Res 2000;25(2):247-55.

[82] Doble BW, Woodgett JR. GSK-3: tricks of the trade for a multi-tasking kinase. J Cell Sci 2003;116(7):1175-86.

[83] Hooper C, Killick R, Lovestone S. The GSK3 hypothesis of Alzheimer's disease. J Neurochem 2008;104(6):1433-9.

[84] Watson GS, Craft S. Insulin resistance, inflammation, and cognition in Alzheimer's Disease: lessons for multiple sclerosis. J Neurol Sci 2006;245(1-2):21-33.

[85] Farooqui AA. In: Lipid mediators and their metabolism in the brain. New York: Springer 2011; pp. 327.

[86] Rovio S, Kåreholt I, Helkala EL, *et al.* Leisure-time physical activity at midlife and the risk of dementia and Alzheimer's disease. Lancet Neurol 2005;4(11):705-11.

[87] Adlard PA, Perreau VM, Pop V, Cotman CW. Voluntary exercise decreases amyloid load in a transgenic model of Alzheimer's disease. J Neurosci 2005;25(17):4217-21.

[88] de la Monte SM, Wands JR. Alzheimer's disease is type 3 diabetes-evidence reviewed. J Diabetes Sci Technol 2008;2(6):1101-13.

[89] Spitzer P, Klafki HW, Blennow K, *et al.* cNEUPRO: Novel Biomarkers for Neurodegenerative Diseases. Int J Alzheimers Dis 2010; pii: 548145.

[90] European Medicines Agency/CHMP. Qualification opinion of novel methodologies in the predementia stage of Alzheimer's disease: cerebro-spinal fluid related biomarkers for drugs affecting amyloid burden 2011. Available from: http://www.ema.europa.eu/docs/en_GB/document_library/Regulatory_and_procedural_guideline/2011/05/WC500106357.pdf [cited: 5th Jul 2012].

[91] Jack Jr CR, Lowe VJ, Weigand SD, *et al.* Serial PIB and MRI in normal, mild cognitive impairment and Alzheimer's disease: implications for sequence of pathological events in Alzheimer's disease. Brain 2009;132(5):1355-65.

[92] Schenk D, Barbour R, Dunn W, *et al.* Immunization with amyloid-beta attenuates Alzheimer-disease-like pathology in the PDAPP mouse. Nature 1999;400(6740):173-7.

[93] Schindowski K, Eckert A, Peters J, *et al.* Increased T-cell reactivity and elevated levels of CD8+ memory T-cells in Alzheimer's disease-patients and T-cell hyporeactivity in an Alzheimer's disease-mouse model: implications for immunotherapy. Neuromolecular Med 2007;9(4):340-54.

[94] Schindowski K, Frohlich L, Maurer K, Muller WE, Eckert A. Age-related impairment of human T lymphocytes' activation: specific differences between CD4(+) and CD8(+) subsets. Mech Ageing Dev 2002;123(4):375-90.

[95] Schindowski K, Leutner S, Muller WE, Eckert A. Age-related changes of apoptotic cell death in human lymphocytes. Neurobiol Aging 2000;21(5):661-70.

[96] Weksler ME. The immunotherapy of Alzheimer's disease. Immun Ageing 2004;1(1):2.

[97] Pifer J, Hennes JL, Lee JM, Witte PL. Age-related differences in the immune response to immunization with human Abeta42 peptide. J Gerontol A Biol Sci Med Sci 2002;57(10):B355-8.

[98] Theofilopoulos AN, Bona CA. In: The molecular pathology of autoimmune diseases. 2nd ed. New York: Taylor & Francis 2002; pp. 1191

[99] Orgogozo JM, Gilman S, Dartigues JF, *et al.* Subacute meningoencephalitis in a subset of patients with AD after Abeta42 immunization. Neurology 2003;61(1):46-54.

[100] Cribbs DH, Ghochikyan A, Vasilevko V, *et al.* Adjuvant - dependent modulation of Th1 and Th2 responses to immunization with β - amyloid. International Immunology 2003;15(4):505-14.

[101] Nicoll JA, Wilkinson D, Holmes C, *et al.* Neuropathology of human Alzheimer disease after immunization with amyloid-beta peptide: a case report. Nat Med 2003;9(4):448-52.

[102] Chauhan NB, Siegel GJ. Intracerebroventricular passive immunization with anti-Abeta antibody in Tg2576. J Neurosci Res 2003;74(1):142-7.

[103] Dodart JC, Bales KR, Gannon KS, *et al.* Immunization reverses memory deficits without reducing brain Abeta burden in Alzheimer's disease model. Nat Neurosci 2002;5(5):452-7.

[104] Schenk DB. Passive immunization treatment of Alzheimer's disease. US Patent US6761888; 2004.

[105] Schenk DB, Bard F, Yednock T. Humanized and chimeric N-terminal amyloid beta-antibodies.US Patent US6750324; 2004.

[106] Holtzman DM, DeMattos R, Bales KR, *et al.* Humanized antibodies that sequester abeta peptide. US Patent US7195761; 2007.

[107] Lemere CA, Maier M, Peng Y, Jiang L, Seabrook TJ. Novel Abeta immunogens: is shorter better? Curr Alzheimer Res 2007;4(4):427-36.

[108] Bard F, Cannon C, Barbour R, *et al.* Peripherally administered antibodies against amyloid beta-peptide enter the central nervous system and reduce pathology in a mouse model of Alzheimer disease. Nat Med 2000;6(8):916-9.

[109] Wilcock DM, Rojiani A, Rosenthal A, *et al.* Passive amyloid immunotherapy clears amyloid and transiently activates microglia in a transgenic mouse model of amyloid deposition. J Neurosci 2004;24(27):6144-51.

[110] Vanmechelen E, Grognet P, Sergeant N, *et al.* Antibodies specific of the beta-amyloid peptides and their uses as diagnostic agents or drugs. US Patent US20110059092; 2008.

[111] Casas C, Sergeant N, Itier JM, *et al.* Massive CA1/2 neuronal loss with intraneuronal and N-terminal truncated Abeta42 accumulation in a novel Alzheimer transgenic model. Am J Pathol 2004;165(4):1289-300.

[112] Frost JL, Liu B, Kleinschmidt M, *et al.* Passive immunization against pyroglutamate-3 amyloid-beta reduces plaque burden in Alzheimer-like transgenic mice: a pilot study. Neurodegener Dis 2012;10(1-4):265-70.

[113] Jawhar S, Wirths O, Bayer TA. Pyroglutamate amyloid-beta (Abeta): a hatchet man in Alzheimer disease. J Biol Chem 2011;286(45):38825-32.

[114] Russo C, Violani E, Salis S, *et al.* Pyroglutamate-modified amyloid beta-peptides--AbetaN3(pE)--strongly affect cultured neuron and astrocyte survival. J Neurochem 2002;82(6):1480-9.

[115] Liu K, Solano I, Mann D, *et al.* Characterization of Abeta11-40/42 peptide deposition in Alzheimer's disease and young Down's syndrome brains: implication of N-terminally truncated Abeta species in the pathogenesis of Alzheimer's disease. Acta Neuropathol 2006;112(2):163-74.

[116] Wittnam JL, Portelius E, Zetterberg H, *et al.* Pyroglutamate amyloid beta (Abeta) aggravates behavioral deficits in transgenic amyloid mouse model for Alzheimer disease. J Biol Chem 2012;287(11):8154-62.

[117] Nussbaum JM, Schilling S, Cynis H, *et al.* Prion-like behaviour and tau-dependent cytotoxicity of pyroglutamylated amyloid-beta. Nature 2012;485(7400):651-5.

[118] Marcello A, Wirths O, Schneider-Axmann T, *et al.* Reduced levels of IgM autoantibodies against N-truncated pyroglutamate Abeta in plasma of patients with Alzheimer's disease. Neurobiol Aging 2011;32(8):1379-87.

[119] Demuth HU, Schilling S, Kleinschmidt M, *et al.* Antibodies specific for N-terminal truncated and pyroglutamate modified Amyloid-beta peptides. US Patent US 8058405; 2009.

[120] Haass C, Selkoe DJ. Soluble protein oligomers in neurodegeneration: lessons from the Alzheimer's amyloid β-peptide. Nature Reviews Molecular Cell Biology 2007;8(2):101-12.

[121] Petkova AT, Ishii Y, Balbach JJ, *et al.* A structural model for Alzheimer's beta -amyloid fibrils based on experimental constraints from solid state NMR. Proc Natl Acad Sci U S A 2002;99(26):16742-7.

[122] Cleary JP, Walsh DM, Hofmeister JJ, *et al.* Natural oligomers of the amyloid-beta protein specifically disrupt cognitive function. Nat Neurosci 2005;8(1):79-84.

[123] Townsend M, Shankar GM, Mehta T, Walsh DM, Selkoe DJ. Effects of secreted oligomers of amyloid beta-protein on hippocampal synaptic plasticity: a potent role for trimers. J Physiol 2006;572(Pt 2):477-92.

[124] Glabe CG. Structural classification of toxic amyloid oligomers. J Biol Chem 2008;283(44):29639-43.

[125] Shankar GM, Li S, Mehta TH, *et al.* Amyloid-beta protein dimers isolated directly from Alzheimer's brains impair synaptic plasticity and memory. Nat Med 2008;14(8):837-42.

[126] Tomic JL, Pensalfini A, Head E, Glabe CG. Soluble fibrillar oligomer levels are elevated in Alzheimer's disease brain and correlate with cognitive dysfunction. Neurobiol Dis 2009;35(3):352-8.

[127] Jin M, Shepardson N, Yang T, *et al.* Soluble amyloid beta-protein dimers isolated from Alzheimer cortex directly induce Tau hyperphosphorylation and neuritic degeneration. Proc Natl Acad Sci U S A 2011;108(14):5819-24.

[128] Villemagne VL, Perez KA, Pike KE, *et al.* Blood-borne amyloid-beta dimer correlates with clinical markers of Alzheimer's disease. J Neurosci 2010;30(18):6315-22.

[129] Dodel R, Balakrishnan K, Keyvani K, *et al.* Naturally occurring autoantibodies against beta-amyloid: investigating their role in transgenic animal and *in vitro* models of Alzheimer's disease. J Neurosci 2011;31(15):5847-54.

[130] Hillen H, Barghorn S, Striebinger A, *et al.* Generation and therapeutic efficacy of highly oligomer-specific beta-amyloid antibodies. J Neurosci 2010;30(31):10369-79.

[131] Barghorn S, Ebert U, Hillen H, *et al.* Anti-Aβ Globulomer antibodies, antigen-binding moieties thereof, corresponding hybridomas, nucleic acids, vectors, host cells, methods of producing said antibodies, compositions comprising said antibodies, uses of said antibodies and methods of using said antibodies. World Patent WO2007062852; 2007.

[132] Barghorn S, Hillen H. New Aβ confomers selective anti-Aβ globulomer monoclonal antibodies. World Patent WO2008067464; 2008.

[133] Hillen H, Barghorn S, Labkovsky B, *et al.* Humanized antibodies which bind to Aβ (1-42) globulomer and uses thereof. World Patent WO2008150946; 2008.

[134] Hillen H, Striebinger A, Giaisi S, *et al.* Aß (X-38. 43) oligomers, and processes, compositions, and uses thereof. World Patent WO2010010469; 2009.

[135] Broisat A, Hernot S, Toczek J, *et al.* Nanobodies targeting mouse/human VCAM1 for the nuclear imaging of atherosclerotic lesions. Circ Res 2012;110(7):927-37.

[136] Park JE, Dorner-ciossek C, Hoerer S, *et al.* A-beta binding polypeptides. US Patent US20120058118; 2012.
[137] Atwal JK, Chen Y, Chiu C, *et al.* A therapeutic antibody targeting BACE1 inhibits amyloid-beta production *in vivo*. Sci Transl Med 2011;3(84):84ra43.
[138] Abbas AK, Lichtman AH. In: Basic immunology : functions and disorders of the immune system. 3rd ed. Philadelphia: Saunders/Elsevier 2009; viii, pp. 312
[139] Murphy K, Travers P, Walport M, Janeway C. In: Janeway's immunobiology. 8th ed. New York: Garland Science 2012; xix, pp. 868
[140] Horejs - Hoeck J, Hren A, Mudde GC, Woisetschläger M. Inhibition of immunoglobulin E synthesis through FcγRII (CD32) by a mechanism independent of B - cell receptor co - cross - linking. Immunology 2005;115(3):407-15.
[141] Roopenian DC, Akilesh S. FcRn: the neonatal Fc receptor comes of age. Nat Rev Immunol 2007;7(9):715-25.
[142] Fu HJ, Liu B, Frost JL, Lemere CA. Amyloid-beta immunotherapy for Alzheimer's disease. CNS Neurol Disord Drug Targets 2010;9(2):197-206.
[143] Bard F, Barbour R, Cannon C, *et al.* Epitope and isotype specificities of antibodies to beta - amyloid peptide for protection against Alzheimer's disease-like neuropathology. Proc Natl Acad Sci U S A 2003;100(4):2023-8.
[144] Ferrer I, Boada Rovira M, Sanchez Guerra ML, Rey MJ, Costa-Jussa F. Neuropathology and pathogenesis of encephalitis following amyloid-beta immunization in Alzheimer's disease. Brain Pathol 2004;14(1):11-20.
[145] Das P, Howard V, Loosbrock N, *et al.* Amyloid-beta immunization effectively reduces amyloid deposition in FcRgamma-/- knock-out mice. J Neurosci 2003;23(24):8532-8.
[146] Tamura Y, Hamajima K, Matsui K, *et al.* The F(ab)'2 fragment of an Abeta-specific monoclonal antibody reduces Abeta deposits in the brain. Neurobiol Dis 2005;20(2):541-9.
[147] Bacskai BJ, Kajdasz ST, McLellan ME, *et al.* Non-Fc-mediated mechanisms are involved in clearance of amyloid-beta *in vivo* by immunotherapy. J Neurosci 2002;22(18):7873-8.
[148] Solomon B, Koppel R, Hanan E, Katzav T. Monoclonal antibodies inhibit *in vitro* fibrillar aggregation of the Alzheimer beta-amyloid peptide. Proc Natl Acad Sci U S A 1996;93(1):452-5.
[149] Legleiter J, Czilli DL, Gitter B, *et al.* Effect of different anti-Abeta antibodies on Abeta fibrillogenesis as assessed by atomic force microscopy. J Mol Biol 2004;335(4):997-1006.
[150] Solomon B, Koppel R, Frankel D, Hanan-Aharon E. Disaggregation of Alzheimer beta-amyloid by site-directed mAb. Proc Natl Acad Sci U S A 1997;94(8):4109-12.
[151] Frenkel D, Katz O, Solomon B. Immunization against Alzheimer's beta -amyloid plaques *via* EFRH phage administration. Proc Natl Acad Sci U S A 2000;97(21):11455-9.
[152] Dodel RC, Du Y, Depboylu C, *et al.* Intravenous immunoglobulins containing antibodies against beta-amyloid for the treatment of Alzheimer's disease. J Neurol Neurosurg Psychiatry 2004;75(10):1472-4.
[153] Taguchi H, Planque S, Nishiyama Y, *et al.* Autoantibody-catalyzed hydrolysis of amyloid beta peptide. J Biol Chem 2008;283(8):4714-22.
[154] Bacher M, Depboylu C, Du Y, *et al.* Peripheral and central biodistribution of (111)In-labeled anti-beta-amyloid autoantibodies in a transgenic mouse model of Alzheimer's disease. Neurosci Lett 2009;449(3):240-5.
[155] Relkin NR, Szabo P, Adamiak B, *et al.* 18-Month study of intravenous immunoglobulin for treatment of mild Alzheimer disease. Neurobiol Aging 2009;30(11):1728-36.
[156] Bacskai BJ, Kajdasz ST, Christie RH, *et al.* Imaging of amyloid-beta deposits in brains of living mice permits direct observation of clearance of plaques with immunotherapy. Nat Med 2001;7(3):369-72.

[157] Oddo S, Caccamo A, Tran L, *et al.* Temporal profile of amyloid-beta (Abeta) oligomerization in an *in vivo* model of Alzheimer disease. A link between Abeta and tau pathology. J Biol Chem 2006;281(3):1599-604.
[158] Oddo S, Vasilevko V, Caccamo A, *et al.* Reduction of soluble Abeta and tau, but not soluble Abeta alone, ameliorates cognitive decline in transgenic mice with plaques and tangles. J Biol Chem 2006;281(51):39413-23.
[159] Brody DL, Holtzman DM. Active and passive immunotherapy for neurodegenerative disorders. Annu Rev Neurosci 2008;31:175-93.
[160] DeMattos RB, Bales KR, Cummins DJ, *et al.* Peripheral anti-A beta antibody alters CNS and plasma A beta clearance and decreases brain A beta burden in a mouse model of Alzheimer's disease. Proc Natl Acad Sci U S A 2001;98(15):8850-5.
[161] Poetsch V, Bennani-Baiti B, Neuhaus W, Muchitsch EM, Noe CR. Serum-derived immunoglobulins alter amyloid beta transport across a blood-brain barrier *in vitro* model. Pharmazie 2010;65(4):267-73.
[162] Sigurdsson EM, Scholtzova H, Mehta PD, Frangione B, Wisniewski T. Immunization with a nontoxic/nonfibrillar amyloid-beta homologous peptide reduces Alzheimer's disease-associated pathology in transgenic mice. Am J Pathol 2001;159(2):439-47.
[163] Sigurdsson EM, Knudsen E, Asuni A, *et al.* An attenuated immune response is sufficient to enhance cognition in an Alzheimer's disease mouse model immunized with amyloid-beta derivatives. J Neurosci 2004;24(28):6277-82.
[164] Sigurdsson EM, Wisniewski T, Frangione B. A safer vaccine for Alzheimer's disease? Neurobiol Aging 2002;23(6):1001-8.
[165] Lemere CA, Spooner ET, LaFrancois J, *et al.* Evidence for peripheral clearance of cerebral Abeta protein following chronic, active Abeta immunization in PSAPP mice. Neurobiol Dis 2003;14(1):10-8.
[166] Dodel R, Hampel H, Depboylu C, *et al.* Human antibodies against amyloid beta peptide: a potential treatment for Alzheimer's disease. Ann Neurol 2002;52(2):253-6.
[167] Bell RD, Sagare AP, Friedman AE, *et al.* Transport pathways for clearance of human Alzheimer's amyloid beta-peptide and apolipoproteins E and J in the mouse central nervous system. J Cereb Blood Flow Metab 2007;27(5):909-18.
[168] Deane R, Du Yan S, Submamaryan RK, *et al.* RAGE mediates amyloid-beta peptide transport across the blood-brain barrier and accumulation in brain. Nat Med 2003;9(7):907-13.
[169] Monro OR, Mackic JB, Yamada S, *et al.* Substitution at codon 22 reduces clearance of Alzheimer's amyloid-beta peptide from the cerebrospinal fluid and prevents its transport from the central nervous system into blood. Neurobiol Aging 2002;23(3):405-12.
[170] Zlokovic BV, Yamada S, Holtzman D, Ghiso J, Frangione B. Clearance of amyloid beta-peptide from brain: transport or metabolism? Nat Med 2000;6(7):718-9.
[171] Deane R, Sagare A, Hamm K, *et al.* IgG-assisted age-dependent clearance of Alzheimer's amyloid beta peptide by the blood-brain barrier neonatal Fc receptor. J Neurosci 2005;25(50):11495-503.
[172] Sagare AP, Deane R, Zetterberg H, *et al.* Impaired lipoprotein receptor-mediated peripheral binding of plasma amyloid-beta is an early biomarker for mild cognitive impairment preceding Alzheimer's disease. J Alzheimers Dis 2011;24(1):25-34.
[173] Sagare AP, Winkler EA, Bell RD, Deane R, Zlokovic BV. From the liver to the blood-brain barrier: an interconnected system regulating brain amyloid-beta levels. J Neurosci Res 2011;89(7):967-8.

[174] Sagare AP, Bell RD, Zlokovic BV. Neurovascular Defects and Faulty Amyloid-beta Vascular Clearance in Alzheimer's Disease. J Alzheimers Dis 2012.

[175] Zipser B, Johanson C, Gonzalez L, *et al.* Microvascular injury and blood–brain barrier leakage in Alzheimer's disease. Neurobiol Aging 2007;28(7):977-86.

[176] Yamada K, Yabuki C, Seubert P, *et al.* Abeta immunotherapy: intracerebral sequestration of Abeta by an anti-Abeta monoclonal antibody 266 with high affinity to soluble Abeta. J Neurosci 2009;29(36):11393-8.

[177] Klyubin I, Walsh DM, Lemere CA, *et al.* Amyloid beta protein immunotherapy neutralizes Abeta oligomers that disrupt synaptic plasticity *in vivo*. Nat Med 2005;11(5):556-61.

[178] Deng J, Hou H, Giunta B, *et al.* Autoreactive-Abeta antibodies promote APP beta-secretase processing. J Neurochem 2012;120(5):732-40.

[179] Pfeifer M, Boncristiano S, Bondolfi L, *et al.* Cerebral hemorrhage after passive anti-Abeta immunotherapy. Science 2002;298(5597):1379.

[180] Racke MM, Boone LI, Hepburn DL, *et al.* Exacerbation of cerebral amyloid angiopathy-associated microhemorrhage in amyloid precursor protein transgenic mice by immunotherapy is dependent on antibody recognition of deposited forms of amyloid beta. J Neurosci 2005;25(3):629-36.

[181] Du Y, Dodel R, Hampel H, *et al.* Reduced levels of amyloid beta-peptide antibody in Alzheimer disease. Neurology 2001;57(5):801-5.

[182] Weksler ME, Relkin N, Turkenich R, *et al.* Patients with Alzheimer disease have lower levels of serum anti-amyloid peptide antibodies than healthy elderly individuals. Exp Gerontol 2002;37(7):943-8.

[183] Britschgi M, Olin CE, Johns HT, *et al.* Neuroprotective natural antibodies to assemblies of amyloidogenic peptides decrease with normal aging and advancing Alzheimer's disease. Proc Natl Acad Sci U S A 2009;106(29):12145-50.

[184] Dodel R, Neff F, Noelker C, *et al.* Intravenous immunoglobulins as a treatment for Alzheimer's disease: rationale and current evidence. Drugs 2010;70(5):513-28.

[185] McLaurin J, Cecal R, Kierstead ME, *et al.* Therapeutically effective antibodies against amyloid-beta peptide target amyloid-beta residues 4-10 and inhibit cytotoxicity and fibrillogenesis. Nat Med 2002;8(11):1263-9.

[186] Hock C, Konietzko U, Papassotiropoulos A, *et al.* Generation of antibodies specific for beta-amyloid by vaccination of patients with Alzheimer disease. Nat Med 2002;8(11):1270-5.

[187] Paul S, Planque S, Nishiyama Y. Immunological origin and functional properties of catalytic autoantibodies to amyloid beta peptide. J Clin Immunol 2010;30 Suppl 1:S43-9.

[188] Johnson-Wood K, Lee M, Motter R, *et al.* Amyloid precursor protein processing and Aβ42 deposition in a transgenic mouse model of Alzheimer disease. Proc Natl Acad Sci U S A 1997;94(4):1550.

[189] Moody's. Credit Outlook: Credit Implications of Current Events: Moody's Investor Service; 2012. Available from: http://www.risk-compliance-association.com/Moodys_to_the_members_of_the_International_Association_of_Risk_and_Compliance_Professionals_July_26_2012.pdf [cited: 19th Jul 2012].

[190] Salloway S, Sperling R, Gilman S, *et al.* A phase 2 multiple ascending dose trial of bapineuzumab in mild to moderate Alzheimer disease. Neurology 2009;73(24):2061-70.

[191] Rinne JO, Brooks DJ, Rossor MN, *et al.* 11C-PiB PET assessment of change in fibrillar amyloid-beta load in patients with Alzheimer's disease treated with bapineuzumab: a phase

2, double-blind, placebo-controlled, ascending-dose study. Lancet Neurol 2010;9(4):363-72.

[192] Chalmers K, Wilcock GK, Love S. APOE epsilon 4 influences the pathological phenotype of Alzheimer's disease by favouring cerebrovascular over parenchymal accumulation of A beta protein. Neuropathol Appl Neurobiol 2003;29(3):231-8.

[193] Boche D, Zotova E, Weller R, *et al.* Consequence of Aβ immunization on the vasculature of human Alzheimer's disease brain. Brain 2008;131(12):3299-310.

[194] Blennow K, Zetterberg H, Rinne JO, *et al.* Effect of immunotherapy with bapineuzumab on cerebrospinal fluid biomarker levels in patients with mild to moderate Alzheimer Disease. Arch Neurol 2012;69(8):1002-10.

[195] Klunk WE, Engler H, Nordberg A, *et al.* Imaging brain amyloid in Alzheimer's disease with Pittsburgh Compound-B. Ann Neurol 2004;55(3):306-19.

[196] Lopez OL, Hamilton R, Ikonomovic M, *et al. In vivo* amyloid deposition and neuropathological findings after humanized amyloid β-specific monoclonal antibodies therapy in a patient with Alzheimer's disease. Alzheimers Dement 2009;5(4):P64-P5.

[197] Johnson J. Johnson & Johnson Announces Discontinuation Of Phase 3 Development of Bapineuzumab Intravenous (IV) In Mild-To-Moderate Alzheimer's Disease: Press release; 2012. Available from: http://www.jnj.com/connect/news/product/johnson-and-johnson-announces-discontinuation-of-phase-3-development-of-bapineuzumab-intravenous-iv-in-mild-to-moderate-alzheimers-disease [cited: 7th Aug 2012].

[198] Salloway S, Sperling R, Raskind M, *et al.* BapineuzumabPhase 3 trials in mild to moderate Alzheimer's disease dementia in apolipoproteinE e4 carriers(Study 302) and non-carriers (Study 301): Clinical Outcomes. Data from Clinical Trials in Alzheimer's Disease; 2012, Oct 29; Monaco, Monaco. Available from: http://www.ctad.fr/07-download/Congres2012/PressRelease/Final-Salloway-CTAD-Presentation-10-29-12.pdf [cited: 5th Nov 2012].

[199] Sperling R, Salloway S, Raskind M, *et al.* Bapineuzumab Phase 3 trials in mild to moderate Alzheimer's disease dementia in apolipoprotein E e4 carriers(Study 302) and non-carriers (Study 301): Safety and PiB PET Amyloid Imaging. Data from: Clinical Trials in Alzheimer's Disease; 2012, Oct 29; Monaco, Monaco. Available from: http://www.ctad.fr/07-download/Congres2012/PressRelease/Final-Sperling-CTAD-Presentation-10-29-12.pdf [cited: 5th Nov 2012].

[200] Sperling R, Salloway S, Raskind M, *et al.* A randomized, double-blind, placebo-controlled clinical trial of intravenous bapineuzumab in patients with Alzheimer's disease who are apolipoprotein E ε4 carriers. Eur J Neurol 2012;19(Suppl 1):SC3012.

[201] Salloway S, Sperling R, Honig L, *et al.* A randomized, double-blind, placebo-controlled clinical trial of intravenous bapineuzumab in patients with Alzheimer's disease who are apolipoprotein E ε4 non-carriers. Eur J Neurol 2012;19(Suppl. 1):SC312.

[202] Seubert P, Vigo-Pelfrey C, Esch F, *et al.* Isolation and quantification of soluble Alzheimer's β-peptide from biological fluids. Nature 1992;359(6393):325-7.

[203] Imbimbo BP, Ottonello S, Frisardi V, *et al.* Solanezumab for the treatment of mild-to-moderate Alzheimer's disease. Expert Review of Clinical Immunology 2012;8(2):135-49.

[204] Samadi H, Sultzer D. Solanezumab for Alzheimer's disease. Expert Opinion on Biological Therapy 2011(0):1-12.

[205] DeMattos RB, Bales KR, Cummins DJ, Paul SM, Holtzman DM. Brain to plasma amyloid-β efflux: a measure of brain amyloid burden in a mouse model of Alzheimer's disease. Science 2002;295(5563):2264-7.

[206] Siemers ER, Friedrich S, Dean RA, *et al.* P4-346: Safety, tolerability and biomarker effects of an Abeta monoclonal antibody administered to patients with Alzheimer's disease. Alzheimers Dement 2008;4(4):T774-T.

[207] Siemers ER, Friedrich S, Dean RA, *et al.* Safety and changes in plasma and cerebrospinal fluid amyloid [beta] after a single administration of an amyloid [beta] monoclonal antibody in subjects with Alzheimer disease. Clinical Neuropharmacology 2010;33(2):67.

[208] Uenaka K, Nakano M, Willis BA, *et al.* Comparison of Pharmacokinetics, Pharmacodynamics, Safety, and Tolerability of the Amyloid β Monoclonal Antibody Solanezumab in Japanese and White Patients With Mild to Moderate Alzheimer Disease. Clinical Neuropharmacology 2012;35(1):25.

[209] Farlow M, Arnold SE, van Dyck CH, *et al.* Safety and biomarker effects of solanezumab in patients with Alzheimer's disease. Alzheimer's & Dementia 2012; 8(4):261–71.

[210] DeMattos RB, Racke M, Gelfanova V, *et al.* Characterization of amino-terminally truncated Abeta-42 peptides in plasma from Alzheimer's patients receiving solanezumab immunotherapy treatment. Alzheimer's and Dementia 2010;6(4):S542-S.

[211] DeMattos RB, Racke MM, Gelfanova V, *et al.* Identification, characterization, and comparison of amino-terminally truncated Aβ42 peptides in Alzheimer's disease brain tissue and in plasma from Alzheimer's patients receiving solanezumab immunotherapy treatment. Alzheimers Dement 2009;5(4):P156-P7.

[212] Siemers E, DeMattos RB, May PC, Dean RA. Role of biochemical Alzheimer's disease biomarkers as end points in clinical trials. Biomarkers 2010;4(1):81-9.

[213] Siemers ER, Dean RA, Lachno D, *et al.* Measurement of cerebrospinal fluid total tau and phospho-tau in phase 2 trials of therapies targeting Aβ. Alzheimers Dement 2009;5(4):P258-P.

[214] Lobello K, Ryan JM, Liu E, Rippon G, Black R. Targeting Beta Amyloid: A Clinical Review of Immunotherapeutic Approaches in Alzheimer's Disease. Int J Alzheimers Dis 2012; Article ID 628070.

[215] Lilly E. Lilly Announces Detailed Results of the Phase 3 Solanezumab EXPEDITION Studies Following a Presentation of the Independent Analyses by the Alzheimer's Disease Cooperative Study (ADCS): Press release; 2012. Available from: http://newsroom.lilly.com/releasedetail.cfm?ReleaseID=702211 [cited: 5th Nov 2012].

[216] Lilly E. Independent analysis of Solanezumab provides evidence that compound may remove amyloid from brain in Alzheimer's disease. Results cause for "cautious" optimism in scientific community. Press Release of Clinical Trial on Alzheimer's disease; 2012, Oct 29; Monaco, Monaco. Available from: http://www.ctad.fr/01-presentation/presentation.asp [cited: 5th Nov 2012]

[217] Genomeweb. Alzheimer's Prevention Study in Genetically Defined High-Risk Patients Includes Roche, Lilly Drugs: Press Release; 2012. Available from: http://www.genomeweb.com/clinical-genomics/alzheimers-prevention-study-genetically-defined-high-risk-patients-includes-roch [cited: 5th Nov 2012].

[218] Knappik A, Ge L, Honegger A, *et al.* Fully synthetic human combinatorial antibody libraries (HuCAL) based on modular consensus frameworks and CDRs randomized with trinucleotides1. J Mol Biol 2000;296(1):57-86.

[219] Bardroff M, Bohrmann B, Brockhaus M, *et al.* Anti-amyloid β antibodies. US Patent US7794719, 2010.

[220] Bardroff M, Bohrmann B, Brockhaus M, *et al.* Anti-amyloid β antibodies and their use. World Patent WO2003070760, 2003.

[221] Bohrmann B, Baumann K, Benz J, *et al.* Gantenerumab: a novel human anti-Aβ antibody demonstrates sustained cerebral amyloid-β binding and elicits cell-mediated removal of human amyloid-β. J Alzheimer Dis 2012;28(1):49-69.

[222] Ostrowitzki S, Deptula D, Thurfjell L, *et al.* Mechanism of amyloid removal in patients with alzheimer disease treated with gantenerumab. Arch Neurol 2012;69(2):198-207.

[223] Alzheimer Research Forum. Aβ Antibody Gantenerumab Clears Plaques 2011. Available from: http://www.alzforum.org/new/detail.asp?id=2933 [cited: 19th Jul 2012].

[224] Morphosys. Morphosys Antibody Reaches Major Milestone in Collaboration with Roche Corporate Media Release2012. Available from: http://www.morphosys.com/pressrelease/morphosys-antibody-reaches-major-milestone-collaboration-roche [cited: 19th Jul 2012].

[225] Glaxo Smith Kline. A randomised, single-blind, placebo-controlled study to investigate the safety, tolerability, immunogenicity, pharmacokinetics and pharmacodynamics of intravenous infusion of GSK933776 in patients with Alzheimer's disease, 2011. Available from: http://www.gsk-clinicalstudyregister.com/result_comp_list.jsp?phase=All&studyType=All&population=All &marketing=No&compound=GSK933776 [cited: 1st Jun 2012].

[226] Glaxo Smith Kline. GSK product development pipeline - February 2012. Available from: http://www.gsk.com/investors/pp_pipeline_standard.htm [cited: 16th Jul 2012].

[227] Ding JD, Lin J, Mace BE, *et al.* Targeting age-related macular degeneration with Alzheimer's disease based immunotherapies: Anti-amyloid-β antibody attenuates pathologies in an age-related macular degeneration mouse model. Vision research 2008;48(3):339-45.

[228] Muhs A, Hickman DT, Pihlgren M, *et al.* Liposomal vaccines with conformation-specific amyloid peptide antigens define immune response and efficacy in APP transgenic mice. Proc Natl Acad Sci USA 2007;104(23):9810.

[229] Poduslo JF, Gilles EJ, Ramakrishnan M, *et al.* HH Domain of Alzheimer's Disease Aβ Provides Structural Basis for Neuronal Binding in PC12 and Mouse Cortical/Hippocampal Neurons. PLoS One 2010;5(1):e8813.

[230] Van der Zee J, Van Swieten P, Aalberse R. Inhibition of complement activation by IgG4 antibodies. Clinical and experimental immunology 1986;64(2):415.

[231] Bruhns P, Iannascoli B, England P, *et al.* Specificity and affinity of human Fcγ receptors and their polymorphic variants for human IgG subclasses. Blood 2009;113(16):3716-25.

[232] Adolfsson O, Pihlgren M, Toni N, *et al.* An Effector-Reduced Anti-β-Amyloid (Aβ) Antibody with Unique Aβ Binding Properties Promotes Neuroprotection and Glial Engulfment of Aβ. J Neurosci 2012;32(28):9677-89.

[233] Carty NC, Wilcock DM, Rosenthal A, *et al.* Intracranial administration of deglycosylated C-terminal-specific anti-Abeta antibody efficiently clears amyloid plaques without activating microglia in amyloid-depositing transgenic mice. J Neuroinflammation 2006;3:11.

[234] The New York Times PB. New Drug Trial Seeks to Stop Alzheimer's Before It Starts 2012. Available from: http://www.nytimes.com/2012/05/16/health/research/prevention-is-goal-of-alzheimers-drug-trial.html [cited: 19th Jul 2012].

[235] La Porte SL, Bollini SS, Lanz TA, *et al.* Structural Basis of C-terminal β– Amyloid Peptide Binding by the Antibody Ponezumab for the Treatment of Alzheimer's Disease. J Mol Biol 2011.

[236] Alzheimer Research Forum. Uppsala: Conference Puts Uppsala on Immunotherapy Map, Part 1 2011. Available from: http://www.alzforum.org/new/detail.asp?id=2821 [cited: 18th Jul 2012].

[237] Hayduck R, Hughes, L., Kalali, A.,. Advancing Clinical Research in Alzheimer's Disease. White Paper from Quintiles 2010. Available from: http://www.quintiles.com/elements/media/white-papers/advancing-clinical-research-alzheimers-disease.pdf [cited: 18th Jul 2012].

[238] Burstein AH, Zhao Q, Ross J, *et al.* Safety and pharmacokinetics following a single 10-minute intravenous infusion of the anti-amyloid mAb ponezumab (PF-04360365) in patients with mild to moderate Alzheimer's disease. Alzheimer's and Dementia 2010;6(4):S585.

[239] Landen JW, Zhao Q, Cohen S, *et al.* Safety and Pharmacokinetics Following a Single Infusion of the Anti-amyloid Monoclonal Antibody Ponezumab (pf-04360365) in Patients With Mild-to-moderate Alzheimer's Disease: Final Results. Alzheimer's and Dementia 2010;6(4):e57.

[240] CBI. Ponezumab (PF-04360365): A novel, potential passive immunotherapy for the treatment of AD 2011. Available from: http://www.cbinet.com/compendiums/PC10188/Presentations/Sprenger_Ken_Presl.pdf [cited: 19th Jul 2012].

[241] Alzheimer Research Forum. Pfizer Halts Development of Aβ Antibody 2011. Available from: http://www.alzforum.org/new/detail.asp?id=2950 [cited: 19th Jul 2012].

[242] Black R, Ekman L, Lieberburg I, *et al.* Immunotherapy regimes dependent on ApoE status. World Patent WO2009052439, 2009.

[243] Gellerfors P, Lannfelt L. Antibodies specific for soluble amyloid beta peptide protofibrils and uses thereof. World Patent WO2005123775, 2005.

[244] Alzheimer Research Forum. Barcelona: Antibody to Sweep Up Aβ Protofibrils in Human Brain 2011. Available from: http://www.alzforum.org/new/detail.asp?id=2757 [cited: 20th Jul 2012].

[245] Eisai. Eisai announces the start of the first clinical study of BAN2401, a novel monoclonal antibody targeting the neurotoxic protofibrils believed to cause Alzheimer's disease: press release; 2010. Available from: http://www.eisai.com/news/enews201044pdf.pdf. [cited: 19th Jul 2012].

[246] Ravetch J, Fukuyama H. Antibodies specific for the protofibril form of beta-Amyloid protein. World Patent WO2009065054, 2009.

[247] Schupf N, Tang MX, Fukuyama H, *et al.* Peripheral Aβ subspecies as risk biomarkers of Alzheimer's disease. Proc Natl Acad Sci USA 2008;105(37):14052.

[248] Sanofi-Aventis. Proceedings of BIO 2009; 2009 May 19, Atlanta, USA. Available from: http://www.virtualpressoffice.com/JPContentAccessServlet?fileContentId=1000000004696&source=sd&showId=1215381716429 [cited: 19th Jul 2012].

[249] Sanofi-Aventis. Sanofi-aventis enters into an Exclusive Global Licensing Agreement with The Rockefeller University 2009. Available from: http://en.sanofi.com/Images/14019_20091203_rockefeller_univ_en.pdf [cited: 19th Jul 2012].

[250] Baurin N, Blanche F, Cameron B, *et al.* assignee. Humanized Antibodies specific to the protofibrillar form of the Beta-Amyloid peptide. World Patent WO2010130946, 2010.
[251] Sanofi-Aventis. Press Release: Strong performance in Q1 2012 including Genzyme contribution 2012. Available from: http://en.sanofi.com/Images/30285_2012-04-27_Results-Q1-2012_Pres.pdf [cited: 20th Jul 2012].
[252] Dunstan R, Bussiere T, Rhodes K, *et al.* Molecular characterization and preclinical efficacy. Alzheimers Dement 2011;7(4):S457.
[253] Nitsch R, Hock C, Esslinger C, *et al.* Method of providing disease-specific binding molecules and targets. World Patent WO2008081008, 2008.
[254] Dunstan R, Bussiere T, Engber T, *et al.* The role of brain macrophages on the clearance of amyloid plaques following the treatment of Tc2576 with BIIB037. Alzheimers Dement 2011;7(4):S700.
[255] Nitsch R, Lindvall O, Biscaro B, Ekdahl C. Method of promoting Neurogenesis. World Patent WO2010004434, 2010.
[256] Neurimmune. Neurimmune Therapeutics Announces Advancement of Alzheimer's Program into Preclinical Development 2009. Available from: http://www.neurimmune.com/index.php?section=news&cmd=details&newsid=7 [cited: 20th Jul 2012].
[257] Barghorn S, Striebinger A, Ebert U, Hillen H. Age dependent A-beta-oligomer formation in Tg2576 mice monitored by using the A-beta-oligomer selective antibody A-887755. Alzheimers Dement 2010;6(4):S238.
[258] Barghorn S, Striebinger A, Giaisi S, *et al.* Abeta-oligomer selective antibody A-887755 exhibits a favorable profile for Alzheimer's disease immunotherapy compared to Abeta-peptide unselective antibodies. Alzheimers Dement 2009;5(4):P424-P.
[259] Relo A, Barghorn S, Ebert U, *et al.* Restoration of home cage activity in Tg2576 mice by immunotherapy with the Ab-oligomer selective antibody A-887755. Alzheimers Dement 2011;7(4):S772.
[260] Alzheimer Research Forum. Paper Alert: Grappling Globulomers Improves Mouse Cognition 2010. Available from: http://www.alzforum.org/new/detail.asp?id=2526 [cited: 20th Jul 2012].
[261] Hamers-Casterman C, Atarhouch T, Muyldermans S, *et al.* Naturally occurring antibodies devoid of light chains. Nature 1993;363(6428):446-8.
[262] Kuang B, King L, Wang HF. Therapeutic monoclonal antibody concentration monitoring: free or total? Bioanalysis 2010;2(6):1125-40.
[263] Wilcock DM, Colton CA. Immunotherapy, vascular pathology, and microhemorrhages in transgenic mice. CNS Neurol Disord Drug Targets 2009;8(1):50.
[264] Ablynx. Neurology 2012. Available from: http://www.ablynx.com/en/research-development/diseases/neurology/ [cited: 19th Jul 2012].
[265] Marketwire. Nanobody for Alzheimer's Disease From The Alliance With Boehringer Ingelheim Moves Closer Towards Phase I Clinical Trials, 2012. Available from: http://www.marketwire.com/press-release/nanobody-alzheimers-disease-from-the-alliance-with-boehringer-ingelheim-moves-closer-brussels-ablx-1648348.htm [cited: 19th Jul 2012].
[266] Wirths O, Erck C, Martens H, *et al.* Identification of Low Molecular Weight Pyroglutamate Aβ Oligomers in Alzheimer Disease. J Biol Chem 2010;285(53):41517-24.
[267] Moir RD, Tseitlin KA, Soscia S, *et al.* Autoantibodies to redox-modified oligomeric Abeta are attenuated in the plasma of Alzheimer's disease patients. J Biol Chem 2005;280(17):17458-63.

[268] Bush A, Cherny R. Neurotoxic oligomers. World Patent WO 2002000245, 2002.

[269] Prana Biotechnology Limited. Annual report 2011, 2011. Available from: http://www.pranabio.com/downloads/Annual Reports/2011/Prana AR 2011 FINAL.pdf [cited: 20th Jul 2012].

[270] Adekar SP, Klyubin I, Macy S, *et al.* Inherent Anti-amyloidogenic Activity of Human Immunoglobulin γ Heavy Chains. J Biol Chem 2010;285(2):1066-74.

[271] Magga J, Puli L, Pihlaja R, *et al.* Human intravenous immunoglobulin provides protection against Aβ toxicity by multiple mechanisms in a mouse model of Alzheimer's disease. J Neuroinflammation;7(1):90.

[272] O'Nuallain B, Hrncic R, Wall JS, Weiss DT, Solomon A. Diagnostic and Therapeutic Potential of Amyloid-Reactive IgG Antibodies Contained in Human Sera. J Immunol 2006;176(11):7071-8.

[273] Fillit H, Hess G, Hill J, Bonnet P, Toso C. IV immunoglobulin is associated with a reduced risk of Alzheimer disease and related disorders. Neurology 2009;73(3):180-5.

[274] Evans J. IVIG Reduced Brain Atrophy in Alzheimer's. Clinical Psychiatry News 2010;38(6):12.

[275] Evans J. IVIG Reduced Brain Atrophy, Improved Cognition in AD. Family Practice News 2010;40(8):47.

[276] Relkin NR. Use of ventricular Enlargement Rate in intravenous Immunoglobulin Treatment of Alzheimer's Disease. World Patent WO2011130355; 2011.

[277] Szabo P, Mujalli DM, Rotondi ML, *et al.* Measurement of anti-beta amyloid antibodies in human blood. J Neuroimmunol 2010;227(1-2):167-74.

[278] Alzheimer Research Forum. Comments on Paper and Primary News: Norman Relkin and Paul Szabo 2011. Available from: http://www.alzforum.org/pap/annotation.asp?powID=116328 [cited: 20th Jul 2012].

[279] Wurster U, Haas J. Passage of intravenous immunoglobulin and interaction with the CNS. J Neurol Neurosurg Psychiatry 1994;57(Suppl):21-5.

[280] Alzheimer's Association. Alzheimer's Treatment Study Reports Three Years with No Decline in Memory and Function at AAIC 2012: AAIC Press Release; 2012. Available from: http://www.alz.org/aaic/tues_1030amct_ivig_trial.asp [cited: 6th Jul 2012].

[281] Kerchner GA, Boxer AL. Bapineuzumab. Expert Opin Biol Ther 2010;10(7):1121-30.

[282] Hofbauer L, Bruckschwaiger L, Butterweck HA, Teschner W. Affinity Chromatography for Purification of IgG from Human Plasma. Available from: http://cdn.intechopen.com/pdfs/33051/InTech-Affinity_chromatography_for_purification_of_igg_from_human_plasma.pdf [cited: 18th Jul 2012].

[283] Du Y, Dodel R. Human beta-amyloid antibody and use thereof for treatment of alzheimer's disease. European Patent EP 1172378 A1, 2001.

[284] Datamonitor. Market and Product Forecasts: Alzheimer's Disease - Multi-billion dollar potential in high-risk pipeline 2011. Available from: http://www.datamonitor.com/store/Product/market_and_product_forecasts_alzheimers_disease_multi_billion_dollar_potential_in_high_risk_pipeline?productid=HC00089-004 [cited: 18th Jul 2012].

[285] Scolnik PA. mAbs: a business perspective. MAbs. 2009;1(2):179-84.

[286] FierceBiotech. Johnson & Johnson and Elan Corporation plc Announce Definitive Agreement for Alzheimer's Immunotherapy Program and Equity Invest. 2009. Available

from: http://www.fiercebiotech.com/press-releases/johnson-johnson-and-elan-corporation-plc-announce-definitive-agreement-alzheimers-imm [cited: 18th Jul 2012].

[287] FierceBiotech. Pfizer and Wyeth: Top 10 deals of 2009. 2010. Available from: http://www.fiercebiotech.com/special-reports/pfizer-and-wyeth-top-10-deals-2009 [cited: 18th Jul 2012].

[288] European Medicines Agency. Guideline on medicinal products for the treatment of Alzheimer's Disease and other dementias. 2008. Available from: http://www.ema.europa.eu/docs/en_GB/document_library/Scientific_guideline/2009/09/WC500003562.pdf [cited: 05th Jul 2012].

[289] European Medicines Agency. Scientific Discussion 2008. Available from: http://www.ema.europa.eu/docs/en_GB/document_library/EPAR_-_Scientific_Discussion/human/000463/WC500058759.pdf [cited: 05th Jul 2012].

[290] World_Health_Organization. Neurological Disorders: Public Health Challenges 206. Available from: http://www.who.int/mental_health/neurology/neurodiso/en/index.html [cited: 19th Jul 2012].

[291] World Health Organization. Alzheimer's Disease: Of Emerging Importance 2002. Available from: http://www.searo.who.int/en/Section1243/Section1310/Section1343/Section1344/Section1355_5309.htm [cited: 05th Jul 2012].

[292] World Health Organization. Alzheimer's disease: the brain killer 2006. Available from: http://www.searo.who.int/en/section1174/section1199/section1567_6740.htm [cited: 20th Jul 2012].

[293] European Medicines Agency. Concept paper on no need for revision of the guideline on medicinal products for the treatment of Alzheimer's disease and other dementias 2012. Available from: http://www.ema.europa.eu/docs/en_GB/document_library/Scientific_guideline/2012/03/WC500124534.pdf [cited: 18th Jul 2012].

[294] Panza F, Frisardi V, Imbimbo BP, *et al.* Monoclonal antibodies against β-amyloid (Aβ) for the treatment of Alzheimer's disease: the Aβ target at a crossroads. Expert Opin Biol Ther 2011(0):1-8.

[295] Xu Y, Oomen R, Klein MH. Residue at position 331 in the IgG1 and IgG4 CH2 domains contributes to their differential ability to bind and activate complement. J Biol Chem 1994;269(5):3469-74.

[296] Reporter P. Alzheimer's Prevention Study in Genetically Defined High-Risk Patients Includes Roche, Lilly Drugs 2012. Available from: http://www.genomeweb.com/clinical-genomics/alzheimers-prevention-study-genetically-defined-high-risk-patients-includes-roch [cited: 31st Oct 2012].

[297] Delacourte A, David J, Sergeant N, *et al.* The biochemical pathway of neurofibrillary degeneration in aging and Alzheimer's disease. Neurology 1999;52(6):1158.

[298] Winblad B, Andreasen N, Minthon L, *et al.* Safety, tolerability, and antibody response of active Aβ immunotherapy with CAD106 in patients with Alzheimer's disease: randomised, double-blind, placebo-controlled, first-in-human study. Lancet Neurol 2012.

Send Orders of Reprints at reprints@benthamscience.net
86 *Frontiers in Clinical Drug Research – Alzheimer Disorders, Vol. 1,* 2013, 86-108

CHAPTER 2

The Development of Multi-Target-Directed Ligands (MTDL) to Treat Alzheimer's Disease

John W. Wright[*] and Joseph W. Harding

Departments of Psychology, Veterinary and Comparative Anatomy, Pharmacology, and Physiology, and Programs in Neuroscience and Biotechnology, Washington State University, Pullman, Washington, USA

Abstract: The loss of neurons and synapses resulting in neurotransmitter dysfunction, along with β-amyloid peptide (Aβ) aggregation and neurofribrillary tangles, have served as the hallmarks of Alzheimer's disease (AD). Alterations associated with neurochemical disturbances of several neurotransmitter systems appear to be the major underling cause of the cognitive and neuropsychiatric symptoms accompanying this disease. Treatments designed to target cholinergic deficiencies have provided only minimal benefit to the large majority of AD patients. As populations live longer there is a persistent and growing need for the development of new and much more efficacious treatment strategies designed to prevent progressive synaptic and neuron losses and encourage neurogenesis and synaptogenesis. There is also the emerging realization that neurodegenerative diseases such as AD, Parkinson's disease, and Amyotrophic Lateral Sclerosis are the result of multiple insults, not one or two, and must be countered with multi-target lead compounds. The development of these new drugs requires a reevaluation of high throughput screens against protein targets in favor of the development of multi-target-directed ligands (MTDL) consisting of a single or hybrid compound capable of influencing several targets and/or systems. The present chapter focuses on this MTDL approach in the development of drugs to treat AD, but recognizes that this drug design strategy is applicable to other clinically significant diseases. This new approach represents a transition away from the "one molecule one target" high throughput screening assays, toward the rational design of drugs. As such it ushers in a new approach to drug development.

Keywords: Aβ aggregate, Acetylcholinesterase inhibitor drugs, Alzheimer's disease, Angiotensin receptor blocking drugs, AT4 receptor subtype, Brain renin-angiotensin system, β-secretase, Cerebral blood flow, c-Met receptor, Dementia, Drug development strategies, MAO inhibitor drugs, Multi-target-directed drugs, Neurogenesis, NMDA receptor antagonist drugs, Oxidative stress, Parkinson's disease-induced dementia, Synaptogenesis.

INTRODUCTION

With the onset of Alzheimer's disease (AD) begins subtle behavioral and memory

***Address correspondence to John W. Wright:** Department of Psychology, Washington State University, Pullman WA, 99164-4820 USA; Email: wrightjw@wsu.edu

Atta-ur-Rahman (Ed)

disturbances that become progressively more severe, culminating in cognitive deterioration, memory loss, behavioral and psychiatric disturbances, and functional impairment [1-3]. At present there are approximately 4.5 million AD patients in the U.S. with the same number of new cases diagnosed yearly worldwide, resulting in an estimated total of 37 million people afflicted with this disease [4-6]. These patients represent approximately 3% of the population between the ages of 65-74 years of age, and nearly 50% of those 85 years and older [7]. The costs associated with treatment and care for these patients has reached in the range of $70 to 150 billion annually in the U.S. alone [8-10]. These numbers do not address the personal hardships suffered by the patient's family members, and other care givers. Thus, research efforts must be directed to stopping and reversing the progression of this disease.

To date the cholinergic hypothesis of memory formation has driven the development of the major marketed drugs in the form of acetylcholinesterase inhibitors (AChEIs: Tacrine®, Donepezil®, Rivastigmine®, and Galantamine®) which will go generic in the near future [11]. These drugs are only marginally helpful with symptoms and do not impact the underlying neuropathology [12]. The FDA approved Namenda® (memantine HCl), an NMDA receptor antagonist designed to limit glutamate excitotoxicity, in 2004 to treat moderate to severe AD patients [13]. Namenda is also limited regarding its ability to slow disease progression and does nothing to stem the neuropathology. More recently research has focused on the accumulation of brain β-amyloid peptide (Aβ) as an important target in the pathogenesis of AD [14]. There may be a link between Aβ accumulation and NMDA receptor over activation in that oxidative stress, plus the elevated intracellular calcium generated due to Aβ accumulation, appear to enhance glutamate mediated neurotoxicity *via* increased NMDA receptor activation [15]. There are many reasons for the lack of an effective therapy for AD including the complexity of the disease process and the resulting inability to identify reliable biomarkers. In addition, it is becoming apparent that AD is multifactorial rather than a single disease [16]. To further complicate diagnosis, those AD criteria behaviors denoting cognitive decline can also result from a number of other clinical conditions including vascular disease [17,18], frontotemporal dementia, Parkinson's disease (PD)-induced dementia, HIV infection [19,20], as well as cumulative oxidative damage and toxicities accompanying normal aging [21]. The ultimate goal of development efforts must be a drug that prevents progressive synapse and neuron loss and reverses this devastating process.

What is a "Multi-Target-Directed Ligand"?

Looking back some years, with the advent of protein expression technologies, molecular cloning, and combinatorial chemistry came the capacity to synthesize large numbers of novel compounds that were subjected to high throughput screening assays against one or more protein targets [22]. Despite the very significant time and resources invested in such drug development programs the rate of discovery has decreased over the last decade [16,23,24]. Thus, during this time span no new efficacious drugs have been developed for any neurodegenerative disease including AD, PD, amyotrophic lateral sclerosis, or ischemic stroke. It is the premise of a growing number of investigators in these important research areas that this failure is not simply a matter of insufficient effort, commitment, or resources, but rather the fault lies with the drug development approaches being employed [25-31]. The new approach being advocated is Multi-Target-Directed Ligands (MTDL).

As mentioned past drug discovery was facilitated by *in vivo* assays that were replaced by the "one target-one disease" approach [22]. This latter strategy makes use of combinatorial chemistry coupled with high throughput screening based on the notion that the more selective the drug for the target the fewer will be the "side effects". It should be realized that some diseases are multifactorial in origin and are best treated *via* a single drug, or drug cocktail, possessing multiple effects. This has led to the most recent theory that there may be advantage in developing a single or hybrid compound capable of influencing multiple targets and/or systems (Fig. **1**). This new approach has stimulated the search for what has been termed, "multimodal" [28,31] or "multi-target-directed" ligands [25,32]. Prior to 2005 such multi-target drugs were developed by a "design in" strategy to a lead compound in order to increase its affinity to a second target, and/or remove those unwanted characteristics, *i.e.* a "design out" strategy [28]. Since 2005 strategies such as modeling, crystallography, and fragment-based technologies have been utilized.

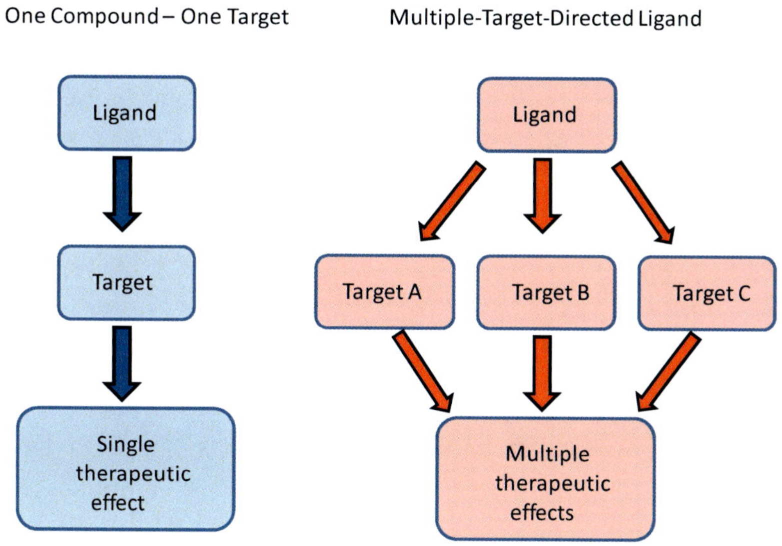

Figure 1: Drug discovery strategies: "One Compound – One Target" represents the frequently used approach of combinatorial chemistry coupled with high throughput screening assays directed at one protein. The "Multiple-Target-Directed Ligand" approach relies on the rational design of drugs capable of influencing several targets. Since the underlying cause of AD is multifactorial this approach may offer an advantage over other drug discovery strategies.

A major obstacle to drug development has been the lack of a comprehensive understanding of the mechanisms contributing to the many histopathological changes seen in AD patients. This complexity of causes underlying brain degeneration at a minimum include the involvement of the cholinergic system, oxidative stress, monoamine oxidase dysfunction, inflammation, cell cycle regulatory failure, β-amyloid aggregation, and NMDA receptor over activation [33-36]. Thus, compounds are being developed to engage more than one of these targets [27]. The starting point for a number of research groups has been the development of dual binding site AChEIs with the goal of preventing Aβ aggregation, and thus improvement in cognitive functioning [37-39].

The present chapter focuses on the MTDL approach in the development of new drugs to treat AD. We describe several systems that show promise in this regard. These include new drugs designed to influence the cholinergic, serotonergic, β-secretase and angiotensinergic systems. Given the complexity of causes underlying AD this development strategy may ultimately prove successful where the "one molecule one target" approach has fallen short. This new approach represents a transition away from high throughput screening assays focused on profiling significant numbers of compounds against large panels of targets, toward the rational design of drugs. Several compounds to be discussed were derived by linking through a spacer two pharmacophores possessing activity directed at two different targets [40-42]. However, the next step of converting or linking a dual-action compound into a triple action ligand promises to be an exceedingly difficult task [43].

Modified Acetylcholinesterase Inhibitor Drugs

Based on the cholinergic hypothesis of memory formation it was originally proposed that drugs designed to inhibit central and peripheral AChE, and serve as a muscarinic M_2 autoreceptor antagonist, would result in facilitated release of ACh (Fig. **2**). Further, AChE binding *via* its peripheral anionic site to the nonamyloidogenic form of Aβ appears to facilitate a conformational shift to the amyloidogenic form [44-46]. Treatment with an AChEI would be expected to neutralize the catalytic site of the enzyme and reduce Aβ peptide aggregation as facilitated by active AChE. The initial compound developed, Caproctamine, is both an AChE inhibitor and an M_2 receptor antagonist [47]. Further structural modifications of this compound have resulted in the synthesis of Momoquin, designed to inhibit AChE and thus Aβ aggregation with the added benefit of also possessing antioxidant activity [37,43,48].

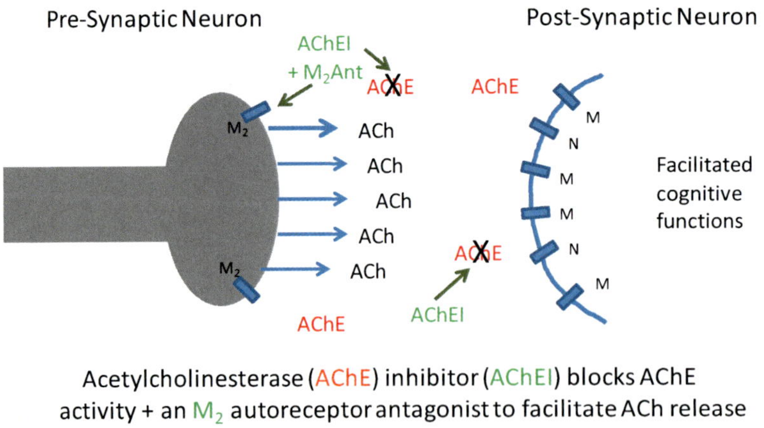

Figure 2: Acetylcholine (Ach) is released from cholinergic neurons, diffuses across the cleft, and binds with muscarinic (M) and nicotinic (N) receptors on the post-synaptic membrane. Acetylcholine esterase (AChE) acts upon Ach to degrade it into the inactive components acetate and choline. Drugs that act as AChE inhibitors (AChEIs) are designed to block degradation and

permit additional Ach to reach M and N receptors. There are also muscarinic (M_2) and nicotinic receptors located on the pre-synaptic membrane. These receptors serve to regulate the synthesis and release of ACh depending upon release rates. An M_2 receptor antagonist acts at the presynaptic membrane to facilitate ACh release.

MTDLs with Antioxidant Plus AChE Inhibitor Properties

Another important target in the treatment of neurodegenerative diseases is the production of reactive oxygen species (ROS) responsible for oxidative stress [49]. Brain tissues from AD patients often show oxidative damage thought to occur due to malfunctioning mitochondria (Fig. **3**). Mitochondria are sensitive to accumulating levels of calcium (also magnesium, zinc, and iron) that can lead to programmed cell death (apoptosis [50]). Thus, treatments directed at sources of ROS production are of potential value in disrupting neurodegeneration. Lipocrine is a protype drug that combines the antioxidant ability of lipoic acid with an AChEI to protect released ACh and discourage Aβ aggregation [37]. Additional modifications have been made to this compound to increase potency and facilitate penetration of the blood-brain barrier (BBB [27,51,52]).

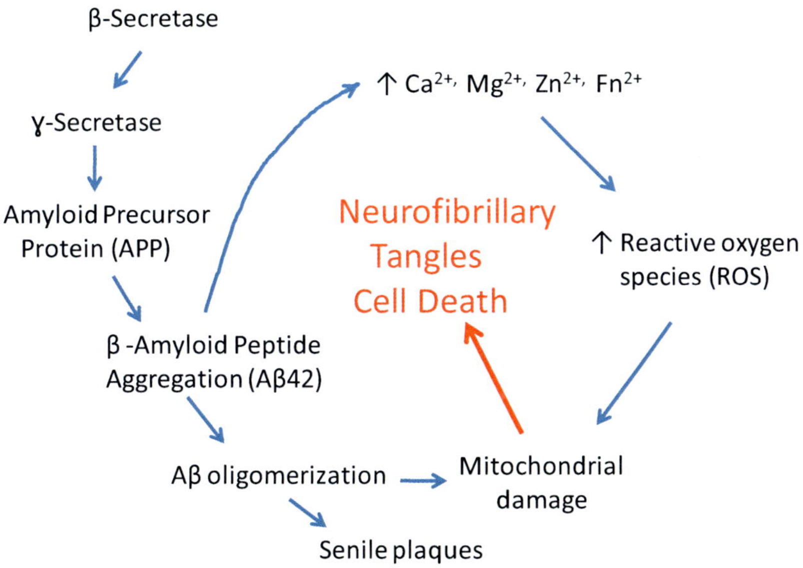

Figure 3: Possible molecular mechanisms contributing to neuronal cell death in Alzheimer's disease. The current notion is that an imbalance occurs between Aβ production and its clearance from neurons. Enhanced β- and ɣ-secretase activity contributes to the increased release of amyloidogenic Aβ42. This produces Aβ oligomers and extracellular senile plaques. There is also convincing evidence supporting a relationship between Aβ42 deposits and increased levels of

metal ions and reactive oxygen species in Aβ oligomerization. Mitochondrial damage appears to be a predisposing factor in the development of neurofribrillary tangles and eventual cell death.

Another approach targets phospholipase A2 which is responsible for clearing membrane phospholipids thus releasing arachidonic acid, a precursor of proinflammatory eicosanoids, and leukotriene A4 hydrolase. Leukotriene A4 hydrolase promotes the synthesis of leukotriene which is a proinflammatory agent present in several inflammatory diseases [28,53]. A dual inhibitor of phospholipase 2 and leukotriene A4 hydrolase has recently been developed [54].

MTDLs Designed to Reduce Aβ Aggregation

Extracellular Aβ plaque accumulation in the brain is a prominent characteristic and appears to be an important causal factor of AD [55-57]. Amyloid precursor protein (APP) acts as the precursor molecule for Aβ42 production. This occurs *via* two proteases, β-secretase and γ-secretase. α-secretase competes with β-secretase for access to APP substrate and is capable of preventing the formation of Aβ by cleaving this peptide into two fragments. Thus, clinical intervention can take the form of inhibitors of β- and γ-secretases, and/or facilitation of α-secretase, in order to reduce Aβ plaque accumulation. Given that elevated levels of β-secretase have been reported in AD patients [58], the major target has been to inhibit β-secretase [59,60]. Several companies have designed β-secretase inhibitors of low molecular weight with acceptable bioavailability [61-63]. Only a few of these compounds have reached clinical trials with the most promising developed by Eli Lilly Pharmaceutical (LY-450139 [25]). However, development of this Lilly compound, Semagacestat, was halted because of unfavorable Phase III clinical trial results [64].

An approach configured to prevent Aβ aggregation induced by AChE is the use of enhanced AChEIs. That is, AChE is thought to facilitate Aβ aggregation by the interaction of its peripheral anionic site with fibrils of the peptide [65]. In this way AChE mediates not only ACh breakdown but also Aβ fibril formation. Thus, efforts have been devoted to developing MTLDs that combine inhibition of AChE with antagonism of muscarinic M2 autoreceptors that facilitate the release of ACh [39,66,67].

MTDLs That Target MAO and AChE

It is generally accepted that the progressive cognitive dysfunction seen in AD is correlated with the loss of cholinergic and glutamatergic neurons. However, damage to noradrenergic and serotonergic systems also contributes to these symptoms. Loss of noradrenergic neurons in the locus coeruleus is associated with depression [68]; while loss of serotonergic neurons in the raphe nucleus is linked with psychosis and depression [69]. Drugs designed to facilitate the activity of biogenic amines are predicted to relieve these symptoms. There is evidence that H_2O_2 is produced accompanying MAO's catalytic activity to deaminate norepinephrine, dopamine, and serotonin. This source of H_2O_2 encourages oxidative stress and may negatively impact vulnerable neurons already compromised by AD [28,70].

Several dual MAO/AChEI compounds have been developed [71]; however, these agents initially revealed low bioavailability and very poor BBB penetrability. Recently, compounds have been synthesized with balanced inhibitory influences on AChE and MAO. Ladostigil shows promise for Parkinson's disease (PD)-related dementia accompanied by depression [72]. The depression accompanying AD has also been treated with inhibitors of serotonin transporters. Attempts have been made to combine serotonin transporter inhibition with AChE inhibition. Although initial results were disappointing [73,74], recent approaches have been more successful [75-77].

The Development of Angiotensin MTDLs

The renin-angiotensin system (RAS) was originally thought to reside only in the periphery and mediate blood pressure and body water regulation. It is now clear that the necessary components of this system are also present in the mammalian brain [78-80]. A role for the brain RAS in cognitive processing, particularly learning and memory, was proposed some time ago and provides justification for the study of components of the RAS as targets for the treatment of AD [81-86]. Our laboratory has devoted much time and effort to understanding the contribution of the brain RAS to memory consolidation, and more recently the potential for these peptides, and their analogues, to be used in the treatment of neurodegenerative diseases.

Angiotensin ligands are derived through the precursor protein angiotensinogen that is acted upon by several enzymatic conversion pathways ([80,87] Fig. **4**).

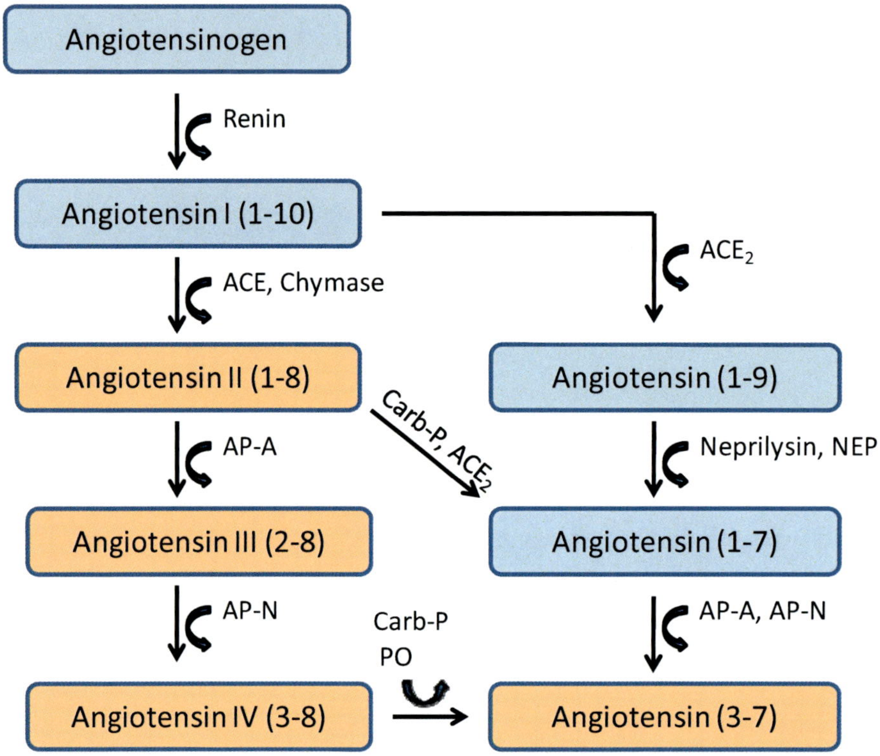

Figure 4: Peptides and enzymes involved in the conversion of angiotensinogen to angiotensins I and II through fragments. The biologically active forms include angiotensins II, III, IV, and (1-7). Abbreviations: ACE; angiotensin-converting enzyme, AP-A; aminopeptidase A, AP-N; aminopeptidase N, Carb-P; carboxypeptidases P, NEP; neutral endopeptidase, PO; propyl oligopeptidase.

First in this sequence is the decapeptide angiotensin I (AngI) formed by renin (EC3.4.23.15) acting at the amino terminal of angiotensinogen. AngI serves as a substrate for angiotensin converting enzyme (ACE: EC 3.4.5.1), a zinc metalloproteases that hydrolizes the carboxy terminal dipeptide His-Leu to form the octapeptide angiotensin II (AngII [88]). AngII is converted to the heptapeptide angiotensin III (AngIII) by glutamyl aminopeptidase A (AP-A: EC 3.4.11.7) that cleaves the Asp residue at the amino terminal [89,90]. AngIII is further converted to the hexapeptide angiotensin IV (AngIV) by aminopeptidase N (AP-N: EC 3.4.11.2) cleavage of Arg at its amino terminal. AngIV can be further converted to

Ang(3-7) by carboxy peptidase P (Carb-P) and propyl oligopeptidase (PO) cleavage of the Pro-Phe bond. Endopeptidases including chymotrypsin, cleave the Val, Tyr, and Ile residues; while dipeptidyl carboxypeptidase cleaves the His-Pro bond, thus reducing AngIV and Ang(3-7) to inactive peptide fragments and amino acid constituents [91-93]. AngII is also converted to Ang(1-7) by Carb-P cleavage of Phe, by ACE_2 [94], or by ACE cleavage of the dipeptide Phe-His from Ang(1-9) [95]. Ang(1-7) is further converted to Ang(2-7) by AP-A acting at the Asp-Arg bond, and to Ang(3-7) by AP-N acting at the Arg-Val bond [96]. AngII and AngIII are full agonists at the AT_1 and AT_2 receptor subtypes [97]. AngIV binds with high affinity and selectivity at the AT_4 receptor subtype [98].

For many years angiotensin peptides shorter than AngIII were considered inactive and of no pharmacological importance. This assumption was based on two findings: 1) AngIV reveals a very poor affinity for the AT_1 and AT_2 sites [98-101]. 2) AngIV, and shorter fragments, are considerably less potent than Ang II and AngIII in eliciting classic angiotensin-dependent functions including the regulation of blood pressure, body water balance, and the release of vasopressin and oxytocin [93,102-105]. Three discoveries changed this assumption. 1) Braszko and colleagues [106] reported that AngIV facilitated acquisition of a conditioned avoidance response in rats. 2) A separate and distinct binding site for AngIV was identified [97] and subsequently classified as the AT_4 subtype ([107] Fig. **5**). This subtype was originally isolated using bovine adrenal membranes [98,101,108,109]. These characterization studies indicated that the AT_4 receptor subtype is distinct from the AT_1 and AT_2 sites given that ligands known to bind to these sites do not bind at the AT_4 site [98]. 3) In contrast with the AT_1 receptor subtype the AT_4 binding site is concentrated in brain regions functionally linked to cognitive processing such as hippocampus, dentate gyrus, neocortex and piriform cortices [86].

The realization that a small peptide activates the AT_4 site, and that the vast majority of small peptide receptors are G protein-linked, suggested that the AT_4 receptor was a serpentine G protein-linked receptor. However, this was not the case because binding to this site was insensitive to guanine nucleotides. In addition, the AT_4 receptor subtype exhibited a molecular weight in the 100+ kDa range as determined by reduced SDS-polyacrylamide gel electrophoresis. An equivalent molecular weight has been reported for this receptor in other bovine tissues including heart, thymus, kidney, bladder, aorta, and hippocampus [110]. Further, Bernier and colleagues [108,111] established a similar molecular weight for the binding subunit of the AT_4 receptor in bovine aortic endothelial cells. The lack of linkage to G proteins was also supported by the observation that GTPγS failed to alter [^{125}I]AngIV binding in rabbit heart [112], guinea pig brain [113], and rat vascular smooth muscle [114]. There is a hypothesis that the AT_4 receptor is the insulin-regulated aminopeptidase (IRAP [115,116]), although this suggestion remains controversial [117]. Recently, our laboratory proposed this site to be the tyrosine kinase receptor c-MET with a 50 kDa α-subunit and a 145 kDa β-subunit [117-121].

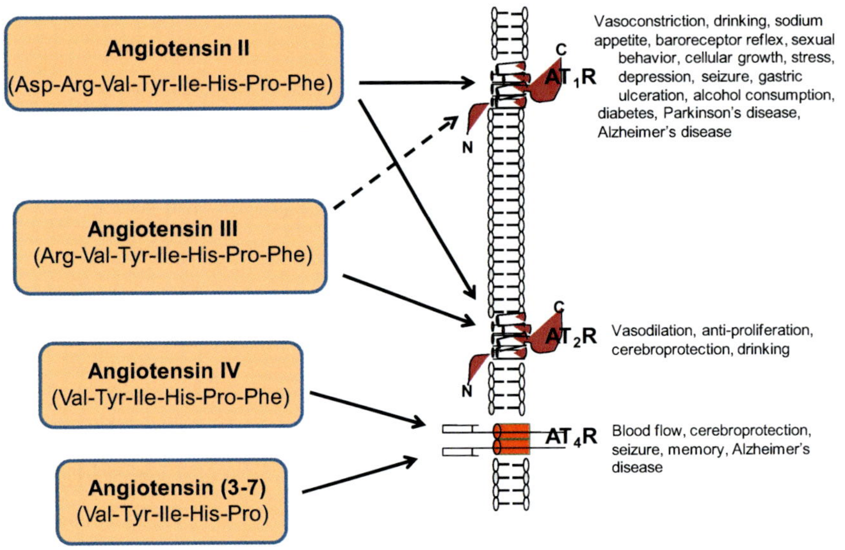

Figure 5: The biologically active forms of angiotensins are highlighted. The receptor proteins where they bind are shown by arrows. The functions mediated by each receptor are indicated.

This AT_4 receptor subtype regulates a number of physiological functions including cerebral blood flow (CBF), neuroprotection, memory consolidation, and synaptogenesis. Our laboratory has reported AngIV-induced dose dependent increases in CBF without significant changes in systemic blood pressure [122]. These effects could not be blocked by AT_1 or AT_2 receptor antagonists, but were prevented by pretreatment with the AT_4 receptor antagonist Divalanal-AngIV

(Divalinal). Related to this, Naveri and colleagues [123] have shown that AngIV infusion restored CBF following subarachnoid hemorrhage. Dalmay *et al.* [124] reported that AngIV infusion following Candesartan pretreatment in the gerbil model of unilateral carotid artery ligation, slightly decreased mortality at post-surgery Day 3, but significantly decreased Lisinopril-induced mortality. These results support the hypothesis that the activation of AT_4 receptors contributes to cerebroprotection. This role for the AT_4 receptor subtype in neuroprotection is consistent with the notion that AngIV increases blood flow by a NO-dependent mechanism [122]. In agreement with this hypothesis recent work by Faure *et al.* [125] has shown that internal carotid artery administration of increasing doses of AngIV significantly decreased mortality and cerebral infarct size in rats 24 hours following embolic stroke due to the intracarotid injection of calibrated microspheres. Pretreatment with Divalinal, or nomga-nitro-L-arginine methyl ester (L-NAME), abolished this protective effect. Sequential cerebral arteriography indicated that AngIV caused the redistribution of blood flow to ischemic areas within a few minutes. Thus, AngIV may yield its cerebral protective effect against acute cerebral ischemia *via* a intracerebro-hemodynamic AT_4 receptor-mediated NO-dependent mechanism.

Recent findings suggest that many of the memory enhancing effects initially attributed to AngII are likely due to the conversion of AngII to AngIV, and it is this peptide acting as an agonist at the AT_4 receptor subtype that is responsible for cognitive facilitation [85,126,127]. This hypothesis is consistent with the finding that angiotensin receptor blockers (ARBs) improve cognition [128-132] presumably due to blockade of the AT_1 receptor subtype thus permitting conversion of excess endogenous AngII to AngIV and Ang(3-7), which then activate the AT_4 receptor (Fig. **4**). This notion is also consistent with the observation that ACE inhibitors enhance cognitive processing in that the resulting increased AngI levels are likely converted to Ang(1-9) and then to Ang(1-7) and Ang(3-7). Both AngIV and Ang(3-7) act as agonists at the AT_4 receptor subtype.

A number of studies indicate that AngIV, and AngIV analogues such as Nle^1-AngIV, can facilitate long-term potentiation, learning, and memory consolidation [106,133-135]. It is noteworthy that studies using various dementia models to test the efficacy of Nle^1-AngIV have demonstrated that acute application reversed deficits initiated by: 1) Treatment with scopolamine [136]; 2) Kainic acid injections into the hippocampus [137]; 3) Perforant path cuts [135]; and 4) Ischemia resulting from transient four vessel occlusion [138]. Consistent with these behavioral and electrophysiological results, brain binding sites for [^{125}I]-AngIV have been autoradiographically localized in structures known to mediate cognitive processing including the neocortex, hippocampus, and basal nucleus of Meynert [3,98,139]. It should be acknowledged that the AT_1 angiotensin receptor subtype may also contribute to the cognitive facilitating effects of AngIV [140].

Although AngIV-like peptides such as Nle^1-AngIV facilitate cognition in animal models of dementia, they possess two characteristics that limit their use as clinically practical therapeutics:1) These compounds are rapidly degraded resulting in short half-lives; and 2) They are too large and hydrophilic to penetrate the BBB. During the past few years our laboratory, and others [141,142], have attempted to overcome these limitations by designing and synthesizing small molecule AngIV analogues with improved stability, smaller size, and hydrophobicity characteristics compatible with BBB permeability. Past studies indicate that the degradation of angiotensin peptides is primarily dependent on the action of aminopeptidases [143]. Thus, we employed several strategies to protect the N-terminal of putative pharmaceuticals. These approaches include the addition of a D-amino acid or non-α-amino acid in the #1 position, acylation of the N-terminal amino acid, and conversion of the 1-2 peptide bond to a reduced-peptide bond. These modifications have recently been shown to provide the desired stability, while maintaining the required cognitive enhancing activity.

As part of our efforts to develop BBB permeable Nle^1-AngIV analogues, we determined the minimum structural features of Nle^1-AngIV capable of supporting pro-cognitive activity [118]. Previous studies established that critical structural information resides at the N-terminal of AngIV. Thus, we sequentially removed amino acids from the C-terminal of Nle^1-AngIV while monitoring pro-cognitive activity. This approach indicated that memory facilitation was achievable with peptides as small as tetra- and tripeptides [118]. Next, we modified these peptides in several ways to enhance stability and hydrophobicity. Included in this group were modified tripeptides protected at both the N- and C-terminals by metabolically resistant constituents. These efforts produced a number of very active molecules with the desired characteristics [11]. Included in this group was the precursor to a BBB permeant molecule with the ability to reverse scopolamine-induced amnesia when delivered peripherally. This compound, N-hexanoic-Tyr-Ile-(6) aminohexanoic amide (Dihexa) has recently been shown to exhibit excellent anti-dementia activity in AD animal models and augment synaptic connections in the hippocampus [144]. Thus, Ang IV and most recently Dihexa, qualify as MTDLs in that they act to facilitate cerebral blood flow, overcome focal and global ischemia and offer neuroprotection, facilitate LTP and memory consolidation. These compounds have also been shown to act as powerful facilitators of synaptogenesis.

CONCLUSION

This chapter proposes that AD drug development efforts move toward a rational approach in the form of Multi-Target-Directed Ligands. There are several reasons to employ this strategy including the fact that the causes of this disease are multifactorial and there is a growing need for drugs designed to attack multiple targets. It is also clear that our present drug development strategies are not working to deliver new efficacious compounds to treat neurodegenerative disorders. It is worthy of note that rationally designed MTDLs should not be confused with "dirty drugs" that often exhibit off target and undesirable side effects. The MTDL approach is presently in its early stages but promises to offer greater selectivity, the engagement of multiple targets with a single drug or combination of drugs, accompanied by fewer unwanted side effects. This drug development strategy is presently being applied to other disorders including depression [145] and schizophrenia [146].

ACKNOWLEDGEMENTS

The research from our laboratory presented in this chapter was supported by the Edward E. and Lucille I. Lainge Endowment for Alzheimer's Research, and funds provided for medical and biological research by the State of Washington Initiative Measure No. 171.

CONFLICT OF INTEREST

The author(s) confirm that this chapter content has no conflict of interest.

REFERENCES

[1] Cummings JL. Alzheimer's disease. N Eng J Med 2004; 351: 56-67.
[2] McKhann GM, Knopman DS, Chertkow H, *et al.* The diagnosis of dementia due to Alzheimer's disease: Recommendations from the National Institute on Aging – Alzheimer's Association workgroups on diagnostic guidelines for Alzheimer's disease. Alzheimers Dement 2011; 7: 263-9.
[3] Wright JW, Harding JW. The brain angiotensin IV/AT_4 receptor system as a new target for the treatment of Alzheimer's disease. Drug Development Research 2009; 70: 472-80.
[4] Mount C, Downton C. Alzheimer disease: Progress or profit? Nat Med 2006: 12: 780-4.
[5] Smith AD. Prevention of dementia: A role for B vitamins? Nutr Health 2006; 18: 25-6.
[6] Yaari R, Corey-Bloom J. Alzheimer's disease. Sem Neurol 2007; 27: 32-41.
[7] Clark CM. Clinical manifestations and diagnostic evaluation of patients with Alzheimer's disease. In: Clark CM, Trajanowski JQ. Eds. Neurodegenerative dementias: Clinical features and pathological mechanisms. New York: McGraw-Hill 2000; pp 95-114.
[8] Isaia G, Bo M, Nobili G, Cappa G, Mondino S, Pilon S, Massaia M. Costs of the in-home patients affected by dementia. Arch Gerontol and Geriatr 2009; 49 (Suppl 1): 147-51.
[9] Maiese K, Chong ZZ, Hou J, Shang YC. New strategies for Alzheimer's disease and cognitive impairment. Oxid Med Cell Longev 2009; 2: 279-89.
[10] Zec.F, Trivedi MA. The effects of estrogen replacement therapy on neuropsychological functioning in postmenopausal women with and without dementia: A critical and theoretical review. Neuropsychol Rev 2002; 12: 65-109.
[11] Wright JW, Harding JW. The brain RAS and Alzheimer's disease. Exp Neurol 2010; 223: 326-33.
[12] Zhu Y, Xiao K, Ma L, *et al.* Design, synthesis and biological evaluation of novel dual inhibitors of acetylcholinesterase and β-secretase. Bioorg Med Chem 2009; 17: 1600-13.
[13] Melnikova I. Therapies for Alzheimer's disease. Nat Rev Drug Discov 2007; 6: 341-2.
[14] Hardy J, Selkoe DJ. The amyloid hypothesis of Alzheimer's disease: Progress and problems on the road to therapeutics. Science 2002; 297: 353-6.
[15] Lipton SA. The molecular basis of memantine action in Alzheimer's disease and other neurologic disorders: Low-affinity, uncompetitive antagonism. Curr Alzheimer Res 2005; 2: 155-65.
[16] Schubert D, Maher P. An alternative approach to drug discovery for Alzheimer's disease dementia. Future Med Chem 2012; 4: 1681-8.
[17] Polidori MC, Pientka L. Bridging the pathophysiology of Alzheimer's disease with vascular pathology: the feed-back, the feed-forward, and oxidative stress. J Alzheimers Dis 2012; 28: 1-9.
[18] Schneider JA, Arvanitakis Z, Bang W, Bennett DA. Mixed brain pathologies account for most dementia cases in community-dwelling older persons. Neurology 2007; 69: 2197-204.
[19] Brayne C, Matthews Fe, Xuereb JH, *et al.* Pathological correlates of late onset dementia in a multicentre, community-based population in England and Wales. Neuropathology Group

of the Medical Research Council Cognitive Function and Ageing Study (MRC CFAS). Lancet 2001; 357: 169-75.

[20] Chen M, Maleski JJ, Sawmiller DR. Scientific truth or false hope: Understanding Alzheimer's disease from an aging perspective. J Alzheimers Dis 2011; 24: 3-10.

[21] Ding Q, Dimayuga E, Keller JN. Proteasome regulation of oxidative stress in aging and age-related disease of the CNS. Antioxidant and Redox Signaling 2006; 8: 163-72.

[22] Swinney DC, Anthony J. How were new medicines discovered? Nat Rev Drug Discov 2011; 10: 507-19.

[23] Munos B. Lessons from 60 years of pharmaceutical innovation. Nat Rev Drug Discov 2009; 8: 959-68.

[24] Paul SM, Myelka DS, Dunwiddie CT, *et al.* How to improve R & D productivity: The pharmaceutical industry's grand challenge. Nat Rev Drug Discov 2010; 9: 203-14.

[25] Bajda M, Guzior N, Ignasik M, Malawska B. Multi-target-directed ligands in Alzheimer's disease treatment. Curr Med Chem 2011; 18: 4949-75.

[26] Boran AD, Iyengar R. Systems approaches to polypharmacology and drug discovery. Curr Opin Drug Discov Dev 2010; 13: 297-309.

[27] Cavalli A, Bolognesi ML, Minarini A, Rosini M, Tumiatti V, Recanatini M, Melchiorre C. Multi-target-directed ligands to combat neurodegenerative diseases. J Med Chem 2008; 51: 347-72.

[28] Costantino L, Barlocco D. Designed multiple ligands: Basic research vs clinical outcomes. Curr Med Chem 2012; 19: 3353-87.

[29] Petrelli A, Giordano S. From single- to multi-target drugs in cancer therapy: When aspecificity becomes an advantage. Curr Med Chem 2008; 15: 422-32.

[30] Puls IN, Eadens M, Messersmith W. Current status of SRC inhibitors in solid tumor malignancies. Oncologist 2011; 16: 56-78.

[31] Van der Schyf CJ, Geldenhuys WJ. Multimodal drugs and their future for Alzheimer's and Parkinson's disease. Int Rev Neurobiol 2011; 100: 107-25.

[32] Huang W, Tang L, Shi Y, Huang S, Xu L, *et al.* Searching for the multi-target-directed ligands against Alzheimer's disease: Discovery of quinoxaline-based hybrid compounds with AChE, H 3R and BACE 1 inhibitory activities. Bioorg Med Chem 2011; 19: 7158-67.

[33] Herrup K, Yang Y. Cell cycle regulation in the postmitotic neuron: Oxymoron or new biology? Nat Rev Neurosci 2007; 8: 368-78.

[34] Mattson MP. Pathways towards and away from Alzheimer's disease. Nature 2004; 430: 631-9.

[35] Petersen RB, Nunomura A, Lee HG, Casadesus G, Perry G, *et al.* Signal transduction cascades associated with oxidative stress in Alzheimer's disease. J Alzheimers Dis 2007; 11: 143-52.

[36] Rosini M, simony E, Bartolini M, Cavalli A, *et al.* Inhibition of acetylcholinesterase, β-amyloid aggregation, and NMDA receptors in Alzheimer's disease: A promising direction for the multi-target-directed ligands gold rush. J Med Chem 2008; 51: 4381-4.

[37] Bolognesi ML, Minarini A, Rosini M, Tumiatti V, Melchiorre C. From dual binding site acetylcholinesterase inhibitors to multi-target-directed ligands (MTDLs): A step forward in the treatment of Alzheimer's disease. Mini-Rev Med Ch 2008; 8: 860-7.

[38] Castro A, Martinez A. Peripheral and dual binding site acetylcholinesterase inhibitors: Implications in treatment of Alzheimer's disease. Mini-Rev Med Chem 2001; 1: 267-72.

[39] Recanatini M, Valenti P. Acetylcholinesterase inhibitors as a starting point towards improved Alzheimer's disease therapeutics. Curr Pharm Des 2004; 10: 3157-66.

[40] Hopkins AL, Mason JS, Overington JP. Can we rationally design promiscuous drugs? Curr Opin Struct Biol 2006; 16: 127-36.

[41] Morphy R, Kay C, Rankovic Z. From magic bullets to designed multiple ligands. Drug Disc Today 2004; 9: 641-51.

[42] Morphy R, Rankovic Z. The physicochemical challenges of designing multiple ligands. J Med Chem 2006; 49: 4961-4970.

[43] Bolognesi ML, Cavlli A, Valgimigli L, Bartolini M, *et al.* Multi-target-directed drug design strategy: From a dual binding site acetylcholinesterase inhibitor to a trifunctional compound against Alzheimer's disease. J Med Chem 2007; 50: 6446-9.

[44] DeFerrari GV, Canales MA, Shin I, Weiner LM, Silman I, Inestrosa NC. A structural motif of acetylcholinesterase that promotes amyloid beta-peptide fibril formation. Biochemistry 2001; 40: 10447-57.

[45] Inestrosa NC, Alvarez A, Perez CA, Moreno RD, Vicente M, Linger C, Casanueva OI, Soto C, Garrido J. Acetylcholinesterase accelerates assembly of amyloid-beta-peptides into Alzheimer's fibrils: Possible role of the peripheral site of the enzyme. Neuron 1996; 16: 881-91.

[46] Rees T, Hammond PI, Soreq H, Younkin S, Brimijoin S. Acetylcholinesterase promotes beta-amyloid plaques in cerebral cortex. Neurobiol Aging 2003; 24: 777-87.

[47] Melchiorre C, Andrisano V, Bolognesi ML, Budriesi R, Cavalli A, Cavrini V, Rosini M, Tumiatti V, Recanatini M. Acetylcholinesterase noncovalent inhibitors based on a plyamine backbone for potential use against Alzheimer's disease. J Med Chem 1998; 41: 186-9.

[48] Bolognesi ML, Rosini M, Andrisano V, Bartolini M, Minarini A, Tumiatte V, Melchiorre C. MTDL design strategy in the context of Alzheimer's disease: From lipocrine to memoquin and beyond. Curr Pharm Des 2009; 15: 601-13.

[49] Loh KP, Huang SH, De Silva R, Tan BK, Zhu YZ. Oxidative stress: Apoptosis in neuronal injury. Curr Alzheimer Res 2006; 3: 327-37.

[50] Aliev G, Palacios HH, Walrafen B, Lipsitt AE, Obrenovich ME, Morales L. Brain mitochondria as a primary target in the development of treatment strategies for Alzheimer disease. Int J Biochem Cell Biol 2009; 41: 1989-2004.

[51] Cheng Y, Feng Z, Zhang QZ, Zhang JT. Beneficial effects of melatonin in experimental models of Alzheimer disease. Acta Pharmacologica Sin 2006; 27: 129-139.

[52] Rodriquez-Franco MI, Fernandez-Bachiller MI, Perez C, Hernandez-Ledesma B, Barolome B. Novel tacrine-melatonin hybrids as dual-acting drugs for Alzheimer disease, with improved acetylcholinesterase inhibitory and antioxidant properties. J Med Chem 2006; 49: 459-62.

[53] Wei D, Jiang X, Zhou L, Chen J, Chen Z, He C, Yang K, Liu Y, Pei J, Lai L. Discovery of multitarget inhibitors by combining molecular docking with common pharmacophores matching. J Med Chem 2008; 51: 7882-8.

[54] Lu JJ, Pan W, Hu YJ, Wang YT. Multi-target drugs: The trend of drug research and development. Plos One 2012; 7: e40262.

[55] Bu G. Apolipoprotein E and its receptors in alzheimer's disease: Pathways, pathogenesis and therapy. Nat Rev Neurosci 2009; 10: 333-44.

[56] Haass C. Amyloid peptide is produced by cultured cells during normal metabolism Nature 1992; 359: 322-5.

[57] Walsh DM, Selkoe DJ. Aβ oligomers – a decadeof discovery. J Neurochem 2007; 101: 1172-84.

[58] Citron M. β-secretase inhibition for the treatment of Alzheimer's disease – promise and challenge. Trends Pharmacol Sci 2004; 25: 59-112.
[59] Leung D, Abbernante G, Fairlie DP. Protease inhibitors: Current status and future prospects. J Med Chem 2000; 43: 305-41.
[60] Al-Tel TH, Semreen MH, Al-Qawasmeh RA, Schmidt MF, El-Awadi R, *et al.* Design, synthesis, and qualitative structure – activity evaluations of novel β-secretase inhibitors as potential Alzheimer's drug leads. J Med Chem 2011; 54: 8373-85.
[61] Charrier N, Clarke B, Cutler L, Demont E, *et al.* Second generation of BACE-1 inhibitors part 3: Towards non hydroxyethylamine transition stte mimetics. Bioorg Med Chem Lett 2009; 19: 3674-8.
[62] Congreve M, Aharony D, Albert J, Callaghan O, *et al.* Application of fragment screening by X-ray crystallography to the discovery of aminopyridines as inhibitors of beta-secretase. J Med Chem 2007; 50: 1124-32.
[63] Zhu Z, Sun Z, Ye Y, Voigt J, *et al.* Discovery of cyclic acylguanidines as highly potent and selective β-site amyloid cleaving enzyme (BACE) inhibitors: Part I – inhibitor design and validation. J Med Chem 2010; 53: 951-5.
[64] http://clinicalrials.gov/ct2/show/NCT01035138 (accessed September 12, 2012).
[65] Alvarez A, Bronfman F, Perez CA, Vicente M, Garrido J, Inestrosa NC. Acetylcholinesterase, a senile plaque component, affects the fibrillogenesis of amyloid-beta-peptides. Neurosc Lett 1995; 201: 49-52.
[66] Castro A, Martinez A. Targeting beta-amyloid pathogenesis through acetylcholinesterase inhibitors. Curr Pharmacol Des 2006; 12: 4377-87.
[67] Holzgrabe U, Kapkova P, Alptuzun V, Scheiber J, Kugelmann E. Targeting acetylcholinesterase to treat neurodegeneration. Exp Opin Therapeut Targets 2007; 11: 161-79.
[68] Gottfries CG. Neurochemical aspects on aging and disease with cognitive impairment. J Neurosci Res 1990; 27: 541-7.
[69] Meltzer CC, Smith G, DeKosky ST, Pollock BG, Mathis CA, *et al.* 3rd serotonin in aging, late-life depression, and Alzheimer's disease: The emerging role of functional imaging. Neuropsychopharmacology 1998; 18: 407-30.
[70] Alper G, Girgin FK, Ozgonul M, Mentes G, Ersoz B. MAO inhibitors and oxidant stress in aging brain tissue. Eur Neuropsychopharmacol 1999; 9: 247-52.
[71] Fink DM, Palermo MG, Bores GM, Huger FP, Kurys BE, Merriman MC, Olsen GE, Petko W, O'Malley GJ. Imino 1,2,3,4-tetrahydrocyclopent[*b*]indole carbamates as dual inhibitors of acetylcholinesterase and monoamine oxidase. Bioorg Med Chem Lett 1996; 6: 625-30.
[72] Youdim MB, Amit T, Bar-Am O, Weinreb O, Yogev-Falach M. Implications of co-morbidity for etiology and treatment of neurodegenerative diseases with multifunctional neuroprotective neurorescue drugs: Ladostigil. Neurotox Res 2006; 10: 181-92.
[73] Hirai K, Kato K, Nakayama T, Hayako H, *et al.* Neurochemical effects of 3-[1-(phenylmethyl)-4-piperidinyl]-1-(2,3,4,5-tetrahydro-1H-1-b enzazepin-8-yl)-1-propanone fumarate (TAK-147), a novel acetylcholinesterase inhibitor, in rats. J Pharmacol Exp Ther 1997; 280: 1261-9.
[74] McKenna MT, Proctor GR, Young LC, Harvey AL. Novel tacrine analogues for potential use against Alzheimer's disease: Potent and selective acetylcholinesterase inhibitors and 5-HT uptake inhibitors. J Med Chem 1997; 40: 3516-23.

[75] Abe Y, Aoyagi A, Hara T, Abe K, Yamazaki R, *et al.* Pharmacological characterization of RS-1259, an orally active dual inhibitor of acetylcholinesterase and serotonin transporter, in rodents: Possible treatment of Alzheimer's disease. J Pharmacol Sci 2003; 93: 95-105.
[76] Cappelli A, Gallelli A, Manini M, Anzini M, *et al.* Further studies on the interaction of the 5-hydroxytryptamine3 (5-HT3) receptor with arylpiperazine ligands. Development of a new 5-HT3 receptor ligand showing potent acetylcholinesterase inhibitory properties. J Med Chem 2005; 48: 3564-75.
[77] Toda N, Tago K, Marumoto S, Takami K, *et al.* A conformational restriction approach to the development of dual inhibitors of acetylcholinesterase and serotonin transporter as potential agents for Alzheimer's disease. Bioorg Med Chem 2003; 11: 4389-415.
[78] De Bundel D, Smolders I, Vanderheyden P, Michotte Y. AngII and AngIV: Unraveling the mechanism of action on synaptic plasticity, memory, and epilepsy. CNS Neurosci Ther 2008; 14: 315-39.
[79] Guo DF, Inagami,T. The genomic organization of the rat angiotensin II receptor AT_{1B}. Biochim Biophysics Acta 1994; 1218,: 91-4.
[80] Speth RC, Karamyan VT. The significance of brain aminopeptidases in the regulation of the actions of angiotensin peptides in the brain. Heart Fail Rev 2008; 13: 299-309.
[81] Gard PR. Angiotensin as a target for the treatment of Alzheimer's disease, anxiety and depression. Exp Opin Ther Targets 2004; 8: 7-14.
[82] Mustafa,T, Lee JH, Chai SY, Albiston AL, McDowall SG., Mendelsohn FA. Bioactive angiotensin peptides: Focus on angiotensin IV. J Renin Angiotensin Aldosterone Syst 2001; 2: 205-10.
[83] von Bohlen und Halbach O. Angiotensin IV in the central nervous system. Cell Tissue Res 2003; 311: 1-9.
[84] Wright JW, Harding JW. 1992. Regulatory role of brain angiotensins in the control of physiological and behavioral responses. Brain Res Rev 1992; 17: 227-62.
[85] Wright JW, Harding JW. 1997. Important roles for angiotensin III and IV in the brain renin-angiotensin system. Brain Res Rev 1997; 25: 96-124.
[86] Wright JW, Harding JW. 2004. The brain angiotensin system and extracellular matrix molecules in neural plasticity, learning, and memory. Prog Neurobiol 2004; 72: 263-93.
[87] Wright JW, Harding JW. 2011. Brain renin-angiotensin – A new look at an old system. Prog Neurobiol 2011; 95: 49-67.
[88] Johnston CI. Biochemistry and pharmacology of the renin-angiotensin system. Drugs 1990; 39: 21-31.
[89] Chauvel EN, Llorens-Cortes C, Coric P, Wilk S, Roques BP, Fournie-Zaluski MC. Differential inhibition of aminopeptidase A and aminopeptidase N by new -amino thiols. J Med Chem 1994; 37: 2950-7.
[90] Ramírez M, Arechage G, García S, Sanchez B, De Gandarias JM. Mn2(+)-activated aspartate aminopeptidase activity, subcellular localization in young and adult rat brain. Brain Res 1990; 522: 165-7.
[91] Banegas I, Prieto I, Vives F, Alba F, de Gasparo M, Segarra AB, *et al.* Brain aminopeptidases and hypertension. J Renin Angiotensin Aldosterone Syst 2006; 7: 129-34.
[92] Reudelhuber TL. The renin-angiotensin system: Peptides and enzymes beyond angiotensin II. Curr Opin Nephrol Hypertens 2005; 14: 155-9.
[93] Unger T, Badoer E, Ganten D, Lang.E, Rettig R. Brain angiotensin: Pathways and pharmacology. Circulation 1988; 77: 140-154.

[94] Ferrario CM, Chappell MD. Novel angiotensin peptides. Cell Mol Life Sci 2004; 61: 2720-7.

[95] Vauquelin G, Michotte Y, Smolders I, Sarre S, Ebinger G, Dupont A, Vanderheyden P. Cellular targets for angiotensin II fragments: pharmacological and molecular evidence. J Renin Angiotensin Aldosterone Syst 2002; 3: 195-204.

[96] Mentlein R, Roos T. Proteases involved in the metabolism of angiotensin II, bradykinin, calcitonin gene-related peptide (CGRP), and neuropeptides Y by vascular smooth muscle cells. Peptides 1996; 17: 709-20.

[97] de Gasparo M, Catt KJ, Inagami T, Wright JW, Unger T. 2000. International Union of Pharmacology. XXIII. The angiotensin II receptors. Pharmacol Rev 2000; 52: 415-72.

[98] Harding JW, Cook VI, Miller-Wing AV, Hanesworth JM, Sardinia MF, *et al.* Identification of an AII (3-8) [AIV] binding site in guinea pig hippocampus. Brain Res 1992; 583: 340-343.

[99] Bennett Jr JP, Snyder SH. Angiotensin II binding to mammalian brain membranes. J Biol Chem 1976; 251: 7423-30.

[100] Glossmann H, Baukal A, Catt KJ. Angiotensin II receptors in bovine adrenal cortex. Modification of angiotensin II binding by guanyl nucleotides. J Biol Chem 1974; 249: 664-6.

[101] Swanson GN, Hanesworth JM, Sardinia MF, Coleman JK, Wright JW. Discovery of a distinct binding site for angiotensin II (3-8), a putative angiotensin IV receptor. Regul Pept 1992; 40: 409-19.

[102] Blair-West JR, Carey KD, Denton DA, Madden IJ, Weisinger RS, Shade RE. Possible contribution of brain angiotensin III to ingestive behaviors in baboons Am J Physiol Regul Integr Comp Physiol 2001; 281: R1633-6.

[103] Fitzsimons JT. The effect on drinking of peptide precursors and of shorter chain peptide fragments of angiotensin II injected into the rat's diencephalon. J Physiol (London) 1971; 214: 295-303.

[104] Tonnaer JA, Wiegant VM, DeJong W, DeWied D. Central effects of angiotensin on drinking and blood pressure: structure-activity relationships. Brain Res 1982; 236: 417-28

[105] Wright JW, Jensen LL, Roberts KA, Sardinia MF, Harding JW. Structure-function analyses of brain angiotensin control of pressor action in rats. Am J Physiol 1989; 257: R1551-7.

[106] Braszko JJ, Kupryszewski G, Witczuk B, Wisniewski K. Angiotensin II (3-8)-hexapeptide affects motor activity, performance of passive avoidance, and a conditioned avoidance response in rats. Neuroscience 1988; 27: 777-83.

[107] de Gasparo M, Husain A, Alexander W, Catt KJ, Chiu AT, *et al.* Proposed update of angiotensin receptor nomenclature. Hypertension 1995; 25: 924-39.

[108] Bernier SG, Bellemare JM, Escher E, Guillemette G. Characterization of AT_4 receptor from bovine aortic endothel photosensitive analogues of angiotensin IV. Biochemistry 1998; 37: 4280-7.

[109] Jarvis MF, Gessner GW, Ly CG. The angiotensin hexapeptide 3-8 fragment potently inhibits [^{125}I] angiotensin II binding to non-AT_1 or -AT_2 recognition sites in bovine adrenal cortex. Eur J Pharmacol 1992; 219: 319-22.

[110] Zhang JH, Hanesworth JM, Sardinia M, Alt JA, Wright JW, Harding JW. 1999. Structural analysis of angiotensin IV receptor (AT_4) from selected bovine tissues. J Pharmacol Exp Ther 1999; 289: 1075-83.

[111] Bernier SG, Servant G, Boudreau M, Fournier A, Guillemette G. 1995. Characterization of a binding site for angiotensin IV on bovin endothelial cells. Eur J Pharmacol 1995; 291: 191-200.

[112] Hanesworth JM, Sardinia JF, Krebs LT, Hall KL, Harding JW. Elucidation of a specific binding site for angiotensin II(3-8), angiotensin IV, in mammalian heart membranes. J Pharmacol Exp Ther 1993; 266: 1036-42.

[113] Miller-Wing AV, Hanesworth JM, Sardinia MF, Wright JW, Speth RC. *et al.* Central angiotensin IV receptors: distribution and specificity in guinea pig brain. J Pharmacol Exp Ther 1993; 266: 1718-26.

[114] Hall KL, Hanesworth JM, Ball AE, Felgenhaner GP, Hosick HL, Harding JW. Identification and chracterization of a novel angiotensin binding site in cultured vascular smooth muscle cells that is specific for the hexapeptide (3-8) fragment of angiotensin II, angiotensin IV. Regul Pept 1993; 44: 225-32.

[115] Albiston AL, McDowall SG, Matsacos D, Sim P, Clune E, Mustafa T. Evidence that the angiotensin IV (AT_4) receptor is the enzyme insulin-regulated aminopeptidase. J Biol Chem 2001; 276: 48623-6.

[116] Chai SY, Yeatman HR, Parker MW, Ascher DB, Thompson PE, *et al.* Development of cognitive enhancers based on inhibition of insulin-regulated aminopeptidase. BMC Neurosci 2008; 9 (Suppl 2): S2-S14.

[117] Wright JW, Yamamoto BJ, Harding JW. Angiotensin receptor subtype mediated physiologies and behaviors: New discoveries and clinical targets. Prog Neurobiol 2008; 84: 157-81.

[118] Benoist CC, Wright JW, Zhu M, Appleyard SM, Wayman GA, Harding JW. Facilitation of hippocampal synaptogenesis and spatial memory by C-terminal truncated Nle^1-angiotensin IV analogues. J Pharmacol Exp Ther 2011; 339: 35-44.

[119] Kawas LH, Yamamoto BJ, Wright JW, Harding JW. Mimics of the dimerization domain of hepatocyte growth factor exhibit anti-Met and anticancer activity. J Pharmacol Exp Ther 2011; 339: 509-18.

[120] Kawas LH, McCoy AT, Yamamoto BJ, Wright JW, Harding JW. Development of angiotensin IV analogs as hepatocyte growth factor/met modifiers. J Pharmacol Exp Ther 2012; 340: 539-48.

[121] Yamamoto BJ, Elias PD, Masino JA, Hudson BD, McCoy AT, *et al.* The angiotensin IV analog Nle-Tyr-Leu-psi-(CH_2-NH_2)3-4-His-Pro-Phe (norleual) can act as a hepatocyte growth factor/c-Met inhibitor. J Pharmacol Exp Ther 2010; 333: 161-73.

[122] Kramár EA, Harding JW, Wright JW. Angiotensin II- and IV-induced changes in cerebral blood flow. Roles of AT1, AT2, and AT4 receptor subtypes. Regul Pept 1997; 68: 131-8.

[123] Naveri L, Stromberg C, Saavedra JM. Angiotensin IV reverses the acute cerebral blood flow reduction after experimental subarachnoid hemorrhage in the rat. J Cereb Blood Flow Metab 1994; 14: 1096-9.

[124] Dalmay F, Pesteil F, Allard J, Nisse-Durgeat S, Fernandez L, Fournier A. Angiotensin IV decreases acute stroke mortality in the gerbil. Hypertension 2001; 14: 56A.

[125] Faure S, Chapot R, Tallet D, Javellaud J, Achard JM, Oudart N. Cerebroprotective effect of angiotensin IV in experimental ischemic stroke in the rat mediated by AT(4) receptors. J Physiol Pharmacol 2006; 57: 329-42.

[126] Braszko JJ, Walesiuk A, Wielgat P. Cognitive effects attributed to angiotensin II may result from its conversion to angiotensin IV. J Renin Angiotensin Aldosterone Syst 2006; 7: 168-74.

[127] Chai SY, Fernando R, Peck G, Ye SY, Mendelsohn FA, Jenkins TA, Albiston AL. The angiotensin IV/AT_4 receptor. CMLS Cell Mol Life Sci 2004; 61: 2728-37.

[128] Basso N, Paglia N, Stella I. Protective effect of the inhibition of the renin-angiotensin system on aging. Regul Pept 2005; 128: 247-52.

[129] Ellul J, Archer N, Foy CM, Poppe M, Boothby H, *et al.* The effects of commonly prescribed drugs in patients with Alzheimer's disease on the rate of deterioration. J Neurol Neurosurg Psychiat 2006; 78: 233-9.

[130] Hajjar IM, Keown M., Frost B. Antihypertensive agents for aging patients who are at risk for cognitive dysfunction. Curr Hypertens Rep 2005; 7: 466-73.

[131] Khachaturian AS, Zandi PP, Lyketsos CG, Hayden KM, Skoog I, *et al.* Antihypertensive medication use and incident Alzheimer disease: the Cache County study. Arch Neurol 2006; 63: 686-92.

[132] Mogi M, Horiuchi M. Effects of angiotensin II receptor blockers on dementia. Hypertens Res 2009; 32: 738-40.

[133] Albiston AL, Pederson ES, Burns P, Purcell B, Wright JW, Harding JW, Mendelsohn FA, Weisinger RS, Chai SY. Attenuation of scopolamine-induced learning deficits by LVV-hermorphin-7 in rats in the passive avoidance and water maze paradigms. Behav Brain Res 2004; 154: 239-43.

[134] Kramár EA, Armstrong DL, Ikeda S, Wayner MJ, Harding JW, Wright JW. The effects of angiotensin IV analogs on long-term potentiation within the CA1 region of the hippocampus *in vitro*. Brain Res 2001; 897: 114-21.

[135] Wright JW, Stubley L Pederson ES, Kramár EA, Hanesworth JM, Harding JW. Contributions of the brain angiotensin IV-AT_4 receptor subtype system to spatial learning. J Neurosci 1999; 19: 3952-61.

[136] Pederson ES, Krishnan R, Harding JW, Wright JW. A role for the angiotensin AT_4 receptor subtype in overcoming scopolamine-induced spatial memory deficits. Regul Pept 2001; 102: 147-56.

[137] Stubley-Weatherly L, Harding JW, Wright JW. Effects of discrete kainic acid-induced hippocampal lesions on spatial and contextual learning and memory in rats. Brain Res 1996; 716: 29-38.

[138] Wright JW, Clemens JA, Panetta JA, Smalstig EB, Weatherly LA, Kramár EA, Pederson ES, Mungall BH, Harding JW. Effects of LY231617 and angiotensin IV on ischemia-induced deficits in circular water maze and passive avoidance performance in rats. Brain Res 1996; 717: 1-11.

[139] Chai SY, Bastias MA, Clune EF, Matsacos DJ, Mustafa T, Lee JH, McDowall SG, Paxinos G, Mendelsohn FA, Albiston AL. Distribution of angiotensin IV binding sites (AT_4 receptor) in the human forebrain, midbrain and pons as visualized by *in vitro* receptor autoradiography. J Chem Neuroanat 2000; 20: 339-48.

[140] De Bundle DD, Demaegdt H, Lahoutte T, Caveliers V, Kersemans K, Ceulemans AG, Vauquelin G, Clinckers R, Vanderheyden P, Michotte Y, Smolders I. Involvement of the AT_1 receptor subtype in the effects of angiotensin IV and LVV-haemorphin 7 on hippocampal neurotransmitter levels and spatial working memory. J Neurochem 2010; 112: 1223-34.

[141] Andersson H, Demaegdt H, Vauquelin G, Lindeberg G, Karlen A, Hallberg M. Ligands to the (IRAP)/AT4 receptor encompassing a 4-hydroxydiphenylmethane scaffold replacing Tyr2. Bioorg Med Chem 2008; 16: 6924-35.

[142] Lukaszuk A, Demaegdt H, Feytens D, Vanderheyden P, Vauquelin G, Tourwe D. The replacement of His(4) in angiotensin IV by conformationally constrained residues provides highly potent and selective analogues. J Med Chem 2009; 52: 5612-8.

[143] Wright JW, Tamura-Myers E, Wilson WL, Roques BP, Llorens-Cortes C, Speth RC, Harding JW. Conversion of brain angiotensin II to angiotensin III is critical for pressor response in rats. Am J Physiol Reg Integr Comp Physiol 2002; 284: R725-33.

[144] McCoy AT, Benoist CC, Wright JW, Kawas LH, Bule J, Zhu M, Appleyard SM, Wayman GA, Harding JW. Evaluation of metabolically stabilized angiotensin IV analogs as pro-cognitive/Anti-dementia agents. J Pharmacol Exp Ther October 2012, Epub ahead of print.

[145] Millan MJ. Multi-target strategies for the improved treatment of depressive states: Conceptual foundations and neuronal substrates, drug discovery and therapeutic application. Pharmacol Ther 2006; 110: 135-370.

[146] Roth BL, Sheffler DJ, Kroeze WK. Magic shotguns *vs.* magic bullets: Selectively non-selective drugs for mood disorders and schizophrenia. Nat Rev Drug Disc 2004; 3: 353-9.

CHAPTER 3

Next Generation Drugs in Alzheimer's Disease Treatment: From Benchtop to Bedside

Rosaria A. Cavallaro[1] and Andrea Fuso[2,*]

[1]Department of Surgery "P. Valdoni" and [2]Department of Psychology - Section of Neuroscience, Sapienza University of Rome, Italy

Abstract: Alzheimer's disease (AD), the most common cause of dementia in the elderly, is a neurodegenerative disorder characterized by severe and progressive cognitive impairment, including impaired decision making and orientation, inability to perform activities of daily living, language impairment and psycho-behavioural disturbances.

Molecular studies based on interdisciplinary approaches, including biochemical investigations, immunohistochemistry, molecular and cell biology, genetic approaches and animal models, have revealed that AD is a multifactorial disease with many different etiopathogenic mechanisms. The two major AD hallmarks are extracellular amyloid beta (Aβ) plaques and intracellular neurofibrillary tangles (NFTs), but additional pathogenic mechanisms have been described, including inflammation, oxidative damage, iron dysregulation, cholesterol metabolism and other amyloid-independent hypotheses.

There are still no effective treatments to prevent, stop or revert AD: the existing treatments only provide symptomatic relief, but poorly affect the progression of the disease.

Current approved therapy is based on the cholinesterase inhibitors and NMDA receptor antagonists. These treatments are symptomatic and drug discovery has been directed, in recent years, to develop "disease modifying drugs" able to counteract the progression of AD by modifying the evolution of disease. Recent advances on diagnostic criteria for the disease based on biological markers, even in a preclinical phase, can facilitate the identification of new effective therapies, impacting on drug development and monitoring disease progression during therapeutic trials.

The aim of this work is to provide an overview on the main therapeutic approaches, summarizing current treatments and new molecules still under clinical development and also discussing new perspectives on therapeutic intervention.

Keywords: Alzheimer's Disease (AD), neurodegeneration, AD treatment, AD prevention, clinical trials.

***Address correspondence to Andrea Fuso:** Department of Psychology - Section of Neuroscience, Sapienza University of Rome, Italy; Tel: +39 0649766601; Fax: +39 0649766600; E-mail: andrea.fuso@uniroma1.it

INTRODUCTION

Alzheimer's Disease (AD) is the main neurodegenerative disease and it is a clinical syndrome leading to dementia; it is characterized by memory impairment in the short and long term and by alterations in one or more of the primary mental activities, such as abstract thinking, critical, language and topographical orientation, in the absence of altered consciousness. AD accounts for 60-80% of cases; it is suggested that this estimate can be false both in negative and positive value since AD is sometimes confused with other causes of dementia and *vice versa*. It was estimated that in 2006 approximately 33.9 million people worldwide had AD [1]. Last release by Alzheimer's Association in 2010 has estimated 3 to 5 million people affected by AD in the U.S. [2]. It is foreseen that the prevalence will increase from three to four times in the next 40 years; this is mainly due to demographic changes and the longer life expectancy [1, 3]. AD prevalence at 60 years of age is 1% and it doubles every 5 years reaching 30-50% for 80 years olds. Geographical differences in AD prevalence indicate that industrialized countries are the most affected (probably as a consequence of the increased average lifespan); for example, the prevalence for the sixty years old group is 6.4% in North America, 5.4% in Western Europe and 4.9% in Latin America. Since the aging population is foreseen to be increased by 250% in industrialized countries by 2030, it is easy to connect the relevance of the parallel aging and dementia increasing. As a matter of fact, it is well documented that rates of dementia exponentially increase with age and, contrarily to the previous knowledge, it is destined to continuously increase also in very olds; recent data show that the incidence of dementia doubles every 5 years, from 0.66/100 persons aged 70 to 74 years to 11.30/100 persons for those aged 90 or more [4].

On the clinical point of view, AD is characterized by progressive neurodegeneration in selected brain regions, including the temporal and parietal lobes and restricted regions within the frontal cortex and the cingulate gyrus, leading to memory loss accompanied by changes in behavior and personality severe enough to affect daily living. The first clinical manifestations consist in memory and cognitive deficits, and alteration of motor and sensorial functions. On the anatomical and pathological point of view, AD is characterized by typical features consistent in the presence of extracellular deposits of β-amyloid protein (senile plaques, SP), intracellular fibrillar deposits of tau protein (neurofibrillary tangles, NFTs) and neuronal death (primarily in cortex, hippocampus, amygdala).

The primary component of SP, the β-amyloid (Aβ) peptide, results from the proteolytic processing of its precursor, the amyloid precursor protein (APP). APP can be processed either by α- and γ-secretases, producing non-amyloidogenic peptides, or by β- and γ-secretase, producing Aβ peptides. γ-secretase activity is exerted by a multiprotein complex that involves Presenilin 1 or 2 (PSEN1 or PSEN2), PEN2, Aph1A and Nicastrin (NCT); β-secretase activity is, on the contrary, exerted by a single protein, the β-site APP cleaving enzyme 1 (BACE1).

The balance between these three different secretase activities appears to be fundamental for the maintenance of physiological levels of non-amyloidogenic *vs.* amyloidogenic fragments. Under physiological conditions, APP is processed to originate two main forms of amyloid protein of 40 (Aβ-40) and 42 (Aβ-42) aminoacids; in these conditions, Aβ-40 is the most abundant form. Aβ-42 has a higher capacity to aggregate and is therefore the major component of SP; as a matter of fact, mutations causing familial AD lead to increased Aβ-42/Aβ-40 ratio [5, 6].

The other characteristic protein deposits present in AD brains, NFTs, are composed of abnormally hyperphosphorylated tau protein, a neuronal microtubule-bound protein whose function is regulated by site-specific phosphorylation. In AD, tau is phosphorylated at multiple sites throughout its 441 aminoacids, but some studies suggest that only specific phosphorylation sites are associated with neurodegenerative features [7]. Tau hyperphosphorylation is one of the earliest events in the progression of AD, preceding the appearance of dystrophic neurites and neurofibrillary tangles [8]. Tau can be hyperphosphorylated by increase of serine/threonine kinases activities or reduced phosphoseryl/phosphothreonyl protein phosphatases activities. Although several protein kinases are known to act on tau [9], glycogen synthase kinase 3β (GSK3β) is the major kinase that phosphorylates tau in the brain. Current evidences suggest that levels and activity of GSK3β are increased in AD brains [10]. GSK3β activity is regulated by site-specific phosphorylation: it was demonstrated that phosphorylation at Ser^{9} inhibits GSK3β whereas phosphorylation at Tyr^{216} is for its activation [11]. Thus, changes in the phosphorylation state of these sites reflect modulation of GSK3β activity. Protein phosphatase 2A (PP2A) is the major brain serine/threonine protein phosphatase that can dephosphorylate tau at multiple sites. PP2A is a multimeric protein complex consisting of a core heterodimer that can be associated with a variety of regulatory B subunits, which confer specificity of functions [12].

AD exists in familial forms, caused by known mutation in key genes, and sporadic forms, largely multifactorial and rather complex. The familial forms, accounting for less than the 5% of the total AD cases, are characterized by early-onset (< 65 years) and rapid progression; for this reason, familial AD is often indicated as EOAD (Early Onset AD). Some of the mutations responsible for EOAD are known and involve few key genes: APP, PSEN1, PSEN2 and ApoE1; nevertheless, about 50% EOAD cases are not explained by these known mutations, suggesting the existence of yet unknown genetic factors [13]. In all the cases of known genes, all the mutations lead to altered APP production and/or processing, in the direction of increased Aβ production and aggregation, causing the dominant Mendelian inheritance pattern observed in the affected families [13, 14].

Despite the dramatic individual impact of EOAD, due to the very rapid and devastating decline of the people affected by the disease, the sporadic forms represent the major relevance of AD on the epidemiological and social point of view. As a matter of fact, most of AD cases (90-95%) are caused by sporadic forms diagnosed in people over 65 years of age. These forms are therefore characterized by late onset (LOAD: Late Onset AD) and slow progression; LOAD has been only recently described as a multifactorial disease [15-17] likely resulting from the interaction among genetic, epigenetic, environmental, and stochastic factors superimposed on the physiological decline of cognitive functions with age [18].

Despite the number of factors possibly involved in LOAD onset and progression, and despite the non-mendelian etiology of the sporadic forms of AD, a genetic basis is however very clear as demonstrated by genetic association studies; as a matter of fact, new candidate genes are continuously added to the lists of genetic variants associated to increased LOAD risk, due to the polymorphisms revealed by recent genome-wide association studies (GWAS) [19]. Meta-analyses of association studies allowed to identify the most relevant genes associated to LOAD risk, as highlighted in the Alzgene database (www.alzgene.org): 1) apolipoprotein E (APOE ε4 allele, OR 1⁄4 3.68); 2) bridging integrator 1 (BIN1 rs744373, OR 1⁄4 1.17); 3) clusterin (CLU rs11136000, OR 1⁄4 0.88); 4) ATP-binding cassette sub-family A member 7 (ABCA7 rs3764650, OR 1⁄4 1.23); 5) complement receptor 1 (CR1 rs3818361, OR 1⁄4 1.17); 6) phosphatidylinositol binding clathrin assembly protein (PICALM rs3851179, OR 1⁄4 0.88); 7) membrane-spanning 4-domains, subfamily A, member 6A (MS4A6A rs610932, OR 1⁄4 0.90); 8) myeloid cell surface antigen CD33 (CD33 rs3865444, OR 1⁄4 0.89); 9) membrane-spanning 4-domains, subfamily A, member 4E (MS4A4E rs670139, OR 1⁄4 1.08); 10) CD2-associated protein (CD2AP rs9349407, OR 1⁄4 1.12). APOE ε4 variant, with an odds ratio of 3.7, remains the main relevant genetic association with LOAD; all the other variants only confer a modest LOAD risk with ORs ranging from 0.9 to 1.2, stressing the existence of complex gene-gene and gene-environment interactions.

The vast majority of the remaining factors associated to LOAD risk can be included under the general class of "environmental factors". Unfortunately, due to the high multifactorial nature of the disease, to the high variability in terms of age at onset and progression rate, and to the high interconnectivity between these factors, studies clearly identifying role, relevance and mechanisms of action of different environmental agents are still lacking or present results often contrasting and inconclusive [18]. However, it is quite accepted the identification of protective and inducing factors. The first group includes low caloric diet, Mediterranean diet, fruit, vegetables and fish consumption, antioxidants intake, physical activity and the preservation of the "so-called" brain reserve through the continue mental exercise and brain stimulation also in the elderly. The second group includes obesity, high fat/cholesterol diet, exposure to pollutants and pesticides, exposure to metal ions, brain trauma and injuries, oxidative stress, hyperhomocysteinemia, repeated viral infections and sedentary lifestyle. Deciphering this complex puzzle is complicated by the fact that exposure to these negative factors can also be encountered in the early life, remaining silent until elderly and senescence [20]. The research aimed at clarify the causes of LOAD and the underlying molecular mechanisms seems still far from the achievement o its scope, nevertheless, a new and promising line if research is focusing on the role of epigenetics. It is clear, indeed, that the environmental factors associated to LOD risk are able to interact with the epigenome: epigenetic changes can act as mediators of environmental stimuli causing the alteration of the expression of genes involved in the pathogenic AD processes [21]. The brain is certainly the human organ most hit by aging: it shows sensitive impairment yet in normal aging, and appears more prone than other organs to the occurrence of aging-related diseases. Neurodegenerative disorders represent the main class of age-associated diseases and, among these, AD represents the most prevalent form of neurodegenerative disease. Moreover, many of these disorders have been recently associated with epigenetic events. For this reasons, it seems of particular relevance the discussion of the epigenetic changes occurring in the brain and observed in adulthood and aging [22].

Thanks to the development of wide-scale analysis techniques and their application to the study of epigenome, it has been possible to demonstrate that DNA methylation in individuals changes over time [23]. Moreover, DNA methylation pattern in human brain is different in different brain areas [24] and dynamically changes during the lifespan [25]. It has also been demonstrated that specific epigenetic changes in LOAD patients can contribute to the onset and development of disease since different loci showed differential methylation stressing a correlation between gene demethylation and LOAD [26]. Specifically, two genes directly involved in AD showed changes in methylation pattern. It was demonstrated that methylation pattern of APP gene promoter was differentially modulated [27] and this region was found hypomethylated in post-mortem brains of AD patients [28]. Besides APP, also PSEN1 has been found hypomethylated both in AD patients [26] and, in our laboratory, in a murine experimental model characterized by unbalanced methylation pathway [29, 30, 31]; in this model, we demonstrated that PSEN1 promoter hypomethylation is responsible for the overexpression of the gene and the induction of several Alzheimer-like features.

The very high social and economic impact of AD on the modern society promoted a very intense and competitive effort not just on the research aimed at the comprehension of the molecular causes of the disease, but also in the finding of an effective therapy. The search for a clinical drug to be used in AD has been mainly directed toward targeting of molecular mechanisms leading to Aβ and Tau processing, modification and accumulation, as illustrated in the next paragraphs. Despite this incredible effort, a therapy able to stop the progression of the disease is still lacking; current therapies are limited to the temporary slowing down of the symptoms. Very recently, the individuation of the above-mentioned environmental risk factors has stressed the idea that a healthy life-style can contribute to the prevention of the disease or, at least, can delay its onset.

Clinical Diagnosis

AD has been considered a "dual clinical-pathological entity", characterized by both progressive cognitive decline and the presence of brain Aβ plaques, neurofibrillary tangles (NFTs) and progressive neuronal loss [32, 33]. Obviously, the latter features could not be observed during life; for this reason AD has primarily evolved as "probable AD", with a probabilistic diagnosis. According to the National Institute of Neurological and Communicative Disease and Stroke (NINCDS)/Alzheimer's Disease and Related Disorders Association (ADRDA) criteria [34], AD could be definite, only at autopsy, as "probable" or "possible".

In the last two decades the researchers focused great efforts to the finding of biomarkers able to show *in vivo* clinical and biological evidence of AD. Together with the presence of hippocampal atrophy on magnetic resonance imaging (MRI), glucose hypo-metabolism on positron emission tomography (PET) scan, biomarkers include abnormal CSF Aβ profile, with reduction of Aβ42 and PET evidence of Aβ aggregates deposition. Furthermore, the association of elevated Tau protein with low Aβ42 levels in CSF is now considered the most informative biomarker of AD diagnosis and might help to predict the progression of patients with MCI to AD. Furthermore, other biomarkers are the presence of a known pathogenic mutation in genes encoding for APP, PSEN1 and PSEN2. Studies performed in carriers of AD genetic factors demonstrated the presence of these biomarkers up to a decade before the clinical onset of AD [35], so they may be very helpful for the early detection of the disease. In light of these new findings, in 2007 new research criteria were proposed [36] and, according to them, the new algorithm used proposed that the diagnosis of AD could be made in the presence of episodic memory impairment and a positive biomarker. This is particularly useful when monitoring the progression of the disease, even at the stage of earliest clinical manifestation, so-called MCI. Today the definition of "AD" refers only to *in vivo* clinical-biological manifestation of disease, while the term "prodromal AD" is used when early symptoms are present, like memory impairment but not affecting daily living, together with a positive biomarker. The term "MCI" should only apply to patients with mild cognitive impairment but negative biomarkers, while the term "preclinical AD" defines subjects that carry a causal mutation, even in absence of clinical manifestation. The discovery of these new criteria and biomarkers is very useful for the development of the new disease modifying drugs, even if these biomarkers need to be prospectively tested for predictive accuracy and specific cut-off values need further validation.

Here will be discussed current approved treatments, new disease modifying compounds under clinical study and future hopes.

TREATMENTS FOR AD

As we have seen before, the term "AD" refers only to *in vivo* clinical-biological manifestation of disease. AD exists in familial forms, with early onset and rapid progression (EOAD, early onset AD) and sporadic forms, with late onset and slow progression (LOAD, late onset AD). The term "MCI" should refer to patients with mild cognitive impairment but without positive biomarkers. Current approved drug therapies for Alzheimer's disease (AD) include four acetylcholinesterase (AchE) inhibitors, used for the treatment of mild to moderate AD, and N-methyl-D-aspartate- NMDA receptor antagonist, for the treatment of moderate to severe cases. These therapies only provide modest symptomatic improvements, without affecting the progression of the disease, though a neuroprotective potential has been proposed. For this reasons, they are called "symptomatic drugs", referring to an agent that is not able to cure or to stop the progression of the disease, affecting the underlying pathogenic mechanisms, but only palliates the severity of symptoms. The effect is usually reversible, so, when the treatment is interrupted, differences between treated and placebo groups are no longer lasting [37-40]. On the basis of recent new findings on AD pathogenesis, in the last 10-15 years drug discovery has been directed to the development of novel therapeutic strategies able to modify the pathological steps leading to AD, with the attempt to counteract the progression of the disease. For this reason, they have been termed "disease-modifying" drugs. It is well known that the patho-physiological process of AD begins many years before clinical diagnosis, therefore the optimal time for applying disease modifying drug therapies may therefore be in the pre-symptomatic stage.

A disease-modifying drug is able to slow the progression of structural damage, to interfere with the pathogenic steps, in attempt to block the course of the disease in the early stages. The effect of this kind of drugs, in contrast with symptomatic drugs, is persistent and continues even after stopping the treatment, being the cumulative pathological changes less severe in the treated group respect to the control (placebo) group. Recently, the National Institute of Ageing revised the criteria for the clinical diagnosis of AD, discovering biomarkers useful to identify early stages of the pathology. This is of a paramount importance to assess the efficacy of modifying drug therapies for AD, impacting on drug development and in future trials, including population susceptible to AD before clinical onset.

On the basis of recent findings on AD pathogenesis, on the new diagnostic criteria and on the definition and validation of new biomarkers [41-44], novel drugs are under development. They include drugs to prevent Aβ aggregation, to reduce Aβ production, to promote Aβ clearance, vaccination, drugs targeting tau phosphorylation and assembly, as well as other approaches.

Up to now no disease modifying drugs are available for AD, even if several have been tested, down to phase III.

CURRENT APPROVED THERAPY

Acethylcholine Agonists

The so called "cholinergic hypothesis of AD", primarily supported by the post-mortem examinations of brains from patients with advanced dementia, is based on the fact that the destruction of cholinergic neurons in the basal forebrain and the resulting deficit in central cholinergic transmission explained most of the cognitive impairment and non-cognitive symptoms observed in AD patients. It was observed and first reported in 1976 [45-47] that in brain tissues from AD patients there is a reduction in the activities of choline acetyltransferase and acetylcholine esterase (AChE), enzymes involved in the synthesis and degradation of the acetylcholine. These findings were confirmed by post-mortem brain analysis of AD patients [48, 49]. This provided the rationale for the development of the AChE inhibitors therapy, together with alternative approaches based on the improvement of the cholinergic transmission by the administration of acetylcholine precursors or the stimulation of presynaptic acetylcholine release or muscarinergic agonists. These alternative approaches were however not successful because of the lack of efficacy or strong side effects [50].

Recent studies challenged the assumption that cholinergic deficits occur in early stages of AD: in fact, it was reported that the activities of AChE and choline acetyltransferase were not reduced in mild AD [51] and that there can be even up-regulation of the cholinergic activity in the early stage of AD [52, 38].

The destiny of the neurotransmitter acetylcholine is, after its release into the synaptic cleft, the hydrolytic degradation by cholinesterases, primarily AChE, but also butyrrylcholinesterase (BChE), that accounts for about 10% of the cholinesterase activity in normal human brain. The first drugs developed for AD were AChE inhibitors, aimed to increase acetylcholine levels in the synaptic cleft, trying to enhance cholinergic transmission ameliorating cholinergic deficits. To date, four AChE inhibitors are approved and commonly used for the treatment of mild to moderate AD: tacrine (First Horizon Pharmaceuticals) was the first drug approved for AD in 1993, but now rarely used for its hepatotoxicity.

The others AChE inhibitors are: donepezil (Pfizer), rivastigmine (Novartis) and galantamine (Janssen) [53]: donepezil is also now approved for the treatment of severe AD. Donepezil and galantamine are selective AChE inhibitors, whether rivastigmine also inhibits BChE [54]. In 2004, a study carried out by the AD2000 Collaborative Group aimed to clarify the efficacy-cost ratio of donepezil treatment, trying to determine whether such treatment produced interesting and useful improvements in disability, dependency, psychological and behavioural symptoms, concluded that the benefits of donepezil treatment were below minimally relevant thresholds and that donepezil is not cost effective [55].

In 2006 the Cochrane Dementia and Cognitive Improvement Group's Specialized Register considered a meta-analysis of 13 randomized, double-blind, placebo-controlled trials with donepezil, galantamine and rivastigmine. They supported the use of the three AChE inhibitors for mild to moderate AD, even the patients who responded to the treatment were not easy to identify prior to treatment. There is no evidence that the treatment is not cost effective. There are slight variations in the mode of action of the three AChE inhibitors, but no differences were reported respect to efficacy. Fewer adverse effects were associated with donepezil compared with rivastigmine, but it may be that galantamine and rivastigmine match donepezil in tolerability if a careful and gradual titration routine over more three months is used [56].

Glutamate Antagonists

The excitotoxicity derived from excessive NMDA receptors activation by glutamate, the most important neurotransmitter in the central nervous system, plays an important role in neuronal death in AD and other neurodegenerative diseases [57, 58]. Glutamate receptors can be divided into metabotropic glutamate receptors, coupled with G-proteins, and ionotropic receptors, that are ligand-gated ion channels. The latter are divided, on the basis of specific synthetic agonists, into NMDA, α-amino-3-hydroxy-5-methyl4-isoxazole-propionate (AMPA) and kainate receptors. Excessive activation of NMDA receptors is thought to be the cause of intracellular calcium increase, associated to events that lead to neurodegeneration: for this reason, NMDA-receptor antagonists may be useful for AD treatment, protecting neurons from glutamate-dependent neurotoxicity.

Molecules as MK-801 or phencyclidine (PCP), potent NMDA-receptor antagonists, probably interfere with normal physiological functions of NMDA receptors, so producing adverse psychotomimetic side effects. Otherwise, Memantine is a non-competitive NMDA receptor antagonist with moderate affinity that protects neurons but does not affect the physiological activity of NMDA receptors [59]. In Europe, memantine was approved in 2002 as a drug for the treatment of "moderately severe to severe AD", and in the 2003 it was approved in the United States for the treatment of moderate to severe AD cases. The Cochrane Collaboration published a review on double-blind, parallel group, placebo-controlled randomized trial on memantine suggested that this drug has beneficial effect on functional decline and cognitive abilities in moderate to severe AD cases, on cognitive function in vascular dementia, and it is well tolerated. The most frequently adverse effects were dizziness, headache and confusion.

Memantine does not interact *in vitro* with AChE inhibitors [60], then another possible approach is the clinical combination of memantine with cholinesterase inhibitors. In 2004 a randomized, double-blind, placebo-controlled clinical trial of patients with moderate to severe AD was published, showing statistically significant benefits of the combination therapy respect to the monotherapy on cognitive function, on activities of daily living, on behavior and clinical global status [61]. Raina and collaborators [62] reported that the treatment with AChE inhibitors and memantine provided statistically significant but only marginal clinical improvement. The use of AChE inhibitors was associated to adverse cholinomimetic effects: nausea and vomiting are the most common, but also dizziness, anorexia and diarrhea were founded. The transdermal administration of rivastigmine can improve tolerability.

NEXT GENERATION DRUGS

Drug Targeting Aβ

Anti-Aβ Aggregation Agents

The primary neurotoxic species in AD are formed by the aggregation of momomeric Aβ species into higher molecular weight oligomers. Preventing the formation of these oligomeric aggregates is an interesting approach to the development of new treatment for AD.

Glycosaminoglycanes (GAGs) bind to soluble Aβ, promoting fibril formation and Aβ deposition. Tramiprosate (3-amino-L-propanesulfonic acid), one of the most studied compounds (Alzhemed^{TM}, Neurochem Inc), is a GAG mimetic that binds to soluble Aβ monomers, inhibiting the formation of Aβ aggregates, enhancing

Aβ clearance. GAG mimetics compete for GAG binding sites, blocking fibril formation and reducing soluble Aβ [63].

After a phase I that demonstrated the tolerability of the drug, a phase II study was conducted in patients with mild to moderate AD, who were randomized to tramiprosate 50,100 and 150 mg twice a day, or placebo. The study showed that tramiprosate 100 or 50 mg reduced CSF Aβ concentration, even no differences in cognitive function between the tramiprosate and placebo group were observed. Similar results, indicating the lack of any significant effect of this drugs, were showed in a phase III study [39, 64]; moreover, the variability among the clinical sites in the trial confounded the observed effects. Another similar trial conducted in Europe has been discontinued. Recent findings showed that tramiprosate stimulates abnormal tau protein aggregation in neuronal cells, suggesting that new potential drugs should be tested on both types of pathology - amyloid and tau related [65].

Colostrinin (O-CLN, ReGen Therapeutics), a proline-rich polypeptide complex derived from sheep colostrum, is another molecule that inhibits Aβ aggregation and neurotoxicity *in vitro* and improves cognitive activity in animal models. After a phase I confirming the good tolerability, the phase II showed a modest improvement in Mini Mental State Evaluation (MMSE) score in patients with mild AD, in the period of 15 months. This effect was not observed in the 15 additional months of treatment.

McLaurin and colleagues in 2000 [66] identified a molecule, scyllo-inositol (scyllo-cyclohexanehexol, AZD 103, ELND-005), which is an orally administered stereoisomer of inositol that crosses the BBB using inositol transporters. Scyllo-inositol can directly bind to Aβ oligomers, stabilizing them, promoting dissociation of Aβ aggregates and thus inhibiting Aβ toxicity. In TgCRND8 mice, a mouse model of AD, the treatment with AZD-103 caused a 25% reduction of Aβ oligomers with concomitant 133% increase of monomeric species, suggesting that this molecule can prevent the oligomerization. A phase II trial on ELND-005 (formerly known as AZD-103) was completed,: it was reported that, in spite of lack of significant clinical improvement, patients receiving 250 mg of ELND-005 showed a reduction of Aβ in CSF. A large-scale and long-term clinical study is now ongoing [39].

Metal ions like Cu^{2+} and Zn^{2+} are involved in the mediation of Aβ aggregation and toxicity [67]: Aβ is easily precipitated by Zn^{2+}. Cu^{2+} and Fe^{3+} ions induce Aβ aggregation, but only in acidic conditions, like those occur in the AD brain [68, 69]. When binding Cu^{2+} and Fe^{3+}, Aβ not only becomes amyloidogenic but also reduces metal ions, generating H_2O_2 and promoting Cu-mediated lipid oxidation, forming the toxic product 4-hydroxyninenal. Chelating agents may be effective in dissolving amyloid deposits. One of these is PBT2, an orally administered 8-hydroxy quinolone with good BBB permeability, which prevents metal–dependent aggregation, deposition and toxicity of Aβ. In AD mouse models it was showed that this drug removes copper and zinc from CSF, promotes Aβ oligomer clearance and restore cognitive abilities [70, 71]. In a recent phase II trial it was showed that patients treated with 250 mg of PBT2 did not reported serious adverse events and they had a significant reduction of CSF Aβ 42 levels, compared with those treated with placebo [72]. No cognitive changes were observed in correlation with CSF Aβ levels, but larger and longer additional trials are needed, since the efficacy on cognitive impairment was restricted to two measures only.

Clioquinol (PBT1, Prana Biotechology), a parent compound of PBT2, was tested in a clinical trial that reported a reduction on the rate of cognitive impairment only in the more severely affected patients but, according to Cochrane collaborative study, it was not clear if clioquinol showed any relevant positive clinical results, also because of the small sample size [73].

There are other small molecules like polyphenolic compounds as curcumin(-) epigallocatechin-3-gallate (EGCG) and grape seed extract, that act attenuating Aβ aggregation. Showing good tolerability, EGCG is now studied in a phase II-III trial [74, 75].

Aβ Immunotherapy

One of the most promising approaches to develop disease-modifying drugs in AD is immunotherapy toward Aβ, direct either against its production, aggregation and deposition [76, 77]. The active immunization by vaccination stimulates an immune response by formation of antibodies against pathogenic forms of Aβ, while passive immunotherapy consists in the administration of antibodies from an exogenous source. Active Aβ immunotherapy was studied and validated since 1999, when Schenk and collaborators [78] demonstrated that immunization using Aβ as an antigen attenuated AD like pathology by removing Aβ in the CNS in a PDAPP transgenic mouse model of AD. This AD transgenic mouse progressively develops many neuropathological AD features in age-related and brain region-specific manner. Generation of Aβ antibodies produced the clearance of cerebral Aβ by microglial-phagocytosis of antibody-opsonized Aβ deposits. Aβ immunization was shown to act on different aspects of the Aβ-associated pathology, preventing development of plaque formation, neuritic dystrophy and astroglyosis in young animals, whereas in old animals vaccination reduces the extent and the progression of AD features [38]. It was also showed that Aβ immunization reduces synaptic degeneration and early tau accumulation [79, 80]. These observations on Aβ immunization have been confirmed in different transgenic mouse models, showing improvement of cognitive deficits, and lowering plaque load in aged non-human primates, which develop some brain amyloid-associated pathology like cerebral amyloid angiopathy. On the basis of these promising preclinical results, a multicentre, randomized, placebo controlled phase II double blind clinical trial of Aβ immunotherapy with Aβ 42 plus adjuvant using the pre-aggregated Aβ peptide AN1792 was started in 2001, on 300 patients. Unfortunately, the trial was prematurely stopped because some patients reported brain inflammation with aseptic meningoencephalitis [81]. Double-blind assessment, maintained for 12 months, showed no differences in cognition between antibody responders and placebo group for the Alzheimer's Disease Assessment Scale cognitive subscale (ADAS-Cog), Disability Assessment for Dementia (DAS), Clinical Dementia Rating (CDR), MMSE, and Clinical Global Impression of Change (CGIC). CSF tau levels were also decreased in a small number of responder patients, but Aβ levels were not affected. Subsequent studies on AN1792 treated patients or transgenic models moderate the negative first results, showing that antibodies against Aβ-related epitopes are able to slow the progression of neuropathological AD features. Further studies provided support for continuing the investigation of antibody treatment in AD [82, 83] until 2003, when 80 patients (or caregivers) included in the phase I of AN1792, gave their consent for long-term follow up and post-mortem neuropathological analysis. It was showed that, in patients treated with vaccine, Aβ accumulation was lower respect to placebo group, but without any improvement in survival or in severity of dementia. Immunization approach was not abandoned but, due to the low responsiveness and the severe adverse effect of active immunization by vaccine, passive immunotherapy has been proposed as an alternative strategy [84-86]. Passive immunotherapy is associated with vasogenic edema and cerebral amyloid angiopathy with micro hemorrhages [87, 88]. The most studied Aβ targeted antibody is the humanized monoclonal antibody bapineuzumab (AAB-001, Wyeth and Elan), that has been tested in a phase II trial in 200 patients with mild to moderate AD, aiming at assessing safety, tolerability and efficacy of multiple ascending doses. Prior of the conclusion of phase II, basing on the accumulated clinical data from phases I, II and III of the patients involved in the original AN1792 trial, in 2007 Wyeth and Elan decided to start a phase III trial of bapineuzumab. Efficacy and safety of this drug seemed to be related to the APOE allele status [89], being APOE e4 carriers more prone to vasogenic edema after the treatment with bapineuzumab. For this reason, the use of the highest dose of this drug was abandoned. Lower doses of bapineuzumab are now used in phase III trials in APOE e4 carriers, whereas slightly higher dose can be used in non-APOE 4 carriers, responding truly well by 78 weeks. In 2010, a study based on imaging analyses on AD patients demonstrated that bapineuzumab administration reduces fibrillar plaque. To date, seven phase III studies with bapineuzumab are ongoing.

Another humanized anti Aβ monoclonal antibody presently under clinical development and studied in three phase III trials is solanezumab. Other antibodies are ACC-001 (Wyeth), that is now studied in two phases II ongoing in Japan and USA, MAB5102A (Genetech, phase I completed), PF-04360365 (Pfizer, phase I completed), R1450 (Hoffman-La Roche, phase I completed), GSK933776A (GlaxoSmithKline) and V950 (Merck, phase I ongoing).

Natural occurring anti-amyloid auto-antibodies, that specifically recognize Aβ and block the toxic effects, have been found in human intravenous immunoglobulines (IVIg), obtained from pooled plasma of healthy blood donors.

A phase III study with IVIg 10 % is now ongoing. It was demonstrated [90] that free human Ig-gamma heavy chains have anti amyloidogenic activity binding to an amyloid, fibril-related conformational epitope, but without affecting Aβ monomers. Moreover, they may cross BBB and are less associated to adverse inflammatory side effects. It is very important to achieve drugs directed to Aβ dimers or toxic oligomers, preserving monomers, probably involved in maintaining learning memory and neuronal survival.

It is also important to underline that passive vaccination is associated to repeated infusions and elevated cost. So, active vaccination has been considered again, despite the unclear results reported so far. Specific antigens have been designed to generate high Aβ antibody titers but without inducing Aβ reactive T-cells. The first compound tested in AD patients is CAD106 (Novartis), now under phase II trial study [39].

Drugs Targeting Aβ Production

It is well known that the generation of Aβ is the result of two sequential cleavages of APP by two proteases, β- and γ-secretases. APP cleavage by α-secretases, the first proteolytic cleavage identified, prevents Aβ generation, since the cleavage site is within the Aβ sequence [91, 92]. Many therapeutic attempts have targeted inhibition of β- and γ-secretases to modulate the production of Aβ.

β-secretase 1 is an aspartyl protease sharing some features with HIV aspartyl proteases. No known mutation in β-secretase encoding gene (BACE1) have been related to familial AD, but levels of this enzyme were found elevated in sporadic AD and might be associated with polymorphism in the promoter region.

Development of β-secretase inhibitors may be difficult because of toxicity related to non-specific effects, having β-secretase also other substrates. Rosiglitazone and pioglitazone, two thiazolidinediones tested for AD in randomized controlled trials, may act in part as suppressors of β-secretase expression. A study on the administration of a β-secretase inhibitor was reported by Chang [93], showing that this drug rescued cognitive impairment and Aβ levels in AD Tg2576 mice, without toxic effects.

It is interesting to underline that rosiglitazone is an antidiabetic drug, with well and precisely identified mechanism of activity defined at molecular levels in diabetes (*i.e.* PPAR gamma binding and subsequent transcription of genes involved in metabolic control). It is difficult to state the supposed beneficial effect in AD even if preclinical effects looked for an effect in AD animal models. In fact, even if rosiglitazone seemed to improve spatial memory learning and to slightly decrease Aβ-42, also inducing insulin-degrading enzyme (IDE), it does not affect plaque burden in Tg2576 mice. Therefore, there is not a testable hypothesis in terms of molecular and cellular mechanisms. IDE is a protease that degrades insulin as well as monomeric Aβ. Several clinical studies were conducted with different doses of rosiglitazone, even analyzing data according to APOe4 allele status, but the last phase III study lacked to show any significant clinical benefit of this drug in all the APOE genetic groups examined. Up to now, a precise biological hypothesis on the mechanism of action of rosiglitazone and related useful biomarker are not yet identified.

Thus, research in BACE inhibitors is still in the preclinical phase.

Some interesting compounds were designed using crystal structure based inhibitor design. Using the Discovery Studio 2.0 software package, the pharmacophore model of arylpiperazine amide derivatives was built and the newly synthesized compounds were tested by experimentally determining their BACE1 inhibitory activities. Among these compounds, there are N-phenylarylamides and N-phenyl-benezenesulfonamide derivatives, neocorlyne [94], epigallocatechin-3-gallate and curcumin [95] and N(4)-substituted piperazine naphtamide derivatives [96].

γ-secretase is a protease complex consisting of four individual proteins: presenilin (the catalytic subunit), nicastrin, APH-1 and PEN-2, with CD147 as negative regulator of the complex [97, 98]. The most known substrate of γ-secretase is APP and Aβ is the product of this cleavage. Development of γ-secretase inhibitors is difficult because there may be potential non-specific effects, being γ-secretase involved in many biologically essential pathways. In fact, it is involved in intra-membranous cleavage of many proteins, like N-cadherin, Notch receptor (involved in differentiation and proliferation of embryonic cells, T-cells and splenic B-cells) and others [99].

Semagacestat (LY 450139, Eli Lilly) was the first non selective γ-secretase inhibitor clinical tested, but two phase III trials were stopped because of severe collateral adverse effects were observed, including hematological, gastrointestinal and skin toxic effects, probably due to the inhibition of Notch signaling pathway [100]. Furthermore, no significant changes in both plasma and CSF Aβ-40 and Aβ-42 levels were showed, neither any amelioration of cognitive status [101]. For this reasons, γ-secretase inhibitors or modulators that do not affect Notch cleavage are now under study. Among these, Begacestat (BMS -708163, Brystol-Myers Squibb, PF-3084014) selectively acts on APP cleavage, 10-100 fold over Notch cleavage and it was tested in a phase I clinical trial showing that it significantly decreased Aβ-40 and Aβ-42 in dose dependent manner. This drug is now under phase II study.

Tarenflurbil (the R-enantiomer of flurbiprofen, MPC-7869; Myriad Pharmaceuticals, Flurizan™) is the first in the new class of selective Aβ-42 lowering agents (SALAs), which specifically modulate γ-secretase activity without affecting Notch or other substrates [102]. Tarenflurbil acts altering γ-secretase conformation by binding to a site other than the catalytic centre, thus avoiding the production of Aβ-42 while producing shorter non-toxic fragments, but without any interference with the other physiologically essential substrates of γ-secretase. It does not act on cyclo-oxygenases 1 and 2 inhibition, like S-flurbiprofen, and its action is not associated with gastrointestinal toxic effects. A study conducted in mice showed that tarenflurbil reduced Aβ plaque burden and prevented behavioral and cognitive impairment [103].

On this basis, a phase I pharmacokinetic study was conducted on healthy, older volunteers and it was demonstrated that the drug is well tolerated, but with no significant effect in Aβ-42 levels in CSF; on the contrary, in plasma the higher drug concentrations were associated to the lower Aβ levels.

Myriad Pharmaceuticals started a large phase II, placebo controlled trial of Flurizan™ of 12 months duration in patients with mild to moderate AD, analyzing the rate of modifying activities of daily living, quantified by the ADCS-ADL (Alzheimer's Disease Cooperative Study-Activities of Daily Living) inventory, global function (by CDR sum of boxes), cognitive function (by ADAS-Cog), extending the treatments for further 12 months. Patients with mild AD and moderate AD responded differently to the treatment: the first group showed lower rates of decline in daily living activities than placebo, whether the same effect was not observed in patients with moderate AD, all patients showing adverse effects like diarrhea, nausea and dizziness. The effect observed in patients with mild AD was confirmed in the tarenflurbil group at 24 months: they showed lower rates of decline for all the analyzed outcomes than the patients in placebo for 0-12 months and tarenflurbil for 12-24 months. Two phase III studies were carried out in USA and in Europe, adding to the primary outcomes even explanatory outcome measures to analyze the physiological, physical and financial impact of AD on caregivers and medical resources. Unfortunately, this definitive phase III trial has failed: in fact, the placebo and treatment curves of the primary outcomes overlapped almost completely, with also increasing frequency of dizziness, anemia, infection and gastrointestinal ulcers in the treated group [104].

Another possible approach on the secretases system is to potentiate APP cleavage by α-secretase activity, thus avoiding or limiting the amyloidogenic pathway. α-secretase can be stimulated by muscarinic acetylcholine-receptor agonists, obtaining the reduction of Aβ levels *in vitro* [105, 106]. The muscarinic acetylcholine-receptor agonist AF267B [107] was tested in triple transgenic mouse, showing improvement in cognitive conditions and reduction in both amyloid and tau pathologies [108, 109]. This suggests that M1 acetylcholine receptor agonist may be useful not only as symptomatic drug but also as progression of disease modifying drug.

Etazolate (EHT 0202, Exonhit Therapeutics) stimulates *in vitro* the neurotrophic non-amyloidogenic pathway inducing the production of sAPP α, with a neuroprotective effect against Aβ-42 toxicity. After a phase I study in healthy volunteers, a phase II clinical trial is now ongoing to analyze the preliminary efficacy on cognitive and behavioral impairment [110].

Drugs Affecting Aβ Production: Other Mechanisms

Other drugs, such as 1,4-dihydropyridine (DHP) L-type calcium channel blockers nilvadipine and amlodipine, used for the treatment of hyperthension, can reduce the risk of developing AD [111, 112], but it seems that this effect is not related to the drug's blood pressure lowering function [113]. Both nilvadipine and amlodipine decrease Aβ production *in vitro*; however, *in vivo,* only nilvadipine reduces Aβ accumulation in a transgenic mouse model of AD, when orally chronically administered. Furthermore, it was demonstrated that nilvadipine treatment slowed cognitive decline in MCI patients with hypertension, being well tolerated and with no dangerous collateral effects related to blood pressure lowering effects [114]. For these reasons, a multicentre phase III clinical trial is now ongoing to assess the efficacy of this drug in AD patients.

Drugs Targeting Tau

The second important neuropathological feature of AD is the presence of intracellular neurofibrillary lesions containing aggregated hyper-phosphorylated forms of the microtubule-associated protein tau. During neurodegeneration, tau is abnormally phosphorylated and this interferes with its normal biological functions, decreasing tau ability to bind to, and to stabilize, microtubules, finally leading to synaptic functions impairment.

Studies on a triple transgenic mouse model of AD [108, 115] indicate that Aβ plaques and NFTs, with hyperphosphorylated tau, lie along the same pathological pathway, although for a long time the roles of Aβ and tau have been considered distinct in the development of AD. Aβ accumulation precedes and promotes tau hyperphosphorylation activating different kinases, like cycline dependent kinase 5 (CDK5) and glycogen synthase kinase 3β (GSK3β) [116]. Studies *in vitro* and *in vivo* demonstrated that tau hyperphosphorylation and its accumulation in NFTs potentiate the toxicity of Aβ, increasing the fragility and vulnerability of neurons [117]. For these reasons, inhibition of tau-related neurodegeneration is a very interesting approach in drug development for AD, pursued by searching for inhibitors of tau phosphorylation and substances preventing tau aggregation and/or improving its disassembly. An important approach is to inhibit the kinases responsible for tau phosphorylation; among the large number of kinases able to phosphorylate the several tau phosphorylation sites, GSK3 was identified as the main enzyme that can represent an interesting therapeutic target. The most studied compound that showed inhibition of GSK3 activity and reduction of tau phosphorylation in AD animal models is lithium, often used as mood stabilizer

[118]. Lithium, furthermore, acts as neuroprotective reducing Aβ production and promoting the release of Tgfβ1 [119]. Despite some studies did not confirm any effect of lithium in cognitive improvement, a recent study showed that lithium treatment was effective to reduce CSF levels of phosphorylated tau, reducing cognitive impairment in patients with amnestic MCI [120]. Another GSK3 inhibitor, NP031112 (NCT00948259) is now under study, and a phase II RCT has been recently completed.

Other compounds, like pyrazolopyrazines, pyrazolopyridines, the aminothiazole AR-A014418 and valproate are under development, the last also studied in the Alzheimer's Disease Cooperative Study (ADCS). This study showed that valproate treatment was not associated with cognitive positive effects but reduced psychosis, results not confirmed by a more recent study in demented patients, that evidenced that valproate is also associated with a lot of adverse effects, like falls, infection and gastrointestinal problems [121].

Another approach is to study compounds that can inhibit tau aggregation, since they could have the potential to protect neurons from degeneration. Thus, some small molecules were screened as tau aggregation inhibitors [122, 123]. Methyl thioninium chloride (MTC), a reducing agent better known as methylene blue, a blue dye used in analytical chemistry and in cellular biology, reduces Aβ oligomerization and interferes with aggregation acting on self-aggregating truncated tau fragments, binding to the domain related to tau phosphorylation [124]. A phase IIb RCT study in MTC (TauRx Therapeutics, Rember™) monotherapy in patients with mild to moderate AD showed cognitive abilities improvement, but a phase III clinical trial validation is needed. One additional problem is that the urine becomes blue, resulting in a lack of blinding.

Drugs Targeting Inflammation/Oxidative Damage

Despite an epidemiologic study of McGeer [125] suggested that long-term use of NSAIDs may be useful in protection against AD development, prospective studies of diclofenac [126], naproxen [127], rofecoxib, celecoxib, napsone, hydroychloroquine and nimesulide [128-131] did not confirm any effect on slowing the progression of cognitive impairment in AD patients. Only indomethacin was able to slow cognitive decline, but it showed severe collateral toxic effects, like gastrointestinal problems [132]. In conclusion, NSAIDs are now not considered as potential therapeutic agents for AD treatment, because they showed many adverse effects (gastrointestinal toxicity, infarction) together with no confirmed efficacy.

High levels of plasma Hcy and oxidative stress, together with low levels of folate, vitamin B6 and vitamin B12 are associated to LOAD [133-136]. A clinical trial named VITAL (VITamins to slow Alzheimer's disease) investigated whether the reduction of Hcy levels by high-dose B vitamins supplementation was able to decrease the rate of cognitive decline in AD patients. The study was conducted for 18 months on 409 patients with mild-to-moderate AD and normal levels of Hcy, folate and vitamin B12, analyzing the variations in the cognitive subscale ADAS-Cog score. The study evidenced that the vitamin supplementation treatment was only able to reduce Hcy levels, without any beneficial effect on the primary the rate of change in ADAS-Cog score, as well on any other secondary measures [137].

Other antioxidants, like mitoquinone (Antipodean Pharmaceuticals), Ginkgo biloba, natural polyphenols like wine, green tea, and vitamin E may be potentially useful in AD treatment, but a clinical trial with Vitamin E and omega-3 fatty acid did not demonstrate any effect as AD therapeutic treatment [138].

Drugs Targeting Cholesterol/Vascular Factors

Hypertension, hyperhomocysteinemia, atherosclerosis and stroke are vascular-related risk factors for the development of AD [139]; hypertension is associated with vascular damage and atherosclerosis, and for this reason it is considered a strong risk factor for AD and vascular dementia, when this conditions occur together [140].

A link between hypercholesterolemia, cardiovascular disease and AD is also suggested. In fact, the brain is the richest organ in cholesterol and a link between the risk of developing AD and ApoE carriers, the major cholesterol transporter in the CNS, is well known.

The treatment with statins, used as cholesterol lowering therapy in patients with cardiovascular problems, was associated to a decrease in the incidence of AD

[141]. Simvastatin, hydrolyzed *in vivo* to generate mevinolinic acid, a high affinity HMG-CoA reductase inhibitor, reduces the levels of mevalonic acid, a precursor of cholesterol. For this reasons, a phase III clinical trial named CLASP is now ongoing, to assess the safety and the efficacy of simvastatin on slowing the progression of AD, measured by ADAS-Cog and other measurements like mental status, behavioral disturbances and quality of life. Another study, named LEADe (Lipitor's Effect in Alzheimer's Dementia), analyzes the potential benefits provided from atorvastatin on mild to moderate AD patients receiving a background of donepezil 10 mg daily. The endpoints were the evaluation of changes in ADAS-Cog and ADCS-CGIC scores. Furthermore, some patients were analyzed by resonance imaging to evaluate the rate of the changes in whole brain and hippocampal volumes. Notwithstanding the promising considerations and the well tolerability of atorvastatin, no effects in all the endpoints analyzed were showed [142].

Other Approaches

AD neurodegeneration and the activation of the amyloidogenic pathway were also associated to nerve growth factor (NGF) pathway impairment. Studies performed in AD animal models showed that targeted delivery of NGF to basal forebrain cholinergic neurons ameliorates cognitive abilities [143] but, unfortunately, protein growth factors do not cross the BBB, making difficult to develop therapies targeting neurotrophic factors. Despite the positive effect on cognition, intracerebroventricular (ICV) infusion NGF showed severe adverse effects. The implant of autologous fibroblasts genetically modified to express NGF into selected brain areas, or the delivery of NGF through viral vectors, have been recently proposed [144, 145]. Recently, a treatment using the stereotactic implantation of a catheter-like device containing NGF-producing cells (NsG0202) demonstrated to be well tolerated [146].

Mitochondrial function impairment is another important factor involved in Aβ toxicity. Latrepirdine, a weak inhibitor of cholinesterases and a low-affinity NMDA receptor antagonist, inhibits mitochondrial permeability transition pores induced by Aβ, stabilizing mitochondria [147, 148]. Clinical trials, however, failed to show any beneficial effect of latrepridine, because of discrepancy emerged form phase II and phase III trials. For this reason, two RCTs are now ongoing trying to evaluate the effect of latrepridine combined with donepezil and memantine (NCT00829374 and NCT00912288).

Another potential target for the study and development of new AD drugs is the receptor for advanced glycation end-products (RAGE), a trans-membrane protein that mediates the effects of Aβ on microglia, BBB and neurons, potentiating the accumulation of Aβ in CNS by modulating BACE1 and improving the transport of Aβ into the brain [149].

Following the promising results obtained in cell cultures and in animal models [150], a phase II trial with the RAGE antagonist PF04494700 on mild to moderate AD patients was recently completed and final results are expected.

Increasing evidences suggest that AD is associated with a progressive impairment in energy metabolism, with a decrease of brain capacity to utilize glucose and to respond to insulin and insulin-like growth factor (IGF) stimulation [151]. Impairments in brain insulin/IGF signaling lead to increased APP expression and Aβ accumulation. In addition to reduced brain insulin signaling, low CSF-to-plasma insulin ratio was observed in AD patients [152, 153]. Furthermore, intracerebroventricular or intravenous insulin administration improves cognitive performance, probably increasing insulin/IGF signaling in brain. The use of intranasal administration allows insulin-like peptides to follow extracellular pathways to the brain within 15 minutes. For these reasons, after a pilot study, Crafts and colleagues [154] tested in a randomized, double blind, placebo-controlled trial the effects of intranasal insulin administration in 104 amnestic MCI and AD patients, finding stabilization or improvement of cognitive functions, and cerebral glucose metabolism.

These results suggest that the administration of intranasal insulin may be a novel approach to the treatment of neurodegenerative disorders.

Finally, a non pharmacological treatment has been recently proposed, consisting in a deep brain stimulation (DBS) of memory circuits, and recent phase I trial showed that patients receiving continuous stimulation for 12 months can revert glucose utilization impairment, also with a decrease in the rate of cognitive decline [155].

PREVENTION OF ALZHEIMER'S DISEASE

A different point of view about the interaction between the environment and the organism is represented by the "stress" concept and its role in the dementia; in this case the word "stress" has to be intended in its wider significance. According to this view, any change or stimulus deriving from outside the organism (that means deriving from the environment) represents a stressor able to induce a response in the individual aimed at coping with external stimuli or changes. Increasing evidence has been accumulated about the role of stress as an important challenge to the onset and progression of AD [156]. Interestingly, the heterogeneous nature of LOAD is only partly explained by the brain propensity to accumulate aberrantly processed, misfolded and aggregated oligomeric structural proteins, including Aβ peptides and hyperphosphorylated tau.

Since specific environmental stimuli have been indicated for their association with LOAD risk and as probably causative of the disease, it is possible to hypothesize that appropriate changes and interventions in lifestyle oriented to correct the stress caused by these factors can also prevent or counteract the associated detrimental effects and the onset or progression of the disease.

For example, different studies suggested that alterations of energy metabolism leading to visceral fat accumulation promote insulin resistance and a chronic inflammatory status that have been recognized as important cofactors in LOAD onset and progression. Therefore, preventive strategies should include actions oriented to the reduction of abdominal fat deposits, through a proper nutrition tailored to the individual needs [157]. In this sense, the new technologies for analysis of body composition that determine fat mass, fat free mass, total body water, intra- and extra-cellular water, mineral, metabolism and inflammatory status in the body (BIA-ACC and TomEEx devices) can represent a valid help [158. 159]. These are non-invasive and low cost technologies that can allow acquiring information on the cited body parameters, to formulate appropriate nutritional interventions and to follow up the changes induced in the low level inflammatory status secondary to modifications in lifestyle [160, 161].

Different studies have already reported that decreased LOAD risk is associated with increased intake of antioxidants like vitamins E and C, fruits and vegetables, both from dietary or supplementary sources. As a matter of fact, diet with high anti-oxidants content may reduce inflammation, which is associated with the risk of dementia and positive effects on brain function have been reported for different supplements like fish oil, teas, fruits, folate, spices and vitamins [162, 163]; more indications derive so far from studies in animal models and need to be confirmed in humans [164]. Besides studies aimed at investigating the beneficial role of specific supplements and nutrients, other studies found that adherence to "Mediterranean diet" (*i.e.* a dietary pattern with high intake of fish, fruits, and vegetables rich in antioxidants) was associated with reduced LOAD. On the contrary, diets enriched with saturated fats and cholesterol increased the risk. Fatty acids may also play a part in the synthesis and fluidity of nerve cell membranes, in synaptic plasticity and neuronal degeneration. In addition, oxidative stress is one of the central features in the AD brain. Thus, it may be plausible that supplementation or diet rich in antioxidants such as fruits, vegetables, and vitamins E and C might protect against AD.

B vitamins were particularly investigated in clinical studies because of the known association between high homocysteine (Hcy) and the risk of dementia and AD. Folate (vitamin B9), vitamin B12 and vitamin B6 are cofactors in the reactions responsible for Hcy transformation and removal in the "so called" one-carbon metabolism [165] and they are also negatively associated with dementia risk. Plasma levels of these three vitamins are therefore strictly related to Hcy levels (low B vitamin = hyperhomocysteinemia). After many studies in humans and in animal models, it is now accepted that moderate hyperhomocysteinemia is a potential risk factor for AD, although the possibility that it represents just a marker of the process exists; the association of low B vitamins with AD is still more controversial [166]. However, the majority of retrospective studies evidenced increased Hcy levels in LOAD subjects and prospective studies pointed out that hyperhomocysteinemia is evident before AD development, stressing the idea of a causative function [167]. Different intervention trials, based on B vitamins supplementation and homocysteine lowering, were made but discordant results were obtained. Only one of these studies demonstrated that Hcy-lowering intervention was associated to the improvement of the effects of cholinesterase inhibitors therapy; other studies demonstrated that B vitamin supplementation was able to reduce Hcy levels but did not improve cognitive status or delay cognitive decline [137]. The apparent controversial results obtained after observational and interventional studies in AD patients is probably due to undisclosed biases and methodological troubles occurring in the design of these protocols. The most evident trouble with these trials is that individuals with normal Hcy levels (presumably presenting no alteration in one-carbon metabolism) where enrolled in these trials; for this reason B vitamin supplementation could very unlikely result in evident beneficial effects [167].

Besides the attempt to prevent dementia and AD by interventions on nutritional aspects, the other main area so far explored in the search of a preventive behavior is related to social and physical activity. Indeed, the risk for dementia and AD is also increased in older people with increasing social isolation and less frequent and unsatisfactory contacts with relatives and friends. Several studies indicate that a poor social network or social disengagement is associated with cognitive decline. Rich and large social networks, on the other hand, provide affective and intellectual stimulation that could influence cognitive function and different health outcomes through behavioral, psychological, and physiological pathways [168, 169].

As for the physical activity, it has been shown that low-intensity activity such as walking may reduce the risk of dementia and cognitive decline; more specific experimental studies in animal models have established a direct correlation between physical exercise and neurogenesis, especially in the hippocampus [170]. Physical activity is important in promoting brain plasticity, and it possibly affects the transcription of several gene and neurotrophic factors relevant for the maintenance of cognitive functions. A strong protective effect of regular physical activity in middle age against the development of dementia and AD in late life was reported, especially for persons with the APOE4 allele. Therefore, regular exercise and intellectual stimulation may represent useful intervention able to stimulate maintenance and repair pathways inducing adaptation of cells and the ability to cope with stronger stresses [171].

Very interesting results were obtained from the Nun Study [172-174], a longitudinal study of 678 Catholic sisters 75 to 107 years of age, all members of the School Sisters of Notre Dame congregation. This study evaluated early and middle-life risk factors, cognitive and physical status in old age and post mortem brain neuropathologic analysis, trying to provide information about the relation between healthy ageing and dementia with the degree of pathology and the resistance to its clinical expression. The results of this study suggest that higher levels of healthy aging, based on measures of global cognitive functions, short term memory, activities of daily living and self-rated functions, were strong associated with greater longevity. Furthermore avoiding Alzheimer and stroke neuropathology is critical to the maintenance of healthy ageing, and the presence of both pathologies dramatically decreases the likelihood of healthy ageing.

CONCLUDING REMARKS

Up today, the pharmacological treatments current approved for AD include memantine and cholinesterases inhibitors, which are symptomatic drugs. Despite the new findings on the development of AD and on the mechanism underlying the pathology allowed the discovery of new treatments, none of these new disease-modifying drugs have succeeded in phase III trial. Many factors could explain this failure, like the lack of useful and adequate biomarkers and time course of treatment in relation to the stage of the development of the disease. Furthermore, it is of paramount importance to better understand the relationship between Aβ, tau and other factors involved in the pathogenesis of AD. Up to now, no biomarkers proposed for the early diagnosis has been validate for monitoring treatment, so new biomarkers and outcome measurements able to follow the effect of the new disease modifying drugs should be identified.

At present, the research is focused to develop drugs able to prevent AD. In fact, new approaches aim to analyze the potential neuroprotective effect of the disease modifying treatment in the early pre-symptomatic stages of AD, with the help of biomarkers predicting the progression of the disease.

ACKNOWLEDGEMENT

Declared none.

CONFLICT OF INTEREST

The author(s) confirm that this chapter content has no conflict of interest.

REFERENCES

[1] Brookmeyer R, Johnson E, Ziegler-Graham K, Arrighi HM. Forecasting the global burden of Alzheimer's disease. Alzheimers Dement 2007; 3: 186–91.
[2] Alzheimer's disease facts and figures, 2010. http://www.alz.org/documents_custom/report_alzfactsfi gures2010.pdf
[3] Hasan MK, Mooney RP. The predisposing factors, biological markers, neuroimaging techniques and medical complications associated with Alzheimer's disease. W V Med J. 2011;107(3):26-9.
[4] Hall CB, Verghese J, Sliwinski M, *et al.* Dementia incidence may increase more slowly after age 90: results from the Bronx aging study. Neurology. 2005;65: 882–886.
[5] Findeis MA. The role of amyloid beta peptide 42 in Alzheimer's disease. Pharmacol Ther 2007;116:266-86.
[6] Nunan J, Small DH. Regulation of APP cleavage by alpha-, beta- and gamma-secretases. FEBS Lett. 2000 Oct 13;483(1):6-10.
[7] Augustinack JC, Schneider A, Mandelkow EM, Hyman BT. Specific tau phosphorylation sites correlate with severity of neuronal cytopathology in Alzheimer's disease. Acta Neuropathol 2002; 103 : 26-35.
[8] Braak H, Braak E, Grundke-Iqbal I, Iqbal K. Occurrence of neurophil threads in the senile human brain and in Alzheimer's disease: a third location of paired helical filaments outside of neurofibrillary tangles and neuritic plaques. Neurosci Lett 1986; 65: (351–355.
[9] Ferrer I, Gomez-Isla T, Puig B, *et al.* Current advances on different kinases involved in tau phosphorylation, and implications in Alzheimer's disease and tauopathies. Curr Alzheimer Res 2005; **2:** 3-18.
[10] Plattner F, Angelo M, Giese KP. The roles of cyclin-dependent kinase 5 and glycogen synthase kinase 3 in tau Hyperphosphorylation. J Biol Chem 2006; 281:, 25457-25465.
[11] Frame S, Cohen P, Biondi RM. A common phosphate binding site explains the unique substrate specificity of GSK3 and its inactivation by phosphorylation. Mol Cell 2001; **7:**1321-1327.
[12] Janssens V, Goris J. Protein phosphatase 2A: a highly regulated family of serine/threonine phosphatases implicated in cell growth and signalling. Biochem J 2001;, **353:** 417-439.
[13] Ertekin-Taner N. Genetics of Alzheimer's disease: a centennial review. Neurol Clin 2007;25:611-7.
[14] Bentahir M, Nyabi O, Verhamme J, *et al.* Presenilin clinical mutations can affect gamma secretase activity by different mechanisms. J Neurochem 2006;96:732-42.

[15] Chopra K, Misra S, Kuhad A. Current perspectives on pharmacotherapy of Alzheimer's disease. Expert Opin Pharmacother 2011; 12: 335-50..

[16] Iqbal K, Grundke-Iqbal I. Alzheimer's disease, a multifactorial disorder seeking multitherapies. Alzheimers Dement 2010; 6: 420-4.

[17] Ballard C, Gauthier S, Corbett A, Brayne C, Aarsland D, Jones E. Alzheimer's disease. Lancet 2011; 377: 1019-31.

[18] Migliore L, Coppede` F. Genetics, environmental factors and the emerging role of epigenetics in neurodegen- erative diseases. Mutat Res 2009;667:82-97.

[19] Bertram L, McQueen MB, Mullin K, Blacker D, Tanzi RE. Systematic meta-analyses of Alzheimer disease genetic association studies: the AlzGene database. Nat Genet 2007;39:17-23.

[20] Lahiri DK, Maloney B, Zawia NH. The LEARn model: an epigenetic explanation for idiopathic neurobiological diseases. Mol Psychiatry 2009;14:992-1003.

[21] Fuso A. The "golden age" of DNA methylation in neurodegenerative diseases. Clin Chem Labor Med 2013; In Press.

[22] Mastroeni D, Grover A, Delvaux E, Whiteside C, Coleman PD, Rogers J. Epigenetic mechanisms in Alzheimer's disease. Neurobiol Aging 2011;7:1161-80.

[23] Bjornsson HT, Sigurdsson MI, Fallin MD, *et al.* Intra-individual change over time in DNA methylation with familial clustering. J Am Med Ass 2008; 299;:2877-83.

[24] Siegmund KD, Connor CM, Campan M, *et al*, DNA methylation in the human cerebral cortex is dynamically regulated throughout the life span and involves differentiated neurons. PLoS ONE 2007;2:e895.

[25] Wang SC, Oelze B, Schumacher A. Age-specific epigenetic drift in late-onset Alzheimer's disease. PLoS ONE 2008;3:e2698.

[26] Rogaev EI, Lukiw WJ, Lavrushina O, Rogaeva EA, St George-Hyslop PH. The upstream promoter of the beta-amyloid precursor protein gene (APP) shows differential patterns of methylation in human brain. Genomics 1994;22:340-7.

[27] Tohgi H, Utsugisawa K, Nagane Y, Yoshimura M, Genda Y, Ukitsu M. Reduction with age in methylcytosine in the promoter region -224 approximately -101 of the amyloid precursor protein gene in autopsy human cortex. Mol Brain Res 1999;70: 288-92.

[28] West RL, Lee JM, Maroun LE. Hypomethylation of the amyloid precursor protein gene in the brain of an Alzheimer's disease patient. J Mol Neurosci 1995;6:141-6.

[29] Fuso A, Nicolia V, Cavallaro RA, *et al.* B-Vitamin deprivation induces hyperhomocysteinemia and brain S-adenosylhomocsyteine, depletes brain S-adenosylmethionine, and enhances PSEN1 and BACE expression and Amyloid-β deposition in mice. Mol Cell Neurosci 2008;37:731-46

[30] Fuso A, Nicolia V, Cavallaro RA, Scarpa S. DNA methylase and demethylase activities are modulated by one-carbon metabolism in Alzheimer's disease models. J Nutr Biochem 2011;122:242-51.

[31] Fuso A, Nicolia V, Pasqualato A, Fiorenza MT, Cavallaro RA, Scarpa S. Changes in Presenilin 1 gene methylation pattern in diet-induced B vitamin deficiency. Neurobiol Aging 2011;32:187-99.

[32] Ballard C, Day S, Sharp S, Wing G, Sorensen S. Neuropsychiatric symptoms in dementia: importance andtreatment considerations. Int Rev Psychiatry 2008; 20: 396–404.

[33] Hardy J. The amyloid hypothesis for Alzheimer's disease: a critical reappraisal. J Neurochem 2009; 110: 1129–34.

[34] McKhann G, Drachman D, Folstein M, Katzman R, Price D,Stadlan EM. Clinical diagnosis of Alzheimer's disease:report of the NINCDS-ADRDA Work Group under the auspices of Department of Health and Human Services Task Force on Alzheimer's Disease. Neurology 1984; 34: 939–944.

[35] Sperling RA, Aisen PS, Beckett LA, *et al.* Toward defining the preclinical stages of Alzheimer's disease: recommendations from the National Institute on Aging-Alzheimer's Association workgroups on diagnostic guidelines for Alzheimer's disease. Alzheimers Dement 2011; 7: 280–92.

[36] Dubois B, Feldman HH, Jacova C, *et al.* Research criteria for the diagnosis of Alzheimer's disease: revising the NINCDS-ADRDA criteria. Lancet Neurol 2007; 6(8):734–746.

[37] Galimberti D, Scarpini E. Progress in Alzheimer's Disease. J Neurol 2012; 259: 201-211.

[38] Klafki HW, Staufenbiel M, Kornhuber J, Wiltfang J. Therapeutic approaches to Alzheimer's Disease. Brain 2006; 129: 2840-2855.

[39] Galimberti D, Scarpini E. Disease-modifying treatments for Alzheimer's disease. Ther Adv Neurol Disord 2011: 4(4): 203-216.

[40] Salomone S, Caraci F, Leggio GM,Fedotova J, Drago F. New pharmacological strategies for treatment of Alzheimer's disease: focus on disease modifying drugs. Br J Clin Pharmacol 2011; 73(4): 504-517.

[41] Albert MS, DeKosky ST, Dickson D, *et al.* The diagnosis of mild cognitive impairment due to Alzheimer's disease: recommendations from the National Institute on Aging-Alzheimer's Association workgroup. Alzheimers Dement2011; 7: 270–9.

[42] Sperling RA, Aisen PS, Beckett LA, *et al.* Toward defining the preclinical stages ofAlzheimer's disease: recommendations from the National Institute on Aging-Alzheimer's Association workgroups on diagnostic guidelines for Alzheimer's disease. Alzheimers Dement 2011; 7: 280–92.

[43] Siemers ER. How can we recognize 'disease modification'effects? J Nutr Health Aging 2009; 13: 341–3.

[44] Cummings JL. Biomarkers in Alzheimer's disease drug development. Alzheimers Dement 2011; 7: e13–44.

[45] Bowen DM, Smith CB, White P, Davison AN. Neurotransmitter-related enzymes and indices of hypoxia in senile dementia and other abiotrophies. Brain 1976; 99:459-496.

[46] Davies P, Maloney AJ. Selective loss of central cholinergic neurons in Alzheimer's disease. Lancet 1976; 2: 1403.

[47] Perry EK, Perry RH, Blessed G, Tomlinson BE. Necropsy evidences of central cholinergic deficit in senile dementia. Lancet 1977; 1: 189.

[48] Perry EK. The cholinergic hypothesis-ten years on. Nr Med Bull 1986; 43: 63-9

[49] Whitehouse PJ, Martino AM, Antuono PG, *et al.* Nicotinic acetylcholine binding sites in Alzheimer's disease. Brain Res 1986; 371: 146-51.

[50] Doody RS, Stevens JC, Beck C, *et al.* Practice parameter: management of dementia (an evidence-based review). Reports of the Quality Standards Subcommettee of the American Academy of Neurology. Neurology 2001; 56: 1154-66.

[51] Davis KL, Mohs RC, Marin D, *et al.* Cho*liner*gic markers in elderly patients with early signs if Alzheimer's disease. JAMA 1999: 281: 1401-6.

[52] Frolich L. The cholinergic pa*tholo*gy in Alzheimer's disease- discrepancies between clinical experience and pathophysiological findings. J Neural Transm 2002; 109: 1003-13.

[53] Farlow M. A clinical overview of cholinesterase inhibitors in Alzheimer's disease. Int Psychogeriatr/IPA 2002 (Suppl 1): 93-126.

[54] Scarpini E, Scheltens P, Feldman H. Treatment of Alzheimer's disease: current status and new perspectives. Lancet Neurol 2003; 2: 539-47.
[55] Courtney C, Farrell D, Gray R, *et al.* Long –term donepezil treatment in 565 patients with Alzheimer's disease (AD 2000): randomised double-blind trial. Lancet 2004; 363(9427): 2105-2115.
[56] Birks J. Cholinesterase inhibitors for Alzheimer's disease. Cochrane Database Syst rev 1 2005, CD005593.
[57] Bleich S, Romer K, Wiltfang J, Kornhuber J. Glutamate and the glutamate receptors system: a target for drug action. Int J Geriatr Psychiatry 2003; 18: S33-40.
[58] Hynd MR, Scott HL, Dodd PR. Glutamate-mediated excitotoxicity and neurodegeneration in Alzheimer's disease. Neurochem Int 2004; 45: 583-95.
[59] Sonkusare SK, Kaul CL, Ramarao P. Dementia of Alzheimer's disease and other neurodegenerative disorders – memantine, a new hope. Pharmacol Res 2005; 51: 1-17.
[60] Wenk GL, Quack G, Moebius HJ, Dansyz W. No interaction of memantine with acetylcholinesterase inhibitors approved for clinical use. Life Sci 2000; 66: 1079-83.
[61] Tariot PN, Farlow MR, Grossberg GT, Graham SM, McDonald S, Gergel I. Memantine treatment in patients with moderate to severe Alzheimer disese already receiving donepezil: randomized controlled trial. JAMA 2004; 291: 317-24.
[62] Raina P, Santaguida P, Ismaila A, *et al.* Effectiveness of cholinesterease inhibitors and memantine for treating dementia: evidence review for clinical practice guideline. Ann Inter Med 2008; 148: 379-397.
[63] Gervais F, Chalifour R, Garceau D, *et al.* Glycosaminoglycan mimetics: a therapeutic approach to cerebral amyloid angiopathy. Amyloid 2001; 8 (Suppl 1); 28-35.
[64] Gauthier S, Aisen PS, Ferris SH, *et al.* Effect of tramiprosate in patients with mild-to-moderate Alzheimer's disease: exploratory analyses of the MRI sub-group of the Alphase study. J Nutr Health Aging 2009; 13: 550-7.
[65] Santa-Maria I, Hernandez F, Del Rio J, Moreno FJ, Avila J., Tramiprosate, a drug of potential interest for the treatment of Alzheimer's disease, promotes abnormal aggregation of tau. Mol Neurodegener 2007; 2: 17.
[66] McLaurin J, Golomb R, Jurewicz A, Antel JP, Fraser PE,. Inositol stereoisomers stabilize an oligomeric aggregate of Alzheimer amyloid β peptide and inhibit Aβ-induced toxicity. J Biol Chem 2000; 275: 18495-18502.
[67] Atwood CS, Moir RD, Hunag X, *et al.* Dramatic aggregation of Alzheimer Aβ by Cu(II) is induced by conditions representing physiological acidosis. J Biol Chem 1998; 273: 12817-26.
[68] Bush AI, Pettingel WH, Multhaup G, *et al.* Rapid induction of Alzheimer Aβ formation by zinc. Science 1994; 265: 1464-1467.
[69] Bush AI. Drug development based on the metal hypothesis of Alzheimer's disease. J Alzheimers Dis 2008;15: 223-240.
[70] Faux NG, Ritchie CW, Gunn A, *et al.* PTB2 rapidly improves cognition in Alzheimer's disese: additional phase II analyses. J Alzheimers Dis 2010; 20: 509-16.
[71] Adlard PA, Cherny RA, Finkelstein DI, *et al.* Rapid restoration of cognition in Alzheimer's transgenic mice with 8-hydroxy quinoline analogs is associated with decreased interstitial Abeta. Neuron 2008; 59: 43-55.
[72] Lannfelt L, Blennow K, Zettrebreg H,,*et al.* Safety, efficacy, and biomarker findings of PBT2 in targeting Aβ as modifying therapy for Alzheimer's disease: a phase IIa, double-blind, randomized, placebo-controlled trial. Lancet Neurol 2008; 7: 779-786.

[73] Jenagartnam L, McShane R. Clioquinol for the treatment of Alzheimer's Disease. Cochrane Database Syst Rev 2006; 25 CD005380.

[74] Amijee H, Scopes DI. The quest for small molecules as amyloid inhibiting therapies for Alzheimer's disease. J Alzhimers Dis 2009; 17:33-47.

[75] Mandel SA, Amit T, Kalfon L, Reznichenko L,Weinreb O,Youdim MB. Cell signaling pathways and iron chelation in the neurorestorative activity of green tea polyphenols: specil reference to epigallocatechin gallate (EGCG). J Alzheimers Dis 2008; 15: 211-22.

[76] Town T. Alternative Aβ immunotherapy approaches for Alzheimer's disease. CNS Neurol Disord Drug Targets 2009; 8: 114–27.

[77] Lemere CA. Developing novel immunogens for a safe and effective Alzheimer's disease vaccine. Prog Brain Res 2009; 175: 83–93.

[78] Schenk D, Barbour R, Dunn W *et al.* Immunization with amyloid-beta attenuates Alzheimer-disease-like pathology in the PDAPP mouse. Nature 1999; 400: 173–7.

[79] Lombardo JA, Stern EA, McLellan ME *et al.* Amyloid-beta antibody treatment leads to rapid normalization of plaque-induced neuritic alterations. J Neurosci 2003; 23: 10879-83.

[80] Oddo S, Billings L, Kesslak JP, Cribbs DH, Laferla FM. Aβ immunotherapy leads to clearance of early, but not late, hyperphosphorylated tau aggregates *via* the proteasome. Neuron 2004; 43: 321-32.

[81] Gilman S, Koller M, Black RS, *et al.* Clinical effects of Aβ immunization (AN1792) in patients with AD in an interrupted trial. Neurology 2005; 64: 1553-62.

[82] Hock C, Nitsch R. Clinical observations with AN1792 using TAPIR analyses. Neurodegener Dis 2005; 2: 273-276.

[83] Holmes C, Boche D, Wilkinson D, *et a*l. Long-term effect of Aβ 42 immunisation in Alzheimer's disease: follow-up of a randomised, placebo-controlled phase I trial. Lancet 2008; 372: 216-223.

[84] Bard F, Cannon C, Barbour R, *et al.* Peripherally administered antibodies against amyloid beta-peptide enter the central nervous system and reduce pathology in a mouse model of Alzheimer disease. Nat Med 2000; 6: 916–9.

[85] DeMattos RB, Bales KR, Cummins DJ, Dodart JC, Paul SM, Holtzman DM. Peripheral anti-A beta antibody alters CNS and plasma A beta clearance and decreases brain A beta burden in a mouse model of Alzheimer's disease. Proc Natl Acad Sci U S A 2001; 98: 8850–5.

[86] Salloway S, Sperling R, Gilman S, *et al.* A phase 2 multiple ascending dose trial of bapineuzumab in mild to moderate Alzheimer disease. Neurology 2009; 73: 2061–70.

[87] Pfeifer LA,White LR, Ross GW, Petrovitch H, Launer LJ. Cerebral amyloid angiopathy and cognitive function: the HAAS autopsy study. Neurology 2002; 58: 1629–34.

[88] Morgan D. Immunotherapy for Alzheimer's disease. J Intern Med 2011; 269: 54–63.

[89] Panza F, Frisardi V, Imbimbo BP, *et al.* Anti-beta-amyloid immunotherapy for Alzheimer's disease: focus on bapineuzumab. Curr Alzheimer Res 2011; 8: 808-17.

[90] Adekar SP, Klyubin I,Macy S, *et al.* Inherent anti-amyloidogenic activity of human immunoglobulin gamma heavy chains. J Biol Chem 2010; 285: 1066–74.

[91] Esch FS, Keim PS, Beattie EC, *et al.* Cleavage of amyloid beta peptide during constitutive processing of its precursor. Science 1990; 248: 1122-4.

[92] Sisodia SS, Koo EH, Beyreuter K, Unterbeck A, Price DL. Evidence that beta-amyloid protein in Alzheimer's disease is not derived by normal processing. Science 1990; 248: 492-5.

[93] Chang WP, Huang X, Downs D, *et al.* Beta-secretase inhibitor GRL-8234 rescues age-related cognitive decline in APP transgenic mice. FASEB J 2011; 25: 775–84.

[94] Choi YH, Yon GH, Hong KS. *et al. In vitro* BACE-1 inhibitory phenolic components from the seeds of *Psoralea Corylifolia*. Planta Med 2008; 74: 1405-1408.

[95] Shimmyo Y, Kihara T., Akaike A, Niidome,T, Sugimoto H. Epigallocatechin-3-gallate and curcumin suppress amyloid beta-induced beta-site APP cleaving enzyme-1 upregulation. Neuroreport 2008; 19: 1329-1333.

[96] Laras Y, Garino C, Dessolin J, *et al.* New N(4)-substituted piperazine naphthamide derivatives as BACE-1 inhibitors. J Enzyme Inhib Med Chem 2009; 24: 181-187.

[97] Kaether C, Haass C, Steiner H. Assembly, trafficking and function of gamma-secretase. Neurodegener Dis 2006; 3: 275–83.

[98] De Strooper B. Aph-1, Pen -2 and nicastrin with presenilin generate an active gamma-secretse comolex. Neuron 2003; 38: 9-12.

[99] Tomita T. Secretase inhibitors and modulators for Alzheimer's disease treatment. Expert Rev Neurother 2009; 9: 661–79.

[100] Henley DB,May PC, Dean RA, Siemers ER. Development of semagacestat (LY450139), a functional gamma-secretase inhibitor, for the treatment of Alzheimer's disease. Expert Opin Pharmacother 2009; 10: 1657–64.

[101] Siemers E, Skinner M, Dean RA, *et al.* Safety, tolerability and changes in amyloid βconcentrations after administration of a γ-secretase inhibitor in volunteers. Clin Neuropharmacol 2005; 28: 126-132.

[102] Weggen S, Eriksen JL, Das P, *et al.* A subset of NSAIDs lower amylodogenic Aβ 42 indipendnetly of cyclooxygenase activity. Nature 2001; 414: 212-216.

[103] Kukar T, Prescott S, Eriksen JL, *et al.* Chronic administration of R-flurbiprofen attenuates learning impairments in transgenic amyloid precursor protein mice. BMC Neurosci 2007; 8: 54.

[104] Green RC, Schneider LS, Amato DA, *et al.* Effect of tarenflurbil on cognitive decline and activities of daily living in patients with mild Alzheimer disease: a randomized controlled trial. JAMA 2009; 302: 2557–64.

[105] Hung AY, Haass C, Nitsch RM, *et al.* Activation of protein kinase C inhibits cellular production of amyloid beta-protein. J Biol Chem 1993; 268: 22959-62.

[106] Wolf BA, Wertkin AM, Jolly YC, *et al.* Muscarinic regulation of Alzheimer's disease amyloid precursor protein secretion on amyloid beta-protein production in human neuronal N2TN cells. J Biol Chem 1995; 270: 4916-22.

[107] Fisher A. Therapeutic strategies in Alzheimer's disease: M1 muscarinic agonists. Jpn J Pharmacol 2000; 834: 101-12.

[108] Oddo S, Caccamo A, Shepherd JD, *et al.* Triple-transgenic model of Alzheimer's disease with plaques and tangles: intracellular Aβ and synaptic dysfunction. Neuron 2003; 39: 409–21.

[109] Caccamo A, Oddo S, Billings LM, *et al.* M1 receptors play a central role in modulating AD-like pathology in transgenic mice. Neuron 2006; 49: 671-82.

[110] Marcade M, Bourdin J, Loiseau N, *et al.* Etazolate, a neuroprotective drug linking GABA (A) receptor pharmacology to amyloid precursor protein processing. J Neurochem 2008; 106(1): 392-404.

[111] Forette F, Seux ML, Staessen JA, *et al.* The prevention of dementia with antihypertensive treatment: new evidence from the Systolic Hypertension in Europe (Syst-Eur) study. Arch Intern Med 2002; 162: 2046–52.

[112] Hanon O, Forette F. Prevention of dementia: lessons from SYST-EUR and PROGRESS. J Neurol Sci 2004; 226: 71–4.

[113] Paris D, Bachmeier C, Patel N, *et al.* Selective antihypertensive dihydropyridines lower Aβ accumulation by targeting both the production and the clearance of Abeta across the blood-brain barrier. Mol Med 2011; 17: 149–62.

[114] Kennelly SP, Abdullah L, Paris D, *et al.* Demonstration of safety in Alzheimer's patients for intervention with an anti-hypertensive drug nilvadipine: results from a 6-week open label study. Int J Geriatr Psychiatry 2011; 26: 1038–45.

[115] Blurton-Jones M, Laferla FM. Pathways by which Abeta facilitates tau pathology. Curr Alzheimer Res 2006; 3: 437–48.

[116] Ittner LM, Gotz J. Amyloid-beta and tau – a toxic pas de deux in Alzheimer's disease. Nat Rev Neurosci 2011; 12: 65–72.

[117] Gong CX, Iqbal K. Hyperphosphorylation of microtubule-associated protein tau: a promising therapeutic target for Alzheimer disease. Curr Med Chem 2008; 15: 2321–8.

[118] Tariot PN, Aisen PS. Can lithium or valproate untie tangles in Alzheimer's disease? J Clin Psychiatry 2009; 70: 919–21.

[119] Wada A, Yokoo H, Yanagita T, Kobayashi H. Lithium: potential therapeutics against acute brain injuries andchronic neurodegenerative diseases. J Pharmacol Sci 2005; 99: 307–21.

[120] Forlenza OV, Diniz BS, Radanovic M, Santos FS, Talib LL, Gattaz WF.Disease-modifying properties of long-term lithium treatment for amnestic mild cognitive impairment: randomised controlled trial. Br J Psychiatry 2011; 198: 351–6.

[121] Lonergan E, Luxenberg J. Valproate preparations for agitation in dementia. Cochrane Database Syst Rev 2009; (3): CD003945.

[122] Chirita C, Necula M, Kuret J. Ligand-dependent inhibition and reversal of tau filament formation. Biochemistry 2004; 5: 677-85.

[123] Pickhardt M, Gazova Z, von Bergen M, *et al.* Anthraquinones inhibit tau aggregation and dissolve Alzheimer's paired helical filaments *in vitro* and in cells. J Biol Chem 2005; 280: 3628-35.

[124] Oz M, Lorke DE, Petroianu GA. Methylene blue and Alzheimer's disease. Biochem Pharmacol 2009; 78: 927–32.

[125] McGeer PL, Schulzer M, McGeer EG. Arthritis and anti-inflammatory agents as possible protective factors for Alzheimer's disease: a review of 17 epidemiologic studies. Neurology 1996; 47: 425-432.

[126] Scharf S, Mander A, Ugoni A, Vajda F, Christophidis N. A double-blind, placebo Controlled trial of diclofenac/misoprostol in Alzheimer's disease. Neurology 1999; 53: 197-201.

[127] Aisen PS, Schafer KA, Grundman M *et al.* Effects of rofecoxib or naproxen vs placebo on Alzheimer's disease progression: a randomized controlled trial. JAMA 2003; 289: 2819-2826.

[128] Soininen H, West C, Robbins J, Niculescu L. Long-term efficacy and safety of celecoxib inAlzheimer's disease. Dement Geriatr Cogn Disord 2007; 23: 8-21.

[129] Eriksen JL, Sagi SA, Smith TE, *et al.* NSAIDs and enantiomers of flurbiprofen target gamma-secretase and lower Aβ 42 *in vivo*. J Clin Invest 2003, 112: 440-449.

[130] Aisen PS, Marin DB, Brickman AM, Santoro J, Fusco M. Pilot tolerability studies of hydroxychloroquine and colchicine in Alzheimer disease. Alz Dis Assoc Disord 2001; 15(2):96–101

[131] Aisen PS, Schmeidler J, Pasinetti GM Randomized pilot study of nimesulide treatment in Alzheimer's disease. Neurology 2002; 8(7):1050–1054

[132] Rogers J, Kirby LC, Hempelman, SR, *et al.* Clinical trial of indomethacin in Alzheimer's disease.Neurology 1993; 43: 1609- 1611.

[133] Ansari MA,Scheff SW. Oxidative stress in the progression of Alzheimer disease in the frontal cortex. J Neuropathol Exp Neurol 2010; 69(2):155-167.

[134] Morris MS. Homocysteine and Alzheimer disease. Lancet Neurol 2003; **2**(7):425-428.
[135] Seshadri S, Beiser A, Selhub J *et al.*Plasma homocysteine as a risk factor for dementia and Alzheimer's disease. New Engl J Med 2002; 346(7):476-483.
[136] Troen AM, Shea-Budgell M, Shukitt-Hale B, Smith DE, Selhub J, Rosenberg IH.B-vitamin deficiency causes hyperhomocysteinemia and vascular cognitive impairment in mice. Proc Natl Acad Sci U S A. 2008; 105(34):12474-12479.
[137] Aisen PS, Schneider LS, Sano M, *et al.* Alzheimer Disease Cooperative Study. High-dose B vitamin supplementation and cognitive decline in Alzheimer disease: a randomized controlled trial. JAMA 2008;300:1774-83.
[138] Barten DM, Albright CF. Therapeutic strategies for Alzheimer's disease. Mol Neurobiol 2008; 37: 171-186.
[139] Hooijmans CR, Kiliaan AJ. Fatty acids, lipid metabolism and Alzheimer pathology. Eur J Pharmacol 2008; 585: 176-196.
[140] Skoog I, Lernfelt B, Landahl S, *et al.* 15-year longitudinal study of blood pressure and dementia. Lancet 1996; 347: 1141-1145.
[141] Jick H, Zornberg GL, Jick SS, Seshadri S, Drachman DA. Statins and the risk of dementia. Lancet 2000; 356: 1627-1631.
[142] Feldman HH, Doody RS, Kivipelto M, *et al.* Randomized controlled trial of atorvastatin in mild tomoderate Alzheimer disease: LEADe. Neurology 2010; 74: 956-964.
[143] Cattaneo A, Capsoni S, Paoletti F. Towards non invasive nerve growth factor therapies for Alzheimer's disease. J Alzheimers Dis 2008; 15: 255–83.
[144] Tuszynski MH, Thal L, Pay M, *et al.* A phase 1 clinical trial of nerve growth factor gene therapy for Alzheimer disease. Nat Med 2005; 11: 551–5.
[145] Mandel RJ. CERE-110, an adeno-associated virus-based gene delivery vector expressing human nerve growth factor for the treatment of Alzheimer's disease. Curr Opin Mol Ther 2010; 12: 240–7.
[146] Eriksdotter JM, Linderoth B, Almqvist P, *et al.* Therapy of Alzheimer's disease with NGF. Neurobiol Aging 2010; 31: (Suppl. 1): S9
[147] Su B,Wang X, Bonda D, Perry G, Smith M, Zhu X. Abnormal mitochondrial dynamics – a novel therapeutic target forAlzheimer's disease? Mol Neurobiol 2010; 41: 87–96.
[148] Zhang S, Hedskog L, Petersen CA, Winblad B, Ankarcrona M. Dimebon (latrepirdine) enhances mitochondrial function and protects neuronal cells from death. J Alzheimers Dis 2010; 21: 389–402.
[149] Cho HJ, Son SM, Jin SM, *et al.* RAGE regulates BACE1 and Aβ generation *via* NFAT1 activation in Alzheimer's disease animal model. FASEB J 2009; 23: 2639–49.
[150] Yan SD, Bierhaus A, Nawroth PP, Stern DM. RAGE and Alzheimer's disease: a progression factor foramyloid-beta-induced cellular perturbation? J Alzheimers Dis 2009; 16: 833–43.
[151] de la Monte SM. Contribution of brain insulin resistance and deficiency in amyloid-related neurodegeneration in Alzheimer's disease. Drugs 2012; 72: 49-66.
[152] Reger MA, Watson GS, Green PS, *et al.* Intranasal insulin improves cognition and modulates beta-aamyloid in early AD. Neurology 2008; 70: 440-8.
[153] Craft S, Peskind E, Schwartz MW, Schellenberg GD, Raskind M, Porte D Jr. Cerebrospinal fluid and plasma insulin levels in Alzheimer's disease: relationship to severity of dementia and apolipoprotein E genotype. Neurology 1998;50:164–168.
[154] Craft S, Baker LD, Montine TJ, *et al.* Intranasal insulin therapy for Alzheimer disease and amnestic mild cognitive impairment: a pilot clinical trial. Arch Neurol 2012; 69: 29-38.

[155] Laxton AW, Tang-Wai DF, McAndrews MP, *et al.* A phase I trial of deep brain stimulation of memory circuits in Alzheimer's disease. Ann Neurol 2010; 68: 521–34
[156] Ricci S, Fuso A, Ippoliti F, Businaro R. Stress-induced cytokines and neuronal dysfunction in Alzheimer's disease. J Alzheimers Dis 2012;28:11-24.
[157] Huffman DM, Barzilai N. Role of visceral adipose tissue in aging. Biochim Biophys Acta 2009;1790:1117-23.
[158] Whitney NP, Eidem TM, Peng H, Huang Y, Zheng JC. Inflammation mediates varying effects in neurogenesis: Relevance to the pathogenesis of brain injury and neurodegenerative disorders. J Neurochem 2009;108:1343–59.
[159] Di Rienzo V, Minelli M, Sambugaro R, *et al.* Applicability of extracellular electrical impedance tomography in monitoring respiratory tract inflammation. J Investig Allergol Clin Immunol 2007;17:34-8.
[160] Ellis KJ. Human body composition: *in vivo* methods. Physiol Rev 2000;80:649-80.
[161] Jaffrin MY, Morel H. Body fluid volumes measurements by impedance: A review of bioimpedance spectroscopy (BIS) and bioimpedance analysis (BIA) methods. Med Eng Phys 2008;30:1257-69.
[162] van Praag H. Exercise and the brain: something to chew on. Trends Neurosci 2009;32(5):283-90.
[163] Gómez-Pinilla F. Brain foods: the effects of nutrients on brain function. Nat Rev Neurosci 2008;9:568–78.
[164] Spencer JP. Food for thought: the role of dietary flavonoids in enhancing human memory, learning and neuro-cognitive performance. Proc Nutr Soc 2008;67:238–52.
[165] Finkelstein J.D., Martin J.J. Homocysteine. Int J Bioch Cell Biol 2000;32:385-9.
[166] Seshadri S. Elevated plasma homocysteine levels: risk factor or risk marker for the development of dementia and Alzheimer's disease ? J. Alzheimers Dis 2006;9:393-8.
[167] Van Dam F., Van Gool W.A. Hyperhomocysteinemia and Alzheimer's disease: a systematic review. Arch Gerontol Geriatr 2009;48:425-30.
[168] Businaro R, Ippoliti F, Ricci S, Canitano N, Fuso A. Alzheimer's Disease promotion by obesity-induced mechanisms and advisable guidelines for its prevention. Curr Gerontol Geriat Res 2012; 2012:986823.
[169] Qiu C, Kivipelto M, von Strauss E. Epidemiology of Alzheimer's disease:occurrence, determinants, and strategies toward intervention. Dialogues Clin Neurosci 2009;11:111-28.
[170] Kapogiannis D, Mattson MP. Disrupted energy metabolism and neuronal circuit dysfunction in cognitive impairment and Alzheimer's disease. Lancet Neurol 2011;10:187-98.
[171] Mikhail V. Blagosklonny: hormesis does not make sense except in the light of TOR-driven aging. Aging 2011;3:1015-62.
[172] Snowdon DA. Healthy aging and dementia:findings from the Nun Study. Ann Intern Med 2003; 139: 450-4.
[173] Tyas SL, Snowdon DA, Desrosiers MF, Riley KP, Markesbery WR. Healthy aging in the Nun Study: definition and neuropathologic correlates. Age Aging 2007; 36:650-5.
[174] Tyas SL, Salazar JC, Snowdon DA, *et al.* Transitions to mild cognitive impairments, dementia, and death: findings from the Nun Study. Am J Epidiemol 2007; 165-1231-8.

Send Orders of Reprints at reprints@benthamscience.net
Frontiers in Clinical Drug Research – Alzheimer Disorders, Vol. 1, 2013, 147-182 147

CHAPTER 4

Tau-Directed Therapies for Alzheimer's Disease and Other Tauopathies

Yasumasa Yoshiyama*

Clinical Research Center, Department of Neurology, Chiba East National Hospital, Chiba, Chiba 260-8712, Japan

Abstract: The rapid increase in size of the elderly population has resulted in a dramatic increase in the number of Alzheimer's disease (AD) patients. To improve patients' quality of life and reduce the economic and care burdens, the development of disease-modifying therapies directly related to the pathomechanism of AD is urgently needed. Based on the amyloid hypothesis, decreasing accumulation of Aβ in the brain is supposed to control downstream AD pathogenesis including formation of neurofibrillary tangles (NFTs) and neurodegeneration. However, the results of recent clinical trials in symptomatic AD patients aimed at Aβ removal, suppression of Aβ production or inhibition of Aβ aggregation have failed to demonstrate significant clinical efficacy. Moreover, active immunization with Aβ did not inhibit NFT formation, neurodegeneration or cognitive decline in AD patients even though it reduced Aβ burden in the brain. This suggests that Aβ-targeting therapies could not suppress tau-mediated neurodegeneration and it might autonomously progress, once it has been initiated. Therefore, the development of tau-directed therapies is an important and urgent issue for AD and related tauopathies. This chapter summarizes recent research progress in tau-directed therapies.

Keywords: Alzheimer's disease, tauopathy, tau, neurodegeneration, progressive supranuclear palsy, Pick's disease, Frontotemporal dementia with parkinsonism linked to chromosome 17, corticobasal degeneration, microtubule, GSK-3, neurofibrillary tangles, hyperphosphorylation, aggregation, axonal transport, dementia, microtubule biniding protein, truncation, fibrillization, misfolding, ubiquitination, lysosome, immunotherapy.

INTRODUCTION

Although the pathomechanisms of Alzheimer's disease (AD) are still unclear, β-amyloid protein (Aβ) and tau protein are the most important targets for disease-modifying therapies. However, the results of clinical trials directed at decreasing the Aβ production or Aβ aggregation, for example tramiprosate (Aβ aggregation

***Address correspondence to Yasumasa Yoshiyama:** Clinical Research Center, Department of Neurology, Chiba East National Hospital, Chiba, Chiba 260-8712, Japan; Tel: +81 43-261-5171; Fax: +81 43-268-2613; E-mail: neuroyy@cehpnet.com

Atta-ur-Rahman (Ed)

inhibitor), tarenflurbil (γ-secretase modulator) and semagacestat (γ-secretase inhibitor), are disappointing [1-3]. Moreover, recent clinical trials of active or passive immunization targeting Aβ to clear Aβ from the brain have also been unsatisfactory. Active immunization with Aβ (AN1792) did not show any clinical benefits, although postmortem pathological examinations confirmed the clearance of senile plaques (SPs) from the brain. Interestingly, histopathological studies also showed the progression of neurofibrillary tangle (NFT) formation and neurodegeneration [4]. This indicated that the active immunization with Aβ attained its goal, Aβ removal, but it did not suppress downstream tau-mediated neurodegeneration.

Table 1: Neurodegenerative diseases associated with the pathological aggregation of tau protein

Diseases showing coexistence of tau and other pathologies
Alzheimer's Disease (familial and sporadic)
Down's syndrome
Creutzfeldt-Jakob disease
Multiple system atrophy
Diseases without distinct other pathology
Pick's disease
Frontotemporal dementia with parkinsonism linked to chromosome 17 (FTDP-17)
Argyrophilic grain disease (AGD)
Diffuse neurofibrillary tangles with calcification
Progressive supranuclear palsy (PSP)
Progressive subcortical gliosis
Corticobasal degeneration (CBD)
Dementia pugilistica (chronic traumatic encephalopathy)
Amyotrophic lateral sclerosis/parkinsonism dementia complex of Guam
Subacute sclerosing panencephalitis

Additionally, the results of two recent phase III clinical trials of passive immunizations with monoclonal anti-Aβ antibodies (bapineuzumab and solanezumab) were announced. Unfortunately, they did not meet the primary endpoints, although there might be some potential efficacy of solanezumab in very mild AD patients. These clinical observations seem to dispute the amyloid hypothesis [5]. Meanwhile, the discovery of mutations in the tau gene in frontotemporal dementia and parkinsonism linked to chromosome 17 (FTDP-17) constitutes convincing evidence that the tau protein itself is involved in the pathogenesis of neurodegenerative disorders. Furthermore, although most cases of familial AD (FAD) pathologically exhibit SPs and NFTs in the brain, a few FAD cases show NFTs without SPs [6], indicating that NFTs might be more directly involved in neurodegeneration and clinical symptoms, even in FAD cases. Additionally, there is a group of neurodegenerative disorders called "tauopathies" that show tau pathology in neurons, or sometimes in glial cells (Table **1**). This also suggests that the process of abnormal tau accumulation might be a common pathological mechanism leading to neurodegeneration. Thus, tau-directed therapies have the potential for beneficial effects not only on AD but also on other tauopathies.

PHYSIOLOGICAL FUNCTIONS AND BIOLOGY OF TAU PROTEIN

Although the physiological functions of most abnormally accumulating proteins in neurodegenerative disorders have not yet been elucidated, the main physiological functions of tau protein are well understood. Tau protein was one of the first microtubule-associated proteins (MAPs) to be characterized, and involved in assembly and stabilization of microtubules (MTs). It is predominantly expressed in neurons, but astrocytes and oligodendrocytes also express tau at a lower level. It plays important roles in cell morphology, as well as in cell differentiation and polarization, especially in neurons [7]. Human tau is encoded on chromosome 17q21 and consists of six isoforms generated by alternative splicing [8]. They differ by the presence or absence of two amino-terminal inserts encoded by exons 2 and 3, and by one of the repeats encoded by exon 10 in the carboxy-terminal half [8, 9]. Clone names and their structures are summarized in Table **2**.

Table 2: Tau isoforms

Clone	N-Terminal Inserts and Repeats	Number of Amino Acids (AA)	Molecular Weight (kDa)
htau40	2N4R	441	45.9
htau39	2N3R	410	42.6
htau34	1N4R	412	43.0
htau37	1N3R	381	39.7
htau24	0N4R	383	40.0
htau23	0N3R	352	36.7

Three- and four-MT-binding repeat tau isoforms express at almost similar levels in the normal adult human brain [10]. However, in some tauopathies insoluble tau protein dominantly consists of tree repeat tau (*e.g.* Pick's disease) or four repeat tau (*e.g.* AGD, PSP, CBD). In physiological conditions, the molecular structure of tau is unfolded because of its hydrophilic properties, although it can fold into an ordered β-structure during aggregation in AD and other tauopathies. Mukrasch *et al.* [11] showed the detailed structure of tau protein by NMR spectroscopy. Most chains of the tau protein are unfolded, with a few short and transient elements of secondary structure including α-helices, β-strands and poly-proline helices. This indicates that the tau molecule is natively an unfolded and stable protein.

The most important physiological function of tau protein is thought to stabilize MTs and to promote MT assembly. MTs are tubulin polymers of the cytoskeleton that are involved in intracellular transport by motor proteins and in stabilizing cell shape and mitosis. The exact binding site of tau on MTs has not been clearly demonstrated, although various imaging techniques have been used to clarify the structure and relationship between tau and MTs, because of the loose and natively unfolded structure in tau protein. The presence of three- or four-repeat domains in the tau protein implies that the repeat domain binds to MTs. Thus, the repeats are frequently called MT-binding repeats, although the flanking domains can bind to MTs even in the absence of the those repeats [12, 13]. The stabilization and the promotion of assembly of MTs by tau are thought to improve axonal transport in neurons. However, tau can inhibit the binding of motor proteins to MTs. Interestingly, kinesin binding (anterograde transport) is inhibited at a much lesser concentration of tau than is dynein binding (retrograde transport) [14]. Therefore, the attachment of tau to MTs or detachment of tau from MTs by dephosphorylation or phosphorylation probably controls axonal transport. In normal neurons, tau is distributed in a proximal-to-distal gradient, which might allow kinesin to easily attach to MTs in the proximal region of the axon and detach in the distal part of axon (synapses) [15].

MOLECULAR MECHANISMS OF TAU-MEDIATED NEURODEGENERATION

Although the precise molecular mechanisms of tau-mediated neurodegeneration have not yet been well elucidated, both loss of tau function and gain of toxic tau function are hypothesized to contribute to neurodegeneration.

Loss of Tau Function

Because promoting MT assembly and stabilizing MTs are the most well characterized roles of tau protein, loss of these normal physiological functions of tau protein leads to instability and disassemble of MTs. MT impairments cause functional and structural disturbance in cells. Because phosphorylation of tau reduces its ability to bind to MTs, the balance of phosphorylation and dephosphorylation controls tau attachment to and detachment from MTs. Increased phosphorylation of tau protein, like that observed in AD and other tauopathies, enhances tau detachment from MTs, leading to the disassembly of MTs followed by the breakdown of the MT network. MTs are like train tracks for the transport of vesicles by motor proteins. Thus, an impaired MT network can cause axonal transport insufficiency, resulting in sever functional impairments. Because neurons have long processes, intracellular transport insufficiency causes profound adverse effects on neurons especially in distal portions of neurons, such as the synapse and dendrites.

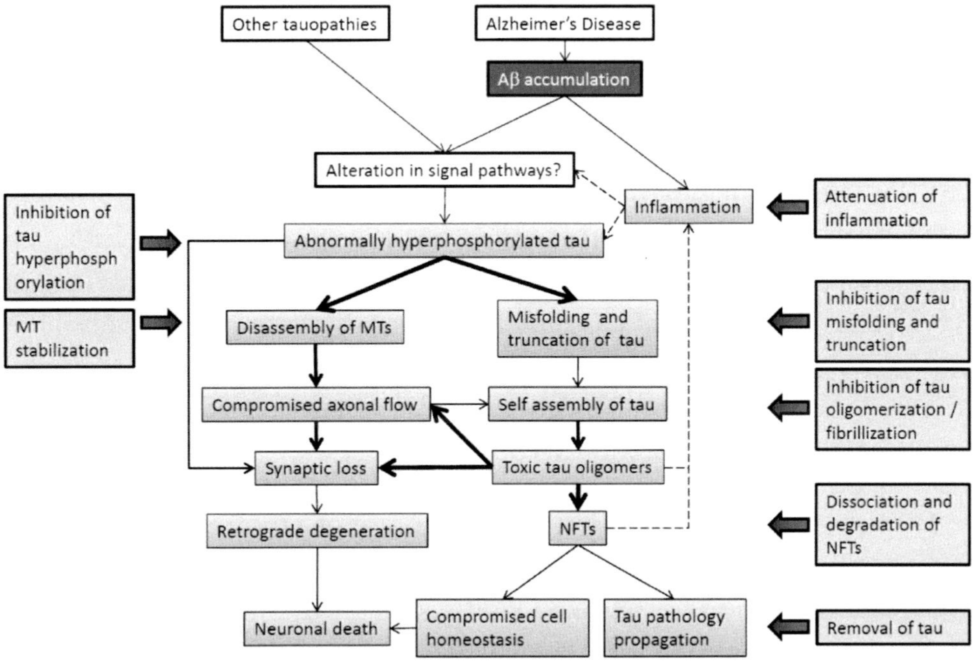

Figure 1: Therapeutic targets relating to possible pathomechanisms in tauopathies (figure partially modified from Yoshiyama *et al.* [16]).

Gain of Toxic Function

Tau aggregation is one of characteristic pathological changes of tau proteins. Thus, tau aggregates including NFTs and neuropil threads seem to be toxic. However, it is still unclear whether accomplished fibrillar tau aggregates like NFTs are toxic or not. They might form by enhancing coaggregation of toxic tau oligomers to escape their toxic effects for cell survival [17]. In either case, toxic functions of pathological tau might be generated in the pathological process from monomeric tau to fibrillar tau aggregates. Furthermore, aggregated tau might induce a clogging of axonal transport, and space-occupying tau aggregates (NFTs) might compromise cell homeostasis. Tau monomers did not inhibit fast axonal transport; however, polymeric tau filaments selectively inhibited anterograde fast axonal transport [18, 19]. Because this effect required an N-terminal amino acid stretch, full-length tau monomers might hide the N-terminal region by global folding, although tau is essentially an unfolded protein [20]. It is suggested that truncation of the amino-terminus in NFTs, as is observed in AD and other tauopathies, might make these toxic effects inert. Meanwhile, overexpression of full-length tau leads to the accumulation of organelles in the perinuclear region in the cells, indicating that the excess tau-binding to MTs itself might disturbs antegrade intracellular transport [21]. Moreover, an increased cytosolic concentration of tau disrupts the Golgi apparatus, leading to fragmentation [22,23]. Fragmentation of the Golgi apparatus in neurons has been observed in AD brains [24]. Although it is still unclear whether abnormal phosphorylation of tau initiates tau aggregation or it results from insolubilization of abnormal tau aggregation. However, abnormal hyperphosphorylation of the tau and mutations might increase cytosolic tau concentrations by reducing tau ability to bind to MTs. Higher concentrations of unbound tau might facilitate the formation of tau oligomer and aggregates, and it also might exhibit more cytotoxic effects on organelles.

Recently, Hoover *et al.* [25] demonstrated that the mislocalization of hyperphosphorylated tau in intact dendritic spines resulted in a disruption of synaptic function through impaired trafficking and synaptic anchoring of glutamate receptors. This suggests that tau phosphorylation plays a critical role in mediating tau mislocalization and subsequent synaptic impairment. Thus, tau phosphorylation itself, but not enhanced tau aggregation, mediates synaptic dysfunction, which may be one of the earliest pathological events in tauopathies [26].

Finally, recent studies have demonstrated the propagation of tau pathology. Previously, analysis of the topographical distribution of neurodegeneration and micromorphologic changes in phosphorylated tau staining suggested anterograde transneuronal mechanisms of neurodegeneration and tau pathology [27, 28]. Cultured cells can take up artificial tau fibrils made from recombinant tau protein [29] or paired helical filaments (PHFs) extracted from AD brain in the medium [30], and those ingested tau fibrils induce tau fibrillization of cytoplasmic tau. Moreover, the transference of tau aggregates from cell to cell was demonstrated by using a co-culture system [29,31]. Clavaguera *et al.* [32] demonstrated that pathological tau aggregates induced the fibrillization of wilt-type tau and spread of tau pathology by using transgenic mouse lines, ALZ17 and P301S tau. Mice form line ALZ17 express the longest isoform of wild-type human tau isoform (htau40) and develop no tau pathology. By contrast, mice from line P301S tau express the 383 amino acid human tau isoform (htau24) with the P301S mutation and develop tau pathology with age. Injection of brain extracts from line P301S tau that contained aggregated tau into the brains of line ALZ17 induced filamentous tau pathology and the spread of tau pathology to other brain regions. This indicates that tau pathology can be transmitted and abnormal tau aggregates can recruit normal tau into aggregates *in vivo*. Recently, de Calignon *et al.* [33] demonstrated the propagation of tau pathology in the brain using a inducible transgenic mouse model in which human tau P301L is primarily directed to be overexpressed in the superficial layers of the medial entorhinal cortex (EC). This mouse model was generated by crossing an inducible tau transgenic mouse line, rTg4510 [34], with a transgenic mouse line expressing the tetracycline transactivator under the control of the neuropsin promoter [35]. The neuropsin gene is primarily expressed in the limbic system of the adult brain [36]. The human tau gene with the P301L mutation in rTg4510 mice is downstream of a tetracycline-operon responsive element and requires the presence of the tetracycline transactivator. Therefore, human tau P301L is expressed primarily in the limbic system in these mice. Immunohistological analysis of tau pathology with mouse age indicated anterograde propagation of tau pathology from the EC to the dentate gyrus, followed by the CA1 and CA3 areas. Because the tau aggregates in this model contained both human and mouse tau, human mutant tau triggered mouse tau into aggregation [33]. Although another paper using similar methods confirmed these observations [37], it is impossible to exclude the possibility that mutant tau expression in non-target areas might induce tau pathology. Because the tetracycline-controlled transcriptional activation system cannot completely inhibit leaky expression, the lower levels of mutant tau expression might tardily induce tau pathology in other regions. Recently, Iba *et al.* reported that synthetic tau fibrils could mediated transmission of tau pathology in a tauopathy mouse model expressing human tau with P301S mutation (PS19) [38]. Interestingly, tau pathology mediated by synthetic tau fibers was more closely resembled AD NFTs. That was more resistant to proteinase K digestion and acetylated compared with tau pathology spontaneously developed in PS19 mice. And an immunohistochemical study using a mouse-tau-specific antibody revealed that tau pathology mediated by synthetic tau fibers consisted of both human and mouse tau but that spontaneously developed in PS19 mice contained only human tau. Moreover, they demonstrated significant tau pathology in the locus ceruleus within a short period of time after hippocampal injection, indicating retrograde transmission of tau pathology. The presumable main pathomechanisms and therapeutic targets for AD and related tauopathies are summarized in Fig. **1**.

TAU-DIRECTED THERAPIES FOR TAU-MEDIATED NEURODEGENERATION

Inhibition of Tau Phosphorylation

NFTs are a diagnostic hallmark of AD, and the brain density of NFTs directly correlates with the degree of dementia in AD patients. Tau is a major component of paired helical filaments (PHF), and tau in PHF has been found to be phosphorylated at over 30 serine/threonine residues [39]. It is still unclear whether hyperphosphorylation of tau protein is causative alteration for AD or just consequences of an altered state of tau protein. However, hyperphosphorylation of tau causes reduction in tau binding ability to MTs, leading to instability and disassemble of MTs. Hyperphosphorylation might induce self-assembly of tau into tangles of paired helical filaments/straight filaments [40]. Thus, reduction in tau phosphorylation, either by inhibiting tau kinases or by activating tau phosphatases is supposed to be the good therapeutic targets.

Inhibition of Tau Kinases

A number of candidate kinases have been proposed to play a role in phosphorylation of tau protein *in vitro* and *in vivo*, and more than 40 phosphorylation sites have been identified in tau form AD brain [41], but the physiological and pathological tau kinases *in vivo* have not been identified yet. Two protein kinases were first identified as responsible for *in vitro* tau phosphorylation. These were initially named tau protein kinases (TPK) -I and -II [42]. Isolation and cloning of these proteins revealed that TPK-I was identical to glycogen synthase kinase 3β (GSK-3β) [43] and TPK-II was a complex composed of two subunits, p35 and cyclin-dependent kinase 5 (CDK5) [44]. In addition to those kinases, cell division cycle (cdc) 2 kinase, mitogen-activated protein kinase Erk1/2, c-Jun N-terminal kinase (JNK) families, casein kinase 1 delta (CK1δ), cAMP-dependent protein kinase A (PKA), calcium/calmodulin-dependent protein kinase II (CaMKII), p38 and MT affinity-regulating kinase 1 (MARK1) have been implicated in tau phosphorylation. GSK-3 and CDK5 are thought to be the most promising therapeutic targets for the treatment of AD in those kinases, because they express at high levels in the brain, and it has been shown that all stages of AD tau pathology associate with GSK-3 and CDK5 [45-47].

GSK-3

GSK-3 is a constitutively partially active serine/threonine kinase that was first isolated and purified as an enzyme capable of phosphorylating and inactivating the enzyme glycogen synthase [48, 49]. Two GSK-3 mammalian genes encode two highly homologous forms of GSK-3, GSK-3α and GSK-3β. Though the structures of these two isoforms are similar, they are not functionally identical, but have overlapping functions. The disruption of the murine GSK-3β gene resulted in embryonic lethality caused by severe liver degeneration during midgestation [50], but GSK-3α knockout mice are viable but show abnormal glucose metabolism [51]. GSK-3β was identified as TPK-I in the 1990s [43], and it can phosphorylate tau at several of the sites found to be phosphorylated in the NFTs of AD patients [52]. Subsequently, more than 50 substrates for GSK-3 have been identified, indicating that GSK-3 is involved in regulating a large variety of cellular processes, including neuronal plasticity, cell growth, cell cycle regulation, apoptosis and inflammation [53]. This is one of the major reasons that it is difficult to develop appropriate inhibitors against GSK-3 to suppress tau phosphorylation without adverse effects. Although numerous GSK-3 substrates have been identified, in many cases the substrates must first be phosphorylated by other kinases. For example, prior phosphorylation by CDK5/p25 enhances phosphorylation of tau at specific sites [43]. This indicates that tau may be both a primed and unprimed substrate of GSK-3. Increased levels of GSK-3β have also been found in AD [54, 55]. Additionally, accumulating evidence suggests that GSK-3 might essentially contribute to the pathogenesis of AD, linking Aβ to tau [56, 57]. Knockdown of GSK-3α expression and activity by intracerebroventricular injections of recombinant adeno-associated virus resulted in a reduction in both SPs and NFTs in the double-transgenic (Tg) AD mouse model (PDAPP × PS19). However, knockdown of GSK-3β expression and activity only modulated NFT formation. These observations emphasize the potential importance of GSK-3α as a possible therapeutic target for pathogenesis of both Aβ and tau in AD [58]. Because GSK-3α is less well-investigated, further research is needed before clinical trials. Aβ leads to increased GSK-3β activity and GSK-3β is essential for Aβ-induced neurotoxicity [59-61]. Additionally, the APP intracellular domain (AICD), which is the C-terminal fragment of APP cleaved by γ-secretase, translocates into the nucleus, resulting in increased GSK-3β expression and activity [62]. GSK-3 might regulate the production of Aβ by impinging on APP cleavage at the γ-secretase site [63, 64]. Meanwhile, injections of an anti-Aβ antibody into cerebral ventricles decreased Aβ and GSK-3β activity in APP_{swe} mice (Tg2576) [65]. Moreover, coexpression of GSK-3β and triple tau mutations (G272V, P301L and R406W) in transgenic mice indicated that overexpression of GSK-3β induced spatial memory impairment, accelerated NFT formation and neurodegeneration [66]. Engle *et al.* [67] conducted two sets of experiments in this double Tg mouse model to analyze the effects of a GSK3 inhibitor lithium on tau pathology. To study the preventive potential of lithium they administrated lithium to 12-month-old Tg mice, which were supposed to show little tau pathology, and in order to analyze the reversal potential of lithium they administrated lithium to 18-month-old Tg mice, which were supposed to show strong tau pathology and tau filaments. Their data indicated that lithium chloride could prevent NFT formation but could not reverse already formed NFTs. Several other studies using tauopathy mouse models have demonstrated that lithium chloride resulted in the inhibition of GSK-3 activity, reductions in both tau hyperphosphorylation and levels of insoluble tau, and improvement of behavioral impairment [67-72]. A small randomized 10-week clinical trial of lithium for mild-to-moderate AD failed to show any beneficial effects on clinical symptoms, GSK-3 activity in lymphocytes or phosphorylated tau in cerebrospinal fluid (CSF) [73]. An open label treatment study with low dose lithium for up to 1 year demonstrated no change in Mini-Mental State Examination (MMSE) [74]. Meanwhile, a 12-month, double-blind trial for amnestic mild cognitive impairment (aMCI) demonstrated a decrease in p-tau in the CSF and better cognitive performance in Alzheimer's Disease Assessment Scale-cognitive ADAS-cog) [75]. Overall, lithium has good tolerability and relatively few side effects. Thus, further larger scale and long-term clinical trials are expected to determine whether lithium has disease-modifying effects and clinical efficacy on AD and other tauopathies. Recently, Palomo *et al.* [76] reported small molecules, 5-imino-1,2,4-thiadiazoles, that inhibit GSK-3 by a unique mechanism of action. These molecules can bind to primed sites of GSK-3 substrates, and inhibit GSK-3 in a substrate-competitive manner. Tideglusib (NP-12) is the only GSK-3 inhibitor for clinical use. It inhibits CSK-3β irreversibly and has the long enzyme half-life [77]. Interestingly, administration of tideglusib (NP-12) improved both tau pathology and Aβ pathology in double Tg mice expressing human mutant APP and tau [78]. GSK-3 inhibitor might have a dual impact on amyloid and tau alterations. However the report of the results of phase II clinical trials of tideglusib for the treatment of mild-to-moderate AD was disappointing. The 26-week ARGO Phase IIb trial did not meet the primary cognitive end-points and two of the secondary end-points.

CDK5

CDK5 inhibitors are expected to prevent tau phosphorylation and consequent pathological events, including NFT formation and neurodegeneration in AD, because CDK5 is, together with GSK-3, one of the major kinases implicated in the abnormal tau phosphorylation in AD brain, and it was identified as TPK-II.

An increase in CDK5 activity is accompanied by the accumulation of p25, which is a truncated form of p35. Inducible transgenic mouse lines overexpressing p25 exhibit neuronal loss, NFT-like pathology and tau hyperphosphorylation accompanied by CDK5 activation [79]. This is direct evidence that CDK5 activation can induce tau pathology and neurodegeneration. However, no CDK5 inhibitors have yet reached clinical trials. The most challenging problem in developing CDK5 inhibitors for clinical usage is to achieve enough specificity among other protein kinases, especially among cyclin-dependent kinases, because CDK5 has high sequence homology and similar three-dimensional structure compared to other CDKs [80]. Recently, it was reported that a novel small-molecule calpain inhibitor, A-705253, inhibited tau phosphorylation by suppressing the calpain-mediated conversion of the CDK5 regulatory protein p35 to p25 *in vitro* [81]. Moreover, the administration of A-705253 decreased the activation of CDK5 and diminished the hyperphosphorylation of tau in 3×TgAD mice *in vivo* [82].

Other Tau Protein Kinases

Tau protein is phosphorylated by both proline-directed and serine/threonine protein kinases (PDPKs), including GSK-3 and CDK5, while other tau kinases including CK1, PKA, CaMKII and MAPKs are non-PDPKs, and phosphorylate tau at only a few sites. However, phosphorylation of tau by these non-PDPKs markedly increases the phosphorylation of tau by PDPKs (GSK-3 and CDK5). There are no tau kinase inhibitors other than GSK-3 inhibitors under clinical trials. It is very challenging to develop inhibitors adequately specific for these tau kinases without adverse effects, because these inhibitors may inhibit the phosphorylation of other substrates important for cell functions.

Activation of Tau Phosphatases

Because most tau phosphorylation sites are serine/threonine residues, serine/threonine protein phosphatases (PPases) play a major role in tau dephosphorylation. The phosphorylation of tau that suppresses its MT binding and assembly activities in the adult mammalian brain is regulated by PP2A, and not by PP2B [83, 84]. Thus, PP2A has now emerged as a key player in tau dephosphorylation [85]. Not only the protein level, but also the mRNA level of

PP2A is reduced in AD brains [86, 87]. PP2A accounts for around 70% of human brain tau phosphatase activity [88], and the activity of PP2A is reduced up to about 20% in AD brain [86, 89-91]. The PP2A core enzyme is composed of a 65 kD scaffold subunit (known as the A or PR65 subunit) and a 36 kD catalytic subunit (the C subunit). To gain full activity toward specific substrates, the PP2A core enzyme interacts with a variable regulatory subunit to form a heterotrimeric holoenzyme. Regulatory subunits belong to four different families of unrelated proteins. Both the B (or PR55) and B' regulatory protein families contain α, β, γ and δ isoforms, with the B' family also including an ε protein. The B″ protein family contains four subunits (PR48, PR59, PR72, PR130), and striatin (PR110) and SG2NA (PR93) belong to the B‴ regulatory protein family [92]. There is no detectable sequence homology among the four families of regulatory subunits. Three major classes of endogenous small molecules, including metal cations, ceramides and polyamines, have been shown to regulate PP2A activity [93]. However, because PP2A plays a role in cellular phospho-regulatory networks that maintain normal cellular functions, indiscriminate activation of PP2A must induce toxic effects on cellular homeostasis. It is difficult and complicated to evaluate and to compare data across studies about PPA2 activation because a variety of assay systems and substrates, and different PP2A subunit composition are employed in those studies. Based on data related to clinical observations or *in vivo* observations for activation of PP2A, some candidates are described below. Folate and methyl-folate promote the methylation of PP2A and the subsequent formation of an active Bα-containing PP2A complex that should lead to increased dephosphorylation of tau [94, 95]. It is also known that blood homocysteine levels are increased in AD patients. Additionally folate deficiency, which leads to raised homocysteine levels [96], is also observed in AD patients [97]. Memantine is a low-affinity voltage-dependent non-competitive antagonist of glutamatergic N-Methyl-D-aspartic acid (NMDA) receptors, and it has been shown to have a modest effect on the cognition of moderate-to-severe AD. Memantine is thought to selectively block the excitotoxic effects associated with abnormal transmission of glutamate, while allowing for the physiological transmission associated with normal cell functioning. Interestingly, memantine recovers the PP2A activity inhibited by the inhibitor of protein phosphatase 2A (IPP2A) *in vitro* [98]. The levels of IPP2A are increased about 20% in AD brains [98]. Memantine treatment for one year resulted in a significant decrease in the levels of phosphorylated tau in the CSF of AD patients [99]. Although the clinical significance is unknown and it is unclear whether the actual PP2A activities were increased in those patients' brains, the effect of memantine that disinhibits the activity of IPP2A against PP2A might be responsible for decreased phosphorylated tau in the CSF. Lastly, sodium selenate acts as a specific agonist for PP2A, significantly boosting phosphatase activity. It induced dephosphorylation of tau, reduced tau insolubility and reduced behavioral impairments as well as prevention of neuronal loss in tauopathy model mice [100, 101]. A neurotoxic assay using primary rat hippocampal slice culture demonstrated no neurotoxic effects of sodium selenite [101]. Although further studies are required to confirm the efficacy and to exclude adverse effects on non-neuronal cells, sodium selenate is a promising candidate for AD treatment.

Modulation of Other Post-Translational Modifications

Several post-translational modifications of tau protein other than phosphorylation, including glycosylation, glycation, acetylation and nitration, have been reported. Because the physiological and pathological significance of these modifications is unclear, they have received much less researcher's attention and are less well investigated than phosphorylation. However, because these modifications could modulate tau function directly or indirectly, these post-translational modifications of tau could be a new therapeutic target of AD therapy, but more research is necessary before they will be the subject of clinical studies.

Tau Glycosylation

Protein glycosylation refers to the enzymatic process that attaches mono- or polysaccharides to proteins. Glycosylation types are classified into *O*-linked, *N*-linked and *C*-linked, according to the identity of the atom of the amino acid which binds the carbohydrate chain. *O*-GlcNAcylation, a type of protein *O*-glycosylation in which the monosaccharide-N-acetylglucosamine (GlcNAc) is attached to serine/threonine residues *via* an *O*-linked glycosidic bond, has received widespread attention, because it regulates the phosphorylation of tau in a site-specific manner both *in vitro* and *in vivo*. Stoichiometric analysis of tau revealed relatively high *O*-GlcNAcylation in tau protein, 4.2 ± 0.9 mol *O*-GlcNAc/mol tau [102]. There is evidence of a reverse relation between the *O*-GlcNAc modification and tau phosphorylation [103]. Inhibition of the hexosaminidase *O*-GlcNAcase (OGA) or increased expression of *O*-β-N-acetylglucosaminyltransferase reduces phosphorylated tau. OGS inhibitors cause a decrease of tau phosphorylation in normal rats [104]. Meanwhile, chronic treatment of a tauopathy mouse model (JNPL3) with an OGA inhibitor increased tau *O*-GlcNAc, inhibited tau aggregate formation and decreased neuronal cell loss. However, increases in tau *O*-GlcNAc did not reduce phosphorylation of tau contrary to expectations [105]. Tau protein from AD brains also show multiple other types of glycosylation, mainly *N*-linked glycosylation. This modification might enhance tau hyperphosphorylation [106].

Tau Glycation

Glycation is a non-enzymatic process of glycosylation. A fraction of PHF-tau from AD brain is highly glycated but soluble tau from either AD or non-demented human brain is not glycated at all. Tau glycation is accompanied by a decrease in the tau binding to MTs *in vitro* [107]. Glycation forms advanced glycation end products (AGEs), which generate oxygen free radicals. AGE immunoreactivity is present in both Aβ plaques and NFTs in patients with AD. AGE-tau generate oxygen free radicals, which activate transcription *via* nuclear factor-kappa B, increasing Aβ production [108].

Tau Acetylation

Recent studies have indicated that acetylation of tau may occur after its phosphorylation and may contribute to tau-mediated neurodegeneration by driving tau polymerization and inhibiting its degradation [109-111]. Increased acetylation of tau protein was observed in tau Tg mice, AD and other tauopathies [109, 111]. Tau acetylation may be regulated mainly by two enzymes, histone acetyltransferase p300, which is responsible for acetylation, and SIRT1, which removes acetylation from tau [109]. SIRT1 levels are decreased in AD brains, and this reduction correlates with the accumulation of hyperphosphorylated tau aggregates [112]. Injection of SIRT1 lentivirus into the hippocampal CA1 region and intracerebroventricular injection of resveratrol, a SIRT1 activator, reduced neurodegeneration in the hippocampus in p25 inducible transgenic mice, a model of AD and tauopathies [113]. Thus, reduction in tau acetylation by inhibiting histone acetyltransferase p300 or by activating SIRT1 is supposed to be a good therapeutic target for AD and relating tauopathies.

Tau Nitration

Increasing evidence indicates that oxidative and reactive oxygen and nitrogen species are implicated in the pathogenesis of neurodegenerative disorders, including AD and other tauopathies. Peroxynitrite, which is a powerful oxidant produced from the reaction of superoxide with nitric oxide, mediates neuronal damage in AD [114]. 3-nitrotyrosine (3-NT) is generated by peroxynitrite, superoxide and nitric oxide. Anti-3-NT polyclonal antibodies revealed immunoreactive 3-NT in hallmark lesions of neurodegenerative diseases such as neurofibrillary tangles (NFTs) in AD [114]. Although nitration of tau has been reported on four residues, a pathological significance of nitration on each residue is supposed to be different. Immunohistological studies using antibodies specific to each nitration site demonstrated that tau nitrated at tyrosine 18 was localized largely in reactive glial cells, and tau nitrated at tyrosine 29 was found in NFTs in AD. Meanwhile, in both corticobasal degeneration (CBD) and progressive supranuclear palsy (PSP), tau nitrated at tyrosine 29 colocalized with some but not all tau inclusions. It was also found in Pick body inclusions in Pick's disease [115, 116].

Stabilization of Microtubules

Stabilization of MTs is a rational therapeutic intervention to compensate for the loss of tau function. Hyperphosphorylation of tau reduces its ability to bind MTs, leading to instability and disassembly of MTs. Actually, axonal transport impairment and MT density loss are observed, especially in the early stages of AD [117, 118]. The Lee and Trojanowski laboratory administered a MT-stabilizing drug to tauopathy model mice expressing the shortest isoform of human wild-type tau (0N3R) [119]. Paclitaxel, which is used for cancer treatment, improved fast axonal transport and MT density compared to vehicle-treated mice, and it markedly improved the motor functions [120]. However, the effect of paclitaxel on tau pathology was uncertain, because this mouse model does not show apparent tau pathology. Next, they used another tauopathy mouse model expressing P301S mutant tau (PS19). This model shows tau pathology and neuronal loss [26]. Because paclitaxel has poor blood-brain barrier permeability, they next used a similar MT-stabilizing drug, epothilone D, which has a favorable pharmacokinetic and pharmacodynamic profile compared to paclitaxel. Treatment with epothilone D reduced forebrain tau pathology and increased hippocampal neuronal integrity, with no dose-limiting side effects, in PS19 mice [121, 122]. Treatment with a very low dose of BMS-241027 (identical to epothilone D) also improved cognitive functions and tau pathology in rTg4510 tau Tg mice [123]. Phase I clinical trials of epothilone D for mild AD, aimed at evaluating its safety, tolerability and effect on CSF biomarkers, have begun. Intranasal administration of another MT stabilizing drug, davunetide [124], for the treatment of tauopathies, including PSP, FTDP-17, CBD and progressive non-fluent aphasia, is undergoing phase II clinical trials. Very recently, the result of the multinational phase II/III randomized, double-blind, placebo-controlled trial of davunetide for PSP treatment was announced. Unfortunately, it failed to demonstrate the efficacy.

Inhibition of Tau Oligomerization/Fibrillization

To prevent the formation of potentially harmful tau oligomers and fibrils would be a promising therapeutic target. Additionally, because the enhanced tau polymerization could recruit normal monomeric tau to tau oligomers and/or fibrils, inhibition of tau polymerization might contribute to maintaining physiological MT functions. High-throughput screening campaigns using *in vitro* assays relying on β-sheet specific dyes, such as thioflavin T or thioflavin S, to monitor the fibrillization reaction to identify aggregation inhibitors have been conducted [125, 126]. These efforts identified a number of compounds that reduced tau fibril formation *in vitro* [125, 126].

Methylene blue is the first tau-aggregation inhibitor reported [127], and this is the only compound that has entered clinical evaluation. It is commonly used in diagnostic procedures for surgeries and endoscopic procedures, but is also used to treat some medical conditions. Prosed/DS™ is combination of drugs including methylene blue, methenamine, phenyl salicylate, benzoic acid and hyoscyamine sulfate, and it is frequently prescribed for treatment of urinary tract infection. Methylene blue is also infrequently used for treatment of malaria, methemoglobinemia and cyanide poisoning. van Bebber *et al.* [128] reported the effects of methylene blue on a zebrafish tauopathy model. Methylene blue reduced insolubility of tau protein by 35%. However, it did not reduce phosphorylation of tau protein, neuronal degeneration and motor deficits contrary to the expectation. Meanwhile, a *C. elegans* tauopathy model improved locomotive deficits by methylene blue treatment [129]. Administration of methylene blue to a tauopathy mouse model, rTg4510, was a good example of the effectiveness of methylene blue to improve tau abnormalities and cognitive deficits in tauopathies [130]. A phase II clinical trial of methylene blue (Rember™) in more than 300 mild-to-moderate AD patients showed a 90% reduction in the rate of disease progression over two years by measured with ADAS-cog and MMSE [131]. This results also supported by functional brain imaging. Methylene blue is expected to inhibit tau polymerization, but it has been reported that methylene blue also inhibits MT polymerization *in vitro* [132]. Therefore, there is some concern about serious adverse effects in the long-term administration. Phase III clinical trials of methylene blue have already started enrolling AD patients. They consist of three different programs for mild-to-moderate AD, mild AD or front temporal dementia (FTD).

Improvement in Degradation of Misfolded Tau/NFTs

Tauopathies, including AD, are characterized by the accumulation of pathological aggregates of misfolded tau. Intracellular levels of misfolded and/or aggregated tau are regulated by a balanced equilibrium between their synthesis and degradation. Thus, enhancement of degradation of tau oligomers and/or fibrils is thought to be another good strategy to reduce harmful levels of pathological or toxic tau species. Two major proteolytic systems contribute to protein degradation inside cells: the ubiquitin-proteasome and the lysosomal (autophagy) systems. The colocalization of ubiquitin with NFTs in AD brains leads to the speculation that the ubiquitin-proteasome system may play a more important role in tau degradation. However, alterations in the lysosomal system have also been demonstrated in AD [133]. Therefore, it remains unclear whether tau is normally degraded by the proteasome system or the lysosomal system.

Ubiquitin-Proteasome System

The chaperone system of the heat shock protein (Hsp) family is an evolutionarily conserved system that is charged with preventing unfolded or misfolded proteins in the cell from aggregating. A major component of the chaperone network is

Hsp70, which folds and degrades nascent and misfolded polypeptides. Hsp70 can work alone or in cooperation with another major chaperone protein, Hsp90. Proteins containing tetratricopeptide repeat (TPR) domains are involved in many protein-protein interactions. In particular, Hsp70 and Hsp90 interact with substrates through TPR domains. Carboxy-terminus of Hsp70-interacting protein (CHIP) is a critical TPR-containing chaperone that is a highly conserved ubiquitin ligase for protein quality control and stress recovery systems. Tau is ubiquitinated and degraded by the cooperation of Hsp70 and CHIP [134]. The levels of Hsp70 and CHIP were found to be increased, and CHIP levels were found to be inversely proportional to the amount of insoluble tau accumulation in AD brains [135]. The Hsp70 family is comprised of 13 members. However, constitutive heat shock cognate (Hsc) 70 and inducible Hsp70 are the predominant family members in the cytoplasm. Although Hsc70 and Hsp70 can bind tau, Hsc70 is much more abundant in the cytosol [136]. Hsc70 facilitates tau-mediated MT polymerization [137]. Hsp70 enhanced tau folding *in vitro*. However, in a hyperphosphorylating environment induced by transfection with GSK-3β, there was reduced tau folding but enhanced Hsc70 binding to tau [138]. Although increasing Hsp levels might be expected to allow more abnormal proteins to bind to Hsps, leading to degradation, Hsps might exert the opposite influence on tau misfolding and fibrillization in the pathological environment [138]. Hsp90 inhibitors enhance the degradation of client proteins bound to Hsp90 and activate heat shock factor 1 (HSF1), which induces other chaperone proteins [139]. A blood-brain barrier-permeable Hsp90 inhibitor EC102 significantly reduced phosphorylated tau levels in tauopathy model mice [140]. Interestingly, the results of binding affinity studies in human AD brain tissue demonstrated Hsp90 extracted from affected region of AD brains had significantly increased binding affinity for EC102. This indicates that lower concentrations of Hsp90 inhibitors could be used for treatment of AD to reduce adverse effects on Hsp90/client complexes in normal tissues [140]. JNPL3 mice overexpressing P301L mutant tau crossed with CHIP knockout mice demonstrated marked phosphorylated tau accumulation throughout the brain. However, deletion of CHIP reduced tau pathology in the brain in JNPL mice [141]. This suggests that hyperphosphorylation of tau alone in the absence of CHP might be insufficient to promote tau fibrils.

Recently, Hsp27, one of the small heat shock proteins, has attracted attention as a new target for tauopathies, because it interacts with unfolding proteins and can disrupt protein aggregation [142]. Atomic force microscopy and dynamic light scattering showed that recombinant wild-type Hsp27 prevented tau filament formation. Extrinsic genetic delivery of Hsp27 to a tauopathy mouse model (rTg4510) using an adeno-asociated viral vector showed that Hsp27 expression reduced neuronal tau accumulation and rescued long-term potentiation (LTP) deficits [143].

Immunophilins are cytoplasmic proteins with chaperone functions. The FK506-binding proteins (FKBP) of the immunophilin family were found to associate with tau. FKBP12 is co-localized with NFTs in AD, and the levels of FKBP12 in AD brains are lower than in normal brains [144]. FKBP51 and FKBP52 are co-chaperones of Hsp90-tau complexes, and they are involved in tau turnover. FKBP51 overexpression increased levels of tau [145], while FKBP52 overexpression decreased levels of tau [146]. Even though FKBP51 and FKBP52 have very similar structures, their opposite effects on tau make it difficult to understand their functions. Although further research is required, analogs of FK51/52 might modulate FKBP activities and reduce misfolded tau without immunosuppression, because FKBP51 and 52 do not bind calcineurin.

Lysosome System

Lysosomes are responsible for the degradation of macromolecules derived from the extracellular space through endocytosis or phagocytosis, as well as from the cytoplasm through autophagy. It could be an appropriate strategy to up-regulate autophagy for degradation of accumulated tau aggregates in neurons. The mammalian target of rapamycin (mTOR) is an inhibitor of autophagy and mTOR activity is inhibited by rapamycin. Thus, rapamycin is an activator of macroautophagy. Rapamycin reduced tau aggregation and neurotoxicity in Tg *Drosophila* expressing P301L mutant tau [147]. Proteasome inhibitors show no effect on the clearance of tau aggregates in an inducible mouse neuroblastoma N2a cell expressing the repeat domain (tau 244–372) of tau with the ΔK280 deletion mutation (TauRDΔK280). While, autophagy inhibitors block the degradation of tau aggregates and enhance neurotoxicity [148]. Subsequently, Dolan *et al.* [149] demonstrated that TauΔC (truncated at Asp^{421}) was degraded significantly faster than full-length tau, and full-length tau was degraded by the proteasome, but TauΔC was degraded predominantly by macroautophagy in inducible neuronal cell lines expressing full-length tau or tauΔC. These results indicate that the lysosomal system plays a major role in degrading pathologically truncated tau, which is presumably more toxic and more amyloidogenic. Administration of trehalose, an activator of autophagy, reduced tau phosphorylation, tau insolubility and neuronal loss in the brains of Tg mice expressing P301S mutant human tau [150].

Suppression of Pathological Tau Expression

PDAPP mice, which show Aβ deposition and cognitive impairments, crossed with tau knockout mice to reduce levels of endogenous tau, showed an improvement in cognitive deficits and reduced lethality, thereby suggesting that tau might be necessary for Aβ-induced neurotoxicity [151]. However, the beneficial effects of reduction in the tau expression have not been confirmed by other studies, and instead deleterious effects of reduced tau protein levels on tau knockout mice have been shown [152, 153]. Thus, it is so far unclear whether a reduction in endogenous wild-type tau expression in the brain can reduce Aβ-induced neurotoxicity. Switching off pathological tau (mutant tau) expression recovered synaptic plasticity, memory impairments and LTP deficits in regulatable tauopathy mouse models. However the levels of neuronal loss was unchanged and NFTs continued to accumulate [34, 154, 155]. Because the switching-off systems used in these experiments cannot completely halt the expression of pathological tau, low expression levels of pathological tau might play a part in neuronal loss and NFT formation. In either case, this suggests that the regulation of DNA transcription of the mutant tau gene could ameliorate or prevent the progression of neurodegeneration and clinical symptoms in FTDP-17 patients. The delivery of small interfering RNA (siRNA) against mutant tau could be a promissing therapy for FTDP-17 patients. One possible way to supply siRNA to the brain is intranasal administration, like the intranasal administration of insulin for treatment for AD [156, 157].

Inhibition of Tau Pathology Propagation

Recently, the concept that tau pathology can trans-synaptically propagate in a cell-to-cell manner has been proposed. Tauopathy mouse brain lysates containing pathological tau induced propagation of tau pathology in recipient mice [32]. Synthetic tau fibril injections into the tauopathy model mouse brains demonstrated transmission of tau pathology [38]. If the propagation of tau pathology plays an important role in tau-mediated neurodegeneration, inhibition of tau pathology propagation should be a good target for treatment. Several *in vitro* and *in vivo* studies have indicated that tau antibodies suppress tau aggregate transmission from cell to cell and those reduce tau pathology.

Tau Immunization

Antibodies against tau are supposed to capture extracellular tau and they could be good therapeutic candidates for inhibition of tau pathology propagation. Indeed Kfoury *et al.* [158] demonstrated that tau aggregates were secreted into the extracellular space, and they transferred and triggered aggregation in recipient cells. Moreover, they also showed that a monoclonal antibody against tau blocked trans-cellular propagation in a dose-dependent manner, although that antibody did not reduce intracellular tau aggregates. Studies of active tau immunotherapy using tau Tg mouse models have shown its potential for treatment of tauopathies. However, a preliminary study using non-Tg mice, female C57BL/6 mice, indicated that inoculation with full-length human recombinant tau emulsified in complete Freund adjuvant and with pertussis toxin induced NFT-like structures, axonal damage, gliosis and mononuclear infiltration as well as cognitive impairments [159]. This observation pointed to the serious risk of tau immunization injuring normal neurons and to inducing encephalitis, especially when targeting normal tau. Moreover, other active immunizations targeting phosphorylated tau also showed unexpected reactions. Autoantibodies recognizing normal tau [160] and astrogliosis [161] were reported. These studies indicate that active immunization may not be able to selectively target pathological tau. To increase target specificity, a passive immunization protocol may be a better therapeutic approach. Boutajangout *et al.* [162] reported that intraperitoneal injections of PHF1, a phosphorylated tau (Ser396/Ser404)-specific antibody, reduced the levels of insoluble pathological tau. Plasma levels of PHF1 were found to be inversely correlated with tau pathology in the brainstem, with a strong trend in the motor cortex, as well as with insoluble tau levels in a tau transgenic mouse model (JNPL3) [162]. Additionally, performance on the traverse beam was better in treated mice. Chai *et al.* [163] used two antibodies, PHF1 and MC1. MC1 is a conformation-dependent antibody that recognizes an early pathological tau conformational change [164]. Both treatments reduced insoluble tau levels and tau pathology in the spinal cord and brainstem [163].

Exosomes [165, 166] and tunneling nanotubes [167, 168] have been proposed to mediate tau-aggregate transmission. Thus, other than antibodies, small molecules that could pass through the blood-brain barrier to target exosomes and/or tunneling nanotubes could be potential therapeutic agents.

Modulation of Inflammatory Responses

Inflammatory responses, including microglial activation and increased levels of inflammatory mediators, are observed in brains from AD [169] and other tauopathies [170, 171]. A positive correlation has also been demonstrated between the NFT burden and microglial activation [172, 173]. The numbers of interleukin (IL)-1β-positive microglia and S100β-positive astrocytes increased progressively with NFT load in the AD brain [174]. Moreover, histopathological analysis in various tau Tg mice have demonstrated that the distribution of activated microglia is linked to the distribution of tau pathology [26, 175-178]. Fibrillar forms of Aβ stimulates the production and secretion of proinflammatory cytokines and production of reactive oxygen and nitrogen radicals in microglia [179]. Meanwhile, it is well known that diabetes mellitus, hypertension and periodontitis, all of which are considered as a chronic inflammatory state, are risk factors for AD [180]. The long-term use of nonsteroidal anti-inflammatory drugs (NSAIDs) reduced the risk of developing AD, but not vascular dementia, in a prospective, population-based cohort study [181]. However, different from other targets based on the molecular pathomechanisms of tau-mediated neurodegeneration discussed above, it is unclear how inflammation enhances tau-mediated neurodegeneration. Recently, the effects of deleting the fractalkine receptor, which resulted in activation of microglia, were reported in a tauopathy model. Tau pathology was exacerbated by the deletion of the fractalkine receptor [182]. This clearly showed that the activation of microglia can enhance tau pathology. On the other hand, the same deletion reduced the amount of amyloid deposits in amyloidogenic transgenic mice (TgCRND8, APPPS1-21) [183, 184].

Therefore, it is important to carefully interpret data on the effects of inflammation in the pathomechanism of AD, because inflammation might have divergent effects on Aβ and tau. Interestingly, microglial activation precedes NFT formation, and administration of an immunosuppressant, FK506, ameliorated microgliosis, synaptic loss, tau pathology and survival durations in a PS19 [26]. Prostaglandin J2, an endogenous product of inflammation, induces caspase-mediated cleavage of tau. This truncated tau might seed aggregates prior to NFT formation [185]. Moreover, chronic administration of donepezil, an acetylcholine esterase inhibitor, ameliorated tau pathology with a reduction in inflammatory responses including microglial activation [186]. On the other hand, anticholinergic drugs accelerated tau pathology, with enhanced inflammatory responses [187]. These observations may relate to anti-inflammatory effects mediated through the peripheral cholinergic system [188]. Several clinical trials targeting inflammation including NSAIDs and COX-2 inhibitors in AD patients have been conducted. The results of these clinical trials have been disappointing, except for a small beneficial effect on reduced AD incidence [189]. Etanercept, an inhibitor of tumor necrosis factor (TNF), which is used for the treatment of rheumatoid arthritis and other inflammatory disorders, is a fusion protein of the TNF receptor and the constant end of the IgG1 antibody. An open-label study of 15 patients treated weekly for a period of six months, perispinal extrathecal administration of etanercept in the posterior neck led to sustained cognitive improvement in mild-to-severe AD [190]. Interestingly, some AD patients showed rapid cognitive improvement following perispinal etanercept administration [191]. This observation suggests that inflammation might directly affect cognitive functions in AD. A phase II, double-blind, placebo-controlled study of 50 mg etanercept given as a once weekly subcutaneous injection in AD patients is ongoing.

DISCUSSION AND CONCLUSION

Although the amyloid hypothesis of AD is widely accepted, it has not been proved that Aβ itself directly relates to AD onset and cognition. Moreover, clinical trials aimed at removal of Aβ accumulation, or suppression of Aβ production or Aβ oligomerization in the brain have failed to demonstrate significant clinical efficacy in symptomatic AD patients. Meanwhile, it was reported that tau pathology progressed autonomously even after Aβ clearance by active immunization [4]. This suggests that Aβ might initiate a pathological circuit of tau-mediated neurodegeneration, but once this cascade is started, Aβ removal cannot stop it. Tau-mediated neurodegeneration seems more closely related to clinical symptoms and their progression. In AD, amnesia is a common initial symptom and is followed by cortical symptoms such as aphasia, apraxia and agnosia. This could be explained by the topographical expansion of tau pathology from the hippocampus to the neocortices along the neuronal network *via* synaptic transmission. Thus, tau-targeted therapies might be necessary to halt or to improve patients' symptoms in AD and other tauopathies.

In conclusion, the compelling reasons mentioned in the previous sections may encourage researchers to attempt to identify drug candidates that can inhibit the gain of toxic functions of tau protein or compensate for the loss of the physiological functions of tau protein. Although tau-targeted therapies are not as advanced as Aβ-targeted therapies, recent advancements in the understanding of tau-mediated neurodegeneration should facilitate the development of new therapies.

CONFLICT OF INTEREST

None declared.

ACKNOWLEDGEMENT

This work was supported by a Grant-in-Aid for Scientific Research (C) grant No 22590948.

DISCLOSURE

Part of information included in this chapter has been previously published in J. Neurol. Neurosurg. Psychiatry doi:10.1136/jnnp-2012-303144.

REFERENCES

[1] Aisen PS, Gauthier S, Ferris SH, *et al.* Tramiprosate in mild-to-moderate Alzheimer's disease - a randomized, double-blind, placebo-controlled, multi-centre study (the Alphase Study). Archives of medical science : AMS. 2011;7:102-11.

[2] Green RC, Schneider LS, Amato DA, *et al.* Effect of tarenflurbil on cognitive decline and activities of daily living in patients with mild Alzheimer disease: a randomized controlled trial. JAMA. 2009;302:2557-64.

[3] Imbimbo BP, Panza F, Frisardi V, *et al.* Therapeutic intervention for Alzheimer's disease with gamma-secretase inhibitors: still a viable option? Expert opinion on investigational drugs. 2011;20:325-41.

[4] Holmes C, Boche D, Wilkinson D, *et al.* Long-term effects of Abeta42 immunisation in Alzheimer's disease: follow-up of a randomised, placebo-controlled phase I trial. Lancet. 2008;372:216-23.

[5] Hardy J, Selkoe DJ. The amyloid hypothesis of Alzheimer's disease: progress and problems on the road to therapeutics. Science. 2002;297:353-6.

[6] Dermaut B, Kumar-Singh S, Engelborghs S, *et al.* A novel presenilin 1 mutation associated with Pick's disease but not ?-amyloid plaques. Annals of Neurology. 2004;55:617-26.

[7] Cleveland DW, Hwo SY, Kirschner MW. Physical and chemical properties of purified tau factor and the role of tau in microtubule assembly. J Mol Biol. 1977;116:227-47.

[8] Goedert M, Spillantini MG, Potier MC, Ulrich J, Crowther RA. Cloning and sequencing of the cDNA encoding an isoform of microtubule-associated protein tau containing four tandem repeats: differential expression of tau protein mRNAs in human brain. EMBO J. 1989;8:393-9.

[9] Neve RL, Harris P, Kosik KS, Kurnit DM, Donlon TA. Identification of cDNA clones for the human microtubule-associated protein tau and chromosomal localization of the genes for tau and microtubule-associated protein 2. Brain Res 1986;387:271-80.

[10] Kosik KS, Orecchio LD, Bakalis S, Neve RL. Developmentally regulated expression of specific tau sequences. Neuron 1989;2:1389-97.

[11] Mukrasch MD, Bibow S, Korukottu J, *et al.* Structural polymorphism of 441-residue tau at single residue resolution. PLoS biology 2009;7:e34.

[12] Mukrasch MD, von Bergen M, Biernat J, *et al.* The "jaws" of the tau-microtubule interaction. J Biol Chem 2007;282:12230-9.

[13] Fauquant C, Redeker V, Landrieu I, *et al.* Systematic identification of tubulin-interacting fragments of the microtubule-associated protein Tau leads to a highly efficient promoter of microtubule assembly. J Biol Chem 2011;286:33358-68.

[14] Dixit R, Ross JL, Goldman YE, Holzbaur EL. Differential regulation of dynein and kinesin motor proteins by tau. Science 2008;319:1086-9.

[15] Kempf M, Clement A, Faissner A, Lee G, Brandt R. Tau binds to the distal axon early in development of polarity in a microtubule- and microfilament-dependent manner. J Neurosci 1996;16:5583-92.

[16] Yoshiyama Y, Lee VM, Trojanowski JQ. Therapeutic strategies for tau mediated neurodegeneration. J Neurol Neurosurg Psychiatry 2012.

[17] Spires-Jones TL, Kopeikina KJ, Koffie RM, de Calignon A, Hyman BT. Are tangles as toxic as they look? J Mol Neurosci 2011;45:438-44.

[18] Morfini G, Pigino G, Mizuno N, Kikkawa M, Brady ST. Tau binding to microtubules does not directly affect microtubule-based vesicle motility. J Neurosci Res 2007;85:2620-30.

[19] LaPointe NE, Morfini G, Pigino G, *et al.* The amino terminus of tau inhibits kinesin-dependent axonal transport: implications for filament toxicity. J Neurosci Res 2009;87:440-51.

[20] Jeganathan S, von Bergen M, Brutlach H, Steinhoff HJ, Mandelkow E. Global hairpin folding of tau in solution. Biochem 2006;45:2283-93.

[21] Stamer K, Vogel R, Thies E, Mandelkow E, Mandelkow EM. Tau blocks traffic of organelles, neurofilaments, and APP vesicles in neurons and enhances oxidative stress. J Cell Biol 2002;156:1051-63.

[22] Yoshiyama Y, Zhang B, Bruce J, Trojanowski JQ, Lee VM. Reduction of detyrosinated microtubules and Golgi fragmentation are linked to tau-induced degeneration in astrocytes. J Neurosci 2003;23:10662-71.

[23] Liazoghli D, Perreault S, Micheva KD, Desjardins M, Leclerc N. Fragmentation of the Golgi apparatus induced by the overexpression of wild-type and mutant human tau forms in neurons. Am J Pathol 2005;166:1499-514.

[24] Gonatas NK, Gonatas JO, Stieber A. The involvement of the Golgi apparatus in the pathogenesis of amyotrophic lateral sclerosis, Alzheimer's disease, and ricin intoxication. Histochem cell biol 1998;109:591-600.

[25] Hoover BR, Reed MN, Su J, *et al.* Tau mislocalization to dendritic spines mediates synaptic dysfunction independently of neurodegeneration. Neuron 2010;68:1067-81.

[26] Yoshiyama Y, Higuchi M, Zhang B, *et al.* Synapse loss and microglial activation precede tangles in a P301S tauopathy mouse model. Neuron 2007;53:337-51.

[27] Su JH, Deng G, Cotman CW. Transneuronal degeneration in the spread of Alzheimer's disease pathology: immunohistochemical evidence for the transmission of tau hyperphosphorylation. Neurobiol Dis 1997;4:365-75.

[28] Hertz L. Is Alzheimer's disease an anterograde degeneration, originating in the brainstem, and disrupting metabolic and functional interactions between neurons and glial cells? Brain Res Brain Res Rev 1989;14:335-53.

[29] Frost B, Jacks RL, Diamond MI. Propagation of tau misfolding from the outside to the inside of a cell. J Biol Chem 2009;284:12845-52.

[30] Santa-Maria I, Varghese M, Ksiezak-Reding H, *et al.* Paired helical filaments from Alzheimer disease brain induce intracellular accumulation of Tau protein in aggresomes. J Biol Chem 2012;287:20522-33.

[31] Guo JL, Lee VM. Seeding of normal Tau by pathological Tau conformers drives pathogenesis of Alzheimer-like tangles. J biol chem 2011;286:15317-31.

[32] Clavaguera F, Bolmont T, Crowther RA, *et al.* Transmission and spreading of tauopathy in transgenic mouse brain. Nature Cell Biology. 2009;11:909-13.

[33] de Calignon A, Polydoro M, Suarez-Calvet M, *et al.* Propagation of tau pathology in a model of early Alzheimer's disease. Neuron 2012;73:685-97.

[34] Santacruz K, Lewis J, Spires T, *et al.* Tau suppression in a neurodegenerative mouse model improves memory function. Science 2005;309:476-81.

[35] Yasuda M, Mayford MR. CaMKII activation in the entorhinal cortex disrupts previously encoded spatial memory. Neuron 2006;50:309-18.

[36] Chen ZL, Yoshida S, Kato K, *et al.* Expression and activity-dependent changes of a novel limbic-serine protease gene in the hippocampus. J Neurosci 1995;15:5088-97.

[37] Liu L, Drouet V, Wu JW, *et al.* Trans-synaptic spread of tau pathology *in vivo*. PLoS ONE 2012;7:e31302.

[38] Iba M, Guo JL, McBride JD, *et al.* Synthetic Tau Fibrils Mediate Transmission of Neurofibrillary Tangles in a Transgenic Mouse Model of Alzheimer's-Like Tauopathy. J Neurosci 2013;33:1024-37.

[39] Hanger DP, Betts JC, Loviny TL, Blackstock WP, Anderton BH. New phosphorylation sites identified in hyperphosphorylated tau (paired helical filament-tau) from Alzheimer's disease brain using nanoelectrospray mass spectrometry. J Neurochem 1998;71:2465-76.

[40] Alonso A, Zaidi T, Novak M, Grundke-Iqbal I, Iqbal K. Hyperphosphorylation induces self-assembly of tau into tangles of paired helical filaments/straight filaments. Proceedings of the National Academy of Sciences of the United States of America. 2001;98:6923-8.

[41] Hanger DP, Anderton BH, Noble W. Tau phosphorylation: the therapeutic challenge for neurodegenerative disease. Trends Mol Med 2009;15:112-9.

[42] Yamaguchi H, Ishiguro K, Uchida T, *et al.* Preferential labeling of Alzheimer neurofibrillary tangles with antisera for tau protein kinase (TPK) I/glycogen synthase kinase-3 beta and cyclin-dependent kinase 5, a component of TPK II. Acta Neuropathologica 1996;92:232-41.

[43] Ishiguro K, Shiratsuchi A, Sato S, *et al.* Glycogen synthase kinase 3 beta is identical to tau protein kinase I generating several epitopes of paired helical filaments. FEBS Lett 1993;325:167-72.

[44] Ishiguro K, Kobayashi S, Omori A, *et al.* Identification of the 23 kDa subunit of tau protein kinase II as a putative activator of cdk5 in bovine brain. FEBS Lett 1994;342:203-8.

[45] Pei JJ, Grundke-Iqbal I, Iqbal K, *et al.* Accumulation of cyclin-dependent kinase 5 (cdk5) in neurons with early stages of Alzheimer's disease neurofibrillary degeneration. Brain Research 1998;797:267-77.

[46] Pei JJ, Braak E, Braak H, *et al.* Distribution of active glycogen synthase kinase 3beta (GSK-3beta) in brains staged for Alzheimer disease neurofibrillary changes. Journal of neuropathology and experimental neurology 1999;58:1010-9.

[47] Kremer A, Louis JV, Jaworski T, Van Leuven F. GSK3 and Alzheimer's Disease: Facts and Fiction. Front Mol Neurosci 2011;4:17.

[48] Embi N, Rylatt DB, Cohen P. Glycogen synthase kinase-3 from rabbit skeletal muscle. Separation from cyclic-AMP-dependent protein kinase and phosphorylase kinase. Eur J Biochem 1980;107:519-27.

[49] Woodgett JR, Cohen P. Multisite phosphorylation of glycogen synthase. Molecular basis for the substrate specificity of glycogen synthase kinase-3 and casein kinase-II (glycogen synthase kinase-5). Biochim Biophys Acta 1984;788:339-47.

[50] Hoeflich KP, Luo J, Rubie EA, *et al.* Requirement for glycogen synthase kinase-3beta in cell survival and NF-kappaB activation. Nature 2000;406:86-90.

[51] MacAulay K, Doble BW, Patel S, *et al.* Glycogen synthase kinase 3alpha-specific regulation of murine hepatic glycogen metabolism. Cell metabolism 2007;6:329-37.

[52] Lovestone S, Reynolds CH, Latimer D, *et al.* Alzheimer's disease-like phosphorylation of the microtubule-associated protein tau by glycogen synthase kinase-3 in transfected mammalian cells. Curr Biol 1994;4:1077-86.

[53] Doble BW, Woodgett JR. GSK-3: tricks of the trade for a multi-tasking kinase. J Cell Sci 2003;116:1175-86.

[54] Pei JJ, Tanaka T, Tung YC, *et al.* Distribution, levels, and activity of glycogen synthase kinase-3 in the Alzheimer disease brain. J Neuropathol Exp Neurol 1997;56:70-8.

[55] Imahori K, Uchida T. Physiology and pathology of tau protein kinases in relation to Alzheimer's disease. J biochem 1997;121:179-88.

[56] Jope RS, Johnson GV. The glamour and gloom of glycogen synthase kinase-3. Trends in Biochemical Sciences 2004;29:95-102.

[57] Giese KP. GSK-3: a key player in neurodegeneration and memory. IUBMB Life 2009;61:516-21.

[58] Hurtado DE, Molina-Porcel L, Carroll JC, *et al.* Selectively Silencing GSK-3 Isoforms Reduces Plaques and Tangles in Mouse Models of Alzheimer's Disease. J Neurosci 2012;32:7392-402.

[59] Takashima A, Noguchi K, Sato K, Hoshino T, Imahori K. Tau protein kinase I is essential for amyloid beta-protein-induced neurotoxicity. Proceedings of the National Academy of Sciences of the United States of America 1993;90:7789-93.

[60] Takashima A, Noguchi K, Michel G, *et al.* Exposure of rat hippocampal neurons to amyloid beta peptide (25-35) induces the inactivation of phosphatidyl inositol-3 kinase and the activation of tau protein kinase I/glycogen synthase kinase-3 beta. Neuroscience Letters 1996;203:33-6.

[61] Hoshi M, Sato M, Matsumoto S, *et al.* Spherical aggregates of beta-amyloid (amylospheroid) show high neurotoxicity and activate tau protein kinase I/glycogen synthase kinase-3beta. Proceedings of the National Academy of Sciences of the United States of America. 2003;100:6370-5.

[62] Kim HS, Kim EM, Lee JP, *et al.* C-terminal fragments of amyloid precursor protein exert neurotoxicity by inducing glycogen synthase kinase-3beta expression. FASEB journal : official publication of the Federation of American Societies for Experimental Biology. 2003;17:1951-3.

[63] Phiel CJ, Wilson CA, Lee VM, Klein PS. GSK-3alpha regulates production of Alzheimer's disease amyloid-beta peptides. Nature 2003;423:435-9.

[64] Jaworski T, Dewachter I, Lechat B, *et al.* GSK-3alpha/beta kinases and amyloid production *in vivo*. Nature 2011;480:E4-5; discussion E6.

[65] Ma QL, Lim GP, Harris-White ME, *et al.* Antibodies against beta-amyloid reduce Abeta oligomers, glycogen synthase kinase-3beta activation and tau phosphorylation *in vivo* and *in vitro*. J Neurosci Res 2006;83:374-84.

[66] Engel T, Lucas JJ, Gomez-Ramos P, *et al.* Cooexpression of FTDP-17 tau and GSK-3beta in transgenic mice induce tau polymerization and neurodegeneration. Neurobiology of Aging 2006;27:1258-68.

[67] Engel T, Goni-Oliver P, Lucas JJ, Avila J, Hernandez F. Chronic lithium administration to FTDP-17 tau and GSK-3beta overexpressing mice prevents tau hyperphosphorylation and neurofibrillary tangle formation, but pre-formed neurofibrillary tangles do not revert. J Neurochem 2006;99:1445-55.

[68] Perez M, Hernandez F, Lim F, Diaz-Nido J, Avila J. Chronic lithium treatment decreases mutant tau protein aggregation in a transgenic mouse model. J Alzheimers Dis. 2003;5:301-8.

[69] Nakashima H, Ishihara T, Suguimoto P, *et al.* Chronic lithium treatment decreases tau lesions by promoting ubiquitination in a mouse model of tauopathies. Acta Neuropathol 2005;110:547-56.

[70] Noble W, Planel E, Zehr C, *et al.* Inhibition of glycogen synthase kinase-3 by lithium correlates with reduced tauopathy and degeneration *in vivo*. Proc Natl Acad Sci U S A 2005;102:6990-5.

[71] Leroy K, Ando K, Heraud C, *et al.* Lithium treatment arrests the development of neurofibrillary tangles in mutant tau transgenic mice with advanced neurofibrillary pathology. J Alzheimers Dis 2010;19:705-19.

[72] Shimada K, Motoi Y, Ishiguro K, *et al.* Long-term oral lithium treatment attenuates motor disturbance in tauopathy model mice: implications of autophagy promotion. Neurobiol Dis 2012;46:101-8.

[73] Hampel H, Ewers M, Burger K, *et al.* Lithium trial in Alzheimer's disease: a randomized, single-blind, placebo-controlled, multicenter 10-week study. J Clin Psychiatry 2009;70:922-31.

[74] Macdonald A, Briggs K, Poppe M, *et al.* A feasibility and tolerability study of lithium in Alzheimer's disease. Int J Geriatr Psychiatry 2008;23:704-11.

[75] Forlenza OV, Diniz BS, Radanovic M, *et al.* Disease-modifying properties of long-term lithium treatment for amnestic mild cognitive impairment: randomised controlled trial. Br J Psychiatry 2011;198:351-6.

[76] Palomo V, Perez DI, Perez C, *et al.* 5-imino-1,2,4-thiadiazoles: first small molecules as substrate competitive inhibitors of glycogen synthase kinase 3. J Med Chem 2012;55:1645-61.

[77] Dominguez JM, Fuertes A, Orozco L, *et al.* Evidence for irreversible inhibition of glycogen synthase kinase-3beta by tideglusib. J Biol Chem 2012;287:893-904.

[78] Sereno L, Coma M, Rodriguez M, *et al.* A novel GSK-3beta inhibitor reduces Alzheimer's pathology and rescues neuronal loss *in vivo*. Neurobiology of Disease 2009;35:359-67.

[79] Cruz JC, Tseng HC, Goldman JA, Shih H, Tsai LH. Aberrant Cdk5 activation by p25 triggers pathological events leading to neurodegeneration and neurofibrillary tangles. Neuron 2003;40:471-83.

[80] Glicksman MA, Cuny GD, Liu M, *et al.* New approaches to the discovery of cdk5 inhibitors. Curr Alzheimer Res 2007;4:547-9.

[81] Nikkel AL, Martino B, Markosyan S, *et al.* The novel calpain inhibitor A-705253 prevents stress-induced tau hyperphosphorylation *in vitro* and *in vivo*. Neuropharmacology 2012;63:606-12.

[82] Medeiros R, Kitazawa M, Chabrier MA, *et al.* Calpain inhibitor A-705253 mitigates Alzheimer's disease-like pathology and cognitive decline in aged 3xTgAD mice. Am J Pathol 2012;181:616-25.

[83] Gong CX, Lidsky T, Wegiel J, *et al.* Phosphorylation of microtubule-associated protein tau is regulated by protein phosphatase 2A in mammalian brain. Implications for neurofibrillary degeneration in Alzheimer's disease. J biol chem 2000;275:5535-44.

[84] Bennecib M, Gong CX, Grundke-Iqbal I, Iqbal K. Inhibition of PP-2A upregulates CaMKII in rat forebrain and induces hyperphosphorylation of tau at Ser 262/356. FEBS Letters 2001;490:15-22.

[85] Xu Y, Chen Y, Zhang P, Jeffrey PD, Shi Y. Structure of a protein phosphatase 2A holoenzyme: insights into B55-mediated Tau dephosphorylation. Mol Cell 2008;31:873-85.

[86] Vogelsberg-Ragaglia V, Schuck T, Trojanowski JQ, Lee VM. PP2A mRNA expression is quantitatively decreased in Alzheimer's disease hippocampus. Experimental Neurology 2001;168:402-12.

[87] Sontag E, Luangpirom A, Hladik C, *et al.* Altered expression levels of the protein phosphatase 2A ABalphaC enzyme are associated with Alzheimer disease pathology. J Neuropathol Exp Neurol 2004;63:287-301.

[88] Liu F, Grundke-Iqbal I, Iqbal K, Gong CX. Contributions of protein phosphatases PP1, PP2A, PP2B and PP5 to the regulation of tau phosphorylation. Euro j neurosci 2005;22:1942-50.

[89] Gong CX, Singh TJ, Grundke-Iqbal I, Iqbal K. Phosphoprotein phosphatase activities in Alzheimer disease brain. J Neurochem 1993;61:921-7.

[90] Matsuo ES, Shin RW, Billingsley ML, *et al.* Biopsy-derived adult human brain tau is phosphorylated at many of the same sites as Alzheimer's disease paired helical filament tau. Neuron 1994;13:989-1002.
[91] Gong CX, Shaikh S, Wang JZ, *et al.* Phosphatase activity toward abnormally phosphorylated tau: decrease in Alzheimer disease brain. J Neurochem 1995;65:732-8.
[92] Lechward K, Awotunde OS, Swiatek W, Muszynska G. Protein phosphatase 2A: variety of forms and diversity of functions. Acta biochimica Polonica 2001;48:921-33.
[93] Voronkov M, Braithwaite SP, Stock JB. Phosphoprotein phosphatase 2A: a novel druggable target for Alzheimer's disease. Future Med Chem 2011;3:821-33.
[94] Sontag E, Nunbhakdi-Craig V, Sontag JM, *et al.* Protein phosphatase 2A methyltransferase links homocysteine metabolism with tau and amyloid precursor protein regulation. J Neurosci 2007;27:2751-9.
[95] Sontag JM, Nunbhakdi-Craig V, Montgomery L, *et al.* Folate deficiency induces *in vitro* and mouse brain region-specific downregulation of leucine carboxyl methyltransferase-1 and protein phosphatase 2A B(alpha) subunit expression that correlate with enhanced tau phosphorylation. J Neurosci 2008;28:11477-87.
[96] Shea TB, Lyons-Weiler J, Rogers E. Homocysteine, folate deprivation and Alzheimer neuropathology. J. Alzh. Dis : JAD 2002;4:261-7.
[97] Kim JM, Stewart R, Kim SW, *et al.* Changes in folate, vitamin B12 and homocysteine associated with incident dementia. J Neurol Neurosurg Psychiatry 2008;79:864-8.
[98] Chohan MO, Khatoon S, Iqbal IG, Iqbal K. Involvement of I2PP2A in the abnormal hyperphosphorylation of tau and its reversal by Memantine. FEBS Lett 2006;580:3973-9.
[99] Degerman Gunnarsson M, Kilander L, Basun H, Lannfelt L. Reduction of phosphorylated tau during memantine treatment of Alzheimer's disease. Dem Geriat Cognit Disord 2007;24:247-52.
[100] van Eersel J, Ke YD, Liu X, *et al.* Sodium selenate mitigates tau pathology, neurodegeneration, and functional deficits in Alzheimer's disease models. Proc Natl Acad Sci U S A 2010;107:13888-93.
[101] Corcoran NM, Martin D, Hutter-Paier B, *et al.* Sodium selenate specifically activates PP2A phosphatase, dephosphorylates tau and reverses memory deficits in an Alzheimer's disease model. J Clin Neurosci 2010;17:1025-33.
[102] Arnold CS, Johnson GV, Cole RN, *et al.* The microtubule-associated protein tau is extensively modified with O-linked N-acetylglucosamine. J Biol Chem 1996;271:28741-4.
[103] Liu F, Iqbal K, Grundke-Iqbal I, Hart GW, Gong CX. O-GlcNAcylation regulates phosphorylation of tau: a mechanism involved in Alzheimer's disease. Proc Natl Acad Sci U S A 2004;101:10804-9.
[104] Yuzwa SA, Macauley MS, Heinonen JE, *et al.* A potent mechanism-inspired O-GlcNAcase inhibitor that blocks phosphorylation of tau *in vivo*. Nat Chem Biol 2008;4:483-90.
[105] Yuzwa SA, Shan X, Macauley MS, *et al.* Increasing O-GlcNAc slows neurodegeneration and stabilizes tau against aggregation. Nat Chem Biol 2012;8:393-9.
[106] Liu F, Zaidi T, Iqbal K, *et al.* Role of glycosylation in hyperphosphorylation of tau in Alzheimer's disease. FEBS Lett 2002;512:101-6.
[107] Ledesma MD, Bonay P, Colaco C, Avila J. Analysis of microtubule-associated protein tau glycation in paired helical filaments. J Biol Chem 1994;269:21614-9.
[108] Yan SD, Yan SF, Chen X, *et al.* Non-enzymatically glycated tau in Alzheimer's disease induces neuronal oxidant stress resulting in cytokine gene expression and release of amyloid beta-peptide. Nat Med 1995;1:693-9.

[109] Min SW, Cho SH, Zhou Y, *et al.* Acetylation of tau inhibits its degradation and contributes to tauopathy. Neuron 2010;67:953-66.
[110] Cohen TJ, Guo JL, Hurtado DE, *et al.* The acetylation of tau inhibits its function and promotes pathological tau aggregation. Nat Commun 2011;2:252.
[111] Irwin DJ, Cohen TJ, Grossman M, *et al.* Acetylated tau, a novel pathological signature in Alzheimer's disease and other tauopathies. Brain 2012;135:807-18.
[112] Julien C, Tremblay C, Emond V, *et al.* Sirtuin 1 reduction parallels the accumulation of tau in Alzheimer disease. J Neuropathol Exp Neurol 2009;68:48-58.
[113] Kim D, Nguyen MD, Dobbin MM, *et al.* SIRT1 deacetylase protects against neurodegeneration in models for Alzheimer's disease and amyotrophic lateral sclerosis. EMBO J 2007;26:3169-79.
[114] Smith MA, Richey Harris PL, Sayre LM, Beckman JS, Perry G. Widespread peroxynitrite-mediated damage in Alzheimer's disease. J Neurosci 1997;17:2653-7.
[115] Reynolds MR, Reyes JF, Fu Y, *et al.* Tau nitration occurs at tyrosine 29 in the fibrillar lesions of Alzheimer's disease and other tauopathies. The Journal of neuroscience : the official journal of the Society for Neuroscience 2006;26:10636-45.
[116] Reyes JF, Geula C, Vana L, Binder LI. Selective tau tyrosine nitration in non-AD tauopathies. Acta Neuropathol 2012;123:119-32.
[117] Cash AD, Aliev G, Siedlak SL, *et al.* Microtubule reduction in Alzheimer's disease and aging is independent of tau filament formation. The American journal of pathology 2003;162:1623-7.
[118] Stokin GB, Lillo C, Falzone TL, *et al.* Axonopathy and transport deficits early in the pathogenesis of Alzheimer's disease. Science 2005;307:1282-8.
[119] Ishihara T, Hong M, Zhang B, *et al.* Age-dependent emergence and progression of a tauopathy in transgenic mice overexpressing the shortest human tau isoform. Neuron 1999;24:751-62.
[120] Zhang B. Microtubule-binding drugs offset tau sequestration by stabilizing microtubules and reversing fast axonal transport deficits in a tauopathy model. Proc Natl Acad Sci 2005;102:227-31.
[121] Brunden KR, Zhang B, Carroll J, *et al.* Epothilone D improves microtubule density, axonal integrity, and cognition in a transgenic mouse model of tauopathy. J Neurosci 2010;30:13861-6.
[122] Zhang B, Carroll J, Trojanowski JQ, *et al.* The microtubule-stabilizing agent, epothilone d, reduces axonal dysfunction, neurotoxicity, cognitive deficits, and Alzheimer-like pathology in an interventional study with aged tau transgenic mice. The Journal of neuroscience : the official journal of the Society for Neuroscience 2012;32:3601-11.
[123] Barten DM, Fanara P, Andorfer C, *et al.* Hyperdynamic Microtubules, Cognitive Deficits, and Pathology Are Improved in Tau Transgenic Mice with Low Doses of the Microtubule-Stabilizing Agent BMS-241027. J Neurosci 2012;32:7137-45.
[124] Bassan M, Zamostiano R, Davidson A, *et al.* Complete sequence of a novel protein containing a femtomolar-activity-dependent neuroprotective peptide. J Neurochem 1999;72:1283-93.
[125] Crowe A, Ballatore C, Hyde E, Trojanowski J, Lee V. High throughput screening for small molecule inhibitors of heparin-induced tau fibril formation. Biochemical and Biophysical Research Communications 2007;358:1-6.

[126] Bulic B, Pickhardt M, Mandelkow EM, Mandelkow E. Tau protein and tau aggregation inhibitors. Neuropharmacology 2010;59:276-89.
[127] Wischik CM, Edwards PC, Lai RY, Roth M, Harrington CR. Selective inhibition of Alzheimer disease-like tau aggregation by phenothiazines. Proc Natl Acad Sci U S A 1996;93:11213-8.
[128] van Bebber F, Paquet D, Hruscha A, Schmid B, Haass C. Methylene blue fails to inhibit Tau and polyglutamine protein dependent toxicity in zebrafish. Neurobiology of Disease 2010;39:265-71.
[129] Fatouros C, Pir GJ, Biernat J, *et al.* Inhibition of tau aggregation in a novel Caenorhabditis elegans model of tauopathy mitigates proteotoxicity. Hum Mol Genet 2012;21:3587-603.
[130] O'Leary JC, 3rd, Li Q, Marinec P, *et al.* Phenothiazine-mediated rescue of cognition in tau transgenic mice requires neuroprotection and reduced soluble tau burden. Molecul Neurodegen 2010;5:45.
[131] Wischik C, Staff R. Challenges in the conduct of disease-modifying trials in AD: practical experience from a phase 2 trial of Tau-aggregation inhibitor therapy. The journal of nutrition, health & aging 2009;13:367-9.
[132] Crowe A, Huang W, Ballatore C, *et al.* Identification of aminothienopyridazine inhibitors of tau assembly by quantitative high-throughput screening. Biochem 2009;48:7732-45.
[133] Nixon RA, Yang DS. Autophagy failure in Alzheimer's disease--locating the primary defect. Neurobiol Dis 2011;43:38-45.
[134] Petrucelli L, Dickson D, Kehoe K, *et al.* CHIP and Hsp70 regulate tau ubiquitination, degradation and aggregation. Human Molecular Genetics 2004;13:703-14.
[135] Sahara N, Murayama M, Mizoroki T, *et al. In vivo* evidence of CHIP up-regulation attenuating tau aggregation. J Neurochem 2005;94:1254-63.
[136] Sarkar M, Kuret J, Lee G. Two motifs within the tau microtubule-binding domain mediate its association with the hsc70 molecular chaperone. J Neurosci Res 2008;86:2763-73.
[137] Dou F, Netzer WJ, Tanemura K, *et al.* Chaperones increase association of tau protein with microtubules. Proc Natl Acad Sci U S A 2003;100:721-6.
[138] Jinwal UK, O'Leary JC, 3rd, Borysov SI, *et al.* Hsc70 rapidly engages tau after microtubule destabilization. J Biol Chem 2010;285:16798-805.
[139] Ciechanover A, Brundin P. The ubiquitin proteasome system in neurodegenerative diseases: sometimes the chicken, sometimes the egg. Neuron 2003;40:427-46.
[140] Dickey CA, Kamal A, Lundgren K, *et al.* The high-affinity HSP90-CHIP complex recognizes and selectively degrades phosphorylated tau client proteins. The Journal of Clinical Investigation 2007;117:648-58.
[141] Dickey CA, Yue M, Lin WL, *et al.* Deletion of the ubiquitin ligase CHIP leads to the accumulation, but not the aggregation, of both endogenous phospho- and caspase-3-cleaved tau species. J Neurosci 2006;26:6985-96.
[142] Stromer T, Ehrnsperger M, Gaestel M, Buchner J. Analysis of the interaction of small heat shock proteins with unfolding proteins. J Biol Chem 2003;278:18015-21.
[143] Abisambra JF, Blair LJ, Hill SE, *et al.* Phosphorylation dynamics regulate Hsp27-mediated rescue of neuronal plasticity deficits in tau transgenic mice. J Neurosci 2010;30:15374-82.
[144] Sugata H, Matsuo K, Nakagawa T, *et al.* A peptidyl-prolyl isomerase, FKBP12, accumulates in Alzheimer neurofibrillary tangles. Neurosci Lett 2009;459:96-9.
[145] Jinwal UK, Koren J, 3rd, Borysov SI, *et al.* The Hsp90 cochaperone, FKBP51, increases Tau stability and polymerizes microtubules. J Neurosci 2010;30:591-9.

[146] Chambraud B, Sardin E, Giustiniani J, *et al.* A role for FKBP52 in Tau protein function. Proc Natl Acad Sci U S A 2010;107:2658-63.

[147] Berger Z, Ravikumar B, Menzies FM, *et al.* Rapamycin alleviates toxicity of different aggregate-prone proteins. Human Molecular Genetics 2006;15:433-42.

[148] Wang Y, Martinez-Vicente M, Kruger U, *et al.* Tau fragmentation, aggregation and clearance: the dual role of lysosomal processing. Human Molecular Genetics 2009;18:4153-70.

[149] Dolan PJ, Johnson GV. A caspase cleaved form of tau is preferentially degraded through the autophagy pathway. The Journal of biological chemistry 2010;285:21978-87.

[150] Schaeffer V, Lavenir I, Ozcelik S, *et al.* Stimulation of autophagy reduces neurodegeneration in a mouse model of human tauopathy. Brain 2012;135:2169-77.

[151] Roberson ED, Scearce-Levie K, Palop JJ, *et al.* Reducing endogenous tau ameliorates amyloid beta-induced deficits in an Alzheimer's disease mouse model. Science 2007;316:750-4.

[152] Dawson HN, Cantillana V, Jansen M, *et al.* Loss of tau elicits axonal degeneration in a mouse model of Alzheimer's disease. Neuroscience 2010;169:516-31.

[153] Lei P, Ayton S, Finkelstein DI, *et al.* Tau deficiency induces parkinsonism with dementia by impairing APP-mediated iron export. Nature Medicine 2012;18:291-5.

[154] Sydow A, Van der Jeugd A, Zheng F, *et al.* Tau-induced defects in synaptic plasticity, learning, and memory are reversible in transgenic mice after switching off the toxic Tau mutant. The Journal of neuroscience : the official journal of the Society for Neuroscience 2011;31:2511-25.

[155] Van der Jeugd A, Hochgrafe K, Ahmed T, *et al.* Cognitive defects are reversible in inducible mice expressing pro-aggregant full-length human Tau. Acta Neuropathol 2012;123:787-805.

[156] Reger MA, Watson GS, Green PS, *et al.* Intranasal insulin administration dose-dependently modulates verbal memory and plasma amyloid-beta in memory-impaired older adults. J Alzheimers Dis 2008;13:323-31.

[157] Reger MA, Watson GS, Green PS, *et al.* Intranasal insulin improves cognition and modulates beta-amyloid in early AD. Neurology 2008;70:440-8.

[158] Kfoury N, Holmes BB, Jiang H, Holtzman DM, Diamond MI. Trans-cellular propagation of Tau aggregation by fibrillar species. J Biol Chem 2012;287:19440-51.

[159] Rosenmann H, Grigoriadis N, Karussis D, *et al.* Tauopathy-like Abnormalities and Neurologic Deficits in Mice Immunized With Neuronal Tau Protein. Arch Neurol 2006;63:1459-67.

[160] Asuni AA, Boutajangout A, Quartermain D, Sigurdsson EM. Immunotherapy Targeting Pathological Tau Conformers in a Tangle Mouse Model Reduces Brain Pathology with Associated Functional Improvements. J Neurosci 2007;27:9115-29.

[161] Bi M, Ittner A, Ke YD, Gotz J, Ittner LM. Tau-targeted immunization impedes progression of neurofibrillary histopathology in aged P301L tau transgenic mice. PLoS ONE 2011;6:e26860.

[162] Boutajangout A, Ingadottir J, Davies P, Sigurdsson EM. Passive immunization targeting pathological phospho-tau protein in a mouse model reduces functional decline and clears tau aggregates from the brain. J Neurochem 2011;118:658-67.

[163] Chai X, Wu S, Murray TK, *et al.* Passive immunization with anti-Tau antibodies in two transgenic models: reduction of Tau pathology and delay of disease progression. J Biol Chem 2011;286:34457-67.

[164] Jicha GA, Weaver C, Lane E, *et al.* cAMP-dependent protein kinase phosphorylations on tau in Alzheimer's disease. J Neurosci 1999;19:7486-94.

[165] Porto-Carreiro I, Fevrier B, Paquet S, Vilette D, Raposo G. Prions and exosomes: from PrPc trafficking to PrPsc propagation. Blood cells, molecules & diseases 2005;35:143-8.

[166] Vella LJ, Sharples RA, Nisbet RM, Cappai R, Hill AF. The role of exosomes in the processing of proteins associated with neurodegenerative diseases. Eur Biophys J 2008;37:323-32.

[167] Gerdes HH. Prions tunnel between cells. Nat Cell Biol 2009;11:235-6.

[168] Gousset K, Zurzolo C. Tunnelling nanotubes: a highway for prion spreading? Prion 2009;3:94-8.

[169] Akiyama H, Barger S, Barnum S, *et al.* Inflammation and Alzheimer's disease. Neurobiol Aging 2000;21:383-421.

[170] Ishizawa K, Dickson DW. Microglial activation parallels system degeneration in progressive supranuclear palsy and corticobasal degeneration. J Neuropathol Exp Neurol 2001;60:647-57.

[171] Gerhard A, Trender-Gerhard I, Turkheimer F, *et al. In vivo* imaging of microglial activation with [11C](R)-PK11195 PET in progressive supranuclear palsy. Mov Disord 2006;21:89-93.

[172] Cras P, Kawai M, Siedlak S, Perry G. Microglia are associated with the extracellular neurofibrillary tangles of Alzheimer disease. Brain Research 1991;558:312-4.

[173] Sheffield LG, Marquis JG, Berman NE. Regional distribution of cortical microglia parallels that of neurofibrillary tangles in Alzheimer's disease. Neuroscience Letters 2000;285:165-8.

[174] Sheng JG, Mrak RE, Griffin WS. Glial-neuronal interactions in Alzheimer disease: progressive association of IL-1alpha+ microglia and S100beta+ astrocytes with neurofibrillary tangle stages. Journal of neuropathology and experimental neurology 1997;56:285-90.

[175] Bellucci A, Westwood AJ, Ingram E, *et al.* Induction of Inflammatory Mediators and Microglial Activation in Mice Transgenic for Mutant Human P301S Tau Protein. Am J Pathol 2004;165:1643-52.

[176] Ikeda M, Shoji M, Kawarai T, *et al.* Accumulation of filamentous tau in the cerebral cortex of human tau R406W transgenic mice. Am J Pathol 2005;166:521-31.

[177] Sasaki A, Kawarabayashi T, Murakami T, *et al.* Microglial activation in brain lesions with tau deposits: comparison of human tauopathies and tau transgenic mice TgTauP301L. Brain Res 2008;1214:159-68.

[178] Zilka N, Stozicka Z, Kovac A, *et al.* Human misfolded truncated tau protein promotes activation of microglia and leukocyte infiltration in the transgenic rat model of tauopathy. J Neuroimmunol 2009;209:16-25.

[179] Cameron B, Landreth GE. Inflammation, microglia, and Alzheimer's disease. Neurobiol Dis 2010;37:503-9.

[180] Profenno LA, Porsteinsson AP, Faraone SV. Meta-analysis of Alzheimer's disease risk with obesity, diabetes, and related disorders. Biol Psychiatry 2010;67:505-12.

[181] in t' Veld BA, Ruitenberg A, Hofman A, *et al.* Nonsteroidal antiinflammatory drugs and the risk of Alzheimer's disease. N Engl J Med 2001;345:1515-21.

[182] Bhaskar K, Konerth M, Kokiko-Cochran ON, *et al.* Regulation of tau pathology by the microglial fractalkine receptor. Neuron 2010;68:19-31.

[183] Liu Z, Condello C, Schain A, Harb R, Grutzendler J. CX3CR1 in microglia regulates brain amyloid deposition through selective protofibrillar amyloid-beta phagocytosis. J Neurosci 2010;30:17091-101.

[184] Lee S, Varvel NH, Konerth ME, *et al.* CX3CR1 deficiency alters microglial activation and reduces beta-amyloid deposition in two Alzheimer's disease mouse models. Am J Pathol 2010;177:2549-62.

[185] Arnaud LT, Myeku N, Figueiredo-Pereira ME. Proteasome-caspase-cathepsin sequence leading to tau pathology induced by prostaglandin J2 in neuronal cells. J Neurochem 2009;110:328-42.

[186] Yoshiyama Y, Kojima A, Ishikawa C, Arai K. Anti-inflammatory action of donepezil ameliorates tau pathology, synaptic loss, and neurodegeneration in a tauopathy mouse model. J Alzheimers Dis 2010;22:295-306.

[187] Yoshiyama Y, Kojima A, Itoh K, Uchiyama T, Arai K. Anticholinergics boost the pathological process of neurodegeneration with increased inflammation in a tauopathy mouse model. Neurobiology of Disease 2012;45:329-36.

[188] Wang H, Yu M, Ochani M, *et al.* Nicotinic acetylcholine receptor alpha7 subunit is an essential regulator of inflammation. Nature 2003;421:384-8.

[189] Breitner JC, Baker LD, Montine TJ, *et al.* Extended results of the Alzheimer's disease anti-inflammatory prevention trial. Alzheimer's & dementia : the journal of the Alzheimer's Association 2011;7:402-11.

[190] Tobinick E, Gross H, Weinberger A, Cohen H. TNF-alpha modulation for treatment of Alzheimer's disease: a 6-month pilot study. MedGenMed 2006;8:25.

[191] Tobinick EL, Gross H. Rapid cognitive improvement in Alzheimer's disease following perispinal etanercept administration. J Neuroinflammation 2008;5:2.

Send Orders of Reprints at reprints@benthamscience.net
Frontiers in Clinical Drug Research – Alzheimer Disorders, Vol. 1, **2013, 183-216** **183**

CHAPTER 5

Different Approaches in the Treatment of Alzheimer's Disease: Comprehensive Insight

Zaciragic Asija*

Department of Physiology, Medical Faculty University of Sarajevo, Bosnia and Herzegovina

Abstract: Alzheimer's disease (AD) is a progressive and most common form of dementia. Pathogenesis of AD is multifactorial and yet fully unknown. Proper and on-time diagnosis of this disease is pivotal but however not easy to achieve. Studies indicate that early diagnosis of AD provides best treatment outcomes. Risks for AD include age, genetic, environmental, vascular, life-style factors and low educational income. Novel findings suggest that inflammation, oxidative stress as well as disturbance of immune system may also play significant role in the development of AD. This disease is neuropathologically characterized by presence of neurofibrillary tangles and senile neuritic plaques, amyloid-β peptid deposits and widespread neuronal loss. Main symptoms of AD are cognitive decline, memory loss, altered behavior and language deficit. As they progress in time, these symptoms lead to severe impairment in daily functioning and in later stages AD patients require total care. The aim of this review is to provide a comprehensive insight into the different, currently available, approaches in the treatment of AD. Furthermore, review presents findings of ongoing clinical trials and drug research studies in advanced stages of development and their application in the prevention and treatment of AD.

Based on currently available FDA guidelines, treatment options for AD include acetycholine esterase inhibitors for mild to moderate cases and memantine, an N-methyl-D-aspartate receptor antagonist, for moderate to severe forms of AD. Alongside with symptomatic drugs, novel disease modifying approaches are currently tested such as amyloid-based therapies, inhibitor and/or modulators of secretases, statins, inhibitors of receptors for advanced glycation end products, aggregation inhibitors, erythropoetin derivatives, peroxisome proliferator-activated receptor-gamma agonists and immunotherapy. Furthermore, therapeutic targets in the treatment of AD include approaches directed against tau protein such as tau protein aggregation inhibitors and inhibitors of tau kinases. Anti-inflammatory drugs and antioxidants are also used as neuroprotective advancements in AD therapy. Finally, alteration of ion homeostasis and neuronal regeneration approaches are currently implemented in the attempt to modify AD progression, stop neurodegeneration or enhance neurogenesis.

***Address correspondence to Zaciragic Asija:** Medical Faculty University of Sarajevo, Cekalusa 90, 71000 Sarajevo, Bosnia and Herzegovina; Tel/Fax: 00 387 33 226 472; Ext. 135; E-mail: zarias@lol.ba

Even though AD treatment options as stated above are numerous,in majority of cases these drugs produce only modest improvements of AD symptoms and none of the available treatment options can cure or stop disease progression. Only full and complete knowledge of its etiology will provide favorable results in AD prevention and therapy.

Keywords: Alzheimer's disease, Alzheimer's disease treatment, Alzheimer's disease clinical trials, symptomatic drugs, disease modifying approaches, amyloid-based therapies, approaches directed against tau protein, neuroprotective advancements in AD therapy, alteration of ion homeostasis and neuronal regeneration approaches.

INTRODUCTION

AD represents neurodegenerative disorder that affects nearly 2% of the population in industrialized countries. In individuals older then 70 years the risk of AD significantly increases but is of note that in older people there can be other causes of memory loss [1].

AD comprises approximately 70 % out of all cases of dementia. Incidence of AD increases with age, doubling every five to ten years. Prevalence also increases exponentially with age, rising from 3% among those 65–74, to almost 50% among those 85 or older. Based on a data from 2010, AD affects 25 million people worldwide. The number of people aged 80 years and older is also expected to double, from 9.3 million in 2000 to 19.5 million in 2030. Globally, the number of older adults (aged 65 years and older) is projected to increase even more dramatically—more than doubling from 420 million in 2000 to 973 million in 2030 [2]. Since advanced age is the most significant factor for AD, medical and other kinds of costs of this disease can not be overlooked.

AD has enormous impact worldwide on the affected individuals, caregivers and overall public health care system. Individuals suffering from AD have significantly shortened life expectancy, their quality of life is drastically decreased and in many cases they require institutionalization caused by their physical disability. Studies have shown that AD patients have two-to fivefold increased risk of death and median survival time for newly diagnosed AD patients is from 3 to 6 years. Shorter survival is often reported for those AD patients that have poorer cognitive function, low education level, and comorbidities such as heart disease, hypertension and diabetes. Furthermore, white race, older age and male sex are associated with shorter life expectancy in individuals that suffer from AD [3]. Alzheimer's Association data from 2009 have revealed that in US the annual costs for patients with AD and other forms of dementia were estimated to be US$148 billion plus US$94 billion of unpaid care service [4]. Having in mind progressive nature of AD and a fact that its adequate treatments are lacking, this disease represents tremendous burden for family members and overall society since AD patients require permanent care and therapy.

Different treatment approaches are currently tested in the treatment of AD. Those mainly include approaches targeted on Aβ that are based on an increase of its clearance, decrease of its production, and other endeavors that aim to ameliorate toxic actions of amyloid cascade. Furthermore, there are approaches that aim to modify actions of various neurotransmitters, tau effects, inflammation, and many others. However, there are many avenues that are tried in the attempt to modify course and progression of AD, still none of the currently used treatment approaches have given optimal and fully successful results. Possible explanations for the lack of desired results may be not enough studies with demonstrated pharmacodynamic activity, treatment of the disease in the late phase of its progression, derisory preclinical models, insufficient number of trials with completed phase III studies, or simply targeting of unimportant mechanisms [5].

At present there are no treatments that can stop or reverse neurodegeneration typical for AD. However, better understanding of AD etiology and pathophysiology will lead to the discovery of effective preventive and therapeutic approaches that will reduce incidence and improve treatment of AD.

The aim of this review is to provide a comprehensive insight into the different, currently available approaches in the treatment of AD. Furthermore, review presents findings of ongoing clinical trials and drug research studies in advanced stages of development and their application in the prevention and treatment of AD.

Incidence and Prevalence of AD

Presently, it is estimated that the number of AD patients worldwide is 35.6 million and it is expected that this number will rise by 2030 to 65.7 million. According to current predictions largest increase of AD cases will be observed in low and middle income countries [6].

Results of recent population-based studies from Europe have revealed that prevalence of AD in people older than 65 was 4.4% and that incidence rate of AD in a same age group was 19.4 per 1000 person-years. Prevalence of AD in United States in people older than 70 is 9.7% and two large community-based studies in US Baltimore and Seattle area have shown that incidence rate for AD in individuals older than 65 was 15.0 per 1000 person-years. Epidemiological studies conducted in developing countries have shown that overall prevalence of AD is 3.4% and generally incidence rate of AD in developing countries is lower than in Europe and North America [3].

Important question that remains to be elucidated regarding better understanding of AD etiology and epidemiology, and projection of its burden to the society is whether the incidence of this disease continues to increase with advancing age or it reaches a plateau at a certain age. Furthermore, findings of numerous studies are suggesting that there are geographical variations in AD incidence rate. For instance, it has been demonstrated that higher incidence rate of AD has been observed in European north-western countries than in southern European countries. Studies have also shown that incidence rate of AD in North America is slightly lower than in Europe [7]. However, possible explanation for observed geographical variations in AD incidence rate may be different approaches in study designs and different procedures in case analysis.

AD Risk Factors

Genetic factors are important determinants of AD development. However, majority of AD cases are sporadic and do not run in families. Molecular genetic analysis has shown that there are many genes that determine one's susceptibility to AD. One of the first susceptibility genes that was identified was apolipoprotein E that has three isoforms (E2, E3, E4). Increased risk of AD was determined in individuals that produce E4 isoform, whereas E2 isoform is associated with lowest AD risk. Even though exact mechanism of E4 isoform action is still not discovered, evidences have shown that E4 may increase the risk of AD by impairing neuronal plasticity, and increasing oxidative stress that lead to the promotion of cerebrovascular pathology. For late onset of AD, another susceptibility locus has been localized on chromosome 10, but responsible gene has not yet been identified [1].

Genetic etiology of AD is in the focus of intensive research and based on current evidence from the genetic point of view AD is a complex disease. Studies have shown that mutations in three genes including presenilin 1, presenilin 2 and amyloid precursor protein (APP) lead to early onset of familial AD that occurs prior to age 65 [8].

Generally, it can be stated that there are two main forms of AD: familial Alzheimer's disease that has genetic background and its characterized with early onset and sporadic form Alzheimer's' disease that is age-depended and more prevalent.

Inherited forms of AD account for up to 10% of all AD cases. Other forms of AD include familial and early-onset AD that may encompass autosomal-dominant form of AD, but as well may include non-dominant causes of AD such as apolipoprotein E4 allele mutation or sporadic form of AD [5]. Studies have demonstrated that mutations on chromosomes 21, 1 and 14 could lead to the development of familial AD and main feature of this AD form is increase in Aß peptide production [9].

Alongside with genetics, key risk factors for AD development are environmental factors including dietary, behavioral and lifestyle factors. However, even though there are numerous evidences that support importance of these factors in AD development strong and definitive recommendations in this area still can not be made.

Besides well established risk factors such as older age and genetic susceptibility, other etiological factors still remain to be fully determined. At present, possible culprits are vascular factors such as cerebrovascular lesions, diminished vascular function, elevated LDL cholesterol, diabetes, obesity, high blood pressure and cigarette smoking. Many authors believe that important determinants in AD development and its clinical manifestations are psychosocial factors namely low education level, insufficient physical activity, lack of mentally simulating exercises and poor social engagement. For factors such as nutritional (dietary), inflammatory or occupational there are still not enough evidence to support their role in AD development [10]. However, results from few investigations point to the significance of immune and metabolic factors in AD pathogenesis. In this line of evidence, a study conducted at our Institute showed increased concentration of C-reactive protein (CRP), a low-grade inflammatory marker, in patients suffering from AD compared to healthy age-matched individuals [11]. Other studies have determined higher values of this marker in midlife and this CRP elevation was associated with increased risk of AD, suggesting that inflammatory mechanisms are linked to dementia, and that they can be measured in blood even before dementia is manifested [12]. In general, role of inflammatory mechanisms in the development of AD is still under intense debate, but results obtained from different studies have shown that products of apoptosis and Aβ aggregates can induce local inflammation by actions of activated astrocytes and microglia that in turn intensify neuronal death. Local inflammatory mediators and neurotoxic factors may increase production of Aβ and therefore contribute to AD pathology [9]. Furthermore, important feature of AD represents molecular events such as reduced neuronal energy production and increased oxidative stress. Novel findings suggest that these molecular events have significant role in the onset, development and overall pathogenesis of AD.

Recent cross-sectional study conducted in USA that included AD patients with different ethnoracial background compared cognitive, mood, and functional impairments, age of AD onset, as well as the frequencies of apolipoprotein E genotypes. Results have shown significant differences among different ethnoracial groups in nearly all tested variables. One of the main conclusions of this study was that Latinos exhibit a more severe profile of AD compared to White non-

Hispanic group and African Americans that could not be accounted by difference in lag time between onset of symptoms and initial evaluation [13].

Pathohistological Characteristics and Symptoms of AD

'Mild cognitive impairment' (MCI) often precedes AD, and by widely accepted definition it represents a transitional state between normal aging and developed dementia. It is characterized by greater cognitive decline then expected for the age of the affected individuals but with still preserved activities of daily functioning. Between 3% and 19% of adults over the age of 65 suffer from this condition, and according to current evidence more than a half of individuals with MCI will develop dementia within 5 years [6]. Significant efforts are made in early diagnosis of this condition and in its proper pharmacotherapy so that development of dementia in MCI suffering patients is prevented.

As there is yet no reliable peripheral biochemical marker for the AD, definitive diagnosis of AD can only be confirmed by postmortem examination of the brain that must contain sufficient number of plaques and tangles. Extracellular deposits of fibrils and amorphous aggregates of amyloid β-peptide (Aβ) form plaques, and intracellular fibrillar aggregates of the microtubule-associated protein tau which exhibit hyperphosphorylation and oxidative modifications form neurofibrillary tangles. Plaques and tangles are present mainly in brain regions involved in learning, memory and emotional behaviors such as the entorhinal cortex, hippocampus, basal forebrain and amygdala. Pivotal characteristic of AD brain is altered proteolytic processing of amyloid precursor protein (APP) that leads to the production and aggregation of neurotoxic forms of Aβ. Studies have shown that affected regions of the brain have reduced number of synapses, and that neurites in these regions are often damaged. It has also been demonstrated that neurons which utilize glutamate or acetylcholine as neurotransmitters are predominantly affected, but neurons that employ serotonin and norepinephrine are also damaged. Furthermore, impaired energy metabolism, increased oxidative stress and disturbed cellular calcium homeostasis characterize neurons that degenerate in AD [14].

Disturbed energy metabolism, ion homeostasis and oxidative damage due to excessive production of highly reactive oxygen species are important features of

AD brain. All of these processes make neurons prone to apoptosis and exitotoxicity. Exposure of synapses to Aβ leads to theirs degeneration and dysfunction that manifests as impaired function of glutamate transporters and membrane ions followed with compromised mitochondrial function [15].

Glucose utilization in AD patients is deficient and studies have shown that this symptom occurs prior to the onset of AD clinical symptoms. There are supporting evidences that insulin resistance, as well as abnormal glucose regulation may be a risk factor for AD. *In vivo* studies on APP mutant mice have revealed that glucose metabolism is decreased in those animals with increased amounts of Aβ in cognition-related brain regions. Furthermore, decreased activities of enzymes that are critical in energy metabolism such as pyruvate dehydrogenase complex, α-ketoglutarate dehydrogenase complex and cytochrome c oxidase have been established [16].

Role of increased oxidative stress in AD has been studied for many years. Studies have shown that oxidatively modified proteins, lipids and DNA in extremely high amounts are present in the cells of AD brain, especially in those regions where plaques and tangles are prominent. Main sources of oxidative stress are Aβ and redox-active metals such as Fe^{2+} and Cu^{+}. It has been suggested that disturbed energy metabolism and increased oxidative stress that increase with advancing age may be promoters of amyloidogenesis since they can provoke altered processing of APP that results in formation and accumulation of neurotoxic forms of Aβ [15].

Essential role in memory and learning abilities play calcium (Ca^{2+}) ions. The dysfunction and death of neuron which is typical for AD brain may result as a consequence of disturbed calcium homeostasis. Numerous findings support the importance of perturbed calcium regulation in the development in both familial and sporadic forms of AD. Studies have shown that tangles in AD brain have increased amounts of Ca^{2+}, as well as hyperactivation of Ca^{2+}-dependent proteases [17]. Altered APP processing is known to destabilize calcium homeostasis. Increase of calcium influx and impairment of calcium pumps may be result of oxidative stress that is induced by Aβ.

Dysfunction of mitochondria is another feature of AD disturbed brain metabolism. Mitochondrial dysfunction manifests itself by decrease in ATP production and increase in reactive oxygen species production. It is believed that it develops as a result of Aβ deleterious actions. Neurons use energy for ion pump and ion channels activity, synaptic transmission and transport. Lack of energy caused by mitochondrial dysfunction significantly contributes to altered brain metabolism. Furthermore, *in vivo* studies have shown that mitochondrial disturbance develops early in the AD animal models, and that it correlates with the extent of cognitive impairment [18].

Clinical evidence of memory loss and impairment of at least one other cognitive domain, with evidence of disturbance in social or occupational function is necessary in order to set the diagnosis of AD [2]. Furthermore, in initial clinical presentation considerable variability may be expected based on the brain regions that are affected.

AD is characterized by memory loss, decrease in cognitive abilities and personality changes. Death and dysfunction on nerve cells that occur due to yet unknown reasons are responsible for these changes. Studies have shown that death of neurons in AD brain occurs in a prolonged period of time, and that relatively small number of neurons dye in a same time. Furthermore, these neurons dye by programmed cell death, apoptosis, although it is believed that this may not be the only form of cell death that is present in AD brain [19].

AD is progressive disorder with long latent period. In many fully manifested AD cases, first symptoms could be described as mild cognitive impairment that is characterized by memory loss, but with relatively yet preserved daily functioning. Loss of memory is mainly manifested as difficulty in retaining of new information. This symptom is followed with changes in mood, misplacing things, disorientation in place and time, and in later disease stages language impairment develops. In severe cases neuropsychiatric conditions such as psychosis, agitation, insomnia, apathy or depression may expend [9]. All of these symptoms progress in time and radically impair patient’s daily activities.

As mentioned before, in majority of AD patients neuropsychiatric symptoms develop gradually. Treatment of these symptoms in demented patients requires careful monitoring, and this is especially important when it comes to therapy with atypical neuroleptics.

TREATMENT OPTIONS FOR AD

Overall, interventions aimed on AD can be divided to primary, secondary and tertiary preventive approaches. The aim of primary AD prevention is to postpone the onset of AD by acting on potential etiological factors including protective and risk factors. Secondary prevention is based on early detection of AD, and tertiary prevention requires appropriate pharmacotherapy and care for AD patients which aims to improve quality of life, stabilize cognitive functions and control neuropsychiatric manifestations [3].

As part of AD primary prevention, few antihypertensive clinical trails have been tested [20,21]. So far, results of those trials are still not in the full accordance, and it seems that longer observation period is needed in order to get complete knowledge on beneficial effects of this class of drugs in the prevention of AD. Similarly, the effects of estrogen and hormonal replacement therapy in the prevention of AD have given discordant results. Moreover, supplementation with vitamin E has failed to demonstrate any benefit in the prevention of AD [22,23].

Furthermore, new strategies in AD primary prevention are currently implemented and can be divided into life-course approach and multidomain interventions. The life-course approach is based on a notion that action of psychosocial, biological, and environmental factors during different lifetime periods including early childhood, midlife, and late life are important in the development of AD. The main goal of this approach is to establish "time windows" when acting of above mentioned factors have the greatest effect on AD development. For example, evidence have shown that low body mass index (BMI) and low blood pressure in elderly are associated with increased risk of AD, whereas high BMI and high blood pressure in middle age are risk factors for the development of AD in late life [24-26].

Most of the AD primary prevention approaches so far have been based on a single intervention in elderly people or in individuals with already decreased cognitive function, and this kind of approach did not yield desirable results. The main purpose of multidomain interventions is based on the attempt to optimally control different risk factors for AD. Currently, several multidomain interventions are planned or ongoing such as the Cognitive Substudy of the Finnish Diabetes Prevention Study, Finnish Geriatric Intervention Study to Prevent Cognitive Impairment and Disability (FINGER), and the Dose-Response to Exercise Training (DR's EXTRA) study [27].

Important, modifiable life-style factor for which it has been confirmed in different studies that it has protective effects against the development of AD is regular physical exercise [9]. As part of AD primary prevention, physical activity starting in midlife is strongly encouraged since it has been reported that it protects against cognitive impairment.

AD is a chronic disorder with long latent period. Pathological processes that characterize typical AD neurodegeneration begin long before clinical manifestation of this disease. It is believed that early detection of AD followed with appropriate intervention in early stages of disease may result in a better AD treatment. Efforts are made in AD secondary prevention to discover useful markers that may help in the detection of individuals in preclinical AD phase [28]. Different biochemical, clinical and neuroimaging markers are tested with this purpose and results of these studies yet remain to be fully appreciated.

Main goal of AD tertiary prevention is to provide appropriate treatment that will lead to the improvement of quality of life and to decrease overall disability of AD patients. Standard approach in the AD treatment aims to improve clinical symptoms of this disease, including neuropsychiatric and cognitive disturbances [29]. Currently, many other treatment approaches are tested that will be discussed in detail in this review.

A. Symptomatic Treatment of AD

Neurotransmitter acetylcholine plays pivotal role in cognition and memory. Hydrolysis of this neurotransmitter is catalyzed by Acetylcholinesterase (AChE) which is one of the most essential enzymes in the family of serine hydrolases. Studies have shown that AD is associated with cholinergic deficiency, hence inhibition of AChE biological activity in order to increase the brain acetylcholine level is the therapeutic strategy in the treatment of AD with promising potential [30].

Cognitive dysfunction is linked to the loss of cholinergic function. Drugs such as nicotinic and muscarinic agonists in last two decades are widely used in the treatment of AD patients. However, results of clinical trials with these agents have shown that they may halt cognitive decline for up to 3-6 months, but their use can not stop general disease progression nor it can modify disease duration [2].

One of the oldest hypotheses that try to explain development of AD is "cholinergic hypothesis". It is based on a notion that AD develops as a result of cholinergic neuron degeneration in the basal forebrain which is followed with disturbed, and in many cases complete loss of cholinergic transmission in cerebral cortex and other brain regions. Above mentioned hypothesis gradually led to an introduction of acetylcholinesterase inhibitors as a class of drugs that is currently used in the treatment of mild to moderate AD cases. Partly beneficial, symptomatic effects of these drugs that potentiate central cholinergic function confirm cholinergic hypothesis to a certain extent. At present, four of the drugs from this class have been approved by Food and Drug Administration (FDA) in United States: tacrine, rivastigmine, donepezil, and galantamine. Main action of all of these drugs is inhibition of acetylcholinesterase, an enzyme that degrades acetylcholine, and by doing so they prolong actions of acetylcholine. Observations from human studies have shown that these drugs do stabilize memory and tend to make cognitive decline more gradual, but their effects are limited in time, and may last up to one or at most few years [31].

Acetycholine Esterase Inhibitors

Acetylcholinesterase inhibitors (ACIs) are used for mild to moderate forms of AD and they may slow progression of this disease. Main mechanism of ACIs action is that they decrease production of amyloidogenic compound and Aβ [32,33]. The first widely used ACI was Tacrine, and a 30-week randomized control clinical trial showed a significant dose depended improvement in cognitive function accomplished with its use in AD patients. However, subsequent studies have shown that administration of Tacrin is related with significant side-effects such as hepatotoxicity, and these findings have limited its use in AD treatment. Donepezil, galantamine and rivastigmine are second generation cholinergics with longer halflives, greater efficacy and fewer side effects. A meta-analysis of 29 randomized placebo-controlled trials has reported that the use of ACIs significantly improves activity of daily living in AD patients [34]. In general, attending physicians proscribe ACIs at low doses to minimize adverse effects such as nausea, vomiting, diarrhea, dyspepsia and facial flushing, and then continue to increase dose up to the maximum tolerated dose [35].

Mehta *et al.* [36] aimed to investigate whether the use of AChIs in USA patients with AD vary by ethnicity. The results of their study have shown that compared with white patients with AD, minority patients had 40% lower odds of AChI use (odds ratio 0.6, 95% confidence interval: 0.5 to 0.7), even after adjusting for covariates such as severity of disease, comorbid diseases, health insurance status and demographics.

Although ACIs are in the last few decades the only class of drugs that have shown beneficial effects in causing an increase of cognitive function in AD patients, a randomized double-blind trial conducted by Courtney *et al.* [37] did not confirm these findings. It seems possible that during preclinical and clinical phases of ACIs development and testing, pharmacokinetics and administered dosage of ACIs were not optimal when used in AD treatment. Certain authors believe that the used dosage of 10 mg daily should be increased if adverse effects permit [38]. In a line with these suggestions, Farlow *et al.* [39] tested effectiveness and tolerability of high-dose (23 mg/d) *vs.* standard-dose (10 mg/d) of donepezil in moderate to severe AD in a 24-week, randomized, double-blind study. The authors reported that donepezil 23 mg/d was associated with greater benefits in cognition compared with donepezil 10 mg/d, but in a same time they suggest that obtained observations require additional studies for confirmation.

Bohnen *et al.* [40] investigated *in vivo* cortical acetylcholinesterase (AChE) activity and cognitive effects in subjects with mild Alzheimer's disease prior to and after 12 weeks of donepezil therapy. Results of their study have demonstrated that donepezil induced only modest inhibition of cortical AChE activity, and that the degree of this inhibition correlates with changes in attentional and executive functions in patients with mild AD.

The use of ACIs in a form of transdermal patch is currently tested and results of The Investigation of Transdermal Exelon® in Alzheimer's Disease (IDEAL) Study in AD patients with Mini-Mental Status Exam (MMSE) scores of 10–20 have shown that rivastigmine patch has similar efficacy compared to high dose capsules. Furthermore, butyrylcholinesterases, compounds that are capable of blocking the nonsynaptic (nonspecific) cholinesterases are presently investigated [41].

N-methyl-D-Aspartate Receptor Antagonist/Blocker of Glutamate Gated NMDA Channels (Memantine)

One of the principal findings of AD brain is excess activation of NMDA receptors by glutamate that increases vulnerability of neurons and finally results in their degeneration. N-methyl-D-aspartate (NMDA) receptor agonists are another class of symptomatic drugs that are used in the treatment of moderate to severe AD cases. The only representative drug from this class is memantine. Its use was in 2003 approved for the AD treatment in USA and in Europe. Main mechanism of memantine's action is that it provides protection against the neurodegenerative effects of glutamate, an excitatory neurotransmitter. However, results of studies that investigated efficacy of memantine's use in AD treatment are evasive. A randomized clinical trial in AD patients with severe form of disease has reported that use of memantine significantly decreases deterioration [42]. Conversely, doubleblind, placebo-controlled trial did not confirm these findings [43]. Adverse effects associated with the use of memantine include dizziness, and more rarely hallucinations and confusion. Furthermore, duration of favorable actions of memantine is still not determined.

Recent study conducted by Martinez-Coria *et al.* [44] in transgenic mice have shown that memantine improves cognitive function and reduces Alzheimer's – like neuropathology. Furthermore, obtained results suggest that use of mementine induces disease modification and prevents synaptic dysfunction in AD brain.

Numerous studies have demonstrated that administration of memantine prevents neuronal death by excitotoxic mechanisms. Results of a meta-analysis of all clinical trials and pivotal studies conducted between 1992 and 2008 have shown that AD patients receiving memantine had significant improvements in both the Neuropsychiatric Inventory (NPI) and in the Behavioral Rating Scale for Geriatric Patients (BGP) compared to those on placebo. An important finding of this meta-analysis was that the use of memantine leads to a significant delay in the development of behavioral symptoms in AD patients [45].

In a placebo-controlled trial in 166 AD patients with a Mini-Mental State Examination (MMSE) score < 10, administration of 10 mg/day (half of the currently recommended dosage) of memantine for 12 weeks led to significant improvement in global and functional parameters. However, the measurements for behavior were not statistically significant [46].

Study by Rosi *et al.* [47] reported that low doses of memantine reduce neuroinflammation, attenuates cognitive deficits induced by lipopolysaccharide and returns Arc gene expression to control levels. Moreover, authors imply that memantine when given early during the development of AD pathology confers cognitive and neuronal protection while indirectly prevents pathological microglial activation.

Findings from the survey of hospice medical directors published in 2009 have shown that AChIs and/or NMDA receptor agonists are prescribed for patient with advanced dementia, and their use is associated with clinical benefit [48].

A 24-week trial in 322 patients with MMSE scores from 5 to 14 has shown that concomitant use of donepezil and memantine may have beneficial effects on cognition, behavior and activities of daily living [49]. Based on current evidence, the transition from AChIs to memantine is well tolerated, but better results in AD treatment are achieved when memantine is used in a combination with AChIs.

B. Disease Modifying Approaches Targeting Aβ in AD Treatment

Amyloid-Based Therapies

According to "amyloid hypothesis" excess in Aβ production is the underlying cause of neuronal dysfunction and death. This hypothesis is based on the amyloid cascade of events that start with increased production of Aβ, followed with its aggregation that leads to plaques and neurofibrillary tangles formation and tau phosphorylation [50]. All of these processes ultimately cause significant cognitive decline and other AD-related symptoms. In recent years, the so called disease-modifying approach in AD treatment aimed to decrease or prevent Aβ production, but majority of clinical trials that tested drugs with such mechanism of action have failed.

Although much of the research is dedicated to the discovery of AD etiology, full mechanisms of its development are not clear. One of the main features of AD is disturbed metabolism of Aβ. Novel therapeutic approaches in AD treatment are based on the intent to decrease Aβ production and aggregation, and to increase its clearance. It is widely believed that this kind of approach will prevent further neuronal loss, and will lead to the improvement of cognitive abilities in AD patients [51]. Although previous disease modifying approaches based on the attempt to decrease Aß production and aggregation were not successful in phase 3 clinical trials, several trials are currently ongoing and their results are waited with anticipation.

Inhibitors and/or Modulators of Secretases

As part of anti-amyloid based therapies inhibitors/or modulators of APP secretases are used. Studies so far have shown that the decrease in Aβ production can be achieved by suppression of APP expression, up-regulation of α-secretase, or reduction in either β- or γ-secretase. Main goal of this kind of approach is to direct processing of APP to non-amyloidogenic pathway. However, majority of agents that are used as secretases inhibitors and/or modulators are not specific only for APP and studies so far have shown that their use may have significant adverse effects. Furthermore, recent findings from animal studies have demonstrated that employment of γ-secretases may induce harmful effects in spleen, gastrointestinal system and thymus. Majority of current efforts are focused on interventions that modify or modulate activity of α-, β- or γ-secretases [52].

Another important group of secretases inhibitors that are widely studied in animal models are β-site amyloid precursor protein-cleaving enzyme (BACE) inhibitors. Animal studies have shown that the use of BACE 1 (β-secretase) led to an improvement of cognitive function and a decrease in Aβ production without inflammatory side-effects [53]. Based on current observation, the use of BACE inhibitors is not associated with adverse effects, but they have not yet been tested in clinical trials.

β- and γ-secretases cleave APP, and as a result Aβ peptide, that is known by its neurotoxicity, is generated. Inhibitors of these secretases were tested in animal studies and have demonstrated that their use can lower the levels of Aβ in the brain. The first selective γ-secretase inhibitor that reached phase III testing for the treatment of AD is LY450139 (Semagacestat). Results of phase II study that lasted 14 weeks and included 51 AD patients have shown that this compound decreases the level of Aβ peptide in the blood but not in CSF, and certain safety concerns such as 3 possible drug rashes and 1 transient bowel obstruction were observed [54]. In general, studies in humans have also shown that the use of LY450139 leads to the dose dependent decrease of Aβ, but eosinophilia and GI side effects were observed. Two phase III clinical trials were conducted that aimed to estimate efficacy and safety of LY450139 compared to placebo in patients with mild to moderate AD. However, in 2010 it was announced that phase III trials failed since this compound performed worse then placebo [55].

In patients with mild to moderate AD tarenflurbil, a γ-secretase modulator was tested, and since this drug does not interfere with γ-secretase substrates, no significant adverse effects were noted. However, results have shown that tarenflurbil is effective in mild form of AD where improvements of cognitive functions and activities of daily living were observed, but in moderate AD these beneficial effects were not noted [56]. However, results of phase III trial that lasted 18 months showed no difference between placebo- and tarenflurbil-treated AD patients. One of the possible explanations of these observed results was that this agent has insufficient potency in CNS penetration [57].

Aggregation Inhibitors

Novel findings suggest that inhibition of Aβ aggregation may present important feature of AD treatment, since numerous studies have shown that aggregation of Aβ is particularly toxic and it causes significant damage in the brain of AD patients. A drug with possible anti-Aβ aggregation effects that has been tested in clinical trials is tramiprosate. This drug belongs to a glycosaminoglycan (GAG) mimetics and it has been demonstrated that it has potentials in the reduction of Aβ deposition in the brain [9]. Results of phase II studies of tramiprosate use in AD treatment were encouraging, but unfortunately were not confirmed by phase III findings in USA. In November 2007, the sponsor stopped further tramiprosate development as a pharmaceutical, as well as the EU phase III trial. One of the reasons why the sponsor decided to halt tramiprosate development was the observation that this drug in fact promotes the aggregation of tau, which is known as primary component of neurofibrillary tangles [58].

Colostrinin is a praline-rich polypeptide complex derived from sheep colostrum. *In vitro* and *in vivo* studies have shown that main mechanism of this drug action is that it inhibits Aβ aggregation and neurotoxicity, and improves cognitive function. Findings from Phase I studies on humans have reported that colostrinin is well tolerated with the dosage of 200 μg daily or 100 μg every other day. Results from Phase II studies have demonstrated modest improvement in MMSE score for patients with mild AD but obtained effect was not sustained when the drug was given in a longer period of time. Generally, beneficial effects of colostrinin are more prominent in patients with mild than in those with moderate AD [56].

Statins

Results of numerous studies have pointed out to the important relation between elevated cholesterol level and amyloid deposition. These findings led to the introduction of statins use as potent protective factors against the development of AD, although full mechanisms of their effects are still not completely understood. One possible explanation of their action may be their ability to increase activity of α-secretase which cleaves APP into soluble products. By this effect statins may prevent the production of Aβ [59].

Studies have shown that statins have potency to decrease Aβ levels [60], but later investigations have reported that this protective effect of statins may be expected only if they are prescribed in early stages of life [61,62]. Furthermore, numerous epidemiological studies have demonstrated that use of statins significantly reduces the risk of AD development [63,64]. However, two large controlled clinical trials have failed to confirm these previous findings of statins protective effects. According to current scientific notion, statins may slow progression of neurodegeneration, but they can not reverse degeneration of neurons once it has begun.

A 6-month study found that probucol, a cholesterol-lowering drug, reduced levels of Aβ in CSF and stabilized cognitive deterioration in patients with mild to moderate AD [65]. However, results of short-term studies failed to demonstrate significant ability of statins to reduce Aβ [66]. Furthermore, results of a phase III trial (N = 400) of simvastatin monotherapy were not encouraging [67]. It is of note that although numerous studies and trials did not show beneficial effects of statins use in the AD treatment, their use as protective agents in AD prevention seems promising and is currently being investigated.

Inhibitors of Receptors for Advanced Glycation End Products

Advanced Glycation End Products (AGEs) have been implicated as proinflammatory mediators that can increase formation of free radicals. Multiple cell types of the brain on their cell surface have receptor for AGEs. It has been demonstrated that in the AD brain, this receptor is up-regulated in microglia and astrocytes of hippocampus. Furthermore, it has been shown that amyloid β has the ability to bind to these receptors which may trigger inflammatory cascade that in final stages leads to cell death. Clinical studies are currently ongoing that are testing the effects of AGEs receptors blockage. These studies are continuation of preclinical studies that have observed decreased inflammation and plaque formation when binding between AGEs receptors and amyloid was blocked [68].

It has been reported that the receptor for advanced glycation end products (RAGE) has an ability to bind to Aβ and mediates its transport across the blood brain barrier. These findings led to a conclusion that inhibitors of RAGE may be used as potential treatment option in the AD therapy. Currently, a phase II safety and efficacy trial is ongoing that aims to investigate potency of PF04494700 (formerly TTP488), RAGE inhibitor, in patients with mild to moderate AD [69].

Erythropoetin Derivatives

It is long known that erythropoietin has neuroprotective and neurotrophic assets. Novel findings have demonstrated that erythropoietin possesses protective effects against traumatic brain injury and cerebral ischemia. Furthermore, studies have shown that its use can prevent inflammatory cells activation and neuronal degeneration. These observations may lead to the use of erythropoietin derivatives in the treatment of AD, since *in vitro* studies have shown that they exert protective effects against Aβ toxicity [9]. However, it is of note that use of erythropoetin derivatives is not without a risk. Primarily, this agent is not well tolerated in all individuals, especially those that suffer from hypertension, cancer, or congestive heart failure. Moreover, it has been observed that its use is linked with development of vascular stenosis. Finally, important concern related to administration of erythropoetin derivatives is that it may provoke cancer progression [70].

Peroxisome Proliferator-Activated Receptor-Gamma Agonists

Novel findings suggest that insulin-related abnormalities in CNS may have significant role in cognitive decline since it has been shown that insulin signaling is important in memory processes and may take part in APP and Aβ regulation. Based on these presumptions it is supposed that the treatment of insulin resistance may reduce the risk of AD development or even delay it [71]. It has been reported that proinflammatory cytokines may cause insulin resistance in adipose tissue, skeletal muscles and liver by inhibition of insulin signaling [72]. Furthermore, in the post mortem brain, and plasma of AD patients over expression of acute phase reactants and pro-inflammatory cytokines is detected [6]. It is of note that in the AD brain accumulation of Aβ may be a result of an impairment of insulin-degrading enzymes. Several drugs such as rosiglitazone and pioglitazone, the peroxisome proliferator–activated receptor-γ agonists, which are primarily used in the treatment of diabetes were tested as possible treatment options in the AD therapy. Results of clinical trials that tested use of these drugs in AD treatment of diabetic and non-diabetic patients are fairly encouraging and have reported that their use may lead to the reduction in Aβ production, inflammation and in plaque deposition. However, results from a recent randomized, double-blind, placebo-controlled Phase III Study in which rosiglitazone was used as monotherapy in mild-to-moderate AD did not show its significant effects on global functioning and cognition compared to placebo [73].

Immunotherapy

The use of vaccines in the treatment of AD is still rather controversial. *In vivo* studies on transgenic mouse models of AD have demonstrated that active immunization with Aβ or passive immunization with anti-Aβ antibodies has led to the reduction of Aβ and tau-related pathologies. Furthermore, this kind of therapeutic approach resulted in the improvement of synaptic plasticity and in the protection against synaptic degeneration, and with the significant improvement in the learning abilities [9]. However, results of Holmes *et al.* [74] have shown that the use of immunotherapy did not improve overall survival of AD patients, nor it has prevented progression of neurodegeneration.

Although principle of active immunization that targets Aβ raised widespread interest, studies that investigated AN1792, a first-generation Aβ vaccine, were rather disappointing since a major safety concern was development of aseptic meningoencephalitis. However, despite these observations there are many other ongoing active immunization approaches such as CAD106 and ACC-001, a second-generation vaccines that are tested in clinical trails [75,76].

Earlier studies have shown that the removal of excessive amounts of Aβ can be triggered by reactive T cell response. These findings led to an introduction of active immunization strategies in animal models of AD. These animal studies have reported that principle of active immunization may result in the prevention of astrogliosis, Aβ plaque formation, neuritic dystrophy and finally in a reduction of Aβ plaques [77]. Studies in humans with the immunogen AN-1792 were not encouraging since Phase IIa clinical trials had to be terminated because of significant side effects such as meningoencephalitis, white matter abnormalities, and lymphocytic pleocytosis [78].

Currently many vaccines that target humoral response are tested in the attempt to avoid T cell mediated immune response. Studies in animals have shown significant reduction of Aβ in the brain after nasal immunization with AdPEDI-(Aβ1–6) [79]. Decline of Aβ and promotion of humoral immunity were also achieved with the administration of Aβ derivatives in alum adjuvant [80].

Several humanized monoclonal antibodies that target Aβ are in clinical development. Bapineuzumab as monoclonal Aβ peptide antibody is the only one that has reached phase III. Use of Bapineuzumab as shown by findings of phase II trials has numerous beneficial effects but also certain side-effects [81]. Results of phase III of Bapineuzumab use that studied its effects on mild to moderate cases of AD have shown that no treatment benefit was apparent, whereas several serious side effects occurred more frequently in the treatment group compared to placebo [82], and in 2012 Pfizer Inc. and Johnson & Johnson announced that Bapineuzumab failed in two major trials.

C. Disease Modifying Approaches Targeting Tau Protein

Alongside with amyloid hypothesis, another hypothesis that is founded on the role of hyperphosphorylation of Tau proteins in the development of AD has many supporters in scientific community. Since the therapeutic approach that targets Aβ did not result in significant improvements in the treatment of AD patients, many efforts are now being made that are based on possible inhibition of Tau protein pathology [9].

The tau protein is a major component of neurofibrillary tangles. In AD brain, tau protein is a subject of phosphorylation, and studies have shown that this form of phosphorylated tau is essentially neurotoxic [2]. Much of the investigations are now focused on tau metabolism, its phosphorylation and dephosphorylation, all of which led to an introduction of agents that primarily are trying to decrease and ameliorate tau phosphorilation in an attempt to successfully treat AD.

Two major kinds of inhibitors are currently tested with this purpose: tau protein aggregation inhibitors and inhibitors of tau kinases.

Tau Protein Aggregation Inhibitors

One of the features of AD is the tendency of tau protein to aggregate. Moreover, this aggregation is closely associated with AD progression. Drugs that inhibit aggregation of tau protein would certainly have important benefit in the treatment of AD, and their main mechanism of action would be based on their potency to disaggregate tau filaments. Studies have shown that potency to block tau fibrillization exert agents such as phenothiazines, polyphenols and porphyrins [83].

Results of animal studies have shown that phenothiazines have potency to block neurodegeneration and ameliorate cognitive deficit. Phase II study reported that use of phenothiazines as part of AD treatment attenuates cognitive decline [84]. Results of Phase III study must still be anticipated since sponsors are required for the implementation of testing.

Furthermore, novel findings suggest that similar properties of tau aggregation inhibition have pyrimidotriazine, benzofuran and quinoxaline inhibitors [85].

Inhibitors of Tau Kinases

Numerous kinases such as protein kinase A, protein kinase C, casein kinase-1, glycogen synthase kinase 3 (GSK-3), cyclin-dependent protein kinase-5 and others have been implicated to be up-regulated in AD [86]. Majority of studies so far investigated possibility of GSK-3 inhibition, and several drugs were tested with this purpose including: paullones, maleimides, lithium chloride, thiazoles indirubins, and thiadiazolidinones [84]. However, the results of their administration are far from promising since kinases tend to have pleiotropic actions in numerous cells, and so far none of the tested inhibitors have been specific enough for the functions of kinase they aimed to inhibit.

Overall, it can be concluded that there are significant obstacles in the development of effective tau kinase inhibitors, primarily because current drugs are mainly directed to the common ATP binding sites, and present results have shown that kinase selectivity is hard to achieve [85].

D. Neuroprotective Advancements in AD Therapy

Majority of disease-modifying drugs are targeted against Aß and tau protein since they are main features of AD pathology. However, there are also some other alternative approaches that are directed against secondary mechanisms that are also thought to be important in the AD pathogenesis. Such alternative neuroprotective approaches include those directed against inflammation as well as against oxidative stress.

Anti-Inflammatory Drugs

Results of numerous studies have confirmed that neuroinflammation is associated with neurodegeneration. Furthermore, studies have shown that in the AD brain there is evidence of complement components activation that is followed with inflammation of astrocytes and microglia. These observations introduced long-term use of anti-inflammatory drugs as possible preventive approach against the development of AD. Earlier studies have shown that use of ibuprofen may decrease production of Aβ [87], and reduce activation of microglia [88]. However, clinical trials that tested use of naproxen, the non-selective NSAID, and the use of rofecoxib, the selective cyclooxygenase (COX)-2 inhibitor in the treatment of mild to moderate AD have not reported favorable result in terms of reduction of cognitive decline [89].

It has been observed that patients with rheumatoid arthritis that are treated with anti-inflammatory drugs have lower prevalence of AD. These observations led to a conclusion that anti-inflammatory drugs may have protective effects against AD development. Studies have shown that use of anti-inflammatory drugs reduce risk of AD, and improves performance on Mini-Mental Status Examination test [90]. These results suggest important role of inflammatory processes in AD etiopathogenesis. In accordance with this concept, it has been demonstrated that an innate immune system response is related to Aβ deposition and plaque formation. This innate immune response includes secretion of pro-inflammatory cytokines, complement activation, apoptosis mediated by excretion of nitric oxide, and secretion of pro-inflammatory cytokines. Anti-inflammatory drugs, especially non-steroidal anti-inflammatory drugs (NSAIDs) may reduce risk of AD through some of their mechanisms of actions that may decrease Aβ production by down-regulation of pro-inflammatory signals. Results of Baltimore Longitudinal Study of Aging have shown that use of NSAID may reduce risk of AD and this effect primarily depends on the duration of NSAID use [91]. Other clinical studies and trials have also reported protective effects of NSAID against AD development [92,93]. In a recent Cache County Study on Memory Health and Aging that was a longitudinal, population-based study it was reported that use of NSAID may prevent cognitive decline if its use begins in midlife. However, these findings were not confirmed by randomized controlled trials [94,95].

Results of few prospective studies have shown that NSAIDs reduce the risk of incident AD [96-98]. However, Alzheimer's Disease Anti-inflammatory Prevention Trial (ADAPT), the first prospective primary prevention trial that investigated effects of naproxen and celecoxib treatment in at-risk subjects, failed to demonstrate a protective benefit against AD after 2 years of observation [99]. Interestingly, after 5 years of observation, a protective benefit was observed but these results could be confounded by higher morbidity and mortality in tested group [100].

According to current guidelines, the use of NSAID and cyclooxygenase-II inhibitors is not recommended in AD prevention or treatment.

Antioxidants

Results of numerous investigations have demonstrated that markers of oxidative stress may precede pathological lesions in AD, including neurofibrillary tangles and senile plaques. These findings led to a notion that oxidative stress has a significant role in processes of aging and in the pathogenesis of AD [101,102], which resulted in an introduction of antioxidants as possible therapeutical option in the AD prevention.

Major action of antioxidants is based on their ability to decrease formation of ROS. Mainly, they are not intended to be used in the treatment of AD but as a mean of its prevention. Compounds with proven antioxidant properties include: vitamin C, vitamin E, estrogen, curcumin, soy meal, flavonoids, and gingo bilboa extract [9]. Studies so far have shown that vitamin E may delay time of AD clinical worsening, and this vitamin in combination with vitamin C has been shown to decrease AD incidence and prevalence [103].

Results of a clinical trial that was conducted in 5 US academic centers between the years 2000 and 2008 in 3069 elderly community-dwelling volunteers did not provide strong evidence that Gingo Biloba has a role in AD prevention [104].

Overall, *in vitro* studies and animal models have demonstrated beneficial effects of antioxidants in AD setting, however results from clinical studies are not that encouraging [105].

E. Alteration of Ion Homeostasis and Neuronal Regeneration Approaches

Many authors believe that alteration of ion homeostasis plays pivotal role in AD pathogenesis. Increase of cerebral levels of ions such as Copper (Cu), Iron (Fe) and Zinc (Zn) has been reported in patients with AD. It has been demonstrated that these ions when present in excess can promote tau- hyperphosphorylation and also have the ability to promote Aß neurotoxicity. Moreover, these ions have potency to induce production of ROS, which in turn may lead to synaptic and neuronal loss. These observations led to the understanding that modulation of metal ions, restoration of metal homeostasis, and blockage of their interaction with Aß could be used as one of the possible avenues in the treatment of AD. Agents used so far with this purpose include metal complexes, metal-protein attenuating compounds, and metal chelators [106].

Cu2(+)-mediated and Zn2(+)- mediated toxic oligomerization of Aβ is often seen in AD. PBT2 is a metal-protein attenuating compound and as such affects this process of metal alteration. Results of an investigation completed in 2007 that tested its use in AD treatment were encouraging. However, this study lasted only 12 weeks [107].

Loss of neurons and their synapses is a major feature of AD that is accompanied with memory impairment and other AD related symptoms. For many decades it is known that best possible approach in the treatment of AD would be the one that induces regeneration of lost neurons and generation of new ones. *In vitro* and *in vivo* studies conducted in the past decade have shown that bone marrow multipotent stem cell have the ability to differentiate into neurons and glia. Furthermore, it has been observed that administration of cytokines or Hematopoietic Growth Factor (HGF) increases mobilization of bone marrow-derived microglia into the brain, neurogenesis, and promotes anti-apoptosis [108]. The use of these agents has still not been introduced as part of standard treatment for AD patients, but needless to say their utilization can be expected soon enough.

CONCLUSIONS

Increasing prevalence and severity of AD worldwide make necessary the need for the development of better treatment approaches for this disease.

Since symptomatic drugs that are used in AD treatment have fairly limited benefits, many disease-modifying approaches are currently tested often with disappointing results. However, certain strategies are encouraging and it may be perceived that in future some of the tested compounds will lead to a breakthrough in AD treatment. This is particularly true in the case of those approaches that are trying to fully solve etiology and pathophysiology of AD that still remains an enigma, although there are many novel insights in this area.

Disease-modifying interventions aim to slow or delay AD, but can not reverse already existing neuronal death. In general, it can be stated that the basis of disease modifying treatment is founded on the attempt to delay or stop disease progression by interrupting underlying pathophysiological processes in AD development. These approaches try to decrease plaque formation, increase its clearance, and decrease Aβ aggregation. In contrast, current symptomatic treatment targets symptoms of cognitive dysfunction and provides only temporary relief for some of the AD patients, but does not alter or alter minimally the course of AD. It is important to note that since it is well documented that elderly population is particularly susceptible to adverse drug effects, primary concern of novel disease modifying treatments should be proper testing of safety and efficacy of different classes of disease modifying drugs used for the AD treatment.

Xiong *et al.* [109] suggests that in designing future therapeutic trials novel approaches should be implemented such as randomized start and randomized withdrawal designs in which sample size and the time of treatment switch should be pivotal design parameters.

Saxena [50] in his recent report suggests interesting Bioenergetics Paradigm for AD that is based on the "energy deprivation" model of AD. The authors point to the possible new therapeutics that would target enzymes such as pyruvate dehydrogenase with allosteric modulation to increase its activity; glycogen synthase kinase 3 β and β-secretase for beta-site APP that should be modulated with selective inhibitors. It can be said with the certainty that this and other new ideas will eventually lead to production of new agents that will be efficient in AD treatment.

According to current notion, best possible approach in the attempt to fight against development of AD and its devastating symptoms is a multidisciplinary approach in which scientists of all profiles including not only physicians and pharmacists, but as well molecular biologists, biochemists, pathologists, and others will join their efforts that will eventually and certainly in time lead to a significant progress in AD prevention and treatment. Finally, it is important to emphasize that even though current treatments can not cure AD, compliance and optimal pharmacotherapy are vital in its treatment alongside with the necessity of early and proper disease diagnosis.

ACKNOWLEDGEMENT

Declared none.

CONFLICT OF INTEREST

The author(s) confirm that this chapter content has no conflict of interest.

REFERENCES

[1] Mattson M.P. Pathways Towards and Away from Alzheimer's Disease. Nature 2004; 430 (7000): 631–639.
[2] Castellani RJ, Rolston RK, Smith MA. Alzheimer Disease. Dis Mon 2010;56(9): 484–546.
[3] Qiu C, Kivipelto M, von Strauss E. Epidemiology of Alzheimer's disease: occurrence, determinants, and strategies toward intervention. Dialogues Clin Neurosci 2009;11:111-128.
[4] Alzheimer's Association. (2009) Alzheimer's Disease Facts and Figures.(www.alz.org/national/documents/report_alzfactsfigures2009.pdf.)
[5] Bateman RJ, Aisen PS, De Strooper B, Fox NC, Lemere CA, Ringman JM, Salloway S, Sperling RA, Windisch M, Xiong C. Autosomal-dominant Alzheimer's disease:a review and proposal for the prevention of Alzheimer's disease. Alzheimer's Research & Therapy 2011,3:1; 1-13.
[6] Saunders C, Spencer JPE. Metabolic and immune risk factors for dementia and their modification by flavonoids: New targets for the prevention of cognitive impairment? Nutrition and Aging 2012;1:69–88.
[7] Wimo A, Winblad B, Agüero-Torres H, von Strauss E. The magnitude of dementia occurrence in the world. Alzheimer Dis Assoc Disord 2003;17:63-67.
[8] Williamson J, Goldman J, Marder KS. Genetic Aspects of Alzheimer Disease. Neurologist 2009; 15(2): 80–86.
[9] Khairallah MI, Aal Kassem LA. Alzheimer's disease: Current Status of Etiopathogenesis and Therapeutic Strategies. Pak J Biol Sci 2011;14(4):257-272.
[10] Fratiglioni L, von Strauss E, Qiu CX. (2008) Epidemiology of the dementias of old age. In: Dening T, Jacoby R, Oppenheimer C, Thomas A, eds. The Oxford Textbook of Old Age Psychiatry. 4th ed. New York, NY: Oxford University Press; pp 391-406.

[11] Zaciragic A, Lepara O, Valjevac A, Arslanagic S, Fajkic A *et al.* Elevated serum C-reactive protein concentration in Bosnian patients with probable Alzheimer's disease. J Alzheimers Dis 2007;12:151-156

[12] Schmidt R, Schmidt H, Curb JD, Masaki K, White LR, Launer LJ. Early inflammation and dementia: a 25-year follow-up of the Honolulu-Asia Aging Study. Ann Neurol 2002;52:168-174.

[13] Gartenberg, Gartenberg Livney M, Clark CM, Karlawish JH, Cartmell S, Negrón M, Nuñez-Lopez J, Xie SX, Entenza-Cabrera F, Vega IE, Arnold SE. Ethnoracial Differences in the Clinical Characteristics of Alzheimer Disease at Initial Presentation at an Urban Alzheimer's Disease Center. Am J Geriatr Psychiatry 2011;19(5): 430–439.

[14] Blass JP. Brain metabolism and brain disease: is metabolic deficiency the proximate cause of Alzheimer dementia? J Neurosci Res 2001;66:851–856

[15] Butterfield D.A., Drake J, Pocernich C *et al.* Evidence of oxidative damage in Alzheimer's disease brain: central role for amyloid beta-peptide. Trends Mol Med 2001; 7:548–554

[16] Dodart J.C., Mathis C, Bales K.R., Paul S.M., Ungerer A. Early regional cerebral glucose hypometabolism in transgenic mice overexpressing the V717F beta-amyloid precursor protein. Neurosci Lett 1999;277:49–52.

[17] Mattson MP, Chan SL. Neuronal and glial calcium signaling in Alzheimer's disease. Cell Calcium. 2003;34:385–397.

[18] Hauptmann S, Scherping I, Drose S, Brandt U, Muller W.E. *et al.* Mitochondrial dysfunction: An early event in Alzheimer pathology accumulates with age in AD transgenic mice. Neurobiol Aging 2009;30:1574-1586.

[19] Mattson MP. Apoptosis in neurodegenerative disorders. Nat Rev Mol Cell Biol 2000; 1:120–129.

[20] Forette F, Seux ML, Staessen JA, *et al.* Prevention of dementia in randomized double-blind placebo-controlled Systolic Hypertension in Europe (Syst-Eur) trial. Lancet 1998;352:1347-1351

[21] Tzourio C,Anderson C, Chapman N, *et al.* Effects of blood pressure lowering with perindopril and indapamide therapy on dementia and cognitive decline in patients with cerebrovascular disease. Arch Intern Med 2003;163:1069-1075.

[22] Shumaker SA, Legault C, Kuller L, *et al.* Conjugated equine estrogens and incidence of probable dementia and mild cognitive impairment in postmenopausal women: Women's Health Initiative Memory Study. JAMA 2004; 291:2947-2958.

[23] Heart Protection Study Collaborative Group. MRC/BHF Heart Protection Study of antioxidant vitamin supplementation in 20,536 high-risk individuals: a randomised placebo-controlled trial. Lancet 2002;360:23-33.

[24] Whalley LJ, Dick FD, McNeill G. A life-course approach to the aetiology of late-onset dementias. Lancet Neurol 2006;5:87-96.

[25] Fitzpatrick AL, Kuller LH, Lopez OL, *et al.* Midlife and late-life obesity and the risk of dementia: cardiovascular health study. Arch Neurol 2009;66:336-342.

[26] Qiu C, Winblad B, Fratiglioni L. The age-dependent relation of blood pressure to cognitive function and dementia. Lancet Neurol 2005;4:487-499.

[27] Kivipelto M, Ngandu T, Laatikainen T, Winblad B, Soininen H, Tuomilehto J. Risk score for the prediction of dementia risk in 20 years among middle aged people: a longitudinal, population-based study. Lancet Neurol 2006;5:735-741

[28] Blennow K, de Leon MJ, Zetterberg H. Alzheimer's disease. Lancet 2006; 368:387-403.

[29] Jelic V, Kivipelto M, Winblad B. Clinical trials in mild cognitive impairment: lessons for the future. J Neurol Neurosurg Psychiatry 2006;77:429-438.

[30] Lu S-H, Wu JW, Liu H-L, Zhao J-H, Liu K-T, Chuang C-K, Lin H-Y, Tsai W-B, HoY. The discovery of potential acetylcholinesterase inhibitors: A combination of pharmacophore modeling, virtual screening, and molecular docking studies. Journal of Biomedical Science 2011;18:8.

[31] Salloway S, Correia S. Alzheimer disease: Time to improve its diagnosis and treatment. Cleve Clin J Med 2009;76:49-58.

[32] Giacobini E, Mori F, Lai CC. The effect of cholinesterase inhibitors on the secretion of APPS from rat brain cortex. Ann N Y Acad Sci 1996;777:393–398.

[33] Mori F, Lai CC, Fusi F, Giacobini E. Cholinesterase inhibitors increase secretion of APPs in rat brain cortex. Neuroreport 1995;6:633–636.

[34] Trinh NH, Hoblyn J, Mohanty S, Yaffe K. Efficacy of cholinesterase inhibitors in the treatment of neuropsychiatric symptoms and functional impairment in Alzheimer disease: a meta-analysis. JAMA 2003;289:210–216.

[35] Mulugeta E, Karlsson E, Islam A, Kalaria R, Mangat H, Winblad B, *et al.* Loss of muscarinic M4 receptors in hippocampus of Alzheimer patients. Brain Res 2003;960:259–262.

[36] Mehta KM, Yin M, Resendez C, Yaffe K. Ethnic differences in acetylcholinesterase inhibitor use for Alzheimer disease. Neurology 2005;65(1):159–162.

[37] Courtney C, Farrell D, Gray R, *et al.* Long-term donepezil treatment in 565 patients with Alzheimer's disease (AD2000): randomized double-blind trial. Lancet 2004;363:2105–2115.

[38] Herholz K. Action of cholinesterase inhibitors in patients' brains. J Neurol Neurosurg Psychiatry 2005;76:305.

[39] Farlow MR, Salloway S, Tariot PN, Yardley J, Moline ML, Wang Q, Brand-Schieber E, Zou H, Hsu T, Satlin A. Effectiveness and Tolerability of High-Dose (23 mg/d) *Vs.* Standard-Dose (10 mg/d) Donepezil in Moderate to Severe Alzheimer's Disease: A 24-Week, Randomized, Double-Blind Study. Clin Ther 2010;32(7):1234-1251.

[40] Bohnen NI, Kaufer DI, Hendrickson R, Ivanco LS, Lopresti BJ, Koeppe RA, Meltzer CC, Constantine G, Davis JG, Mathis CA, DeKosky ST, Moore RY. Degree of inhibition of cortical acetylcholinesterase activity and cognitive effects by donepezil treatment in Alzheimer's disease. J Neurol Neurosurg Psychiatry 2005;76(3):315-319.

[41] Kamal MA, Klein P, Yu QS, Tweedie D, Li Y, Holloway HW, *et al.* Kinetics of human serum butyrylcholinesterase and its inhibition by a novel experimental Alzheimer therapeutic, bisnorcymserine. J Alzheimers Dis 2006;10:43–51.

[42] Reisberg B, Doody R, Stoffler A, Schmitt F, Ferris S, Mobius HJ. Memantine in moderate-to-severe Alzheimer's disease. N Engl J Med 2003;348:1333–1341.

[43] van Dyck CH, Tariot PN, Meyers B, Malca Resnick E. A 24-week randomized, controlled trial of memantine in patients with moderate-to-severe Alzheimer disease. Alzheimer Dis Assoc Disord 2007;21:136–143.

[44] Martinez-Coria H, Green KN, Billings LM, Kitazawa M, Albrecht M, Rammes G, CG Parsons, Gupta S, Banerjee P, LaFerla FM. Memantine Improves Cognition and Reduces Alzheimer's-Like Neuropathology in Transgenic Mice. Am J Pathol 2010, 176:870–880.

[45] Thomas SJ, Grossberg GT. Memantine: a review of studies into its safety and efficacy in treating Alzheimer's disease and other dementias. Clinical Interventions in Aging 2009;4:367–377.

[46] Winblad B, Poritis N. Memantine in severe dementia: results of the 9M-Best Study (Benefit and efficacy in severely demented patients during treatment with memantine. Int J Geriatr Psychiatry 1999;14(2):135–146.

[47] Rosi S, Ramirez-Amaya V, Vazdarjanova A, Esparza EE, Larkin PB, Fike JR, Wenk GL, and Barnes CA. Accuracy of hippocampal network activity is disrupted by neuroinflammation: rescue by Memantine. Brain 2009;132:2464–2477.

[48] Shega JW, Ellner L, Lau DT, Maxwell TL. Cholinesterase Inhibitor and N-Methyl-D-Aspartic Acid Receptor Antagonist Use in Older Adults with End-Stage Dementia: A Survey of Hospice Medical Directors. Journal of Palliative Medicine 2009;12(9):779-783.

[49] Tariot PN, Farlow MR, Grossberg GT, Graham SM, McDonald S, Gergel I. Memantine treatment in patients with moderate to severe Alzheimer disease already receiving donepezil: a randomized controlled trial. JAMA 2004;291:317–324.

[50] Saxena U. Bioenergetics breakdown in Alzheimer's disease: targets for new therapies. Int J Physiol Pathophysiol Pharmacol 2011;3(2):133-139.

[51] Walker L.C., Ibegbu C.C., Todd C.W., Robinson H.L., Jucker M., LeVine H., Gandy S. Emerging prospects for the disease-modifying treatment of Alzheimers disease. Biochem Pharmacol 2005;69:1001-1008.

[52] Creed MC, Milgram NW. Amyloid-modifying therapies for Alzheimer's disease: therapeutic progress and its implications. AGE 2010; 32:365–384.

[53] Chang WP, Downs D, Huang XP, Da H, Fung KM, Tang J. Amyloid-beta reduction by memapsin 2 (beta-secretase) immunization. FASEB J 2007;21:3184–3196.

[54] Fleisher AS, Raman R, Siemers ER, *et al.* Phase 2 safety trial targeting amyloid beta production with a gamma-secretase inhibitor in Alzheimer disease. Arch Neurol 2008;65:1031–1038.

[55] Hopkins C.R. Chemical Neuroscience Molecule Spotlight on ELND006: Another γ-Secretase Inhibitor Fails in the Clinic. ACS Chem Neurosci. 2011; 2(6): 279–280.

[56] Christensen DD. Changing the course of Alzheimer's disease: anti-amyloid disease modifying treatments on the horizon. Prim Care Companion J Clin Psychiatry 2007;9:32–41.

[57] Koo E. Gamma-secretase modulators, nonsteroidal anti-inflammatory drugs, and Alzheimer's disease: a review and update. Alzheimer Dement 2008;4(Suppl 2):T102.

[58] Sabbagh MN. Drug Development for Alzheimer's Disease: Where Are We Now and Where Are We Headed? Am J Geriatr Pharmacother 2009;7(3):167–185.

[59] Pedrini S, Carter T.L., Prendergast G, Petanceska S, Ehrlich M.E, Gandy S. Modulation of statin-activated shedding of Alzheimer APP ectodomain by ROCK. PLoS.Med 2005;2:e18-e18.

[60] Fassbender K, Simons M, Bergmann C, Stroick M, Lutjohann D, Keller P, *et al.* Simvastatin strongly reduces levels of Alzheimer's disease beta-amyloid peptides Abeta 42 and Abeta 40 *in vitro* and *in vivo*. Proc Natl Acad Sci U S A 2001;98:5856–5861.

[61] Li G, Higdon R, Kukull WA, Peskind E, Van Valen Moore K, Tsuang D, *et al.* Statin therapy and risk of dementia in the elderly: a community-based prospective cohort study. Neurology 2004;63:1624–1628.

[62] Li G, Kukull WA, Peskind E, McCormick W, Bowen JD, Teri L, *et al.* Differential effect of statins on risk of AD by age, sex and APOE genotype: findings from a community-based prospective cohort study. Alzheimer Dis Assoc Disord 2006;20:S103–S104.

[63] Yaffe K, Barrett-Connor E, Lin F, Grady D. Serum lipoprotein levels, statin use, and cognitive function in older women. Arch Neurol 2002;59:378–384.
[64] Rockwood K, Kirkland S, Hogan DB, MacKnight C, Merry H, Verreault R, *et al.* Use of lipid lowering agents, indication bias, and the risk of dementia in community-dwelling elderly people. Arch Neurol 2002;59:223–227.
[65] Poirier J. Apolipoprotein E, cholesterol transport and synthesis in sporadic Alzheimer's disease. Neurobiol Aging 2005;26:355–361.
[66] Lukiw WJ. Cholesterol and 24S-hydroxycholesterol trafficking in Alzheimer's disease. Expert Rev Neurother 2006;6(5):683–693.
[67] Sano M. Multi-center, double blind, placebo controlled trial of simvastatin to slow the progression of Alzheimer's disease. Alzheimer Dement 2008;4(Suppl 2):T200.
[68] Chen X, Walker D.G., Schmidt A.M., Arancio O, Lue L.F., Yan S.D. RAGE: A potential target for Abeta-mediated cellular perturbation in Alzheimers disease. Curr Mol Med 2007;7:735-742.
[69] Yan SF, Ramasamy R, Schmidt AM. Soluble RAGE: Therapy & Biomarker in Unraveling the RAGE Axis in Chronic Disease and Aging. Biochem Pharmacol 2010; 79(10):1379–1386.
[70] Maiese K, Chong ZZ, Shang YC. Raves and risks for erythropoietin. Cytokine Growth Factor Rev 2008;19(2):145–155.
[71] Watson GS, Craft S. The role of insulin resistance in the pathogenesis of Alzheimer's disease: implications for treatment. CNS Drugs 2003;17(1):27–45.
[72] Zaciragic A. Uloga inflamacije masnog tkiva u nastanku inzulinske rezistencije. In: Ascic-Buturovic B *et al.* (ed) Predijabetes, i dijabetes melitus tip 2. Institut za naucnoistrazivacki rad i razvoj KCUS, 2009. Sarajevo, pp 27-30.
[73] Gold M, Alderton C, Zvartau-Hind M, Egginton S, Saunders AM, Irizarry M, Craft S, Landreth G, Linnamägi U, Sawchak S. Rosiglitazone Monotherapy in Mild-to-Moderate Alzheimer's Disease: Results from a Randomized, Double-Blind, Placebo-Controlled Phase III Study. Dement Geriatr Cogn Disord 2010;30:131–146.
[74] Holmes C, Boche D, Wilkinson D, Yadegarfar G, Hopkins V *et al.* Long-term effects of Abeta 42 immunization in Alzheimer's disease: Follow-up of a randomized, placebo-controlled phase I trial. Lancet 2008;372:216-223.
[75] ClinicalTrials.gov. Safety and tolerability of repeated subcutaneous injections of CAD106 in mild Alzheimer's patients. ClinicalTrials.gov
[76] ClinicalTrials.gov. Study evaluating safety, tolerability, and immunogenicity of ACC-001 in subjects with Alzheimer's disease. ClinicalTrials.gov
[77] Schenk D, Barbour R, Dunn W, Gordon G, Grajeda H, Guido T, *et al.* Immunization with amyloid beta attenuates Alzheimer-disease-like pathology in the PDAPP mouse. Nature 1999;400:173–177.
[78] Orgogozo JM, Gilman S, Dartigues JF, Laurent B, Puel M, Kirby LC, *et al.* Subacute meningoencephalitis in a subset of patients with AD after Abeta42 immunization. Neurology 2003;61:46–54.
[79] Kim HD, Tahara K, Maxwell JA, Lalonde R, Fukuiwa T, Fujihashi K, *et al.* Nasal inoculation of an adenovirus vector encoding 11 tandem repeats of Abeta1-6 upregulates IL-10 expression and reduces amyloid load in a Mo/Hu APPswe PS1dE9 mouse model of Alzheimer's disease. J Gene Med 2007;9:88–98.

[80] Asuni AA, Boutajangout A, Scholtzova H, Knudsen E, Li YS, Quartermain D, *et al.* Vaccination of Alzheimer's model mice with Abeta derivative in alum adjuvant reduces A beta burden without microhemorrhages. Eur J Neurosci 2006;24:2530–2542.

[81] Grundman M, Black R. Clinical trials of bapineuzemab, a beta amyloid targeted immunotherapy in patients with mild to moderate Alzheimer's disease. Alzheimer Dement 2008;4(Suppl 2):T166.

[82] Castellani RJ, Perry G. Pathogenesis and Disease-modifying Therapy in Alzheimer's Disease: The Flat Line of Progress. Arch Med Res. 2012 ;43(8):694-698.

[83] Brunden KR, Ballatore C, Crowe A, Smith III AB, Lee VM-Y, Trojanowski JQ. Tau-Directed Drug Discovery for Alzheimer's Disease and Related Tauopathies: A Focus on Tau Assembly Inhibitors. Exp Neurol 2010;223(2):304–310.

[84] Noble W, Pooler AM, Hanger DP. Advances in tau-based drug discovery. Expert Opin Drug Discov 2011;6(8): 797–810.

[85] Brunden KR, Trojanowski JQ, Lee VM-Y. Advances in Tau-focused drug discovery for Alzheimer's disease and related tauopathies. Nat Rev Drug Discov 2009;8(10): 783–793.

[86] Dolan PJ, Johnson GVW. The role of tau kinases in Alzheimer's disease. Curr Opin Drug Discov Devel 2010; 13(5): 595–603.

[87] Weggan S, Eriksen J.L., Das P. Sagi S.A., Wang R. *et al.* A subset of NSAIDs lower amyloidogenic Abeta42 independently of cyclooxygenase activity. Nature 2001;414:212-216.

[88] Lim G.P., Yang F, Chu T, Chen P, Beech W. *et al.* Ibuprofen suppresses plaque pathology and inflammation in a mouse model for Alzheimer's disease. J. Neurosci 2000;20:5709-5714.

[89] Aisen P.S., Schafer K.A., Grundman M, Pfeiffer E, Sano M, *et al.* Effects of rofecoxib or naproxen *vs.* placebo on Alzheimers disease progression: A randomized controlled trial. J Am Med Assoc 2003;289:2819-2826.

[90] Rich JB, Rasmusson DX, Folstein MF, Carson KA, Kawas C, Brandt J. Nonsteroidal anti-inflammatory drugs in Alzheimer's disease. Neurology 1995;45:51–55.

[91] Stewart WF, Kawas C, Corrada M, Metter EJ. Risk of Alzheimer's disease and duration of NSAID use. Neurology 1997;48:626–632.

[92] Rogers J, Kirby LC, Hempelman SR, Berry DL, McGeer PL, Kaszniak AW, *et al.* Clinical trial of indomethacin in Alzheimer's disease. Neurology 1993;43:1609–1611.

[93] Andersen K, Launer LJ, Ott A, Hoes AW, Breteler MM, Hofman A. Do nonsteroidal anti-inflammatory drugs decrease the risk for Alzheimer's disease? The Rotterdam Study. Neurology 1995;45:1441–1445.

[94] Thal LJ, Ferris SH, Kirby L, Block GA, Lines CR, Yuen E, *et al.* A randomized, double-blind, study of rofecoxib in patients with mild cognitive impairment. Neuropsychopharmacology 2005;30:1204–1215.

[95] Wyss-Coray T. Inflammation in Alzheimer disease: driving force, bystander or beneficial response? Nat Med 2006;12:1005–1015.

[96] Cornelius C, Fastbom J, Winblad B, *et al.* Aspirin, NSAIDs, risk of dementia, and influence of the apolipoprotein E epsilon 4 allele in an elderly population. Neuroepidemiology 2004;23(3):135–143.

[97] Lindsay J, Laurin D, Verreault R, *et al.* Risk factors for Alzheimer's disease: a prospective analysis from the Canadian Study of Health and Aging. Am J Epidemiol 2002;156(5):445–453.

[98] in t' Veld BA, Ruitenberg A, Hofman A, *et al.* Nonsteroidal antiinflammatory drugs and the risk of Alzheimer's disease. N Engl J Med 2001;345(21):1515–1521.

[99] ADAPT Research Group. Lyketsos CG, Breitner JC, Green RC, Martin BK, Meinert C *et al.* Naproxen and celecoxib do not prevent AD in early results from a randomized controlled trial. Neurology 2007;68(21):1800–1808.

[100] Breitner JCS for the ADAPT study group. Onset of Alzheimer's dementia occurs commonly without prior cognitive impairment: results from the Alzheimer's Disease Anti-inflammatory Prevention trial (ADAPT). Alzheimer Dement 2008;4(Suppl 2):T130–T131.

[101] Castellani RJ, Harris PL, Sayre LM, Fujii J, Taniguchi N, Vitek MP, *et al.* Active glycation in neurofibrillary pathology of Alzheimer disease: N(epsilon)-(carboxymethyl) lysine and hexitollysine. Free Radic Biol Med 2001;31:175–180.

[102] Nunomura A, Perry G, Aliev G, Hirai K, Takeda A, Balraj EK, *et al.* Oxidative damage is the earliest event in Alzheimer disease. J Neuropathol Exp Neurol 2001;60:759–767.

[103] Zandi PP, Anthony JC, Khachaturian AS, Stone SV, Gustafson D, Tschanz JT, *et al.* Reduced risk of Alzheimer disease in users of antioxidant vitamin supplements: the Cache County Study. Arch Neurol 2004;61:82–88.

[104] DeKosky ST, Williamson JD, Fitzpatrick AL, *et al.* Ginkgo Evaluation of Memory (GEM) Study Investigators. Ginkgo biloba for prevention of dementia: a randomized controlled trial. JAMA 2008;300:2253–2262.

[105] Sarkaki A, Amani R, Badavi M, Moghaddam A.Z., Aligholi H, Safahani M, Haghighizadeh M.H. Pre-treatment effect of different doses of soy isoflavones on spatial learning and memory in an ovariectomized animal model of Alzheimer’s disease. Pak J Biol Sci 2008;11:1114-1119.

[106] Biran Y, Masters C.L., Barnham K.J., Bush A.I., Adlard P.A. Pharmacotherapeutic targets in Alzheimer’s disease. J Cell Mol Med 2009; 13:61-86.

[107] Lannfelt L, Blennow K, Zetterberg H, *et al.* PBT2-201-EURO study group. Safety, efficacy, and biomarker findings of PBT2 in targeting Abeta as a modifying therapy for Alzheimer's disease: a phase IIa, double-blind, randomized, placebo-controlled trial. Lancet Neurol 2008;7:779–786.

[108] Sanchez-Ramos J, Song S, Cao C, Arendash G. The potential of hematopoietic growth factors for treatment of Alzheimer’s disease. A mini-review. BMC Neurosci 2008;9:S3-S3.

[109] Xiong C, van Belle G, Miller JP, Morris JC. Designing clinical trials to test disease-modifying agents— application to the treatment trials of Alzheimer's disease. Clin Trials 2011;8(1):15–26.

Send Orders of Reprints at reprints@benthamscience.net
Frontiers in Clinical Drug Research – Alzheimer Disorders, Vol. 1, 2013, 217-242

CHAPTER 6

Alzheimer's Disease and Dementia: A Midlife Approach to Treatment is Needed

Angela Clifford, Jennifer Stock, Stephan Bandelow, Tri Budi Rahardjo and Eef Hogervorst*

SSEHS Loughborough University and Center for Aging Studies Universitas Indonesia, Depok, Indonesia

Abstract: Dementia is a global problem and in this chapter we investigate modifiable protective and risk factors for dementia. The main question is when to intervene with preventative measures and which focus points should have priority to direct public health policies. Risk factors for vascular disease are also risk factors for dementia and should probably be addressed in midlife at the latest. We review evidence for the most optimal time of intervention for prevention of dementia when discussing protective and risk factors, with a focus on physical activity and cardiovascular risk factors. We also discuss other factors that may influence the relationship between physical activity and cognitive functioning and the implications for theory and practice.

Keywords: Alzheimer, risk factors, midlife, treatment, prevention.

1. INTRODUCTION

Dementia is a global problem and no current pharmacological treatment is effective long-term. In this chapter, we investigate modifiable protective and risk factors for dementia and discuss the most optimal time for interventions to delay dementia onset. Physical activity has been shown to benefit cognitive functioning and while this may be partially through improving general health, other mechanisms appear to be involved as health does not fully mediate effects of physical activity on cognition. There is growing evidence that engaging in long-term physical activity behaviours in midlife is more effective than in older age when dementia pathology is increasingly present. Finally, the presence of moderating variables may influence the findings of exercise interventions,

***Address correspondence to E. Hogervorst:** SSEHS, Brockington Building, Loughborough University, Loughborough LE11 3TU, UK; Tel: + 44 1509 223020; Fax: + 44 1509 223940; E-mail: e.hogervorst@lboro.ac.uk

Atta-ur-Rahman (Ed)

potentially masking the true effect of physical activity when unaccounted for. These findings have theoretical implications for the field of dementia prevention and recommendations for future research are discussed in this chapter.

1.1. Description of Dementia and Alzheimer's Disease

The term 'dementia' refers to a group of acquired cognitive symptoms that impact on activities of daily living and that are most commonly seen in those over the age of 65 years. The idea of age-associated cognitive decline has been documented since Greco-Roman times [1]. However, we now know that there are several types of dementia, each with their own aetiologies, symptoms and progression. Some forms of dementia can be at least partially reversed or halted by removing the cause, such as vitamin deficiencies, endocrine disease, electrolyte imbalance, lead intoxication and alcoholism. Other types of dementia are progressive and non-reversible, showing gradual decline until death. Such dementias include Lewy Body dementia, Frontotemporal dementia, dementia in Parkinson's disease and Creutzfeld-Jakob disease. Alzheimer's disease (AD) accounts for around 50-70% of diagnosed cases of progressive dementia [2] and often co-occurs with Vascular Dementia (VaD). Dementia can thus be considered an umbrella term for a range of disorders of which AD is the most common.

Many of the difficulties experienced in AD are with memory, both verbal and non-verbal. AD often becomes apparent due to a progressive decline in remembering new information, with more severe cases experiencing difficulties recalling events from many years previously [3]. For example, individuals may fail to recognise familiar people such as family and friends, forget recent events, and get lost in familiar surroundings. Not all symptoms of AD are related to memory, however, and other cognitive domains are usually also affected. These include reduced executive function abilities which manifest as difficulty in planning activities, such as making a meal, washing oneself and playing a game, and making judgements such as dressing appropriately for the occasion. Attention and language can also be affected, causing difficulties with following a conversation, remembering words and using the correct word in a sentence. Although these deficits are common in AD, they do vary widely between individuals and as such different cases can present with very different symptom profiles.

The heterogeneity of symptoms means that diagnosis of AD is not straightforward, especially since many older individuals experience some decline in memory ability and changes in brain activity as part of the normal aging process [4]. In order to be diagnosed as having AD, a memory deficit should be accompanied by impairment in at least one other cognitive domain (*e.g.* language, object recognition, motor functions) that interferes with the ability to carry out daily tasks (Diagnostic and Statistical Manual of Mental Disorders 4th Edition; (DSM-IV)) [3]. Memory impairments should normally be present for at least 6 months before a diagnosis can be made to also help rule out potentially reversible syndromes, such as transient ischemic attack, brain trauma or hydroencephalus (International Statistical Classification of Disease and Related Health Problems 10th Revision (ICD-10)) [5]. Cognitive symptoms should also occur without impaired awareness or altered consciousness (*i.e.* delirium) but may be accompanied by mood changes, such as apathy or irritability [5]. In addition, the diagnostic procedure may be supported by neuroimaging and analysis of biomarkers of the disease [6, 7]. Those with a memory complaint who do not reach the criteria for probable AD [7] may be diagnosed with possible AD or Mild Cognitive Impairment (MCI) [8], which is characterised by a memory impairment that does not affect activities of daily living. Approximately 64% of people who are diagnosed with MCI go on to be diagnosed with AD after 2 years [9]. However, some patients with MCI never develop AD, and it remains unclear whether MCI represents a transitional stage to AD or whether it is an independent disorder that makes an individual vulnerable to further cognitive decline.

An analysis of neuropathology reveals characteristic features of AD that are being increasingly considered as potential biomarkers. A loss of brain volume in the medial temporal lobe, especially in the hippocampus, temporal poles and temporoparietal junction, is seen in over 70% of AD patients [10]. These brain areas are thought to contribute to memory function and increased atrophy in these areas is associated with poor memory performance in AD as well as in healthy individuals [11-13]. At a cellular level, AD patients tend to show increased levels of intracellular neurofibrillary tangles (NFTs), which consist of hyperphosphorylated tau protein, and extracellular protein plaques containing high levels of Amyloid-β (Aβ). Tau protein is important for microtubule structure and stability. Aβ is a peptide formed from the cleavage of Amyloid Precursor Protein whose own function is currently unclear but which is thought to be involved in neurotransmission [14]. Excess insoluble Aβ in the brain causes deposits (plaques) on the outside of the neuron, affecting neuronal quality and function. Although NFTs and Aβ plaques are seen in healthy aging [15], they are generally seen in excessive amounts in those with AD and abnormal levels can be measured in cerebrospinal fluid some years before the onset of symptoms [10]. The dominant theory suggests that these features cause dementia symptoms [16]. However, it remains controversial whether Aβ plaques cause AD or are simply a feature of it, especially since research into pharmaceutical treatments that target these features has to date been relatively unsuccessful (*e.g.* bapineuzumab [17], tarenflurbil [18] and semagacestat [19]).

1.2. Non-Modifiable Risk Factors: Genetics and AD Risk

Despite extensive research using humans as well as mouse, rat and fly models, the cause of AD is currently unknown. Several genes have been identified that appear to increase an individual's risk of developing late onset AD (early onset AD has a clear familial genetic component but is relatively rare and is not the topic of this chapter). One of the better-known genes is called *Apolipoprotein E* (*ApoE*), where possession of two epsilon-4 alleles increases risk of AD in comparison to possession of just one, which itself increases risk of AD compared to possession of no epsilon-4 alleles [20, 21]. In contrast, the epsilon-2 allele has been shown to be protective against AD [20]. *ApoE* has been linked to lipid transport and the breakdown of Aβ [22, 23]. The epsilon-4 allele appears to be less effective at this breakdown process than other variants [23] resulting in higher levels of amyloid plaques. Other genes implicated in AD susceptibility such as *CLU*, *BIN1* and *PICALM* are involved in mitochondrial energy metabolism and stress resistance, lipid metabolism, endocytosis and neuroinflammation [24-26]. However, these genes do not appear to predict all AD cases [27] and so it is likely that external factors, including environment and chance, play a large part. Lifestyle behaviours may thus significantly contribute to an individual's risk of developing the disorder and are an ideal avenue for research into preventative strategies.

1.3. Modifiable Protective Factors: Lifestyle, Physical Activity and Prevention of Dementia

Physical activity refers to any type of sustained movement, including day-to-day activity such as walking as well as structured exercise such as sport and fitness classes. Physical activity has long been recommended to promote physical health in the elderly because it helps to maintain a healthy weight, lowers cholesterol and blood pressure (BP), and reduces risk of disorders such as diabetes, heart disease and cancer [28]. Even a small amount of physical activity can have a huge impact on health. For example, Wen *et al.* (2011) showed that all-cause mortality over 8 years was reduced by 14% in those exercising to the equivalent of just 15 minutes per day compared to those who were inactive [29]. However, less than 40% of middle- to older-aged adults meet the recommended amount of moderate physical activity (which is 30 minutes, at least 5 times per week) [28]. Physical activity can be low cost, there is choice of different activities, and it is accessible to most regardless of ability. Encouraging an active lifestyle in the elderly is therefore a sensible intervention that is worthy of consideration. Moreover, there is a social component to group activities, which may add to their appeal and possibly their effectiveness (psychosocial mediators of the effect of physical activity on cognitive function are discussed briefly in this chapter but are described in more detail elsewhere [30]).

1.4. Mechanisms for Protective Effects of Physical Activity on Brain Function

Physical activity seems to be an attractive intervention for Alzheimer's disease but an equally important question is whether there are any theoretical mechanisms through which physical activity may impact on cognitive functioning. There are many plausible mechanisms that have been suggested, which are described below and shown in Fig. **1**.

1.4.1. Direct Causal Effects and the Association of Physical Activity with Cognitive Reserve

Physical activity is believed to increase cell proliferation and promote the maturation of cells to fully integrated neurons [31, 32]. Van der Borght, Havekes, Bos, Eggen, & Van der Zee (2007) found that running increased the number of young neurons in the mouse dentate gyrus [33]. While previously controversial, it is now generally accepted that neurogenesis occurs in two regions within the adult brain. One is the subventricular zone, the other the subgranular zone of the dentate gyrus in the hippocampus [34], an area implicated in memory [35] that is particularly affected in AD [36]. Reduced neurogenesis is not considered to be directly or solely responsible for AD but neurogenesis is known to decline in older age [37] and preventing cells from dividing impairs memory performance [38], potentially exacerbating the effects of dementia pathology. In addition, AcetylCholinesterase Inhibitor (AChEI) treatments – the primary treatment for AD that can temporarily help to slow symptom progression – increase the number of young cells and neuroblasts (cells that are mature but not yet integrated in neural networks), found *in vivo* and *in vitro* [39].

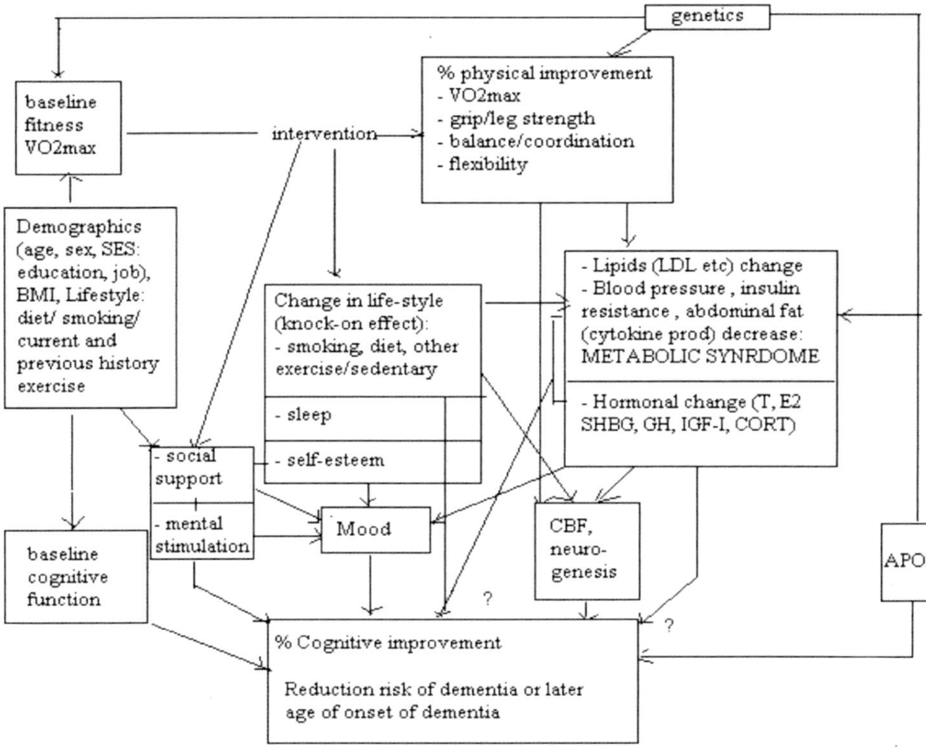

Figure 1: Theoretical mechanisms for protective effects of physical activity on cognitive function[1].

[1]Figure developed by Prof. E. Hogervorst.

Levels of neurotrophins increase after physical activity, such as Insulin-Like Growth Factor-1 (IGF-1) [40]. Exercise induced IGF-1 is associated with increased neurogenesis [41]. This may reflect a downstream process as IGF-1 induces Brain-Derived Neurotrophic Factor (BDNF), which helps to promote the survival of neural stem cells [34, 42]. BDNF supports cholinergic neurons and it is secreted by pyramidal neurons in the hippocampus [43]. Physical activity is also associated with increased levels of Granulocyte Colony Stimulating Factor (G-CSF) [44] in the cingulate and prefrontal cortices in older human adults [44]. G-CSF helps to inhibit proinflammatory cytokines, such as IL1, TNFα and IFN-γ, which are released by microglia as part of an immune response to plaque deposition, worsening dementia pathology through a cyclical relationship [45]. Direct causal effects of physical activity on cognition are therefore plausible and this makes it important to evaluate physical activity itself as an intervention for AD. However, new cells take time to become functional and integrated into the dentate gyrus [46], yet memory and attention improvements have been demonstrated after an acute bout of physical activity [47]. Neurogenesis is therefore not the only potential causal pathway through which physical activity can influence cognitive functioning.

Physical activity that increases heart rate encourages the flow of blood and oxygen around the body including the brain [48-50], preventing loss of tissue through hypoxia to which the hippocampus is thought to be particularly sensitive [51]. Brain volume is at least maintained in both gray and white matter areas after long-term physical activity [52] and perhaps particularly the right anterior frontal cortex [44] and the hippocampus [53], both of which are involved in memory. Direct neurological effects of physical activity are purported to affect cognitive functioning through what has come to be known as "cognitive reserve". Supporters of this theory suggest that the brain has the capacity to function relatively normally in the presence of disease pathology [54, 55]. This is evidenced by the observation that degree of AD pathology at post-mortem is not always indicative of decreased cognitive performance [56]. Increased cognitive reserve is not thought to significantly extend life expectancy but rather maintains cognitive abilities for longer during the lifetime of the individual [57]. Although some accounts of cognitive reserve suggest that it is a predefined threshold, others consider cognitive reserve as adaptive and the proposed impact of physical activity on cognitive reserve is largely based on neural networks being modifiable. While education and childhood IQ (which are associated with early acquired cognitive reserve) predicted age at onset of both hypertension and dementia [58], education in most models remains independent of other risk and protective factors, suggesting scope for later-life interventions based on this reserve model.

1.4.2. Association with Cardiovascular Risk Factors

Healthy lifestyles may influence cognition through their effects on cardiovascular risk factors, which are also risk factors for AD. For example, the Cardiovascular Risk Factors, Aging, and Dementia (CAIDE) study followed 1449 participants from around the mid 40th decade of their lives. The study showed that being obese (Body Mass Index [BMI] >30kg/m^2), having high Blood Pressure (BP) and high total cholesterol in midlife doubled the risk of AD 18 years later [59]. When all three factors occurred together, the risk for AD was increased by a factor 6. This was independent of age, education and reported fat intake or physical activity. Similarly, increased BP in mid-life (5th decade of life) was associated with an increased risk of AD 25 years later in untreated individuals in the Honolulu-Asia Aging Study [60]. In the Kaiser Permanente of Northern California Medical Group study, elevated cholesterol levels at 40-45 years of age were associated with a doubled risk of AD and VaD 30 years later [61]. In addition, being overweight or obese at 40-45 years of age was associated with a 2- to 3-fold increased risk of AD later in life [62]. The Swedish Adoption/Twin Study Aging study also suggested that high BMI at around age 40 years was associated with lower cognitive ability after 25 years and faster cognitive decline [63]. Physical activity reduces fat mass and lowers blood pressure and cholesterol levels [58], and it is thus a cost effective way of reducing cardiovascular and dementia risk in midlife.

1.4.3. Association with an Overall Healthy Lifestyle and Diet

Active lifestyles may appear to influence cognitive functioning because they are accompanied by other healthy behaviours that exert an independent effect on cognition. For example, certain diets help to protect against cardiovascular risk factors and they may also have direct neurological effects. A prospective study of older adults found a positive relationship between monounsaturated and polyunsaturated fatty acid intake and global cognition over 8.5 years [64]. The Mediterranean diet is low in fat but does contain high levels of antioxidants, omega-3 fatty acids and vitamin B12, which are thought to be neuroprotective. Adherence to the Mediterranean diet was associated with a 24% reduced risk of AD after adjustment for age, sex, ethnicity, education, ApoE genotype, caloric intake, smoking, comorbity index and BMI [65]. Other research suggests that caloric restriction may be more important than specific diets in affecting cognitive performance. Witte, Fobker, Gellner, Knecht and Flöel (2009) showed a significant improvement in memory scores in 50 healthy older adults who adhered to a calorie-restricted diet over three months compared to those who continued with their normal diet [66].

The associations between physical activity, diet and cognitive functioning may also be linked to insulin sensitivity. Insulin resistance is a risk factor for AD even in those who do not have diabetes [67, 68]. Increasing insulin resistance through feeding a high fat diet can lead to impaired ability on task switching in rats [69] and reduced long-term potentiation in the dentate gyrus and impaired Morris Water Maze performance [70]. In Witte *et al.*'s study of human elderly, significant relationships were found between increased insulin sensitivity and improvements in memory [66]. Drugs for diabetes have also been found to improve or maintain cognitive ability [71]. Type 2 diabetes is particularly associated with lifestyle factors, including diet and physical activity, and reducing insulin resistance through a healthy lifestyle may help to prevent cognitive decline and AD. It could thus be suggested that those who choose to engage in physical activity will also choose healthier diets and evidence exists for this confounding relationship. The effect of physical activity on cognitive functioning may be due to joint associations with other health behaviours also, such as avoidance of tobacco and illicit drug use.

1.4.4. Physical Activity may be Most Important in Midlife

Despite evidence for strong associations in midlife, cardiovascular risk factors have been shown to change in the years before dementia becomes clinically apparent, possibly due to dysfunctional regulation in the central nervous system caused by pathological changes seen in AD. For instance, a meta-review of the literature [72] showed that high BP in mid-life is associated with AD, whereas low BP in older age is associated with AD. In cross-sectional case control studies in those with established dementia, BP is often found to be lower [73]. In a longitudinal study, high BP was found to be a risk factor in mid-life for later life dementia but was seen to drop in the years preceding the clinical onset of dementia [74]. Similarly, being overweight or obese in later life (>60 years of age) has been associated with a decreased risk of dementia [75] despite mid-life obesity being associated with increased risk. This has important implications for treatment; two Cochrane reviews found no evidence that reducing cholesterol through the use of statins or reducing BP with anti-hypertensive medications lowered risk of AD and dementia in individuals above the age of 70 years [76, 77]. The neurological benefits of physical activity also suggest the benefit of midlife intervention as exercise does not appear to have an effect on Aβ once plaques and memory impairments are already present [78]. These data thus suggest a need for early midlife (rather than later life) interventions, as these risk factors tend to become unstable in later life. In addition, in midlife after menopause, levels of sex steroids (both estrogens and testosterone) drop dramatically in women and men also start to show a gradual decline [58]. Physical activity affects levels of sex hormones which may themselves influence cognitive functioning. For example, free testosterone, which decreases with age [79], affects many factors implicated in pathological brain aging and reduced levels have been associated with dementia risk [80]. Estrogen also has various potential protective effects on brain function [81]. Part of the effect of exercise may thus be mediated by its effects on sex steroids but this remains to be further investigated.

1.4.5. Association with Psychological Mediators

Physical activity is not carried out independently from the surrounding environment and there is also great variation in specific features of exercise interventions. These variations include: the environment (outdoors or indoors, in a gym/laboratory setting or at home), the level of social interaction (alone or in a group, supervised by an instructor or self-led), the purpose of the exercise (targets and goals set or exercise with no feedback for individuals to see their progression) and mental engagement (a mentally passive exercise of walking on a treadmill or playing a sport that requires mental effort, attention and decision making). These varying interactions between the physical activity environments and specific features may have different psychological impacts on individuals, which could impact upon the success of the intervention. Intervention and observational studies in non-exercise contexts have shown that psychological factors, including better mood [82], social engagement [83], mental stimulation [84], positive attitudes to ageing [85] and positive self-perceptions [86, 87] were associated with improvements in health and longevity, as well as cognitive performance. Therefore, the psychological impact of exercise needs to be considered in this context due to this potentially mediating relationship [30].

Potential mechanisms through which psychological mediators could impact on the effectiveness of exercise interventions have also been discussed in more detail elsewhere [30]. The psychological mediators discussed include self-efficacy, perceived control manifesting in a self-fulfilling prophecy, self-perceptions, attitudes to ageing, causal attributions to memory problems and mood. For example, using targets and goal setting during an exercise intervention increases the scope for improvements in self-efficacy through sense of achievement and mastery experiences. This is due to the progression of individual's strength and fitness being recorded and assessed in relation to whether goals were achieved and therefore improvements are more apparent. It was suggested that by improving self-efficacy in one domain (*e.g.* physical ability through exercise intervention), core beliefs of what older adults believe that they are capable of and the degree to which their decline is inevitable and out of their control are challenged [30]. Therefore, more positive core beliefs about ageing could manifest better cognitive performance through greater effort to remember things, employing memory strategies and an increased interest in paying attention to things that they believe they can remember. Further research is needed to understand the magnitude of the effect of psychological mediators in relation to physiological mechanisms and other factors discussed in this chapter.

1.5. Moderators of the Association of Physical Activity and Cognitive Functioning

Physical activity presents as a potential lifestyle intervention with evidence pointing towards the need for midlife intervention before dementia pathology is present. There appear to be multiple mechanisms through which physical activity can affect cognitive performance and reduce the risk of dementia through associations with other lifestyle behaviours, cardiovascular risk factors and cognitive reserve. In practice, it is likely that the beneficial effects of physical activity and cognitive function are due to a combination of factors, both direct and indirect. However, despite these mechanisms and a wide range of studies reporting associations between increased physical activity and cognitive functioning, there are possibly as many published studies that have failed to replicate these findings and show little or no discernible effect of physical activity. A large review of exercise studies for cognitive decline and dementia, which incorporated intervention and observational studies in healthy older adults, cohort studies of dementia risk and intervention studies in elderly with dementia, attempted to qualitatively examine some of the reasons for these discrepancies in findings [88]. This review identified certain variables that may act as moderating factors influencing the relationship between physical activity and cognitive functioning, with the association presenting for specific samples or situations only. This potentially distorts the true magnitude of the effect of physical activity on cognitive functioning, which could have huge implications for theory and practice. These variables are discussed in detail below.

1.5.1. Moderators of Effects of Physical Activity on Cognition: Tests and Functions

A clinical diagnosis of AD is ultimately the most desirable outcome measure when studying disease prevention, but the lengthy follow-ups required to conduct adequate prospective studies into AD mean that cognitive testing is often used as an alternative measure to assess short-term changes in brain function. However, the type of cognitive measurement used in observational studies of healthy adults and in Randomised Controlled Trials (RCTs) seems to have a strong effect on the findings of those studies. Although Hall, Smith and Keele (2001) [89] and Colcombe and Kramer (2003) [90] suggested that the benefits of physical activity were restricted to complex executive function processes, a review of the literature suggests that studies that used a measure of memory, especially the digit span, as the outcomes were more likely to find a significant benefit of physical activity compared to those using a measure of executive function [88]. Measures of memory such as digit span, word list learning and word generation are sensitive to AD, and so the cognitive benefits of physical activity in middle- and older-age adults appear to be particularly relevant to AD. This is also supported by previous studies that have found reduced risk of AD in those who are physically active [91-95].

Although it has been observed that some tests of executive function are sensitive to AD [96-99], performance on these tasks appears to be less affected by physical activity [88]. Indeed, the authors recently conducted an RCT of resistance training in twenty middle-aged adults. After 12 weeks of exercise, memory (as measured by the Hopkins Verbal Learning Task and a word generation task) was improved but executive function ability was not [100]. It has been observed that AD patients with early executive function deficits are less likely to respond to AChEI medication, and poor executive performance may thus be indicative of early or particularly severe pathology [101] that requires more targeted interventions. However, the finding of a lack of an association between physical activity and executive function remains controversial [102]. A longitudinal study showed that young-elderly with high fitness maintained executive function abilities over 6 years while those with low baseline fitness showed decline [103]. Rather than being all-or-nothing, these differences may reflect temporal differences – memory may show immediate benefits of physical activity, while executive function may require longer-term interventions to show an effect.

The observed differences in exercise's effects between memory and executive function can be explained by the neurogenesis hypothesis described previously. Neurogenesis is currently thought to occur mostly in the hippocampus [34], an area that is important for memory function [35]. Since physical activity is thought to have an impact on neurogenesis, memory processes are therefore more likely to benefit compared to functions mediated by other cortical areas. However, the brain is highly interconnected and the prefrontal cortex has connections with many other regions including the medial temporal lobe, which contains the hippocampus [104]. Prolonged physical activity over many years may thus impact on executive function abilities through subsequent effects of stimulating the hippocampus on the rest of the brain. Executive function may also benefit from improved angiogenesis after physical activity, possibly due to the increased availability of IGF-1 [105, 106] or other direct vascular effects, such as enhanced perfusion leading to reduced risk for hypoxia [48-50]. Neuroimaging methods may help to sequence changes in these regions at various points through an exercise intervention to determine whether temporal differences are seen in the effect of physical activity on different cognitive abilities.

1.5.2. Moderators of Effects of Physical Activity on Cognition: Types of Exercise

Different studies have investigated the impact of different types of exercise, which makes it difficult to determine what type of physical activity is most effective at improving cognitive functioning [88]. Aerobic activity and resistance training both appear to have a strong effect on memory and even low-intensity exercise has been observed to improve memory performance [100, 107, 108]. These effects are likely to be due to the range of physiological effects that aerobic activity has on the body that are also implicated in cognitive performance, such as improved lung and vascular function, although improvements in VO_2 max and other physiological and/or health measures have not always been associated with improvements in cognitive functioning [107, 109, 110]. Resistance training is of interest as it does not focus on an aerobic component *per se*, but rather on physiological effects that come from strength changes which may be associated with the changes in sex steroids and subsequent improved cognition mentioned earlier. For example, Williams and Lord (1997) [111] found that changes in muscle strength predicted cognitive change over 12 months (although the limitations of this association are discussed later). However, the effects of resistance training on cognitive performance are also not necessarily associated with changes in strength. In the authors' RCT, changes in grip strength following the intervention were not correlated with memory improvements [100]. Further research should thus consider the influence of other neurophysiological factors, such as growth or sex hormones as well as psychological mediators [30], as possible pathways between physical activity and cognitive functioning. It is currently unclear whether the effects of aerobic and non-aerobic activity are additive or would interact to produce optimal benefits, and interventions using both, one and neither of the two types of activity should be trialled to investigate this.

In contrast, there is little evidence that yoga and flexibility-type exercises have any benefit to cognition. Our recent RCT showed benefits to memory from resistance training but not flexibility exercises that were matched for time commitment [100]. It could be argued that participants may not enjoy a particular intervention and so may not engage as strongly as they do with an intervention they enjoy; however, this null finding has been observed throughout the literature [88]. As the observed benefits of flexibility exercise and yoga often include improved mood, it suggests that mood is not a strong mediator of the association between physical activity and cognition, although this requires further validation. However, as balance may benefit from yoga exercises, which could lead to reduced risk for falls, this may have a long-term benefit as the risk for dementia increases after falls. Flexibility could be assessed in future longitudinal studies as an additional component to an exercise intervention to help reduce risk of dementia further.

1.5.3. Moderators of Effects of Physical Activity on Cognition: Gender

Certain groups may benefit more from physical activity than others, with most of the effects of physical activity being seen in women but not in men. This has been observed in several longitudinal studies of physical activity and dementia risk [92, 112, 113]. A direct comparison of effect sizes for the association of sport with memory in an elderly Indonesia cohort (n=719) supported this, showing a significantly stronger effect in women compared to in men [114]. These differences thus appear to be relatively robust and universal across cultures, and they are perhaps due to the range of physiological mechanisms through which physical activity affects cognitive functioning. Sex steroids affect brain function through multiple mechanisms and women show a sharp decline in both estrogens and testosterone after the menopause around the average age of 50 years [81]. For example, changes in testosterone following resistance training may benefit women more so than men who already have higher levels of testosterone. In addition, there is evidence that estrogen increases BDNF, as does physical activity as described previously [34]. When relatively high levels of estrogen and physical activity co-occur, they may interact such that the benefit on cognition is increased compared to when they occur alone [115]. Due to the range of physiological and psychological mechanisms possible, different types of physical activity may be beneficial to men and women separately.

Studies of men did show some benefit of physical activity [94, 108, 116] and so it should not be considered a female-only intervention. However, it is clear that sex should be considered as a moderating factor rather than a covariate in this relationship. Studies in this field should always compare the effect size between men and women to determine the true effect of the intervention in both groups as the inclusion of men in the analysis has been seen to attenuate the effect size in women, masking the true benefit. In practice, exercise interventions could be particularly targeted at women in middle-age and could be incorporated into existing healthcare advice for this population.

1.5.4. Moderators of Effects of Physical Activity on Cognition: Genetics

Previous literature has suggested that baseline risk of AD, as indicated by genetic status, might mediate the relationship between physical activity and cognitive ability [91, 93]. However, few studies had looked at this and conflicting findings have meant that it is not clear whether the benefit is greatest in those who are ApoE4 positive or negative, or whether this is simply a chance finding from repeated analysis. Our randomised trial of resistance training found a possible effect of education, with those with more years of education showing a faster cognitive improvement after physical activity. As discussed previously, low education level is a risk factor for AD, suggesting that groups with additional risk factors for AD and cognitive decline may be less responsive to short-term treatment interventions. However, it is at present unclear whether they may benefit from longer-term interventions. Any risk factors for AD should therefore be considered as additional moderating factors and interactions between risk factors and intervention group should be analysed to fairly evaluate the intervention's effectiveness. This does have implications for those at high risk for AD, suggesting that they may be less responsive to normal interventions, and this group should receive specific attention in future research.

1.5.5. Moderators of Effects of Physical Activity on Cognition: Mental Activity

Ultimately, the discrepancies in previous research may be due to physical activity alone not being sufficient to maintain cognitive health. Physical activity is thought to promote the release of neural stem cells [43] but there is little evidence that it promotes the survival of these cells. Instead, it may be that mental stimulation and learning is required to integrate these cells into existing networks; if they are not recruited into existing networks and used, they might not offer benefit to cognitive performance. This would also indicate that physical activity with a mental or social component will be more beneficial compared to physical activity that occurs alone. Indeed, completing both physical and mental activities have been shown to create significant improvement in memory performance compared to completing only one or neither activity over 8 weeks [117]. Rather than being seen as two different interventions, physical and mental stimulation may be considered as complementary; the success of a physical activity intervention against cognitive decline may depend upon the mental stimulation that participants receive outside of the study, while research into the benefits of mental stimulation may be most successful if participants are also physically active. Therefore, physical activity may be important but not sufficient for prolonged cognitive health. Although this is speculation, building upon the current literature to combine these interventions may help to investigate these claims further.

1.6. Limitations of Current Research

Although prospective studies are useful for assessing the risk of developing AD over a given timescale based on a set of baseline predictors, these studies take a long time to conduct and it is difficult to control for variables such as type of activity, intensity and mental stimulation. In addition, although we predict that physical activity in midlife before pathology begins is most important, it may not be possible to measure this using cognitive tasks in healthy adults. Although it has been observed previously that physical activity leads to short-term improvements in cognitive function, the ultimate effect of long-term physical activity may be maintenance of function. Giving the brain the resources to adapt to pathology does not mean that it will function more efficiently when already performing at its optimum. Healthy, high functioning middle-aged adults are therefore unlikely to show much observable benefit of most physical activity until they reach older age. One way to get around this problem is to use MRI in healthy middle-aged adults during an RCT. This type of study would be able to assess functional changes that may occur after long-term physical activity, even if these changes do not show up on tests of memory. Another option is to study MCI patients as part of an RCT. Although MCI patients may already show dementia pathology, their brains might still have the capacity to adapt and show improvements on cognitive tasks after an intervention. Although not all MCI patients go on to develop AD, the difference in conversion rate to AD could be used to evaluate the intervention.

CONCLUSION

Alzheimer's disease and other types of age-related cognitive impairment is a growing problem around the world but there is evidence that physical activity is a potential lifestyle intervention that can help to lower an individual's risk of developing symptoms of the disease. If these findings can be replicated in larger controlled studies then they would support government funded exercise programmes for the elderly and suggest that a simple home based and cheap resistance exercise programme should be further developed to enable the full benefits and participation in those who find aerobic activity especially difficult (*e.g.* those with asthma, Chronic Obstructive Pulmonary Disease *etc.*). However, physical activity should not be considered a cure for Alzheimer's disease or a solution to extinguish the presence of dementia. Not only is there currently little evidence that physical activity influences memory in those already affected by cognitive impairment [88], the theories presented in this chapter propose that the effects of physical activity work on an individual level, lessening risk rather than preventing completely. Various factors appear to mediate or moderate these findings and it is important to consider their influence when evaluating the effectiveness of physical activity as an intervention against AD and other age-related cognitive impairments. Thus, physical activity is likely to have its limitations (especially counteracting the impact of genetics) if used independently without consideration of other risk factors. Prevention of dementia should focus on overall lifestyle change including smoking cessation and dietary alterations.

ACKNOWLEDGEMENT

This chapter was prepared during Angela Clifford's doctoral studentship which was funded by Loughborough University, UK.

CONFLICT OF INTEREST

The author(s) confirm that this chapter content has no conflict of interest.

REFERENCES

[1] Karenberg A, Forstl H. Dementia in the Greco-Roman world. Journal of the Neurological Sciences 2006; 244(1-2): 5-9. doi:10.1016/j.jns.2005.12.004

[2] Hardy J. Amyloid, the presenilins and Alzheimer's disease. Trends in Neurosciences 1997; 20(4): 154-159.

[3] American Psychiatric Association. Diagnostic and statistical manual of mental disorders (4th ed.). Washington, D.C.: American Psychiatric Press 1994.

[4] Bishop NA, Lu T, Yankner BA. Neural mechanisms of ageing and cognitive decline. Nature 2010; 464(7288): 529-535. doi:10.1038/nature08983

[5] World Health Organisation. Tenth revision of the international classification of diseases and related health problems (ICD-10). Geneva: WHO 1992.

[6] Dubois B, Feldman HH, Jacova C, *et al.* Revising the definition of Alzheimer's disease: A new lexicon. The Lancet Neurology 2010; 9(11): 1118-1127. doi:10.1016/S1474-4422(10)70223-4

[7] McKhann GM, Knopman DS, Chertkow H, *et al.* The diagnosis of dementia due to Alzheimer's disease: Recommendations from the National Institute on Aging- Alzheimer's Association workgroups on diagnostic guidelines for Alzheimer's disease. Alzheimer's and Dementia 2011; 7(3): 263-269. doi:10.1016/j.jalz.2011.03.005

[8] Petersen RC, Smith GE, Waring SC, Ivnik RJ, Tangalos EG, Kokmen E. Mild cognitive impairment: Clinical characterization and outcome. Archives of Neurology 1999; 56(3): 303-308.

[9] Geslani DM, Tierney MC, Herrmann N, Szalai JP. Mild cognitive impairment: an operational definition and its conversion rate to Alzheimer's disease. Dement Geriatr Cogn Disord 2005; 19(5-6): 383-389. doi: 10.1159/000084709

[10] Dubois B, Feldman HH, Jacova C, *et al.* Research criteria for the diagnosis of Alzheimer's disease: revising the NINCDS-ADRDA criteria. The Lancet Neurology 2007; 6(8): 734-746. doi:10.1016/S1474-4422(07)70178-3

[11] Golomb J, Kluger A, de Leon MJ, *et al.* Hippocampal formation size in normal human aging: a correlate of delayed secondary memory performance. Learning & Memory 1994; 1(1): 45-54.

[12] Lind J, Larsson A, Persson J, *et al.* Reduced hippocampal volume in non-demented carriers of the apolipoprotein E ε4: Relation to chronological age and recognition memory. Neurosci Lett 2006; 396(1): 23-27. doi:10.1016/j.neulet.2005.11.070

[13] de Toledo-Morrell L, Dickerson B, Sullivan MP, Spanovic C, Wilson R, Bennett DA. Hemispheric differences in hippocampal volume predict verbal and spatial memory performance in patients with Alzheimer's disease. Hippocampus 2000; 10(2): 136- 142.

[14] Priller C, Bauer T, Mitteregger G, Krebs B, Kretzschmar HA, Herms J. Synapse formation and function is modulated by the amyloid precursor protein. The Journal of Neuroscience 2006; 26(27): 7212-7221. doi:10.1523/JNEUROSCI.1450- 06.200

[15] Hyman BT, Phelps CH, Beach TG, *et al.* National institute on Aging–Alzheimer's association guidelines for the neuropathologic assessment of Alzheimer's disease. Alzheimer's and Dementia 2012; 8(1): 1-13. doi:10.1016/j.jalz.2011.10.007

[16] Mohajeri MH, Leuba G. Prevention of age-associated dementia. Brain Research Bulletin 2009; 80(4-5): 315-325. doi:10.1016/j.brainresbull.2009.06.014

[17] Johnson & Johnson [webpage on the internet]. Johnson & Johnson Announces Discontinuation Of Phase 3 Development of Bapineuzumab Intravenous (IV) In Mild- To-

Moderate Alzheimer's Disease [Updated 6th August 2012; cited 10th Aug 2012]. Available from: http://www.jnj.com/connect/news/all/johnson-and-johnson- announces-discontinuation-of-phase-3-development-of-bapineuzumab-intravenous- iv-in-mild-to-moderate-alzheimers-disease

[18] Myriad Genetics [webpage on the internet]. Myriad Genetics Reports Results of U.S. Phase 3 Trial of Flurizan™ in Alzheimer's Disease [Updated 30th June 2008; cited 10th Aug 2012]. Available from: http://investor.myriad.com/releasedetail.cfm?ReleaseID=325471

[19] Eli Lilly [webpage on the internet]. Lilly Halts Development of Semagacestat for Alzheimer's Disease Based on Preliminary Results of Phase III Clinical Trials [Updated 17th Aug 2010; cited on 10th Aug 2012]. Available from: http://newsroom.lilly.com/releasedetail.cfm?releaseid=499794

[20] Corder EH, Saunders AM, Strittmatter WJ, *et al.* Gene dose of apolipoprotein E type 4 allele and the risk of Alzheimer's disease in late onset families. Science 1993; 261(5123): 921-923. doi:10.1126/science.8346443

[21] Tsai MS, Tangalos EG, Petersen RC, *et al.* Apolipoprotein E: risk factor for Alzheimer disease. Am J Hum Genet 1994; 54(4): 643-649.

[22] Raber J. AR, apoE, and cognitive function. Hormones and Behavior 2008; 53(5): 706- 715. doi:10.1016/j.yhbeh.2008.02.012

[23] Baum L, Chen L, Ng HK, Pang CP. Apolipoprotein E isoforms in Alzheimer's disease pathology and etiology. Microscopy Research and Technique 2000; 50(4): 278-281. doi:10.1002/1097-0029(20000815)

[24] Harold D, Abraham R, Hollingworth P, *et al.* Genome-wide association study identifies variants at CLU and PICALM associated with Alzheimer's disease. Nature Genetics 2009; 41(10): 1088-1093. doi:10.1038/ng.440

[25] Hu X, Pickering E, Liu YC, *et al.* Meta-analysis for genome-wide association study identifies multiple variants at the BIN1 locus associated with late-onset Alzheimer's disease. PloS One 2011; 6(2): e16616. doi:10.1371/journal.pone.0016616

[26] Liang WS, Reiman EM, Valla J, *et al.* Alzheimer's disease is associated with reduced expression of energy metabolism genes in posterior cingulate neurons. Proceedings of the National Academy of Sciences 2008; 105(11): 4441-4446. doi:10.1073/pnas.0709259105

[27] Rocchi A, Pellegrini S, Siciliano G, Murri L. Causative and susceptibility genes for Alzheimer's disease: a review. Brain Res Bull 2003; 61(1): 1-24. doi:10.1016/S0361-9230(03)00067-4,

[28] Haskell WL, Lee IM, Pate RR, *et al.* Physical activity and public health: Updated recommendation for adults from the American College of Sports Medicine and the American Heart Association. Medicine & Science in Sports & Exercise 2007; 39(8): 1423-1434. doi:10.1249/mss.0b013e3180616b27

[29] Wen CP, Wai JPM, Tsai MK, *et al.* Minimum amount of physical activity for reduced mortality and extended life expectancy: a prospective cohort study. The Lancet 2011; 378(9798): 1244-1253. doi:10.1016/S0140-6736(11)60749-6

[30] Stock J, Clifford A, Hogervorst E. Exercise Interventions to Improve Cognitive Performance in Older Adults – Potential Psychological Mediators to Explain Variation in Findings. European Neurological Review 2012; 7(2): 107-112

[31] van Praag H, Christie BR, Sejnowski TJ, Gage FH. Running enhances neurogenesis, learning, and long-term potentiation in mice. Proceedings of the National Academy of Sciences of the United States of America 1999; 96(23): 13427-13431.

[32] Kim S, Ko I, Kim B, *et al.* Treadmill exercise prevents aging-induced failure of memory through an increase in neurogenesis and suppression of apoptosis in rat hippocampus. Experimental Gerontology 2010; 45(5): 357-365. doi:10.1016/j.exger.2010.02.005

[33] Van der Borght K, Havekes R, Bos T, Eggen BJL, Van der Zee EA. Exercise improves memory acquisition and retrieval in the Y-maze task: Relationship with hippocampal neurogenesis. Behavioral Neuroscience 2007; 121(2): 324. doi:10.1037/07357044.121.2.324

[34] Zhao C, Deng W, Gage FH. Mechanisms and functional implications of adult neurogenesis. Cell 2008; 132(4): 645-660. doi:10.1016/j.cell.2008.01.033

[35] Diana RA, Yonelinas AP, Ranganath C. Imaging recollection and familiarity in the medial temporal lobe: A three-component model. Trends in Cognitive Sciences 2007; 11(9): 379-386. doi:10.1016/j.tics.2007.08.001

[36] Rombouts SA, Barkhof F, Witter MP, Scheltens P. Unbiased whole-brain analysis of gray matter loss in Alzheimer's disease. Neuroscience Letters 2000; 285(3): 231- 233. doi:10.1016/S0304-3940(00)01067-3

[37] Verret L, Trouche S, Zerwas M, Rampon C. Hippocampal neurogenesis during normal and pathological aging. Psychoneuroendocrinology 2007; 32(Supplement 1): S26- S30. doi:10.1016/j.psyneuen.2007.04.014

[38] Shors TJ, Miesegaes G, Beylin A, Zhao M, Rydel T, Gould E. Neurogenesis in the adult is involved in the formation of trace memories. Psychol.Rev 1995; 102: 101-130.

[39] DeCarolis NA, Eisch AJ. Hippocampal neurogenesis as a target for the treatment of mental illness: A critical evaluation. Neuropharmacology 2010; 58(6): 884-893. doi:10.1016/j.neuropharm.2009.12.013

[40] Nakajima S, Ohsawa I, Ohta S, Ohno M, Mikami T. Regular voluntary exercise cures stress-induced impairment of cognitive function and cell proliferation accompanied by increases in cerebral IGF-1 and GST activity in mice. Behavioural Brain Research 2010; 211(2): 178-184. doi:10.1016/j.bbr.2010.03.028

[41] LLorens-Martín M, Torres-Alemán I, Trejo JL. Exercise modulates insulin-like growth factor 1-dependent and -independent effects on adult hippocampal neurogenesis and behaviour. Molecular and Cellular Neuroscience 2010; 44(2): 109-117. doi:org/10.1016/j.mcn.2010.02.006

[42] Bothwell M. Functional interactions of neurotrophins and neurotrophin receptors. Annual Review of Neuroscience 1995; 18(1): 223-253.

[43] Schaeffer EL, Novaes BA, da Silva ER, Skaf HD, Mendes-Neto ÁG. Strategies to promote differentiation of newborn neurons into mature functional cells in Alzheimer brain. Progress in Neuro-Psychopharmacology and Biological Psychiatry 2009; 33(7): 1087-1102. doi:10.1016/j.pnpbp.2009.06.024

[44] Flöel A, Ruscheweyh R, Kruger K, *et al.* Physical activity and memory functions: Are neurotrophins and cerebral gray matter volume the missing link? NeuroImage 2010; 49(3): 2756-2763. doi:10.1016/j.neuroimage.2009.10.043

[45] Nichol KE, Poon WW, Parachikova AI, Cribbs DH, Glabe CG, Cotman CW. Exercise alters the immune profile in Tg2576 Alzheimer mice toward a response coincident with improved cognitive performance and decreased amyloid. J Neuroinflammation 2008; 5: 13. doi:10.1186/1742-2094-5-13

[46] Zhao C, Teng EM, Summers Jr RG, Ming G, Gage FH. Distinct morphological stages of dentate granule neuron maturation in the adult mouse hippocampus. The Journal of Neuroscience 2006; 26(1): 3-11. doi: 10.1523/JNEUROSCI.3648-05.2006

[47] Hogervorst E, Bandelow S, Clifford A, Stock J. Acute physical activity and cognitive performance - a comparison of cycling, dancing and Sudoku. Unpublished raw data 2010.
[48] Querido JS, Sheel AW. Regulation of cerebral blood flow during exercise. Sports Medicine (Auckland, N.Z.) 2007; 37(9): 765-782.
[49] Ogoh S. Autonomic control of cerebral circulation: Exercise. Medicine and Science in Sports and Exercise 2008; 40(12):2046-2054. doi:10.1249/MSS.0b013e318180bc6f
[50] Rhyu IJ, Bytheway JA, Kohler SJ, *et al.* Effects of aerobic exercise training on cognitive function and cortical vascularity in monkeys. Neuroscience 2010; 167(4): 1239-1248. doi:10.1016/j.neuroscience.2010.03.003
[51] Di Paola M, Caltagirone C, Fadda L, Sabatini U, Serra L, Carlesimo GA. Hippocampal atrophy is the critical brain change in patients with hypoxic amnesia. Hippocampus 2008; 18(7): 719-728. doi: 10.1002/hipo.20432
[52] Colcombe SJ, Erickson KI, Scalf PE, *et al.* Aerobic exercise training increases brain volume in aging humans. The Journals of Gerontology Series A: Biological Sciences and Medical Sciences 2006; 61(11): 1166.
[53] Yuede CM, Zimmerman SD, Dong H, *et al.* Effects of voluntary and forced exercise on plaque deposition, hippocampal volume, and behavior in the Tg2576 mouse model of Alzheimer's disease. Neurobiology of Disease 2009; 35(3): 426-432. doi:10.1016/j.nbd.2009.06.002
[54] Nithianantharajah J, Hannan AJ. The neurobiology of brain and cognitive reserve: Mental and physical activity as modulators of brain disorders. Progress in Neurobiology 2009; 89(4): 369-382.doi:10.1016/j.pneurobio.2009.10.001,
[55] Stern Y. What is cognitive reserve? Theory and research application of the reserve concept. Journal of the International Neuropsychological Society 2002; 8(3): 448- 460. doi:10.1017.S1355617701020240
[56] Katzman R, Aronson M, Fuld P, *et al.* Development of dementing illnesses in an 80- year-old volunteer cohort. Ann Neurol 1989; 25(4): 317-324.
[57] Marioni RE, van den Hout A, Valenzuela MJ, Brayne C, Matthews FE. Active cognitive lifestyle associates with cognitive recovery and a reduced risk of cognitive decline. Journal of Alzheimer's Disease 2012; 28(1): 223-230. doi:10.3233/JAD-2011-110377
[58] Hogervorst E, Clifford A, Stock J, Xin X, Bandelow S. Exercise to prevent cognitive decline and Alzheimer's disease: For whom, when, what, and (most importantly) how much? Journal of Alzheimer's Disease and Parkinsonism 2012; 2: e117. doi:10.4172/2161-0460.1000e117
[59] Kivipelto M, Ngandu T, Fratiglioni L, *et al.* Obesity and vascular risk factors at midlife and the risk of dementia and Alzheimer disease. Archives of Neurology 2005; 62(10): 1556-1560.
[60] Launer LJ, Ross GW, Petrovitch H, *et al.* Midlife blood pressure and dementia: The Honolulu-Asia Aging Study. Neurobiology of Aging 2000; 21(1): 49-55.
[61] Solomon A, Kivipelto M, Wolozin B, Zhou J, Whitmer RA. Midlife serum cholesterol and increased risk of Alzheimer's and vascular dementia three decades later. Dementia and Geriatric Cognitive Disorders 2009; 28(1): 75-80. doi:10.1159/000231980
[62] Whitmer RA, Gunderson EP, Quesenberry CP Jr, Zhou J, Yaffe K. Body mass index in midlife and risk of Alzheimer disease and vascular dementia. Current Alzheimer Research 2007; 4(2): 103-109. doi:10.2174/156720507780362047
[63] Dahl A, Hassing LB, Fransson E, *et al.* Being overweight in midlife is associated with lower cognitive ability and steeper cognitive decline in late life. The Journals of

Gerontology Series A: Biological Sciences and Medical Sciences 2009; 65(1): 57-62. doi:10.1093/gerona/glp035

[64] Solfrizzi V, D'Introno A, Colacicco AM, *et al*. Dietary fatty acids intake: Possible role in cognitive decline and dementia. Experimental Gerontology 2005; 40(4): 257-270. doi:10.1016/j.exger.2005.01.001

[65] Scarmeas N, Stern Y, Tang MX, Mayeux R, Luchsinger JA. Mediterranean diet and risk for Alzheimer's disease. Annals of Neurology 2006; 59(6): 912-921. doi:10.1002/ana.20854

[66] Witte AV, Fobker M, Gellner R, Knecht S, Flöel A. Caloric restriction improves memory in elderly humans. Proceedings of the National Academy of Sciences 2009; 106(4): 1255. doi:10.1073/pnas.0808587106

[67] Rönnemaa E, Zethelius B, Sundelöf J, *et al*. Impaired insulin secretion increases the risk of Alzheimer disease. Neurology 2008; 71(14): 1065-1071. doi:10.1212/01.wnl.0000310646.32212.3a

[68] Luchsinger JA, Tang MX, Shea S, Mayeux R. Hyperinsulinemia and risk of Alzheimer disease. Neurology 2004; 63(7): 1187-1192. doi:10.1212/01.WNL.0000140292.04932.87

[69] McNeilly AD, Williamson R, Sutherland C, Balfour DJ, Stewart CA. High fat feeding promotes simultaneous decline in insulin sensitivity and cognitive performance in a delayed matching and non-matching to position task. Behav Brain Res 2011; 217(1): 134-41. doi:10.1016/j.bbr.2010.10.017

[70] Stranahan AM, Norman ED, Lee K, *et al*. Diet-induced insulin resistance impairs hippocampal synaptic plasticity and cognition in middle-aged rats. Hippocampus 2008; 18(11): 1085. doi: 10.1002/hipo.20470

[71] Craft S, Baker LD, Montine TJ, *et al*. Intranasal insulin therapy for Alzheimer disease and amnestic mild cognitive impairment: A pilot clinical trial. Archives of Neurology 2012; 69(1): 29. doi:10.1001/archneurol.2011.233

[72] Qiu C, Winblad B, Fratiglioni L. The age-dependent relation of blood pressure to cognitive function and dementia. Lancet Neurology 2005; 4(8): 487-499. doi:10.1016/S1474-4422(05)70141-1

[73] Hogervorst E, Ribeiro HM, Molyneux A, Budge M, Smith AD. Plasma homocysteine levels, cerebrovascular risk factors, and cerebral white matter changes (leukoaraiosis) in patients with Alzheimer disease. Archives of Neurology 2002; 59(5): 787.

[74] Skoog I, Gustafson D. Update on hypertension and Alzheimer's disease. Neurological Research 2006; 28(6): 605-611. doi:10.1179/016164106X130506

[75] West NA, Haan MN. Body adiposity in late life and risk of dementia or cognitive impairment in a longitudinal community-based study. The Journals of Gerontology Series A: Biological Sciences and Medical Sciences 2009; 64A(1): 103-109. doi: 10.1093/gerona/gln006

[76] McGuinness B, Craig D, Bullock R, Passmore P. Statins for the prevention of dementia. Cochrane Database Syst Rev 2009; 15(2): CD003160. doi:10.1002/14651858.CD003160.pub2

[77] McGuinness B, Todd S, Passmore P, Bullock R. Blood pressure lowering in patients without prior cerebrovascular disease for prevention of cognitive impairment and dementia. Cochrane Database Syst Rev 2009; 7(4): CD004034. doi:10.1002/14651858.CD004034.pub3

[78] Richter H, Ambrée O, Lewejohann L, *et al*. Wheel-running in a transgenic mouse model of Alzheimer's disease: protection or symptom? Behav Brain Res 2008; 190(1): 74- 84.

[79] Shkurnikov MU, Donnikov AE, Akimov EB, Sakharov DA, Tonevitsky AG. Free testosterone as marker of adaptation to medium-intensive exercise. Bulletin of Experimental Biology and Medicine 2008; 146(3): 354-357. doi:10.1007/s10517- 008-0292-2

[80] Hogervorst E, Bandelow S, Combrinck M, Smith AD. Low free testosterone is an independent risk factor for Alzheimer's disease. Experimental Gerontology 2004; 39(11-12): 1633-1639. doi:10.1016/j.exger.2004.06.019

[81] Hogervorst E, Henderson VW, Gibbs RB, Brinton RD. Hormones, Cognition and Dementia: State of the Art and Emergent Therapeutic Strategies. Cambridge, UK: Cambridge University Press 2009.

[82] Ganguli M, Du Y, Dodge HH, *et al.* Depressive symptoms and cognitive decline in late life: a prospective epidemiological study. Arch Gen Psychiatr 2006; 63: 153–60.

[83] Saczynski JS, Pfeifer LA, Masaki K, *et al.* The effect of social engagement on incident dementia. Am J Epidemiol 2006; 163(5): 433–40.

[84] Salthouse TA. Mental exercise and mental aging. Perspect Psychol Sci 2006; 1(1): 68– 87.

[85] Langer EJ, Rodin J. The effects of choice and enhanced personal responsibility for the aged: A field experiment in an institutional setting. J Pers Soc Psychol 1976; 34(2): 191–8.

[86] Levy BR, Slade MD, Kunkel SR, Kasl SV. Longevity increased by positive self-perceptions of aging. J Pers Soc Psychol 2002; 83(2): 261-70.

[87] Levy BR, Slade MD, Kasl SV. Longitudinal benefit of positive self-perceptions of aging on functional health. J Gerontol B Psychol Sci Soc Sci 2002; 57(5): 409-17.

[88] Clifford A, Bandelow S, Hogervorst E. The effects of physical exercise on cognitive function in the elderly: a review. In: Q. Gariépy & R. Ménard. Handbook of Cognitive Aging: Causes, Processes and Effects. New York: Nova Science Publishers 2009; pp. 109-150.

[89] Hall CD, Smith AL, Keele SW. The impact of aerobic activity on cognitive function in older adults: a new synthesis based on the concept of executive control. European Journal of Cognitive Psychology 2001; 113(1/2): 279-300.

[90] Colcombe SJ, Kramer AF. Fitness effects on the cognitive function of older adults: a meta-analytic study. Psychological Science 2003; 14(2): 125-130.

[91] Rovio S, Kareholt I, Helkala E, *et al.* Leisure-time physical activity at midlife and the risk of dementia and Alzheimer's disease. Lancet Neurology 2005; 4: 705-711.

[92] Laurin D, Verreault R, Lindsay J, MacPherson K, Rockwood K. Physical activity and risk of cognitive impairment and dementia in elderly persons. Archives of Neurology 2001; 58: 498-504.

[93] Podewils LJ, Guallar E, Kuller LH, *et al.* Physical activity, APOE genotype, and dementia risk: findings from the Cardiovascular Health Cognition Study. American Journal of Epidemiology 2005; 161: 639-651.

[94] Abbott RD, White LR, Webster Ross G, Masaki KH, Curb JD, Petrovitch H. Walking and dementia in physically capable elderly men. JAMA 2004; 292(12): 1447-1453.

[95] Lindsay J, Laurin D, Verreault R, *et al.* Risk factors for Alzheimer's disease: a prospective analysis from the Canadian study of health and aging. American Journal of Epidemiology 2002; 156(5): 445-453.

[96] Bondi MW, Serody AB, Chan AS, *et al.* Cognitive and neuropathologic correlates of SCWT color-word test performance in Alzheimer's disease. Neuropsychology 2002; 16(3): 335.

[97] Amieva H, Lafont S, Rouch-Leroyer I, *et al.* Evidencing inhibitory deficits in Alzheimer's disease through interference effects and shifting disabilities in the SCWT test. Archives of Clinical Neuropsychology 2004; 19(6): 791-803.

[98] Kramer AF, Erickson EI, Colcombe SJ. Exercise, cognition, and the aging brain. Journal of Applied Physiology 2006; 101: 1237-1242.

[99] Berardi AM, Parasuraman R, Haxby JV. Sustained attention in mild Alzheimer's disease. Developmental Neuropsychology 2005; 28(1): 507-537.

[100] Clifford A, Ferguson R, Bandelow S, Hogervorst E. A Randomised Controlled Trial of Resistance Exercise on Cognition in Healthy Middle-Aged Adults. In: Clifford, A. Physical Activity and Cognition in the Elderly. Doctoral dissertation. Loughborough University, Loughborough, UK 2012.

[101] Parasuraman R, Greenwood PM, Sunderland T. The apolipoprotein E gene, attention, and brain function. Neuropsychology 2002; 16(2): 254-74.

[102] Eggermont LHP, Milberg WP, Lipsitz LA, Scherder EJA, Leveille SG. Physical Activity and Executive Function in Aging: The MOBILIZE Boston Study. J Am Geriatr Soc 2009; 57(10): 1750–1756.

[103] Barnes DE, Yaffe K, Satariano WA, Tager IB. A longitudinal study of cardiorespiratory fitness and cognitive function in healthy older adults. Journal of the American Geriatrics Society 2003; 51(4): 459-465.

[104] Miller EK, Cohen JD. An integrative theory of prefrontal cortex function. Annu Rev Neurosci 2001; 24: 167-202.

[105] Cotman CW, Berchtold NC, Christie LA. Exercise builds brain health: key roles of growth factor cascades and inflammation. Trends Neurosci 2007; 30(9): 464-72.

[106] van Praag H, Shubert T, Zhao C, Gage FH. Exercise Enhances Learning and Hippocampal Neurogenesis in Aged Mice. J Neurosci 2005; 25(38): 8680–8685.

[107] Tsutsumi T, Don BM, Zaichkowsky LD, Delizonna LL. Physical fitness and psychological benefits of strength training in community dwelling older adults. Applied Human Science 1997; 16(6): 257-266.

[108] Cassilhas RC, Viana VAR, Grassmann V, *et al.* The impact of resistance exercise on the cognitive function of the elderly. Medicine and Science in Sports and Exercise 2007; 39(8): 1401-1407.

[109] Blumenthal JA, Madden DJ. Effects of aerobic exercise training, age, and physical fitness on memory-search performance. Psychology and Aging 1988; 3(3): 280-285.

[110] Madden DJ, Blumenthal JA, Allen PA, Emery CF. Improving aerobic capacity in healthy older adults does not necessarily lead to improved cognitive performance. Psychology and Aging 1989; 4(3): 307-320.

[111] Williams P, Lord SR. Effects of group exercise on cognitive functioning and mood in older women. Australian and New Zealand Journal of Public Health 1997; 21(1): 45- 52.

[112] Ho SC, Woo J, Sham A, Chan SG, Yu ALM. A 3-year follow-up study of social, lifestyle and health predictors of cognitive impairment in a Chinese older cohort. International Journal of Epidemiology 2001; 30: 1389-1396.

[113] Sumic A, Michael YL, Carlson NE, Howieson DB, Kaye JA. Physical activity and the risk of dementia in oldest old. Journal of Aging and Health 2007; 19(2): 242-259.

[114] Clifford A, Rahardjo TB, Bandelow S, Hogervorst E. The relationship between physical activity and cognition in an elderly Indonesian cohort. In: Clifford A. Physical Activity and Cognition in the Elderly. Doctoral dissertation. Loughborough University, Loughborough, UK 2012.

[115] Berchtold NC, Kesslak JP, Pike CJ, Adlard PA, Cotman CW. Estrogen and exercise interact to regulate brain-derived neurotrophic factor mRNA and protein expression in the hippocampus. Eur J Neurosci 2001; 14(12): 1992-2002.
[116] Shay KA, Roth DL. Association between aerobic fitness and visuospatial performance in healthy older adults. Psychology and Aging 1992; 7(1):15-24.
[117] Fabre C, Masse-Biron J, Chamari K, Varray A, Mucci P, Prefaut C. Evaluation of quality of life in elderly health subjects after aerobic and/or mental training. Archives of Gerontology and Geriatrics 1999; 28: 9-22.

Send Orders of Reprints at reprints@benthamscience.net
Frontiers in Clinical Drug Research – Alzheimer Disorders, Vol. 1, 2013, 243-264

CHAPTER 7

Neurogenic Molecules Modulating Cell Survival Pathways, Amyloid Beta, Tau Proteins, and Metal Dyshomeostasis: An Alternative Hope for Alzheimer's Therapy

Hailin Zheng[1,*], Mati Fridkin[2] and Moussa B.H. Youdim[3]

[1]Department of Medicinal Chemistry, Intra-cellular Therapies Inc. 3960 Broadway, New York, NY 10032; [2]Department of Organic Chemistry, The Weizmann Institute of Science, Rehovot 76100, Israel and [3]Eve Topf and USA National Parkinson Foundation Centers of Excellence for Neurodegenerative Diseases and Department of Pharmacology, Technion-Rappaport Family Faculty of Medicine Haifa, 31096, Israel, and Department of Biology, Yonsei World Central University, Seoul, South Korea

Abstract: Compelling evidence indicates that Alzheimer's disease (AD) has multiple causes, and its effective treatment should fully address the multi factor nature of the disease. Accordingly, to achieve better efficacy, recent drug development for AD therapy has been moving from single-targeting to multi-targeting, drugs which simultaneously modulates several key disease-related/pathways. In this chapter, we explore some examples of neurogenic molecules which can reduce amyloid beta (Aβ) production, prevent oxidative stress and tau hyperphosphorylation, restore metal dyshomeostasis and also possess neuroprotective/neurorestorative effects. We also discuss the diverse mechanisms contributing to the biological activities of these neurogenic molecules, including regulation of Aβ processing and tau phosphorylation; activation of protein kinase signaling pathways; and modulation of cell survival genes and proteins.

Keywords: Abeta-lowering agent, antioxidant, ladostigil, rasagiline, methylene blue, M30, multitarget drugs, Alzheimer's disease, Parkinson's disease, metal chelators, neurogenic, amyloid beta, tau hyperphosphorylation, monoamine oxidase inhibitors, acetylcholinesterase inhibitors, neuroprotective, neurorestorative, disease modifying drugs, Alzheimer's therapy, neurodegeneration and metal dyshomeostasis.

INTRODUCTION

Alzheimer's disease (AD) is a fatal complex neurodegenerative disease affecting

***Address correspondence to Hailin Zheng:** Department of Medicinal Chemistry, Intra-cellular Therapies Inc. 3960 Broadway, New York, NY 10032; Tel: 1-212-923-3344 ext. 211; Fax: 1-212-923-3388; E-mails: hailin20070109@hotmail.com and hzheng@intracellulartherapies.com

Atta-ur-Rahman (Ed)

36 million people in the world as of 2010, with one new AD case being reported every 7 seconds globally [1]. However, the current FDA-approved anti-AD drugs, which include four acetylcholinesterase inhibitors (AChEIs) and one N-methyl D-aspartate (NMDA) antagonist, offer short-term symptomatic benefit without apparent slowing down the disease progression. The annual cost for AD treatment is estimated to be about 172 billion dollars in USA with expectedly soaring to 1.1 trillion dollars by the year of 2050 [1]. This imposes huge social and economic burden on our society, and calls for new treatments that can prevent or delay the disease onset, slow or even halt the progression.

In fact, there are numerous drug candidates currently under active development as potential new treatments for AD. These drugs can be divided into three major categories: 1) drugs targeting the Aβ peptide; 2) drugs targeting the tau protein; 3) drugs targeting metal ions, oxidative stress, and other signaling pathways related to AD pathogenesis [2, 3]. However, in the past 10 years, the major drugs under development were designed to target the Aβ peptide because Aβ is considered to be the central to the cognitive deficits in AD. It is believed that inhibition of Aβ formation and aggregation, and clearance of Aβ deposition should slow down or even halt the disease progression. Several such Aβ-targeted drugs have recently completed large scale clinical trials, but none of them has shown efficacy in improving the cognitive deficits in AD patients, let alone the disease-modifying effects. These disappointed outcomes have raised questions regarding the validity of targeting only Aβ as an effective treatment for AD [4, 5].

In fact, recent AD treatment has been moving forward to combination therapy, where AChE inhibitors can be given together with NMDA antagonist memantine for greater efficacy with significant slowdown in the rate of disease progression over monotherapy [6]. Drug development for AD have also seen a shift from designing highly selective ligands to multitarget ligands [4, 7]. In this chapter, we will focus on selected multitarget ligands (Table **1**) for AD to demonstrate the drug design strategies and recent development in term of preclinical and clinical studies. These new ligands were designed not only to target Aβ but also other key AD-related targets, and thus provide an alternative hope for effective treatment of AD in the future.

Table 1: Examples of multitarget compounds tested in preclinical/clinical trials for AD therapy and their mechanisms of action.

Name and Structure		Mechanisms of Action	Phase
M30	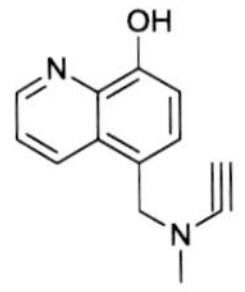	Aβ–lowering agent Antioxidant, Metal chelator Brain-selectiveMAO A/B inhibitor Neuroprotective and neurogenic agent	preclinical
Ladostigil	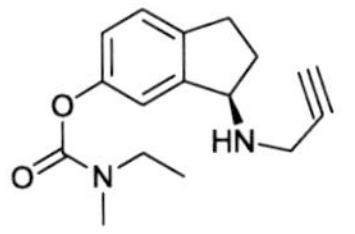	ChE inhibitor, antidepressant Brain-selective MAO A/B inhibitor Neuroprotective/neurotrophic agent	2
LMTX	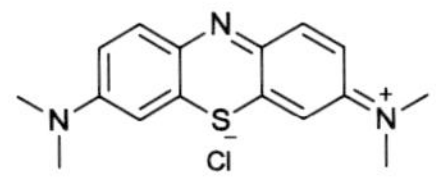	Tau-aggregation inhibitor Aβ oligomerization inhibitor mitochondria-targeting agent ChE and MAO A/B inhibitor	2/3
Davunetide	NAPVSIPQ eight amino acid peptide	Aβ–lowering agent Microtubule stabilizing agent Neurotrophic/neurogenic agent	2
Posiphen	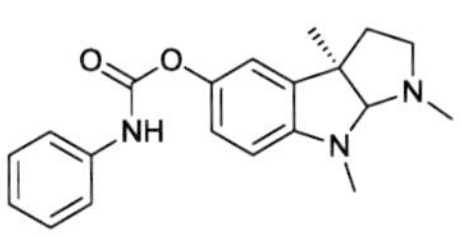	Inhibit or of the synthesis of APP, Tau and α-Synuclein	1
PBT2	structure not disclosed to date	Inhibitor of tau hyperphosphorylation, Aβ aggregation and oligomerization Antioxidant, Metal chelator Neuroprotective/neurotrophic agent	2

M30: A METAL/MAO-FOCUSED MULTI TARGET LIGAND

M30 (5-[N-methyl-N-propargylaminomethyl]-8-hydroxyquinoline) is a new metal/MAO (monoamine oxidase)-focused multitarget ligand (Table **1**), which possesses brain- selective MAO A and B inhibitory activity and does not potentiate the cardiovascular effect of tyramine [8]. M30 was rationally designed based on a series of scientific discoveries. It has been established that AD is a multifactorial syndrome with multiple dysfunctional pathways involved in its pathogenesis. Accordingly the effective treatment of AD is unlikely achieved by any drug acting on a single pathway or target. Multitarget ligands that simultaneously modulate several key disease-related targets/pathways may offer great hope for fighting this devastating disease. Some of the key disease-related targets/pathways include metal dyshomeostasis, oxidative stress, Aβ production, tau hyperphosphorylation, and MAO dysfunction in the brain [2, 7]. Oxidative stress has been shown to be one of the earliest, if not the earliest, change that occurs in AD pathogenesis, as extensive oxidative damage was observed in the brains of persons with mild cognitive impairment (MCI), a condition that often precedes AD [9, 10]. It is known that iron and copper ions can aggravate oxidative stress by producing reactive oxygen species (ROS) through the Fenton reaction. Iron, copper and zinc ions were found to be accumulated in and around Aβ plaques, and/or neurofibrillary tangles (NFTs), with 3/5-fold increase in the AD patient brains compared to those in age-matched controls [11]. These metal ions favor protein misfolding and aggregation (formation of Aβ plaques and NFTs). For example, it was shown that Zn^{2+} induced Aβ aggregation rapidly, even at low physiological concentrations (submicromolar); Cu^{2+} and Fe^{3+} promoted Aβ aggregation at mildly acidic conditions [12]; Cu^{2+} interacted with Aβ, thus promoting tau hyperphosphorylation and subsequent neurofibrillary tangles (NFTs) formation and Fe^{3+}, which is accumulated in neurons with NFTs, bound to hyperphosphorylated tau and induced its aggregation [13]. Metals (Cu, Fe, and Zn) ions also mediate other targets central to AD pathogenesis in different levels: a) iron can activate an iron-responsive element (IRE-type II) in the 5' untranslated region (UTR) of APP mRNA, resulting in enhancement of endogenous Alzheimer's APP translation and subsequent Aβ formation [14]; b) Cu could interact with BACE1 and potentially regulate β-cleavage of APP, thus altering APP processing; c) Zinc can inhibit α-secretase cleavage, increase β- and γ-secretase cleavage of APP and promote Aβ generation [12].

MAO is another potential target for the development of anti-AD drugs. It has been shown that MAO-B activity increases with age and aging is the major non-genetic risk factor for the development of late-onset sporadic AD [15, 16]. MAO-B activity can rise up to 3-fold in the temporal, parietal, and frontal cortex in AD patients compared with controls, and its activity is particularly high in plaque-associated astrocytes in AD brains [17]. This increase in MAO-B activity accelerates the oxidative deamination of neurotrans-mitters, such as dopamine (DA), resulting in enhanced production of free radicals causing oxidative stress [15]. In addition to contributing to oxidative stress, the increased activity of MAO can lead to increased metabolism of neurotransmitters such as dopamine, noradrenaline, and serotonin in the CNS. These neurotransmitters have been found to decrease in PD/AD patients compared to age-matched control [18]. Therefore MAO inhibitors have the potential not only to suppress oxidative stress to interfere with Aβ formation and tau hyperphosphorylation, but also to restore the levels of neurotransmitters. In fact, selegiline, a selective MAO-B inhibitor, has been shown to significantly improve learning and memory deficits in animal models associated with AD and to slow the disease progression in AD patients [19].

Based on this reasoning, M30 was designed to simultaneously modulate: metal dyshomeostasis, oxidative stress, Aβ production, tau hyperphosphorylation, and MAO dysfunction in the brain. M30 has been successfully developed by incorporating the antioxidant chelating moiety into the anti-PD drug rasagiline (Fig. **1**), which contains a key pharmacophore, the propargylamine moiety. Rasagiline is a new generation of MAO-B inhibitor, approved by FDA as monotherapy or adjunct therapy to levodopa for early and late PD patients. In addition to the symptomatic benefits in treating PD patients, rasagiline was recently reported to be the first drug with an apparent disease-modifying effect, at a dose of 1 mg per day [20]. This disease-modifying effect is likely due to the interaction of rasagiline with an array of neuroprotective/neurorescue pathways, which are mediated by its key pharmacophore, the propargylamine moiety [21].

Studies have demonstrated that M30 possess a wide range of biological activities both *in vitro* and *in vivo*. *In vitro*, M30 exhibited moderate chelating affinity for metal (Fe, Cu, and Zn) ions with highly inhibitory action against iron-induced mitochondrial membrane lipid peroxidation, comparable to that of DFO (deferoxamine), a prototype iron chelator [22]. In pancreatic beta cell models, M30 markedly and dose-dependently inhibited H_2O_2-induced cytotoxicity associated with decreased intracellular ROS formation and increased catalase activity. When tested in rat brain homogenates, M30 showed irreversible inhibition of both MAO-A and B (IC_{50} values of 0.037 ± 0.02 μM and 0.057 ± 0.01 μM for MAO-A and -B, respectively) [22]. *In vivo*, M30 was found to possess brain selective inhibition for MAO-A and -B activity with relatively poor inhibition of these enzymes in the liver and small intestine [23]. This suggest that M30 may also have antidepressant activity by increasing brain levels of dopamine (DA), noradrenaline (NA), and serotonin (ST) with limited potentiation of tyramine pressor effect. Indeed, recent studies have shown that rats treated with M30 following oral administration of tyramine showed limited potentiation of blood pressure, similar to that of moclobemide, a selective reversible MAO-A inhibitor used as an anti-depressant in the clinic [8].

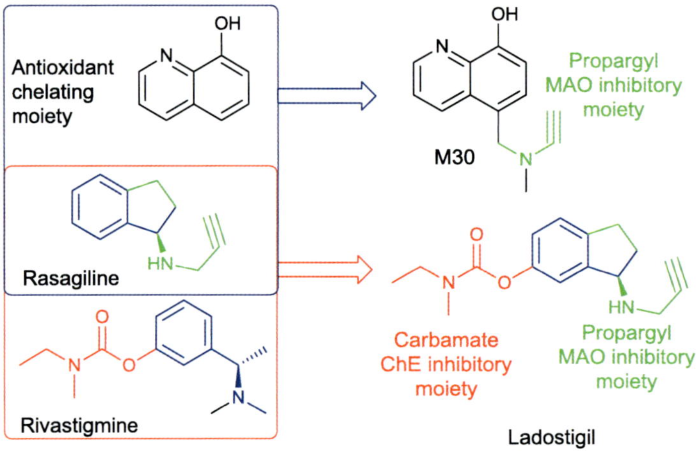

Figure 1: Design strategy of M30 and Ladostigil: M30 was designed by combining two pharmacophores in a single molecule: the antioxidant chelating moiety (black part) and the propargylamine moiety (green part) from the anti-PD drug rasagiline. Ladostigil was developed by introduction of carbamate moiety (red part) from rivastigmine into rasagiline.

The regulatory effects of M30 on APP and Aβ were demonstrated both *in vitro* and *in vivo*. *In vitro*, M30 induced a significant down-regulation of membrane-associated holo-APP levels in the mouse hippocampus and SH-SY5Y neuroblastoma cells [24] [25]; it also markedly lowered the secreted levels of Aβ in the conditioned medium and cellular β-CTF (β-C-terminal fragment) in CHO/ΔNL cells stably transfected with the APP "Swedish" mutation, and significantly increased the release of soluble APPα into the conditioned medium [24-27]. *In vivo*, chronic treatment of aged mice with M30 (1 and 5 mg/kg; four times weekly for 6 months) showed a significant positive impact on neuropsychiatry functions and cognitive age-related impairment;

in addition, M30 also significantly decreased cerebral Aβ level and iron accumulation [28]. In another study, chronic administration of M30 was showed to attenuate cerebral Aβ pathology and behavioral deficits in an APP /PS1 transgenic mice model of AD [29].

M30 exhibited neurogenesis activity in various cell culture and animal models. For example, attenuation of cell death induced by serum deprivation and 6-hydroxydopamine in rat pheochromocytoma (PC12) cells, [22]; elevation of the expression levels of the transcripts of BDNF and growth-associated protein-43 (GAP43) in primary cortical cells [27]; protection against hydrogen peroxide or 3-morpholinosydnonimine- induced neurotoxicity in mouse NSC-34 motor neuron cells [26]. *In vivo* studies, real-time RT-PCR revealed that M30 increased mRNA expression levels of BDNF in the cortex and striatum and glial cell-derived neurotrophic factor (GDNF) in the hippocampus and spinal cord of mouse brain [30].

NEW ChE/MAO INHIBITOR LADOSTIGIL

Degeneration of cholinergic neurons in the basal forebrain and the related loss of cholinergic neurotransmission in the cerebral cortex and other areas are associated with cognitive dysfunction in AD patients. Accordingly increase in cholinergic levels should improve the cognitive function, and indeed, the first drug tacrine (AChEI) approved for AD therapy is working by increasing cholinergic levels. Now there are five FDA-approved drugs for AD therapy, four of which are AChEIs. AChEIs have proven safe and efficacious in the clinic to improve cognition and global behaviors in AD patients, although the effects are symptomatic with no significantly affecting the disease progression. However, recent achievements in our understanding of the role of AChE in AD pathogenesis have highlighted the possibility of AChE as a potential target to modify AD progression. Apart from acting to hydrolyze the neurotransmitter acetylcholine, AChE has been shown to involves in the development of a vicious cycle of Aβ and hyperphosphorylated tau protein (P-tau) dysregulation. Studies have shown that AChE is consistently increased in regions around Aβ plaques and NFT at all stages of AD; It colocalizes with Aβ in senile plaques and serves to accelerate Aβ aggregation into amyloid fibrils, and increase Aβ neurotoxicity. In addition, AChE can induce Aβ aggregation through its catalytic site close to the peripheral anionic binding site (PAS) [31]. Moreover, Aβ was found to regulate AChE expression, assembly and glycosylation in cell culture, transgenic mice and the brain of AD patients; it interacts with AChE to form AChE-Aβ complexes, which are more neurotoxic than Aβ fibrils alone both *in vitro* (PC12 cells and primary retina cells) [32, 33] and *in vivo* (rat hippocampus) [34]. Further studies indicated that P-tau can trigger an increase in AChE expression: for example, overexpression of mutant P-tau increased AChE activity in the brain of Tg-VLW mice, and VLW tau overexpression in SH-SY5Y cells increased AChE activity levels [35].

Based on these findings, recent development of new AChEIs has been shifted to designing multitarget ligands with diverse mechanisms of action including inhibition of both AChE and BuChE, modulation of APP processing and P-tau dysregulation. Ladostigil (Fig. **1**) may be considered one of such new AChEIs, which is currently in Phase II clinical trials for treating MCI associated with early stages of AD. Like other AChEIs ladostigil targets cholinesterase to relieve the disease symptom in AD patients. But unlike other drugs, ladostigil acts *via* additional novel mechanisms and provides potential to improve the behavioral and psychological symptoms of dementia, such as depression and anxiety. Moreover, ladostigil (Table **1**) has the potential to slow the disease progression and to modify the pathology associated with the disease [36].

In vitro studies, Ladostigil was found to inhibit ChE (AChE and BuChE) with 100 times more potent against AChE than BuChE; *in vivo*, oral administration of ladostigil to rats in doses of 35-100 mg/kg was shown to inhibit brain ChE by 25-40% and antagonize scopolamine-induced spatial memory impairments. In addition, ladostigil given orally (75 mg/kg rat, daily for 2 weeks) displayed antidepressant-like activity with significant reduction of immobility in the forced swim test [37]. It was found that ladostigil was metabolized *in vivo* into a potent MAO A/B inhibitor, resulting in ~ 70% inhibition of the enzymes in the brain, but without significant effect on these enzymes in the intestine. The brain selective inhibition of MAO A/B renders it a potentially safer antidepressant than other irreversible MAO-A inhibitors. Indeed, in Phase I and Phase IIa clinical trials, Ladostigil proved to be pharmacologically active, safe and well tolerated. Moreover, Ladostigil also exhibited potent neuroprotective activities *in vitro* and in various neurodegenerative rat models. For example, it improved cognitive functions in aged monkeys, restored spatial memory deficits induced by the cytochrome oxidase inhibitor sodium azide, and prevented memory impairment induced by streptozotocin (STZ) in rats. The neuroprotective activities of ladostigil have been shown to involve several molecular targets and pathways including regulation of APP processing, modulation of cell survival genes and proteins, and activation of protein kinase C (PKC) and mitogen- activated protein kinase (MAPK) pathways [38].

DAVUNETIDE: A TAU-FOCUSED NEUROPROTECTIVE PEPTIDE

AD is multifactional disease, caused probably by complex interactions among multiple factors. It is associated not only with the abnormal accumulation of Aβ, but also with neurofibrillary tangles (NFTs). NFTs are composed predominately of tau protein, an essential protein for neuronal shape and axonal transport, which can binds to microtubules and acts to stabilize neuronal structure and function. In AD patients, Tau proteins become abnormally hyperphosphorylated and accumulate as paired helical filaments that in turn aggregate into NFTs. Formation of NFTs results in loss of microtubule/tau function and appearance of toxic species, both of which lead to cell death associated with neurodegerative diseases such as AD and progressive supranuclear palsy (PSP). In early AD and MCI patients, NFTs were observed to be significantly increased in all four ventromedial temporal lobe structures, as compared to normal cognition controls. The increase in NFTs (and neuronal loss) can be correlated well with cognitive loss and the disease progression and severity in AD patients. By contrast, Aβ plaques, the supposed critical factor in AD pathogenesis, have poor clinical pathological correlations with the extent of cognitive impairment in AD patients. Moreover, Aβ clearing is not necessarily accompanied by cognitive improvement [39-41]. In addition, studies have shown that tau malfunction is able not only to independently induce neurodegeneration even in the absence of Aβ deposits or other pathological events [42, 43], but also act as a key mediator of neurodegeneration in response to other upstream events, including Aβ induced toxicity. For example, neurons expressing either mouse or human tau proteins exhibited degeneration in the presence of Aâ, while tau-depleted neurons showed no signs of degeneration in the presence of Aâ [44]. And, reduction of endogenous tau was found to prevent behavioral deficits in transgenic mice expressing human amyloid precursor protein, without altering their high Aβ levels [45]. Recent studies also demonstrated that both Aβ-associated volume loss and clinical decline in older humans at risk for dementia occurred only in the presence of phospho-tau (p-tau) [46].

Collectively, these studies substantiate the notion that tau-mediated neurodegeration play a key role in AD pathogenesis, and targeting tau, as well as Aβ, may represent a more effective strategy for AD therapy.

Davunetide is a Tau-focused multitarget ligand developed by Allon Therapeutics. It is an eight amino acid peptide (NAPVSIPQ) derived from activity-dependent neuroprotective protein (ADNP), a glial cell mediator of vasoactive intestinal peptide (VIP) induced neuroprotection (Table **1**). Davunetide is now under Phase 2/3 study for the treatment of neurodegenerative disease such as AD and PSP [47]. It has shown neuroprotection against a variety of insults *in vitro*, including oxidative stress, beta amyloid toxicity and excitotoxicity. For example, Davunetide protected cerebral cortical neurons against the major AD toxic peptide Aβ (1-42) with maximal protection observed at concentrations of 10-15 to 10-10 mol/l [48]. It also acted as neurotrophic agent, stimulating neurite elongation and synapse formation in rat hippocampal and cortical cultures [49]. *In vivo* experiments, Davunetide significantly ameliorated the complex array of injuries elicited by head trauma, reduced mortality and facilitated neurobehavioral recovery ($P < 0.005$) in a mouse model of closed head injury [50]. It also protected the brain against ischemic injury by reducing apoptotic death [51]. In a transgenic mouse model of AD, Intranasal davunetide administration was found to reduce both Aβ accumulation and tau hyperphosphorylation at early pathological stage [52]; in a tauopathy mouse model, davunetide increased soluble tau, and reduced tau hyperphosphorylation, and protected memory [53].

In phase 1 clinical studies, davunetide was found to penetrate cells and cross the BBB after nasal or system administration. Phase 2 clinical studies indicated that twice daily intranasal administration of davunetide was safe and well-tolerated in amnestic mild cognitive impaired patients, with the rate of adverse effects equally distributed between the treated groups and the placebo. Significant improvement was observed in short-term recall and working memory after 12 weeks of intranasal administration with davunetide in MCI subjects; but there was no significant improvement on the tests that involved executive functions, such as planning, cognitive flexibility and abstract thinking [47].

METHYLENE BLUE: AN OLD DRUG WITH MULTIPLE MECHANISMS OF ACTION

Methylene blue (MB, Table **1**), which has been described as "the first fully synthetic medicine", is an old drug with diverse applications, including as medicines, staining agents, and as a lead for drug development. It has been used as antimalarial, antidepressant, antiseptic agents, and currently MB is approved by FDA for treating methemoglobinemia, urinary tract infections, and other diseases [54]. The beneficial effects of MB in the treatment of cognitive diseases occurring in psychoses have been known for a long time, but only in recent years has MB again attracted scientific attention as a potential drug to slow down cognitive decline in neurodegenerative diseases such as AD and PSP [55].

In phase 2 clinical studies on 321 patients with mild or moderate AD, MB (trade name rember® by TauRx Therapeutics) given orally 60 mg three times per day, was found to significantly reduce the rate of disease progression by 80% over 50 weeks measured by Alzheimer's Disease Assessment Scale (ADAS-COG); it also showed an improvement on cognitive function in Mini Mental State Examination (MMSE) [56].

However, the high dose (100 mg) of rember® did not show significant clinical benefit. It was reported that a formulation defect delayed MB release from high-dose capsules and prevented absorption. Accordingly the second generation rember® named LMTX® was developed and now is now under Phase 3 testing for treating Pick's Disease, a type of Frontotemporal Dementia (FTD), and will be heading toward two Phase 3 AD trials next year. LMTX® has the same active ingredient (MB) and mode of action as rember®, but with improved bioavailability and tolerability [57].

It is believed that MB acts by reducing levels of aggregated or misfolded Tau proteins, which are associated with the progressive neurodegeneration in AD and other neurodegenerative diseases. MB was the first Tau-aggregation inhibitor reported to selectively inhibit AD-like tau aggregation more than a decade ago [58]. It disaggregated paired helical filaments (PHFs), permitting the release of the core-tau monomer, with activity at submicromolar concentrations in cell-free and cellular assays [59]. In a subsequent study, MB was shown to inhibit heparin-induced assembly of tau into filaments *in vitro*, with an IC_{50} value in the low micromolar range [58]. In a more recent study, MB was further shown to inhibit the formation of Tau oligomers and prevent their conversion to PHFs *in vitro*. *In vivo* with transgenic mouse models of AD, MB was found to dissolve PHFs isolated from the brain of mice expressing truncated Tau and significantly reduce the counts of Tau-positive neurons in hippocampus and entorhinal cortex [60]. However, other modes of action for MB in AD must be considered as well because MB is a really "dirty drug" with an amazing number of different targets. In addition to modulation of tau pathology, MB also influences numerous enzymes and signaling pathways closely related to the progression of AD, including Aβ peptide, mitochondrial function and various neurotransmitter systems (serotonergic, cholinergic, and glutamatergic).

MB and Amyloid Plaques

Among the multiple Aβ aggregated species identified, Aβ oligomers appears to be highly toxic species. Therefore, selective inhibition of Aβ oligomer formation has emerged as an attractive strategy for AD therapy. MB was found to inhibit $A\beta_{42}$ oligomerization *in vitro*, with concomitant promotion of fibrilization in a dose-dependent manner [61]. *In vivo* with 3xTg-AD mice, chronic dietary MB treatment was shown to reduce soluble $A\beta_{40}$ and $A\beta_{42}$ levels and improves learning and memory deficits. This effect may be mediated by an increase in proteasome function in the brain [62].

MB and Mitochondrial Function

Mitochondrial dysfunction such as changes in amount of respiratory enzymes, or the enzyme activities could lead to oxidative stress and increased apoptosis, which are associated with AD pathogenesis [63]. MB can penetrate through cellular and mitochondrial membranes, accumulates within mitochondria and improves mitochondrial functions. *In vitro*, MB was shown to enhance mitochondrial complex IV by 30% and increase cellular oxygen consumption by 37–70% in normal human fibroblasts (IMR90) at nM levels [64]. *In vivo*, administration of MB increased mitochondrial complex IV activity by 100% and 50% in the brains and hearts of old mice, respectively [65].

MB and Neurotransmitter Systems

MB can affect various neurotransmitters in the brain, such as monoamine, choline, and glutamate. For example, it inhibited both MAO A and MAO B with IC_{50} values of 164±8 nm and 5.5±1.7 μM, respectively, when measured in recombinant human liver enzyme [66]. *In vivo*, systemic administration of MB was found to significantly increase the extracellular levels of 5-HT and DA in the rat brain. This increase can be explained by the inhibition of MB on MAO A, an enzyme that is responsible for 5-HT metabolism [67]. The inhibition of MB on AChE activity has been known for many decades. Early studies revealed that MB competitively inhibited AChE activity in cow erythrocytes with an IC50 of 11 μM [68]. In a more recent report, the inhibition of MB on the activity of different types of cholinesterases was studied. The results of the studies indicated that MB concentration-dependently inhibited the esterase activity of human plasma, human pseudocholinesterase and bovine acetylcholinesterase with IC50 values of 1.05±0.05 μmol l^{-1}, 5.32±0.36 μmol l^{-1} and 0.42±0.09 μmol l^{-1}, respectively [69]. In addition, other studies found that MB inhibition of electric eel AChE (K_i =17 nM) was mediated by a single site, whereas the inhibition of of human BuChE (K_i = 0.4μM) was mediated through multiple binding sites [70].

POSIPHEN: AN INHIBITOR OF THE SYNTHESIS OF APP, TAU AND α-SYNUCLEIN

Posiphen (Table **1**) is a multitarget agent derived from physostigmine, with activity against the synthesis of APP, Tau and α-Synuclein. It is now being developed by QR Pharma as a potential disease modifying drug for AD therapy. Posiphen is a chirally pure (+) enantiomer of anti-AChE inhibitor phenserine, but with no direct anti-AChE activity [71]. In cultured human neuroblastoma cells, Posiphen was found to lower sAPP and Aβ levels by acting on the 5' UTR of APP mRNA and inhibiting its translation into APP protein [72]. *In vivo*, Posiphen, given orally to mice (7.5–75 mg/kg daily, 21 consecutive days), dose-dependently reduced brain levels of APP and $A\beta_{42}$, with 40% and 60 % reduction for both biomarkers at the lower dose of 25 mg/kg and at the two higher doses of 50mg/kg and 75 mg/kg, respectively [72].

In phase I studies, Posiphen has proven well tolerated by healthy male and female volunteers at single doses from 10 to 80 mg and in MCI patients at multiple doses (up to 4 × 60 mg daily × 10 days). There was a small but statistically insignificant difference from placebo regarding gastrointestinal side effects and dizziness. In these studies, Posiphen was also found to lower the CSF levels in MCI patients of sAPPα, sAPPβ, t-τ (total tau) to those determined in healthy (Posiphen naive) volunteers, after 10 days of treatment (4 × 60 mg/day). In addition, there were no dramatic differences between Posiphen-treated MCI subjects and healthy volunteers regarding the CSF levels of sAPPα, sAPPβ, t-τ (total tau) [73].

Together, the available data demonstrate that Posiphen is a multitarget agent that can effectively lower the CSF levels of sAPPα, sAPPβ, t-τ (total tau), the 2 key targets for AD therapy, and may therefore modify the disease progression.

PBT2: A METAL-FOCUSED MULTI TARGET LIGAND

PBT2 (Table **1**) is termed as a metal protein attenuating compound (MPAC), which is derived from clioquinol (CQ), the prototype of the first generation MPACs. MPACs differ from traditional chelators in their intermediate reversible affinity towards metals and their ability to reach specific intracellular compartments to target the harmful 'up stream' metal-protein reactions. Rather than systemic binding and removal of metals from tissues (with potential systemic toxicity), MPACs restore metal dyshomeostasis, block free and reactive coordination positions of protein bound metals, and abolish specific and deleterious metal–protein interactions. Therefore, they hold promise as anti-AD drugs for slowing and/or preventing the disease progression.

Studies have shown that treatment with CQ (20 mg/kg daily for 12 weeks, Tg2576 APP transgenic mice) significantly reduced the insoluble Aß levels in the brain extracts and the amyloid deposits in the brain sections [74]. In a pilot Phase II double-blind, placebo controlled trial, CQ given orally (125 - 375 mg/twice daily over 36 weeks) showed activity against the rate of cognitive decline in moderately severe AD patients, with no signs of severe adverse effects [75]. The subsequent phase II/III studies, however, were stalled, reportedly due to the difficulty in purifying CQ in larger scale for such trials. A second generation MPAC or PBT2 was then developed and now is under Phase 2 clinical trials for AD treatment. In the double blind clinical trials, PBT2 showed safe and well tolerated in 78 mild AD patients, when given randomly in multi-center at doses either 50 mg or PBT2 250 mg once a day for 12 weeks. There was no serious side effects observed in any PBT2 treated patient; but there was a dose-dependent and reduction of CSF Aâ levels seen at the 250mg dose. Moreover, cognitive executive function as measured by the NTB (neuropsychological test battery) was significantly improved for patients at the 250mg dose [76].

Multiple mechanisms of action may contribute to the clinical benefits of PBT2 (metal chelator or MAPC) in AD patients. The possible mechanisms include: metal chelating and antioxidant properties, reduction of Aβ and tau levels, and ionophore activity to restore metal dyshomeostasis and activate neuroprotective pathways.

Metal Chelating and Antioxidant Properties

It is known that Aβ binds copper and zinc selectively in AD tissue, and Aβ:Cu complexes are a catalytic source of H_2O_2 [77, 78]. PBT2 is a metal chelator with metal affinity similar to its predecessor CQ, and could act like CQ to capture Cu ion from Aβ:Cu complexes and form redox inert complexes, preventing H_2O_2 formation. Indeed, studies have demonstrated that Both PBT2 and CQ coordinated Cu and Zn in a 2:1 ratio and significant reduced H_2O_2 formation mediated by Aβ:Cu complexes *in vitro*, with PBT2 providing a significantly greater inhibition than CQ ($p = 0.01$) [79].

Reduction of Interstitial $A\beta_{ISF}$ and Phosphorylated Tau Levels *In Vivo*

High levels of Aβ and phosphorylated tau play a role in cognitive deficit of AD patients. Recently, PBT2 was showed to inhibit Zn-induced formation of protease resistant Aβ aggregates and promote Aβ degrada-tion by preventing Aβ:Zn interactions. *In vivo* studies, PBT2 was found to markedly decrease the soluble interstitial brain Aβ and rapidly restore the cognitive function in Alzheimer's transgenic mice. Moreover, PBT2 also significantly reduced insoluble and soluble phosphorylated tau in APP/PS1 mice and insoluble total tau in the Tg2576 mice [79].

Ionophore Activity

Intracellular Cu stores are deficient in AD brain although Cu (and Zn) are indeed markedly enriched in Aβ plaque. This unbalanced Cu homeostasis could promote Aβ production with a causative or diagnostic link to AD [80]. Ionophores are any molecules, as of drugs, which can facilitate the transport of metal ions across the cell membrane and restore unbalanced Cu (and other metals) homeostasis. Studies have shown that different classes of ionophores including CQ can increase intracellular metal levels (copper and zinc), resulting in substantial loss of secreted Aβ, and improve cognitive function in AD model mice [81, 82]. In fact, copper was found to stimulate the pathway of APP secretion in cells and greatly reduce Aβ levels *in vitro*. And also *in vivo*, Cu treatment was shown to lead to significant decrease in Aβ production in APP23 transgenic mice, which is associated with an increase in brain Cu level and restoration of intracellular Cu dyshomeostasis [80, 83]. Recent studies suggested that PBT2 may also act as an ionophore to exert its beneficial effects on AD patients. It has been found that PBT2 can translocate extracellular Zn and Cu ions into cells, leading to an increase in bioavailable Zn and Cu levels within the cells. The increased Cu and Zn levels initiate several neuroprotective signaling cascades, including change in cellular levels of GSK3α/β, inhibition of the phosphatase calcineurin, and increased phosphorylation of other calcineurin substrates such as cAMP response element binding protein and Ca^{2+} /calmodulin-dependent protein kinase [84].

CONCLUSION

The current AD therapies (AChEIs and memantine) have been successful in improving cognition and global functions in AD patients, although none of them exhibits significant disease-modifying effects. The development of disease-modifying drugs for AD has faced a great challenge, as up to date, none of the highly anticipated anti-amyloid drugs, which were designed to modify the disease progression, shows efficacy in Phase 3 clinical trials. The complexity of AD pathogenesis may be a reasonable explanation for this failure. In addition, the fact that most of the relevant nerve cells are already dead in AD patients when treatment begins may also be responsible. Many anti-amyloid drugs, despite being high potency in modulating a single disease related target, have a limited ability to rescue the dying neurons and to regenerate new neurons. In this chapter, we have discussed some potential drugs for AD, which were designed to address the complexity of AD pathogenesis. These drugs can act on multiple targets *via* diverse mechanisms of action, with potential to improve the cognitive and behavior symptoms in AD, as well as modify the course of disease progression. Some of them also demonstrate neuroprotective activity to rescue the dying neurons and neurogenic activity to regenerate new neurons. These drugs may offer an alternative hope for AD treatment.

ACKNOWLEDGEMENT

This chapter was prepared during Angela Clifford's doctoral studentship which was funded by Loughborough University, UK.

CONFLICT OF INTEREST

The author(s) confirm that this chapter content has no conflict of interest.

REFERENCES

[1] 2012 Alzheimer's disease facts and figures. Alzheimers Dement 2012; 8: 131-68.

[2] Huang Y, Mucke L. Alzheimer mechanisms and therapeutic strategies. Cell 2012; 148: 1204-22.

[3] Corbett A, Ballard C. New and emerging treatments for Alzheimer's disease. Expert Opin Emerg Drugs 2012; 17: 147-56.

[4] Van der Schyf CJ. The use of multi-target drugs in the treatment of neurodegenerative diseases. Expert Rev Clin Pharmacol 2011; 4: 293-8.

[5] Gotz J, Ittner A, Ittner LM. Tau-targeted treatment strategies in Alzheimer's disease. Br J Pharmacol 2012; 165: 1246-59.

[6] Atri A, Shaughnessy LW, Locascio JJ, Growdon JH. Long-term course and effectiveness of combination therapy in Alzheimer disease. Alzheimer Dis Assoc Disord 2008; 22: 209-21.

[7] Cavalli A, Bolognesi ML, Minarini A, Rosini M, Tumiatti V, Recanatini M, *et al.* Multi-target-directed ligands to combat neurodegenerative diseases. J Med Chem 2008; 51: 347-72.

[8] Gal S, Abassi ZA, Youdim MB. limited potentiation of blood pressure in response to oral tyramine by the anti-Parkinson brain selective multifunctional monoamine oxidase-AB inhibitor, M30. Neurotox Res 2010; 18: 143-50.

[9] Williams TI, Lynn BC, Markesbery WR, Lovell MA. Increased levels of 4-hydroxynonenal and acrolein, neurotoxic markers of lipid peroxidation, in the brain in Mild Cognitive Impairment and early Alzheimer's disease. Neurobiol Aging 2006; 27: 1094-9.

[10] Keller JN, Schmitt FA, Scheff SW, Ding Q, Chen Q, Butterfield DA, *et al.* Evidence of increased oxidative damage in subjects with mild cognitive impairment. Neurology 2005; 64: 1152-6.

[11] Lovell MA, Robertson JD, Teesdale WJ, Campbell JL, Markesbery WR. Copper, iron and zinc in Alzheimer's disease senile plaques. J Neurol Sci 1998; 158: 47-52.

[12] Duce JA, Bush AI. Biological metals and Alzheimer's disease: Implications for therapeutics and diagnostics. Prog Neurobiol 2010.

[13] Yamamoto A, Shin RW, Hasegawa K, Naiki H, Sato H, Yoshimasu F, *et al.* Iron (III) induces aggregation of hyperphosphorylated tau and its reduction to iron (II) reverses the aggregation: implications in the formation of neurofibrillary tangles of Alzheimer's disease. J Neurochem 2002; 82: 1137-47.

[14] Rogers JT, Randall JD, Cahill CM, Eder PS, Huang X, Gunshin H, *et al.* An iron-responsive element type II in the 5'-untranslated region of the Alzheimer's amyloid precursor protein transcript. J Biol Chem 2002; 277: 45518-28.

[15] Fowler JS, Volkow ND, Wang GJ, Logan J, Pappas N, Shea C, *et al.* Age-related increases in brain monoamine oxidase B in living healthy human subjects. Neurobiol Aging 1997; 18: 431-5.

[16] van der Flier WM, Scheltens P. Epidemiology and risk factors of dementia. J Neurol Neurosurg Psychiatry 2005; 76 Suppl 5: v2-7.

[17] Saura J, Luque JM, Cesura AM, Da Prada M, Chan-Palay V, Huber G, *et al.* Increased monoamine oxidase B activity in plaque-associated astrocytes of Alzheimer brains revealed by quantitative enzyme radioautography. Neuroscience 1994; 62: 15-30.

[18] Reinikainen KJ, Soininen H, Riekkinen PJ. Neurotransmitter changes in Alzheimer's disease: implications to diagnostics and therapy. J Neurosci Res 1990; 27: 576-86.

[19] Sano M, Ernesto C, Thomas RG, Klauber MR, Schafer K, Grundman M, *et al.* A controlled trial of selegiline, alpha-tocopherol, or both as treatment for Alzheimer's disease. The Alzheimer's Disease Cooperative Study. N Engl J Med 1997; 336: 1216-22.

[20] Olanow CW, Rascol O, Hauser R, Feigin PD, Jankovic J, Lang A, *et al.* A double-blind, delayed-start trial of rasagiline in Parkinson's disease. N Engl J Med 2009; 361: 1268-78.

[21] Weinreb O, Amit T, Bar-Am O, Youdim MB. Rasagiline: a novel anti-Parkinsonian monoamine oxidase-B inhibitor with neuroprotective activity. Prog Neurobiol 2010; 92: 330-44.

[22] Zheng H, Gal S, Weiner LM, Bar-Am O, Warshawsky A, Fridkin M, *et al.* Novel multifunctional neuroprotective iron chelator-monoamine oxidase inhibitor drugs for neurodegenerative diseases: *in vitro* studies on antioxidant activity, prevention of lipid peroxide formation and monoamine oxidase inhibition. J Neurochem 2005; 95: 68-78.

[23] Gal S, Zheng H, Fridkin M, Youdim MB. Novel multifunctional neuroprotective iron chelator-monoamine oxidase inhibitor drugs for neurodegenerative diseases. *In vivo* selective brain monoamine oxidase inhibition and prevention of MPTP-induced striatal dopamine depletion. J Neurochem 2005; 95: 79-88.

[24] Avramovich-Tirosh Y, Reznichenko L, Mit T, Zheng H, Fridkin M, Weinreb O, *et al.* Neurorescue activity, APP regulation and amyloid-beta peptide reduction by novel multi-functional brain permeable iron- chelating- antioxidants, M-30 and green tea polyphenol, EGCG. Curr Alzheimer Res 2007; 4: 403-11.

[25] Avramovich-Tirosh Y, Amit T, Bar-Am O, Zheng H, Fridkin M, Youdim MB. Therapeutic targets and potential of the novel brain- permeable multifunctional iron chelator-monoamine oxidase inhibitor drug, M-30, for the treatment of Alzheimer's disease. J Neurochem 2007; 100: 490-502.

[26] Kupershmidt L, Weinreb O, Amit T, Mandel S, Carri MT, Youdim MB. Neuroprotective and neuritogenic activities of novel multimodal iron-chelating drugs in motor-neuron-like NSC-34 cells and transgenic mouse model of amyotrophic lateral sclerosis. FASEB J 2009; 23: 3766-79.

[27] Avramovich-Tirosh Y, Bar-Am O, Amit T, Youdim MB, Weinreb O. Up-regulation of hypoxia-inducible factor (HIF) -1alpha and HIF-target genes in cortical neurons by the novel multifunctional iron chelator anti-Alzheimer drug, M30. Curr Alzheimer Res 2010; 7: 300-6.

[28] Kupershmidt L, Amit T, Bar-Am O, Youdim MB, Weinreb O. Neuroprotection by the multitarget iron chelator M30 on age-related alterations in mice. Mech Ageing Dev 2012; 133: 267-74.

[29] Kupershmidt L, Amit T, Bar-Am O, Youdim MB, Weinreb O. The novel multi-target iron chelating-radical scavenging compound M30 possesses beneficial effects on major hallmarks of Alzheimer's disease. Antioxid Redox Signal 2012; 17: 860-77.

[30] Kupershmidt L, Weinreb O, Amit T, Mandel S, Bar-Am O, Youdim MB. Novel molecular targets of the neuroprotective/neurorescue multimodal iron chelating drug M30 in the mouse brain. Neuroscience 2011.

[31] Verhoeff NP. Acetylcholinergic neurotransmission and the beta-amyloid cascade: implications for Alzheimer's disease. Expert Rev Neurother 2005; 5: 277-84.

[32] Alvarez A, Alarcon R, Opazo C, Campos EO, Munoz FJ, Calderon FH, *et al.* Stable complexes involving acetylcholinesterase and amyloid-beta peptide change the biochemical properties of the enzyme and increase the neurotoxicity of Alzheimer's fibrils. J Neurosci 1998; 18: 3213-23.

[33] Munoz FJ, Inestrosa NC. Neurotoxicity of acetylcholinesterase amyloid beta-peptide aggregates is dependent on the type of Abeta peptide and the AChE concentration present in the complexes. FEBS Lett 1999; 450: 205-9.

[34] Reyes AE, Chacon MA, Dinamarca MC, Cerpa W, Morgan C, Inestrosa NC. Acetylcholinesterase-Abeta complexes are more toxic than Abeta fibrils in rat hippocampus: effect on rat beta-amyloid aggregation, laminin expression, reactive astrocytosis, and neuronal cell loss. Am J Pathol 2004; 164: 2163-74.

[35] Silveyra MX, Garcia-Ayllon MS, de Barreda EG, Small DH, Martinez S, Avila J, *et al.* Altered expression of brain acetylcholinesterase in FTDP-17 human tau transgenic mice. Neurobiol Aging 2012; 33: 624 e23-34.

[36] Weinreb O, Amit T, Bar-Am O, Youdim MB. Ladostigil: a novel multimodal neuroprotective drug with cholinesterase and brain-selective monoamine oxidase inhibitory activities for Alzheimer's disease treatment. Curr Drug Targets 2012; 13: 483-94.

[37] Weinstock M, Bejar C, Wang RH, Poltyrev T, Gross A, Finberg JP, *et al.* TV3326, a novel neuroprotective drug with cholinesterase and monoamine oxidase inhibitory activities for the treatment of Alzheimer's disease. J Neural Transm Suppl 2000: 157-69.

[38] Weinreb O, Amit T, Bar-Am O, Yogev-Falach M, Youdim MB. The neuroprotective mechanism of action of the multimodal drug ladostigil. Front Biosci 2008; 13: 5131-7.

[39] Hardy J. The amyloid hypothesis for Alzheimer's disease: a critical reappraisal. J Neurochem 2009; 110: 1129-34.

[40] Extance A. Alzheimer's failure raises questions about disease-modifying strategies. Nat Rev Drug Discov 2010; 9: 749-51.

[41] Green RC, Schneider LS, Amato DA, Beelen AP, Wilcock G, Swabb EA, *et al.* Effect of tarenflurbil on cognitive decline and activities of daily living in patients with mild Alzheimer disease: a randomized controlled trial. JAMA 2009; 302: 2557-64.
[42] Goedert M, Jakes R. Mutations causing neurodegenerative tauopathies. Biochim Biophys Acta 2005; 1739: 240-50.
[43] von Bergen M, Barghorn S, Li L, Marx A, Biernat J, Mandelkow EM, *et al.* Mutations of tau protein in frontotemporal dementia promote aggregation of paired helical filaments by enhancing local beta-structure. J Biol Chem 2001; 276: 48165-74.
[44] Rapoport M, Dawson HN, Binder LI, Vitek MP, Ferreira A. Tau is essential to beta - amyloid-induced neurotoxicity. Proc Natl Acad Sci U S A 2002; 99: 6364-9.
[45] Roberson ED, Scearce-Levie K, Palop JJ, Yan F, Cheng IH, Wu T, *et al.* Reducing endogenous tau ameliorates amyloid beta-induced deficits in an Alzheimer's disease mouse model. Science 2007; 316: 750-4.
[46] Desikan RS, McEvoy LK, Thompson WK, Holland D, Brewer JB, Aisen PS, *et al.* Amyloid-beta--associated clinical decline occurs only in the presence of elevated P-tau. Arch Neurol 2012; 69: 709-13.
[47] Allon Therapeutics Inc. http://www.allontherapeutics.com/product-development/davunetide/ (Accessed Oct. 16, 2012).
[48] Gozes I, Divinski I, Piltzer I. NAP and D-SAL: neuroprotection against the beta amyloid peptide (1-42). BMC Neurosci 2008; 9 Suppl 3: S3.
[49] Smith-Swintosky VL, Gozes I, Brenneman DE, D'Andrea MR, Plata-Salaman CR. Activity-dependent neurotrophic factor-9 and NAP promote neurite outgrowth in rat hippocampal and cortical cultures. J Mol Neurosci 2005; 25: 225-38.
[50] Beni-Adani L, Gozes I, Cohen Y, Assaf Y, Steingart RA, Brenneman DE, *et al.* A peptide derived from activity-dependent neuroprotective protein (ADNP) ameliorates injury response in closed head injury in mice. J Pharmacol Exp Ther 2001; 296: 57-63.
[51] Leker RR, Teichner A, Grigoriadis N, Ovadia H, Brenneman DE, Fridkin M, *et al.* NAP, a femtomolar-acting peptide, protects the brain against ischemic injury by reducing apoptotic death. Stroke 2002; 33: 1085-92.
[52] Matsuoka Y, Gray AJ, Hirata-Fukae C, Minami SS, Waterhouse EG, Mattson MP, *et al.* Intranasal NAP administration reduces accumulation of amyloid peptide and tau hyperphosphorylation in a transgenic mouse model of Alzheimer's disease at early pathological stage. J Mol Neurosci 2007; 31: 165-70.
[53] Shiryaev N, Jouroukhin Y, Giladi E, Polyzoidou E, Grigoriadis NC, Rosenmann H, *et al.* NAP protects memory, increases soluble tau and reduces tau hyperphosphorylation in a tauopathy model. Neurobiol Dis 2009; 34: 381-8.
[54] Schirmer RH, Adler H, Pickhardt M, Mandelkow E. "Lest we forget you--methylene blue.". Neurobiol Aging 2011; 32: 2325 e7-16.
[55] Gura T. Hope in Alzheimer's fight emerges from unexpected places. Nat Med 2008; 14: 894.
[56] TauRx Therapeutics http://taurx.com/clinical-data.html (Accessed Oct. 16, 2012).
[57] Will Tau Drug Show Its True Colors in Phase 3 Trials? http://www.alzforum.org/new/detail.asp?id=3283 (Accessed Oct. 2, 2012).
[58] Taniguchi S, Suzuki N, Masuda M, Hisanaga S, Iwatsubo T, Goedert M, *et al.* Inhibition of heparin-induced tau filament formation by phenothiazines, polyphenols, and porphyrins. J Biol Chem 2005; 280: 7614-23.

[59] Wischik CM, Edwards PC, Lai RY, Roth M, Harrington CR. Selective inhibition of Alzheimer disease-like tau aggregation by phenothiazines. Proc Natl Acad Sci U S A 1996; 93: 11213-8.
[60] Harrington C, Rickard JE, Horsley D, Harrington KA, Hindley KP, Riedel G, *et al.* O1-06-04: Methylthioninium chloride (MTC) acts as a Tau aggregation inhibitor (TAI) in a cellular model and reverses Tau pathology in transgenic mouse models of Alzheimer's disease. Alzheimer's & Dementia 2008; 4: T120-T1.
[61] Necula M, Breydo L, Milton S, Kayed R, van der Veer WE, Tone P, *et al.* Methylene blue inhibits amyloid Abeta oligomerization by promoting fibrillization. Biochemistry (Mosc) 2007; 46: 8850-60.
[62] Medina DX, Caccamo A, Oddo S. Methylene blue reduces abeta levels and rescues early cognitive deficit by increasing proteasome activity. Brain Pathol 2011; 21: 140-9.
[63] Eckert GP, Renner K, Eckert SH, Eckmann J, Hagl S, Abdel-Kader RM, *et al.* Mitochondrial dysfunction-a pharmacological target in Alzheimer's disease. Mol Neurobiol 2012; 46: 136-50.
[64] Atamna H, Nguyen A, Schultz C, Boyle K, Newberry J, Kato H, *et al.* Methylene blue delays cellular senescence and enhances key mitochondrial biochemical pathways. FASEB J 2008; 22: 703-12.
[65] Atamna, H. and Gharib, A. Methylene Blue Induces Mitochondrial Complex IV and Improves Cognitive Function and Grip Strength in Old Mice. Nova Science Publishers. In: McNeill AS, Ed. "Neurodegeneration: Theory, Disorders and Treatments". 2011 Nova Science Publishers, Inc.; Chapter 3, pp. 63–86.
[66] Ramsay RR, Dunford C, Gillman PK. Methylene blue and serotonin toxicity: inhibition of monoamine oxidase A (MAO A) confirms a theoretical prediction. Br J Pharmacol 2007; 152: 946-51.
[67] Wegener G, Volke V, Rosenberg R. Endogenous nitric oxide decreases hippocampal levels of serotonin and dopamine *in vivo*. Br J Pharmacol 2000; 130: 575-80.
[68] Augustinsson K. Methylene blue as an inhibitor of acetylcholine-esterase. Acta Chem Scand 1950; 4: 536-42.
[69] Pfaffendorf M, Bruning TA, Batnik HD, van Zwieten PA. The interaction between methylene blue and the cholinergic system. Br J Pharmacol 1997; 122: 95-8.
[70] Kucukkilinc T, Ozer I. Multi-site inhibition of human plasma cholinesterase by cationic phenoxazine and phenothiazine dyes. Arch Biochem Biophys 2007; 461: 294-8.
[71] QR Pharma http://www.qrpharma.com/posiphen.html.
[72] Lahiri DK, Chen D, Maloney B, Holloway HW, Yu QS, Utsuki T, *et al.* The experimental Alzheimer's disease drug posiphen [(+)-phenserine] lowers amyloid-beta peptide levels in cell culture and mice. J Pharmacol Exp Ther 2007; 320: 386-96.
[73] Maccecchini ML, Chang MY, Pan C, John V, Zetterberg H, Greig NH. Posiphen as a candidate drug to lower CSF amyloid precursor protein, amyloid-beta peptide and tau levels: target engagement, tolerability and pharmacokinetics in humans. J Neurol Neurosurg Psychiatry 2012; 83: 894-902.
[74] Cherny RA, Atwood CS, Xilinas ME, Gray DN, Jones WD, McLean CA, *et al.* Treatment with a copper-zinc chelator markedly and rapidly inhibits beta-amyloid accumulation in Alzheimer's disease transgenic mice. Neuron 2001; 30: 665-76.
[75] Ritchie CW, Bush AI, Mackinnon A, Macfarlane S, Mastwyk M, MacGregor L, *et al.* Metal-protein attenuation with iodochlorhydroxyquin (clioquinol) targeting Abeta amyloid deposition and toxicity in Alzheimer disease: a pilot phase 2 clinical trial. Arch Neurol 2003; 60: 1685-91.

[76] Faux NG, Ritchie CW, Gunn A, Rembach A, Tsatsanis A, Bedo J, *et al.* PBT2 rapidly improves cognition in Alzheimer's Disease: additional phase II analyses. J Alzheimers Dis 2010; 20: 509-16.

[77] Opazo C, Huang X, Cherny RA, Moir RD, Roher AE, White AR, *et al.* Metalloenzyme-like activity of Alzheimer's disease beta-amyloid. Cu-dependent catalytic conversion of dopamine, cholesterol, and biological reducing agents to neurotoxic H(2)O(2). J Biol Chem 2002; 277: 40302-8.

[78] Huang X, Atwood CS, Hartshorn MA, Multhaup G, Goldstein LE, Scarpa RC, *et al.* The A beta peptide of Alzheimer's disease directly produces hydrogen peroxide through metal ion reduction. Biochemistry (Mosc) 1999; 38: 7609-16.

[79] Adlard PA, Cherny RA, Finkelstein DI, Gautier E, Robb E, Cortes M, *et al.* Rapid restoration of cognition in Alzheimer's transgenic mice with 8-hydroxy quinoline analogs is associated with decreased interstitial Abeta. Neuron 2008; 59: 43-55.

[80] Bayer TA, Schafer S, Simons A, Kemmling A, Kamer T, Tepest R, *et al.* Dietary Cu stabilizes brain superoxide dismutase 1 activity and reduces amyloid Abeta production in APP23 transgenic mice. Proc Natl Acad Sci U S A 2003; 100: 14187-92.

[81] Caragounis A, Du T, Filiz G, Laughton KM, Volitakis I, Sharples RA, *et al.* Differential modulation of Alzheimer's disease amyloid beta-peptide accumulation by diverse classes of metal ligands. Biochem J 2007; 407: 435-50.

[82] Crouch PJ, Hung LW, Adlard PA, Cortes M, Lal V, Filiz G, *et al.* Increasing Cu bioavailability inhibits Abeta oligomers and tau phosphorylation. Proc Natl Acad Sci U S A 2009; 106: 381-6.

[83] Borchardt T, Camakaris J, Cappai R, Masters CL, Beyreuther K, Multhaup G. Copper inhibits beta-amyloid production and stimulates the non-amyloidogenic pathway of amyloid-precursor-protein secretion. Biochem J 1999; 344 Pt 2: 461-7.

[84] Crouch PJ, Savva MS, Hung LW, Donnelly PS, Mot AI, Parker SJ, *et al.* The Alzheimer's therapeutic PBT2 promotes amyloid-beta degradation and GSK3 phosphorylation *via* a metal chaperone activity. J Neurochem 2011; 119: 220-30.

Send Orders of Reprints at reprints@benthamscience.net

Frontiers in Clinical Drug Research – Alzheimer Disorders, Vol. 1, **2013, 265-314** **265**

CHAPTER 8

Cognitive Decline and Alzheimer's Disease: A Central Role for Hypoxia-Obstructive Sleep Apnea and Related Stigmata in Elderly

Mak Adam Daulatzai*

Sleep Disorders Group, EEE/MSE, The University of Melbourne, Parkville, Victoria 3010, Australia

Abstract: Late onset Alzheimer's disease (AD) is the most common cause of dementia in the western world. Available extensive evidence supports the concept that AD is fundamentally a multifactorial syndrome that results in progressive psychiatric and cognitive impairment. Several clinico-pathological and epidemiological data emphasize overlaps of pathological cascades, *e.g.* aging, cerebrovascular pathology and obesity, and their synergistic effects in promoting cognitive and memory decline/AD. In comparison with normal controls, patients with amnestic mild cognitive impairment (MCI) who converted to AD showed cerebral hypoperfusion particularly in the parietal, parahippocampal, inferior temporal and fusiform gyri, whereas those who did not convert yet showed hypoperfusion in the retrosplenial cortex. This suggests that promotion and progression of cognitive decline may precede neuronal damage. The current chapter addresses this important issue and delineates the factors responsible for hypoxia, hypoxemia, cerebral hypoperfusion and compromised blood supply, brain hypometabolism in aging and AD. In the normally aged brain -perfusion, oxygen consumption and glucose metabolism remain quite unchanged from the 3rd to the 7th decade. Thereafter, they decline and lead to relatively lower hypoperfusion, and glucose hypometabolism in aging. This disturbance arises due to snoring and obstructive sleep apnea (OSA) that cause chronic intermittent hypoxia (which is highly prevalent in the obese). OSA could be as much as 50 per hour (or more) – *i.e.* an apnea-hypopnea index of 50. Hence a scenario of chronic hypoxia/hypoxemia may promote cerebral ischemia, changes in oxygen consumption, and glucose hypometabolism -documented in the temporo-parietal cortex of OSA sufferers as well as AD patients. A convergence of "Age" – a proinflammatory state, and several other stigmata due to hypoxia/hypoxemia may upregulate neuroinflammation and enhance oxidative stress. The confluence of the above pathogenic cascades may depress neuronal function by rendering synapses, neurotransmission, and brain metabolism dysfunctional, and thus provoke and enhance cognition impairment. This progressive/evolving neurodegenerative process is also manifested by amyloid angiopathy, and the formation/deposition of amyloid plaques

***Address correspondence to Mak Adam Daulatzai:** Senior Medical Research Fellow, Sleep Disorders Group, EEE/MSE, The University of Melbourne, EEE Building # 193, 3rd Floor, Room 301/2, Parkville, Victoria 3010, Australia; Tel: +61 3 83440042; Fax: +613 93471094; E-mail: makd@unimelb.edu.au

Atta-ur-Rahman (Ed)

and neurofibrillary tangles -the quintessential hallmarks of Alzheimer's dementia. Animal models of intermittent hypoxia have consistently documented upregulation of amyloid beta and neuronal apoptosis -being widespread in key foci of cortex and brainstem.

Keywords: Aging, Alzheimer's disease, Obstructive sleep apnea, hypoxia, hypoxemia, obesity, systemic inflammation, oxidative stress, cardiovascular disease, hypertension, hypoperfusion, hippocampus, hypometabolism, amyloid, neurofibrillary tangles, neurocognitive dysfunction, memory loss, diabetes mellitus, synapse, sensory loss.

1. INTRODUCTION

The most common dementia in older population is the late onset sporadic Alzheimer disease (AD) a progressive neurodegenerative disorder. AD is characterized by progressive loss of memory, especially episodic memory, as well as loss of other cognitive functions. In industrialized countries it is presumed to be responsible for one fourth of deaths. Indeed, AD alone is said to be the seventh-leading cause of death in the United States. This senile dementia is an enormous public health problem. Owing to human longevity, the number of people with dementia will double by 2030, and triple by 2050. The cost of caring for Dementia patients is likely to rise faster than the prevalence. The global impact of dementia would exact a massive toll on the global economy. AD afflicts 5.4 million in the U.S. alone, and its cost is estimated to be about $172 billion annually [1]. These statistics are shocking and currently there are no curative or disease altering therapies. Apart from a huge economic burden, the impact of dementia is immeasurable on the lives of sufferers and their families. This therefore demands a detailed evaluation of possible neuropathogenic factors; efforts to find a cure in the near future are extremely important. Owing to global aging, and an ongoing increase in the incidence of AD, it is essential, that we figure out strategies and therapies which may modify, and delay the progression of this condition – and ultimately cure it. There is an urgent need to thwart this dreadful disease.

The transition from age-associated memory impairment (AAMI) to the dramatic loss of cognitive functions accompanying AD requires progressive development of neocortical and brainstem pathology that triggers neuron death. The selective vulnerability of neurons is reflected in the characteristics of cortical pyramidal neurons that are prone to form NFT. Neuropathologically, AD is characterized by neuronal loss and deposition of NFT and amyloid plaques in the brain [2]. Post-mortem studies have demonstrated that three main stages can be differentiated during the clinical course of AD [3] – entorhinal cortex/parahippocampal region is the earliest to be affected (Braak stages 1 and 2), the pathological alterations spread into additional structures of the medial temporal lobe/limbic system (Braak stages 3 and 4), they then encroach other regions of the temporal, parietal, and frontal cortical lobes [3-6]. Protein misfolding and aggregation play a key role in neurodegenerative disorders such as AD and Parkinson's disease (PD). In AD, two pathological hallmarks, as mentioned above, are amyloid plaques and NFT, composed of β-amyloid (Aβ) and tau proteins respectively [3, 7, 8]. Amyloid plaques correlate poorly with memory decline [9] and immunotherapy-induced plaque removal does not prevent progressive neurodegeneration [10]. This suggests that other therapeutic modalities need to be exploited to ameliorate AD-related neurodegeneration. Plasma levels of Aβ peptides have been studied extensively as blood-based biomarkers for dementia; however, the available data have not produced the conclusion that they are indeed a reliable blood biomarker. Plasma levels of $A\beta_{42}$ alone were not found to be strong predictors of AD, however, the ratio of $A\beta_{42}$: $A\beta_{40}$ may be more useful as predictors of dementia/AD [11]. A lower ratio of $A\beta_{42}$:$A\beta_{40}$ may predict worse cognitive decline, thus possibly reflecting a process of selective $A\beta_{42}$ deposition in insoluble amyloid plaques in the brain [11].

Aβ is generated by the sequential cleavage of the amyloid protein (APP). APP is encoded by a gene located on chromosome 21 [12]. APP can be cleaved by one of two pathways: these are 1. A non-amyloidogenic pathway, and 2. An amyloidogenic pathway. The non-amyloidogenic pathway is the most common; APP is cleaved by the α-secretase enzymes, precluding the formation of Aβ [13]. In the non-amyloidogenic pathway, however, APP is cleaved by BACE1 [14-16], liberating bAPP and a small carboxy terminal fragment termed C99. Subsequently, C99 is further cleaved by the γ-secretase complex (formed by PS1 or PS2, Aph1, Pen2 and nicastrin) generating the peptides Aβ40 and Aβ42 [17-21]. Of the two, Aβ42 is the more amyloidogenic form of Aβ, and it is this species that predominantly accumulates in the amyloid plaques [22]. Aβ oligomers can aggregate to form amyloid fibrils. Aβ fibrils are the major component of the extracellular deposits of amyloid plaques; however, there are data that elucidated the important role of Aβ oligomers also in the AD neuropathogenesis [23-25].

Tau is a microtubule-binding protein with important cellular functions, in terms of regulation of cytoskeletal structure and function [26, 27]. The adult human brain has six different tau isoforms; they differ by the presence of 3 to 4 microtubule binding domains (at the C-terminal), characterized by 18 amino acid repeat sequences that are tubulin binding sites [28]. The six isoforms are generated by the alternative splicing [29-31]. The microtubule-binding properties of tau are mainly regulated by post-translational modifications, including glycation, ubiquitylation, sumoylation, nitration, proteolyis and glycosylation, and indeed phosphorylation at specific serine/threonine sites [32]. It is important to underscore that tau phosphorylation inversely correlates with its ability to bind to microtubules [33]. The tau phosphorylation is a function of the activity of several kinases (and phosphatases) [34-36]. In AD and other tauopathies, tau is abnormally hyperphosphorylated; there is an increase, therefore, in total levels of unbound tau. The unbound tau aggregates to form straight and paired helical filaments -the NFT [33, 37]. Studies have shown a correlation between higher levels of NFT and the memory decline in AD [9, 38]. Interestingly, analogous to amyloid, soluble tau may be more toxic to neurons [39, 42].

Chimpanzees are our closest living relatives. Human lifespan is much longer than the great ape's. Infections in the latter cause high mortality and the apes rarely survive more than 50 years. Chimpanzees in captivity have a much lower incidence of cancer, ischemic heart disease, and neurodegeneration compared to humans. MRI was used to measure the volume of the whole brain, total neocortical gray matter, total neocortical white matter, gray matter and white matter of the frontal lobe, and the hippocampus (in 99 chimpanzee brains from 10 to 51 years of their adult lifespan). These results obtained were compared with measurements of similar brain structure volumes in 87 adult humans (from 22 to 88 years of age). The humans show a decrease in the volume of the above brain structures, when studied over the lifespan; however, no significant age-related changes become evident in the chimpanzees [43]. Undoubtedly, both human and nonhuman primates undergo some functional declines in selective attention, executive functions, and some facets of declarative memory with aging. Despite significant improvement in food and medicine, we still suffer from pathology of aging mediated by malnutrition, infection, inflammation, and neurodegeneration. However, age-related extensive neuronal loss in the hippocampus and entorhinal cortex (memory-related brain regions) has not been found in humans. One important study demonstrated a significant and extensive decline in the number and size of immunolabeled neurons in subcortical cholinergic basal forebrain regions of aged rhesus monkeys (an animal model of human aging) [44]. Cognitive function studies on mice have shown that processes of aging impact brain structures differentially, in that the mice cerebellum shows earlier senescence than the hippocampus. Significant deficits in long-term depression (LTD) appear in cerebellum between 4 and 8 months, whereas long-term potentiation (LTP) and LTD values were similar in 4-to 12-month mice [45]. In humans, the hippocampus undergoes pathological changes particularly in those who progress to AD.

One of the essential pre-requisites for effective lung ventilation is a patent upper airway (UA). The latter is intimately associated with normal physiological functioning of the extrinsic tongue muscle – genioglossus (GG). Alteration in the GG activity occurs when sleep commences; this may lead to airway narrowing and hypoventilation even in normal individuals [46-51]. In obstructive sleep apnea (OSA) patients, however, who already have compromised UA, further sleep-related GG dysfunction causes increased air flow limitation and ventilation cessation [46-53]. OSA is characterized by repetitive collapse of the pharyngeal UA during sleep that results in hypoxia, hypoxemia and hypercapnea. This recurrent blood gas perturbation enhances respiratory drive and causes patency of the UA and arousal. The cyclic intermittent pattern of wakefulness and sleep produces sleep fragmentation, changes in diurnal rhythm, autonomic and hormonal disturbances, and a host of other pathophysiological sequel. These may include hypertension [54-56], myocardial infarction [57, 58], neurocognitive impairment [59], stroke, and sudden death [60, 61]. It is also linked to various other dysfunctions such as memory decline [62-66, 70], and depression [67].

Sleep disordered breathing (SDB) ranges from snoring through increased airway resistance and reduction in airflow (hypopnea) to periods of breathing cessation and airway collapse (apnea). Apnea-hypopnea index (AHI) is the average number of apnea and hypopnea events per hour of sleep. The AHI number reflects the severity of OSA. An AHI of <5 is generally regarded normal; very mild disease is reflected by 5-10 AHI, moderate by 15-30, and severe by an AHI greater than 30 per hour. The classical epidemiological data have characterized hypopnea as a 4% oxygen desaturation utilizing thermisters and/or inductance plethysmography. However, more recently nasal pressure transducers are used in scoring hypopnea. There are several clinical definitions of hypopnea in use, and hence different labs use different criteria owing to a lack of consensus on this issue. Although consequences of hypopnea may seem insignificant to translate in significant pathology, in the long run, however, they are devastating and physiological outcome of hypopneas has been considered in various studies to be similar to that of apnea. This is because the progression of hypopneas results in apnea and the OSA-associated co-morbidities. Sleep-disturbed breathing (SDB) includes hypopneas, obstructive sleep apneas (OSA), and oxygen desaturations; SDB occurs in asymptomatic individuals and increases with age. The gold standard test for evaluating SDB is polysomonography, and OSA is the most frequent type of respiratory disturbance documented. Factors that clearly increase the risk of SDB and OSA include age, obesity, hypertension, structural abnormalities of the UA, sedatives and alcohol use. Some mortality studies have suggested decreased survival in individuals suffering from OSA, due mainly to vascular-related disease.

This chapter will highlight in aged snoring and/or OSA patients the status of our current understanding of pathogenic contributing factors that possibly promote a transition from cognitive decline to AD. The emphasis here, therefore, is on linking OSA and its squeals, *viz.* neuroinflammation, oxidative stress (generation of reactive oxygen species, ROS), coagulation, and metabolic disturbances, in conjunction with dysfunctional brainstem nuclei (such as nucleus tractus solitarius (NTS) and hypoglossal) to triggering of cognitive dysfunction [see 62-66]. The delineation of the above association may lead to a better understanding of the pathogenic factors that underpin cognitive decline; this obviously has important therapeutic implications [68, 69].

2. AGING -PRIMARY RISK FACTOR FOR COGNITIVE IMPAIRMENT

It is well known that healthy aging is associated with structural and physiological decline. Human brain is an organ that is subject to several biological changes in aging. The neuropathological changes due to age-related declines in several sensory systems have been extensively documented. Not uncommonly older patients have multiple impairments involving vision, hearing, mobility; these in conjunction with socioeconomic and nutritional factors possess the potential for enhancing cognitive and memory dysfunctions. Age-associated memory impairment in neurodegenerative disorders is characterized by extensive neuronal apoptosis. However, prior to neuronal death, age-related changes occur in biochemical, physiological, and structural attributes of cortical and brainstem circuits; when dysfunctional, these may mediate neurocognitive functional decline [62-66, 70].

2.1. Aging – A Pro-Inflammatory State

The prevalence of a number of chronic neurological diseases increases with senescence. Aging, therefore, is a major risk factor for the pathogenesis of conditions such as AD and Parkinson's disease (PD). The age-related changes in sensitivity to immune activation in effect commences from the middle age of individuals. Elderly subjects have been extensively studied for age-related inflammatory responses to an immune stimulus [70-73]. The middle-aged subjects show increased sensitivity to acute lipopolysaccharide (LPS) administration; they document enhanced neural and immune responses, decrease in locomotor behavior, and a higher response to a lower dose of endotoxin [74]. Aging as an ongoing immune-challenged condition shows dysregulated immune and inflammatory responses in both humans and animals [75-77]. It impacts general performance and cognitive function in aged subjects owing to exaggerated neuroinflammatory response. During normal aging, microglia become activated and tend to be hyperresponsive to inflammatory stimuli [72, 78, 79, 86]. Accumulation of iron occurs in microglia with aging, and this would stimulate inflammatory processes further [80]. Importantly, there is evidence suggesting that activated microglia may contribute to neurodegeneration [80, 81]. The neurons of aged animals show greater vulnerability to neurotoxins than those of young animals, this may in part be due to enhanced glial activation [82, 83], thus reflecting a correlation between age-dependent hyper-inflammatory response and an increased risk for neurodegeneration. Immune activation in aging has a profound effect on the neurobehavioral response. Following acute LPS administration, middle-age mice are more sensitive to an immune insult than the younger ones; tolerance development to repeated LPS exposure (low dose endotoxin) is also slower in the aged [74]. Similar results were reported in the LPS-stimulated mice in regard to macrophages; the latter from old mice produced significantly more cyclooxygenase activity (as PGE2) than those from young mice [85] (see below). In animals, after a single injection of LPS, the number of c-fos immunoreactive neuronal cells in the paraventricular nucleus (PVN) of the hypothalamus and the central nucleus of the amygdala (CeA) were significantly higher in older mice (12-month) relative to young mice [73, 84]. Assessment of splenic cytokine levels showed that older subjects retain a level of immunoreactivity to the previous LPS exposure, with the IL-1β response showing an apparent slower dissipation [74], and with further LPS exposures, the response becomes additive [74].

Normal aging is a pro-inflammatory state; it is associated with immune alterations in the CNS owing to elevated proinflammatory cytokine release and glial activation. Over-reactive microglia in the aged brain produces a pronounced increase of TNF-α [79; 86]. Furthermore, LPS-stimulated induction of inducible nitric oxide synthase (iNOS) in microglia of aged rats shows resistance to downregulation [87]. The above mentioned reflect alterations in regulatory system of inflammation and an upregulation of NF-κB binding activity [88]. Macrophages from old mice (relative to young) produce more PGE2, contributing to the dysregulation of the immune and inflammatory responses with age. This age-associated increase in LPS-stimulated PGE2 production is due to increased COX activity, higher COX-2 protein and mRNA expression [89]. Available data show enhanced gene expressions of IL-1β, IL-6, TNF-α, COX-2 and iNOS during aging [90]. COX activity and the production of TXA_2 and PGI_2 are also increased during aging. Other pro-inflammatory proteins, such as adhesion molecules *viz.* VCAM-1, ICAM-1, P-, and E-selectin are all upregulated during aging [92], as is NF-κB transcription factor. The activation of NF-κB-dependent genes is involved in the wide-spread systemic inflammatory process [see 66, 92, 93], and pro-inflammatory genes regulate the transcription of several inflammatory mediators, including TNF-α, interleukins (IL-1β, IL-2, and IL-6), and chemokines (IL-8 and RANTES) [90-94].

2.2. Aging and Sensory Dysfunctions

Physiological changes occur in aging; these are slow over time in all body systems in general and sensory modalities in particular. These changes are a function of life events, illnesses, and genetic traits, environmental, socioeconomic, and nutritional factors. Sensory changes include a decline in sight and peripheral vision, hearing, smell, taste, and somatosensory system. The losses are neither rapid nor total, but they do pose a global negative impact *via* affecting sensory-motor reflexes, senile muscle loss, and general health status. Loss of smell and taste *e.g.* affect the nutritional intake in many seniors. A discussion of alterations in sensory system in aging/AD can be found in several papers [62-66, 70, 202]. Olfactory decline is intimately associated with aging and cognitively impaired elderly. The following describes olfactory perturbations in the elderly:

2.3. Olfactory Dysfunction in Aging

There is substantial evidence that olfactory event-related potentials (ERP) change with age [95]; olfactory epithelium and olfactory bulb (OB) undergo neuropathological changes in older humans; these have been copiously documented in AD [100-105]. Direct projections between the olfactory system and the hippocampus, prefrontal cortex, and amygdala are known to exist [96, 97]. Olfactory dysfunction in AD has been repeatedly confirmed by several studies [98-102]. Neuropathological data in the olfactory bulb of AD patients have documented the deposition of amyloid plaques and NFT [103-105]. When compared to healthy controls, patients with mild AD possess NFT pathology [104]. NFT pathology in the OB occurs in the very early Braak stages, even prior to this pathology in the entorhinal cortex [102, 103, 105]. The decrease in OB volume was significantly correlated with a decrease in global cognitive performance [102].

3. OBESITY -RISK FACTOR FOR COGNITIVE IMPAIRMENT

3.1. Obesity Leads to Pro-Inflammatory Pathology

Obesity is a significant health problem throughout the western world placing a major burden on the health care system. 66% of Americans are overweight and 32% meet the WHO criterion for obesity [106]. Obesity has been associated with several serious diseases including diabetes, cardiovascular disorders, and certain cancers [107]. Further, accumulating evidence shows that obesity, associated with chronic low-grade systemic inflammation during mid-and late-life increases the risk for neurodegenerative diseases including AD [108-111].

Several significant data provide substantial evidence that obesity leads to an abnormal activation of neuroendocrine and pro-inflammatory pathways. Activated white adipose tissue increases the synthesis of pro-inflammatory cytokines, such as IL-6, IL-1, IL-8, TNF-α, and IL-18, while regulatory cytokines, such as IL-10, are decreased. In addition adipocytes also produce other inflammation promoting cytokines such as adipokines [112]. Obesity and OSA are associated in terms of airways inflammation [113, 114]. Exhaled nitric oxide (NO) significantly increases in OSA and in obese patients (23.1 and 17.9 p.p.b. respectively) than in healthy subjects (7.2 p.p.b.). OSA and obese patients possess a higher percentage of neutrophils and a lower percentage of macrophages (also found in the induced sputum) compared to healthy subjects [113, 114].

High fat diet has been shown to increase $A\beta_{40}$ and $A\beta_{42}$ burden in the cortex of Tg2576 mice [115] and produce AD-like pathology in 3xTg-AD mice [116]. However, the observed Aβ changes were not correlated with alterations in total APP, C-terminal fragments (α and β cleavage products) of APP or soluble oligomers. This suggests that high-fat diet does not promote the accumulation of Aβ through altered APP metabolism. Several studies highlight that obesity and metabolic disorders are linked to Aβ accumulation in AD [116-118]; it can be hypothesized then, that the increases in weight and brain Aβ following high-fat diet, share common mechanisms. GFAP is a commonly used marker of astrocytic activation; its concentration increases in AD [119], in transgenic mice [120-124], and also in obese mice [58]. This suggests that the upregulation of gliosis in obesity may be independent of the transgene status [125]. This further shows that intake of excessive calories from fat may increase reactive astrocytes and inflammatory changes in the brain, an event also associated with Aβ pathology. Indeed, experimental induction of microgliosis increases Aβ burden [126, 127].

Obesity has been shown to be an independent risk factor for hypertension, cardiovascular disease, and insulin resistance [128-130]. Insulin resistance is also an independent risk factor for hypertension and cardiovascular disease [131]. Meta-analysis of treatment trials suggests that the effects of OSA treatment are modest on blood pressure [132]. Since both snoring and obesity affect blood pressure, either independently or conjointly have a similar impact on blood pressure (albeit of variable degree). Some of the key pathogenic impacts resulting from obesity are oxidative stress, inflammation and sympathetic activation [133-136]. Compared with fat deposits in other regions, abdominal visceral fat has been shown to be a potent risk factor for adverse health consequences -including hypertension, insulin resistance, diabetes and the metabolic syndrome [137-141]. Visceral fat is also a risk factor for snoring compared to BMI-matched controls [142]. Since proinflammatory cytokines are produced by visceral fat [142-144], there will be an enhancement in production of the proinflammatory cytokines in the obese. The highest levels of IL-6, TNF-α, and leptin were found in obese snorers with OSA, intermediate levels in obese subjects without OSA, and the lowest in lean controls [142]. IL-10 is an anti-inflammatory cytokine, and subjects with OSA may have lower IL-10 levels [145]. Obesity is one of the major factors that affect the levels of the inflammatory biomarkers in several diseases including OSA [146-149]. Other inflammatory effects include systemic increases in IL-8, granulocyte chemotactic protein-2 (GCP-2), monocyte chemotactic protein-1 (MCP-1) levels, and inflammatory cell infiltration in the UA occur in snoring obese subjects [150-153]. It is conceivable then that in obese subjects: 1) inflammation of the UA, 2) hypoxia, 3) snoring, 4) the mechanical stress of recurrent negative pressure, 5) obesity, and 6) systemic inflammation, may conjointly cause persistent increases in proinflammatory cytokines, and neuroinflammation owing to prominent overlaps of heterogeneous pathogenic effect of cascades.

Finally, two important points need to be made. 1) Endothelial adhesion molecules like E-selectin play an important role in leukocyte recruitment and development of atherosclerotic plaque. Visceral adipose tissue has been shown to be significantly associated with E-selectin. Not surprisingly, obese subjects possess higher E-selectin and TNF-α activation than the non-obese [154]. 2) There are of course several important negative links between increased body fat and brain functioning in older subjects; one is given below. Extensive neuropsychological testing and high-resolution MRI scanning has shown significant correlation between BMI and perturbation in regional gray and white matter volumes. Higher BMI was associated with decreased gray matter volumes in the left orbitofrontal, right inferior frontal, and right precentral gyri, the parahippocampal, fusiform, and lingual gyri, and right cerebellar regions [155-157].

4. SYNERGISTIC RISK FACTORS FOR COGNITIVE IMPAIRMENT: PATHOLOGY -SNORING AND OBSTRUCTIVE SLEEP APNEA

4.1. Snoring – An Important Risk Factor for Hypertension

Habitual snoring is defined as snoring greater than four times per week. Habitual snorers snore most of the night, and have an AHI of <5. 40% of elderly are habitual snorers, and habitual snoring is an independent risk factor for essential hypertension (EH). Many habitual snorers, similar to OSA patients, do not reduce their blood pressure (BP) during sleep [158]. This is ascribed to the fact that habitual snorers with EH have far greater number of arousals during sleep than "normotensive" simple snorers [159]. In an epidemiological study of female nurses (641 subjects followed up for 4 years, and adjusted for age, sex, BMI, neck and waist circumference, alcohol and cigarette smoking -in Wisconsin Sleep Cohort Study), EH developed in 9.7% in those who had no respiratory stigmata [160, 161]. In a smaller study of 95 males (20-62 year old) 60% had OSA and 33.3% had habitual snoring [162]. A similar odds ratio (OR) was found by Lindberg in normotensives over a 10 year period, EH developed in 12.5% of habitual snorers, and 7.4% of non-habitual snorers; however in younger subjects aged 30-39 years, the odds ratio for developing EH was only 2.6% [163]. In a study of community subjects, (adjusted for age, sex, BMI, and antihypertensive medications) habitual snorers showed higher morning and evening BP (systolic and diastolic) than normal non-apneic occasional simple snorers. Interestingly, in the latter group, after controlling for age, sex, and BMI, an OR for developing EH was 3.8 times higher than in non-apneic non-snorers in one study [164], and 1.6 times in another study [165]. A cohort of 72, 231 females, initially normotensive, were followed up over a period of 8 years [166]. During this follow-up, there were 7,622 incident cases of hypertension. Older age, smoking, body mass index, waist circumference, waist-hip ratio, weight gain, physical inactivity, and sleeping on the back were directly associated with regular snoring. After adjustment for age, body mass index (BMI), waist circumference, and other covariates, snoring was associated with a significantly higher prevalence of hypertension (OR = 1.22 for occasional snorers and1.43 for regular snorers). Compared with non-snorers, 29% occasional snorers, and 55% regular snorers developed EH (after controlling for age, BMI, waist circumference, alcohol and cigarette smoking) [167].

A statistically positive correlation has been documented between BMI and respiratory disturbance index (RDI) in overweight snoring patients -underscoring the correlation between obesity and snoring [200]. However, studies indicate that nocturnal alcohol ingestion affects individuals with a reported history of snoring; they reveal an increase of the apnea-hypopnea-index (AHI) with increasing blood alcohol concentration [168]. Even non-snoring subjects showed a significant increase of hypopneas with nocturnal alcohol ingestion [169]. A great majority of overweight individuals in the community consume alcohol regularly.

4.2. Nose and Upper Airway

Studies have confirmed the presence of inflammation in the uvula mucosa of patients with OSA. The thickness of the lamina propria, plasma cell infiltration and interstitial edema occur in conjunction with soft palate inflammation; these may contribute to UA occlusion observed during sleep in the OSA patients [170]. Furthermore, nasal inflammation is present in OSA sufferers (nonsmoking patients); they possess higher number of polymorphonuclear leukocytes (PMNs) and higher levels of of bradykinin and vasoactive intestinal peptide (VIP) in their nasal lavage [171]. Hence nasal inflammation may play a role in olfactory impairment as well as in UA obstruction in snoring/OSA.

4.3. Obstructive Sleep Apnea and Metabolic Dysfunction

OSA is a well known risk factor for metabolic perturbations and susceptibility to cardiovascular risk and weight gain [172, 173]. Higher leptin (from adipocytes) levels reflect resistance to the normal metabolic effects (appetite suppression and satiety) by this hormone in obese persons [174, 175] and OSA patients [172]. Leptin may predispose not only to increased cardiovascular risk [176] but platelet aggregation also [177]. Independent of body weight, OSA patients are known to possess higher levels of fasting blood glucose, insulin, and glycosylated hemoglobin [178, 179]. The severity of OSA correlates with the degree of insulin resistance [70, 180, 181] and a several-fold increase in the type 2 diabetes [179]. Although CPAP may reduce leptin levels and decrease visceral fat accumulation [182], glucose tolerance is not improved invariably [183]. This aspect is worth noting since impaired glucose tolerance in OSA patients, may be linked to sympathetic activation, leptin resistance, and sleep deprivation [70,184].

4.4. Obstructive Sleep Apnea and Coagulation

A relatively higher level of the primary fibrinolysis-inhibiting plasminogen activator inhibitor (PAI-1) and lower level of the primary fibrin degradation product (D-dimer) across the 24-h in OSA patients reflect decreased fibrinolytic capacity and fibrin degradation, respectively, reflecting evidence for a prothrombotic state in OSA [185]. Elevated nocturnal levels of catecholamines in OSA may enhance platelet aggregability [186, 187]. OSA may be causally related to increased clotting activity owing to the documented increases in fibrinogen [185, 191, 192], blood viscosity [188], and hematocrit [189]. CPAP therapy, however, may reduce platelet aggregability in conjunction with downregulation of catecholamine levels [186, 187, 190], and factor VII clotting activity [193].

4.5. Obstructive Sleep Apnea and Oxidative Stress

There is persuasive evidence from the animal and human studies for an association between hypoxia and upregulation of oxidative damage. Highly reactive free oxygen radicals *i.e.* ROS are known to be generated during intermittent hypoxia and reperfusion in repetitive episodes of nocturnal apnea. In OSA, repeated intermittent arterial oxygen desaturation and reoxygenation – ischemia-reperfusion injury to the vascular wall is instrumental in triggering ROS generation [194-196]. However, CPAP may reduce the level of ROS [195].

4.6. Snoring and Obstructive Sleep Apnea Upregulate Inflammation

20-40% of the general population suffers from snoring. Snoring is considered to be a prodromal stage prior to OSA; however, it may be present in nonapneic snorers also [197]. Snoring and sleep apneas are breathing disorders during sleep. Primary snorers without sleep apnea are labeled 'simple' or 'nonapneic'. The mechanism of snoring is vibration of anatomical structures in the pharyngeal airway. Flutter of the soft palate accounts for the harsh aspect of the snoring sound. The vibrations transmitted to the pharyngeal structures by snoring may span a large range of frequencies, with transmission of high energy. During sleep, the degree of inspiratory airflow reduction is associated with snoring. Snoring is the sound produced by vibration of the soft palate and other soft tissues in the pillars of the oropharynx inlet and usually indicates increased airway resistance. Snoring without arousals is of unknown clinical consequences; however, arousals associated with snoring are potentially important sleep-disordered breathing events. Thus, it is important to determine the snore arousal index (number of events associated with arousal per hour of total sleep time). In the general population major independent risk factors for development of habitual snoring are male gender, age between 40 and 64 years, obesity, use of alcohol or sleep medications, and current cigarette smoking [198, 199]. A deleterious effect of snoring-related vibrations may lead to anatomic lesions of the upper airway mucosa, pharyngeal muscles, and their innervating nerves.

Light microscopy of soft palate and uvula of nonapneic snorers revealed mucous gland hypertrophy with ductal dilation and focal squamous metaplasia, disruption of muscle bundles by infiltrating mucous glands, focal atrophy of muscle fibers, and extensive edema of the lamina propria with vascular dilation [201]. Severe snorers did not differ qualitatively from apneics in the above characteristic changes found; however, some snorers had less extensive changes. A reduction in capillarity of palate muscles is one of the pathophysiological features in long-term non-apneic snorers. This may be due to neuromuscular injury caused by repeated vibratory trauma of the soft palate during snoring [202]. Focal degeneration of myelinated nerve fibers and axons was present (electron microscopy) in both apneics and nonapneic snorers [201]. Definite degenerative changes in myelinated and unmyelinated nerves were confirmed in human uvula by EM in a more recent study;

however, as expected these degenerative changes were present to a lesser degree in cases of nonapneic snorers compared with those of the OSA patients [203]. Other studies found atrophy of the muscle bundles and inflammatory changes noted as lymphocytic infiltrations [204]. Nonapneic patients with habitual snoring possess increased palatopharyngeal abnormalities characteristic of neurogenic lesions, despite fairly normal oximetry findings, in the obese and non-obese patients [205]. This has been interpreted to be due to vibration trauma causing peripheral nerve injury [201, 205, 206]. The vibratory trauma to pharyngeal tissues is correlated with impairing of pharyngeal reflexes [64, 66, 201].

The UA mucosa in snorers is edematous and hyperplastic. A morphometric point counting technique was used to determine the tissue composition; the data showed that nonapneic snorers do possess a significantly greater percentage of intercellular space than controls. Control uvulas contained more muscle compared with those from snorers. Moreover, the covering epithelium was significantly thicker in snorers than in controls. Of note, the uvula morphology was similar in OSA patients and nonapneic snorers. Since the increased amount of intercellular space is the expression of edema, it was hypothesized that these mucosal changes together with hyperplasia of the covering epithelium are due to the pathologic effects of snoring [207]. The levels of proinflammatory cytokines TNF-α and IL-6 are elevated in the uvula of nonapneic snorers, although they are much higher in apneic patients [208]. The effect of vibration was assessed in the rats by measuring the vibration-induced increase in gene expression of the TNF-α and of the neutrophil attractant chemokine macrophage inflammatory protein-2 in the soft-palate tissue. PCR measurement of mRNA showed that vibration induced a significant overexpression of the pro-inflammatory cytokines (relative to control) [209]. In a cell model of snoring-induced airway inflammation, mechanical vibration simulating snoring, triggered an inflammatory cascade in human bronchial epithelial cells, reflected by the increase in IL-8 release mediated by MAPK pathways [210].

4.6. Hypoxia and CPAP therapy

Hypoxia is implicated in the production of inflammatory cytokines, *e.g.* IL-6, TNF-α and increased levels of C-reactive protein (CRP) [210-213]. CRP may play an important role in inhibiting NO synthase [214], and enhancing cell adhesion molecule expression [215] -thereby contribute to cerebrovascular disease. Further, OSA may directly contribute to the expression of adhesion molecules (Ohga *et al.*, 1999); however, CPAP therapy may reduce the level of these inflammatory molecules [216-218].

5. DISCUSSION

5.1. Obstructive Sleep Apnea – Source of Multiple Pathologies

It is essential that this discussion commences with an emphasis on the pathologic attributes of "Snoring/OSA". The deleterious combination of snoring + OSA comprises a major sleep-related disorder in which repetitive partial (hypopneas) or complete (apnea) obstruction of the UA during sleep causes hypoxia, hypoxemia, multiple arousals, sleep fragmentation, genioglossus dysfunction inducing increased respiratory and (consequently promoting) various physiological/metabolic sequels [219]. UA negative pressure is one of the critical factors, and the UA occlusion during sleep is caused by the subatmospheric pharyngeal pressure during respiration, and reduced UA dilator muscle activity [220, 221]. Sleep disordered breathing in this condition encompasses repeated episodes of UA obstruction that ranges from snoring through increased airway resistance and reduction in airflow (hypopnea) to periods of breathing cessation (apnea) and airway collapse. OSA-induced elevated sympathetic activity may contribute to hypertension and other adverse cardiovascular outcomes (Fig. **1**). OSA has been shown to be an important contributory factor that evokes a plethora of dysfunctions contributing to neurocognitive dysfunctions [59, 62-66, 70, 202], hypertension-cardiovascular conditions [56, 222, 223], myocardial infarction [57, 58], stroke, and even sudden death [60, 61]. OSA is considered to be a risk factor for atrial fibrillation [224]. Acute sleep deprivation of healthy young men and women is associated with increased p-wave dispersion on electrocardiography, an electrophysiologic marker for the prediction of paroxysmal atrial fibrillation [225, 226]. In addition, OSA itself has been associated with increased p-wave dispersion and incidence of atrial fibrillation [227, 228]. Sustained sleep deprivation also increases sympathetic (but decreases parasympathetic) activity as measured by heart rate and blood pressure variability [229].

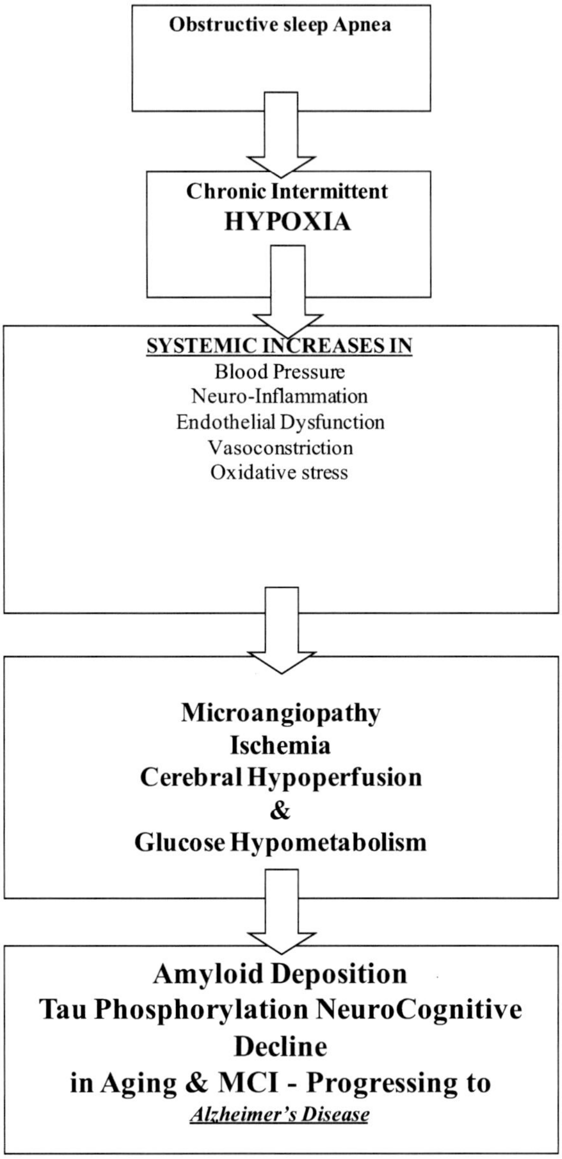

Figure 1: Diagrammatic representation of possible sequence of events commencing from snoring/OSA. Intermittent hypoxia and hypoxemia are precursors of several adverse factors such as hypertension, neuroinflammation, microangiopathy, hypoperfusion, and oxidative stress. These may promote glucose hypometabolism, energy decrease, and amyloid as well as NFT deposition. These pathological stigmata being intimately associated with amyloid plaques and NFT promote neurocognitive decline escalating to AD. AD is due to sustained multifactorial damage, and concomitant salvaging molecular pathways are inefficient and incapable to inhibit/clear the ongoing damage brought about by the heterogensous pathogenic cascades.

5.2. Obstructive Sleep Apnea and Obesity – Harbingers of Synergistic Pathology

Both obesity and OSA have been shown to induce airways inflammation [230]. OSA and obese patients showed a higher percentage of neutrophils and a lower percentage of macrophages in their phlegm [230, 231]. In addition obesity and OSA may promote atrial fibrillation [224]. High fat diet has been shown to increase amyloid burden and produce AD-like pathology in 3xTg-AD mice brain [232]. However, the observed Aβ changes were not correlated with alterations in total APP, C-terminal fragments (α and β cleavage products) of APP or soluble oligomers. This suggests that high-fat diet does not promote the accumulation of Aβ through altered APP metabolism. Several studies highlight that obesity and metabolic disorders are linked to Aβ accumulation in AD [233, 234]; it can be hypothesized then, that the increases in weight and brain Aβ following the high-fat diet, share common mechanisms. GFAP is a commonly used marker of astrocytic activation; its concentration increases in AD [235], in transgenic mice [236-240], and in obese mice [241]. This suggests that the upregulation of gliosis in obesity may be independent of the transgene status [241]. This further shows that intake of excessive calories from fat may increase reactive astrocytes and inflammatory changes in the brain, an event also associated with Aβ pathology [235]. Indeed, experimental induction of microgliosis increases Aβ burden [242, 243].

5.3. Neurocognitive Morbidity in Snoring and Obstructive Sleep Apnea Patients

5.3.1. Oxidative Stress Causes Axonal/Neuronal Degeneration

Normally metabolism of oxygen by cells generates potentially deleterious ROS which is balanced by the oxidant elimination. However, an imbalance between prooxidants and antioxidants results in oxidative stress. Oxidative stress increases in normal aging and is a key event in the pathogenesis of several diseases. An excess generation of vascular ROS has been implicated in various chronic diseases, including hypertension, diabetes, and many other chronic diseases including AD. Oxidative stress is both a cause and a consequence of upregulated circulating inflammatory cytokines (CIC). It is a multisystem phenomenon involving immune system, nervous system and vasculature, and is implicated in inflammation, endothelial dysfunction, and neuronal apoptosis (Fig. **1**). There is strong evidence to support that oxidative stress promotes axonal/neuronal degeneration. The latter is a common pathologic feature in normal aging and in several neurodegenerative diseases including AD [244]. Axonal defects occur prior to neuronal death and correlate closely with functional decline.

In vitro studies have shown that agents of oxidative stress such as H_2O_2 mediate axonal degeneration [245]. Loss of the antioxidant enzyme Cu, Zn-superoxide dismutase (SOD) in mice causes several accelerated age-related pathologies and a 30% reduction in lifespan [246, 247-250]. Similarly, anti-sense knockdown of SOD causes death of motoneurons [251], PC12 neurons [252], and axonal degeneration in *Sod−/−* DRGs in mouse DRG cultures [253]. Superoxide-mediated toxicity may be particularly relevant to the pathogenesis of neuronal atrophy/death in several brainstem nuclei including the multi-functional "nucleus tractus solitarius (NTS)". SOD plays an important role in maintaining neuronal viability under stress and during the aging process. The inflamed NTS neurons may be susceptible to increased superoxide production in the elderly harboring chronic inflammation -and SOD is required for neuronal survival. However, in prolonged neuroinflammation, superoxide radicals may be inefficiently neutralized in the NTS by low SOD enzymes, thus enhancing highly toxic free radicals [254, 255].

Inflammation, hypoxia/hypoxemia, and oxidative stress are deleterious and cause neuronal degeneration in several brainstem foci and the NTS [62-66, 70, 202, 256-260]. The NTS is a hub nucleus of brainstem and has copious afferent and efferent projections that affect central homeostatic control and an array of extensive pathways. Extensive data indicate that the NTS is a key structure for sensory, autonomic and neuroendocrine functions. CNS areas that receive a direct projection from the NTS project back to this nucleus. This nucleus contains an enormous range of neuroactive substances; indeed, most of those identified within the CNS are also found in the NTS, as neurotransmitters and neuromodulators. It is, therefore, tremendously important in subserving key interaction with almost every system of the body by impacting the CNS. Apoptotic neurons (not found in the control animals), were widespread in brainstem regions in hypoxic group of animals; these regions involved pons and medulla including the crucial nuclei in cardiovascular functions *viz.* the nucleus ambiguus, the NTS, and the dorsal motor nucleus of the vagus [261].

5.4. Hypoxia Promotes Neuronal Damage *via* Hypoperfusion, Inflammation and Endothelial Dysfunction

5.4.1. Cerebral Hypoperfusion – Triggered by Obstructive Sleep Apnea

The cerebral blood flow (CBF) is tightly coupled to metabolic demand; however, chronic brain hypoperfusion (CBH) may uncouple blood flow to metabolic activity whereby brain cell function is compromised [62-66, 70, 262]. During cerebral hypoperfusion, CBF decrease is not commensurate with the metabolic demands of the brain. The possibility of aged brain receiving persistent suboptimal CBF, may compromise neurocognitve function over a prolonged period. Thus, unlike an acute ischemic episode/stroke, CBH may promote progressive cognitive deficits, albeit of variable degree depending on the level of pathology. As neurons possess no energy reserves, the performance of cognitive function is critically dependent on the steady delivery of adequate oxygen and glucose in order to produce the main energy source – *viz.* adenosine triphosphate (ATP). In the event of relentless chronic intermittent apnea (CIH/hypoxia) causing CBF interruptions night after night, it can be safely hypothesized that cognitive impairment of aging and mild cognitive impairment (MCI) may develop into AD [62-66, 70, 262].

Vascular pathology in the aged brain and in AD includes cerebral amyloid angiopathy (CAA) [2]. An important question is -what are the factors that enhance amyloid β deposition (amyloid plaque) during aging, even in normal elderly? The microvascular changes encompassing neuroinflammatory and cytokine increases are known to cause pathophysiological abnormalities that perturb metabolic and blood-brain barrier functions, regional cerebral perfusion, and promote neuronal damage. AD-related volume losses have been repeatedly and reproducibly documented in several key brain regions including the hippocampus, the entorhinal cortex, the parahippocampal gyrus, and the parietal cortex [62-66, 70]. It is now well accepted that Fluoro-2-deoxy-D-glucose positron emission tomography (FDG-PET) imaging that reveals glucose metabolic reduction in the parieto-temporal, frontal and posterior cingulate – may be taken as the hallmark of neurodegenerative disease/AD [263]. In fact FDG-PET studies have consistently shown that medial temporal lobe hypometabolism is the most specific and sensitive measure for the identification of MCI [264, 265].

5.4.2. Vascular Endothelial Dysfunction -Triggered by OSA

The hypoxia, hypercapnia, and pressor surges are potent stimuli for the release of vasoactive substances and for impairment of endothelial function during obstructive episodes. Increased levels of endothelin in response to hypoxemia of OSA, may promote sustained vascular changes such as vasoconstriction [266]. Indeed OSA patients free of any other overt cardiac or vascular disease possess impaired endothelial function [267, 268]. Endothelial dysfunction also occurs in subjects having hypertension [62-66, 70, 269], hyperlipidemia [270], diabetes [270-274] or smoking habit [275-277]. OSA patients may possess one or all of the above conditions; therefore, OSA/hypertension alone or in concert with comorbidities may result in overt endothelial dysfunction (Fig. **1**).

Various studies have documented gray matter loss in cognitively relevant brain regions in hypoxia [278, 279]. The cortical neuronal cell damage due to CIH has been repeatedly documented in the brain [280-282]. In a mouse model of OSA, it has been established that there is CIH-mediated cortical neuronal cell death and neurocognitive dysfunction. Exposure to hypoxia may produce hypertension since long-lasting sympathetic activation occurs in humans [283]. Another mechanism analogous to CIH is the ischemic hypoxia/reperfusion-related reoxygenation, where enhanced ROS generation causes damage [284-289]. OSA is associated with reduction of blood flow from the middle cerebral artery to the cortex that may negatively impact cortical neurons [282, 284, 286, 288-291]. Cortical neurons from almost entire cortex project to the NTS reciprocally [65, 66, 70]. Taken together, it is plausible that OSA and the NTS are the perpetrators, in causing microangiopathy, inflammation, hypoperfusion, and oxidative stress; they conceivably contribute to dysfunction in various key CNS foci and promote cognitive decline/AD.

5.4.3. Microglia – Upregulate Proinflammatory Cytokines, Neurotoxic Factors, and Aβ Deposition

Microglia are phagocytic and immunocompetent cells in the CNS involved in the innate defense [292, 304]. Microglia also function as 1) Scavenger cells when the

CNS confronts infection, inflammation, trauma, ischemia and neurodegeneration in the CNS [293], and 2) In the programmed elimination of neural cells [294, 295]. Resting microglia are highly sensitive to the above mentioned alterations/disturbances within their microenvironment; they respond swiftly within 24 hours of stimulus [296]. This may then lead to elevated microglial immunoglobulin IgG reactivity, upregulation of CD1 and cell adhesion molecules, such as lymphocyte function-associated antigen 1 (LFA-1) (CD11a/CD18), intercellular adhesion molecule (ICAM)-1 (CD54), and vascular cell adhesion molecule (VCAM)-1(CD106) [297]. During ongoing noxious stimuli, microglia may adhere to neurons [298]; chemokines such as monocyte chemoattractant protein-1 and interferon (IFN)-inducible protein-10 have been implicated in their neuronal attachment [299, 300]. The phenotypic changes of microglia involve their taking on a macrophage-like appearance [301]. With continuing stimulus, the microglia enhance their functional transformation with upregulation of the major histocompatibility complex (MHC) class two molecules and inflammatory glycoproteins, such as CD40, B7.1 (CD80) and B7.2 (CD86); these provide a powerful stimulus for immune cell activation [300]. Activated microglia are inherently capable of constitutively expressing β2-integrins CD11a, CD11b and CD11c, ICAM1, a ligand for the β2-integrins (FA-1/CD11a/CD18) and mac-1 (CD11b/CD18). Furthermore, complement activation occurring in neuroinflammation, may provide additional ligands for microglial integrins [292, 302, 303]. Despite the above physiological stimulations, microglia can still downregulate the above activation factors, and return to the basal resting state – provided the pathologic stimulus dissipates. However, if the activating stimulus does persist, the above activation markers and cell adhesion molecules remain potentiated; the activated microglia may then cluster and encircle the neurons, become phagocytic [305-307], and cause progressive and irreversible neuronal cell death. This neuronal fate is worsened by the release of chemoattractants by the dying neurons – this leads to higher infiltration of more vigorous and activated microglia in the pathological foci [292, 299].

5.5. Amyloid Deposition, Neurofibrillary Tangles and Inflammation

This section is meant to underscore some salient features rather than being exhaustive on this issue. The cerebral amyloid angiopathy has been documented to be present in the autopsied Alzheimer's disease brain, and ~80% showed Aβ deposits within blood vessels [308-310]. This vascular Aβ correlated with high retention of Pittsburgh Compound B also [311-313]. There is a clinical silent period prior to AD that can last up to two decades, and this duration depends on life style habits, pro/anti inflammatory factors, environmental and genetic factors, and/or cognitive reserve. Thus Aβ deposition occurs at a slow rate indicating that the time window for its accumulation prior to cognitive dysfunction may be quite large. This is correlated with the smoldering inflammatory state due to various inflammatory risk factors/oxidative stress. No correlation was found between age-at-onset of the disease and Aβ burden (amyloid-β-ligand [^{11}C]-labelled Pittsburgh compound-B binding) [314]. Aβ deposition in humans is a temporally later process in aging, possibly as a neuroprotective process for modifying disease progression [315]. The aberrant Aβ activates macrophages, stimulating them to synthesize inflammatory cytokines, chemokines and excitotoxic factors that in turn stimulate neurons to produce more Aβ. This leads to a mismatch between Aβ deposition and its removal by microglia [316, 317], thus causing further neuroinflammation. Additionally, lymphocyte infiltration and complement deposition might further amplify the ongoing inflammatory processes resulting in neuronal death [318-321]. Since several infectious agents have been reported in the CNS of elderly and AD patients [322], Aβ may be synthesized as part of response triggered by the immune system. Furthermore, Aβ is now considered to be an antimicrobial peptide; it may be upregulated in AD as a response to such ongoing stimuli [323]. Older smoking and alcohol consuming obese individuals are more prone to microbial infections. Importantly, various mechanisms that contribute to inflammatory cytokines are upstream to Aβ deposition, they need to be addressed and prevented in aging in order to attenuate cognitive and behavioral dysfunction.

AD is characterized by neurofibrillar degeneration; the latter occurs as intraneuronal NFT. Hyperphosphorylated tau (p-tau) is considered to occupy a central position in AD neuropathogenesis. In Tg4510 mice, p-tau is associated with age-related microglial activation, and in normal aging up to 9 mo CD45+ microglia increase in parallel with tau pathology. Moreover, transient ischemia induced tau hyperphosphorylation and NFT [324]. 1-3 mo old young mice are capable of clearing soluble p-tau; however, in 5.5 mo-old mice insoluble tau aggregates are present [324, 325]. In LPS injected APP mice, activated microglia cleared Aβ [326]. In contrast, LPS-induced microglial activation exacerbated p-tau pathology in Tg4510 mice [325]. Further, peripheral administration of LPS showed no changes in APP processing. Interestingly, p-tau was increased in the hippocampus an important region in AD pathogenesis [126, 327]. Moreover, proinflammatory cytokine IL-1β and IL-6 caused tau phosphorylation in neurons [328, 329]. The above mentioned as well as many evidentiary studies [330-332] support the potential role of inflammatory mediators in enhancing p-tau pathology and underscore that inflammatory stimuli can and do facilitate tau phosphorylation potentiating taupathy, and neuronal degeneration. Such a scenario is very likely to occur in the CNS in general and the NTS in particular, in the elderly and MCI/AD patients who may possess chronic neuroinflammatory state and/or recurring inflammatory insults. Consequently, the dysfunctional NTS may affect functional integrity of a large number of key pathways/synapses/neurons, throughout the CNS including the hypoglossal, locus ceruleus, basal forebrain, and medial temporal lobe regions, and provoke cognitive dysfunction/dementia in the elderly.

5.6. OSA, Obesity and Insulin Resistance

In the current paradigm, apnea-induced hypoxia/hypoxemia has a central role. Several studies suggest this risk factor to be the common denominator in the pathological cascades that promote the onset of hypertension, cardiovascular diseases, and metabolic dysfunctions [62-66, 70]. In obese sleep apneics, visceral fat *e.g.* when significantly greater than obese controls -is strongly associated with insulin resistance and inflammation [178, 333, 334]. Obesity plays a major role in the pathogenesis of sleep apnea; cytokines and insulin resistance are upregulated in excessive daytime sleepiness (EDS) and OSA in humans. In obese persons hypothalamic CRH neuron is hypoactive providing additional evidence on the potential role of central neural mechanisms in mediating depressed ventilation and development of OSA [333, 334]. Thus, obesity *via* inflammation, insulin resistance, visceral adiposity, and hypofunctioning central neural mechanisms (*e.g.* hypothalamic CRH), may synergize in the pathogenesis of sleep apnea, and its associated cardiovascular co-morbidities [334]. The pathogenic role of insulin resistance in OSA can be appreciated from women with the polycystic ovary syndrome (PCOS) (a condition associated with hyperandrogenism and insulin resistance); compared to controls, they are much more likely to possess SDB and daytime sleepiness. Furthermore, interestingly the risk for OSA among nonobese women with PCOS is very low, but the risk for OSA in obese PCOS is high [335]. Recently, two "subtypes" of PCOS have been documented -*i.e.* PCOS with OSA and PCOS without OSA. PCOS women with OSA may be at much higher risk for diabetes and cardiovascular disease than PCOS women without OSA, reflecting that these two subtypes may possess distinct metabolic and endocrine attributes [336]. This aspect was further delineated in that glucose intolerance and insulin resistance were correlated with the presence and severity of OSA [337]. Among PCOS women with normal glucose tolerance, the presence of OSA was associated with a nearly 2-fold higher fasting insulin level and HOMA index. The severity of OSA was highly significant in predicting the fasting concentrations of glucose and insulin as well as HOMA index [338]. Conversely, only the most insulin-resistant subset of obese individuals was found to be at high risk for OSA [339].

Diabetes (type 2, DM) is associated with peripheral insulin resistance, whereas AD is associated with brain insulin resistance and insulin deficiency. In a C57BL/6 mouse model of obesity and DM, high fat diet (HFD) feeding for 16 weeks doubled body weight and caused DM. HFD caused brain insulin resistance also evidenced by reduced insulin receptor binding, and reduced brain weight. However, HFD-fed mouse brains lacked the typical AD characteristics, *i.e.* increases in Aβ or phospho-tau, IGF signaling impairments, or acetylcholine dyshomeostasis. This underscores that obesity and DM may cause brain atrophy with insulin resistance, oxidative stress, and cytoskeleton degradation, but they lack many features that are well known to be associated with AD neuropathology. This clearly indicates that obesity and DM by themselves are not sufficient to trigger neurocognitive impairment of AD [340]. In a study that evaluated the prognostic predictors in AD, the results showed that patients with DM had a 65% reduced risk of fast cognitive decline, compared to AD patients without DM [341].

5.7. Hypoxia and Hypertension in Causing Amyloid Deposition

Renin-angiotensin system (RAS) contributes to cognitive decline and brain damage *via* chronic cerebral ischemia. Chronic cerebral hypoperfusion in

C57BL/6J mice significantly increased brain renin activity and angiotensinogen expression; this was attributed to activated microvessels for the rennin increase and an increase in angiotensinogen in activated astrocytes [342]. Utilizing chimeric transgenic (Tg) mice (having human renin (hRN) and human angiotensinogen (hANG) gene) it was documented that cerebral surface blood flow was greatly reduced in hRN/hANG-Tg mice (20-week-old). Superoxide anion production occurred in the brain of these 10-week-old mice; it further increased with an increase in NADPH oxidase activity. Moreover, the brain of hRN/hANG-Tg mice also had increased expression of p47 (phox) and Nox4 [343], enhanced ischemic brain damage, due mainly to marked reduction of CBF and enhanced oxidative stress [344]. In keeping with epidemiological evidence that hypertension may accelerate the onset and progression of AD, experimental hypertension induced in Tg2576 mice was associated with significantly enhanced accumulation of Aβ peptides in brain. These results indicate that hypertension may promote AD-type Aβ neuropathology in Tg2576 [345]. Additionally, hypoxia/hypoxemia which promotes hypertension also induces Aβ neuropathology in brain. Convincing evidence shows that transient hypoxia can cause an increase in the gene expression of β-secretase (BACE1) – the cleaving enzyme of β-site of amyloid precursor protein [346, 347]. Aβ deposition (that begins insidiously) is an early event in this cascade and may lead to cognitive impairment. The above factors may independently contribute to Aβ accumulation and neurodegeneration. Further, cognitive decline and functional/structural brain changes suggestive of AD may occur when several stigmata such as aging, depression, environmental factors (smoking and alcohol consumption), inflammation, oxidative stress, hypoperfusion, and hypoxia/hypoxemia may converge and keep Aβ production and tau hyperphosporylation upregulated. Finally, the above mentioned pathogenic factors may upregulate each other and fan the fuel of neurocognitive dysfunction.

6. CONTROL OF RISK FACTORS

Based on the above studies/evidence, the risk factors for AD must be attenuated in order to keep cognitive and memory decline under control. Vascular risk factors may significantly increase hypoperfusion and dementia risk, and thus provide important preventive targets. Obviously there is an essential need to control hypertension, diabetes, hyperlipidemia, and malnutrition. However, caution is warranted for aggressively controlling hypertension in a demented patient. In postmenopausal women estrogen replacement therapy has given mixed results in controlling dementia; risks and benefits obviously need to be carefully weighed for each patient. Physically activity in midlife is of importance and may lower the risk of AD; hence regular exercise may enhance protective benefit. Many measures, such as taking anti-inflammatories, probably have a very small benefit over time, so at this point in time, it is difficult to recommend neutraceuticals and anti-inflammatory drugs -since careful clinical trial have not shown their benefit in combating AD. Clinicians should diligently focus on controlling snoring and OSA, and hence lowering cardiovascular risk and a host of other pathogenic factors/cascades discussed above. As detailed earlier, these risk factors are crucially important, and eminently treatable to ameliorate neurocognitive impairment. Our aim is to reduce incidence of this devastating disease, and no doubt exciting new therapies must be on the horizon.

CONCLUSIONS

Aging is said to be the state of accumulation of many forms of biological damage; it is a pro-inflammatory state. Hence, an understanding of the cellular and molecular mechanisms that cause this damage is of crucial importance. A simple and potentially pragmatic pathogenic mechanism(s) is posited here to explain the neuropathogenesis of cognitive decline in aging and AD. Relatively common sources, such as snoring and OSA, potentiate systemic inflammation/ neuroinflammation and are emphasized here to be significant risk factors with the potential to promote hypertension and neurodegenerative diseases in aged individuals, particularly those who are obese. Nocturnal chronic intermittent hypoxia and hypercapnia are cardinal features of OSA. Altered cerebral circulation, brain hypoperfusion/ischemia, and the pathological loss of gray matter are associated with OSA. Hypoxia/hypoxemia may cause hypertension, inflammation and microvascular changes; these may stimulate switching of microglial phenotype to becoming pro-inflammatory. In OSA, the process of vascular (and non-vascular) Aβ deposition may progress due to comorbid disease progression (*i.e.* diabetes, hypertension and obesity – and a combination thereof) over years. The aberrant Aβ activates macrophages, stimulating them to synthesize inflammatory cytokines, chemokines and excitotoxic factors that in turn stimulate microglia and neurons to produce more Aβ. The vicious cycle is perpetuated. These conceivably upregulate physiological dysfunctions in key areas of cortex, and brainstem nuclei -notably, the NTS, locus ceruleus, and the hypoglossal, and their inter-connected pathways. The synergistic role of hypoxia, hypoxemia, (neuro)inflammation, cerebral hypoperfusion and hypometabolism of brain may thus play a critical role in inducing cognitive decline in aging → MCI → AD, and possibly other neurodegenerative diseases.

ACKNOWLEDGEMENT

Declared none.

CONFLICT OF INTEREST

The author(s) confirm that this chapter content has no conflict of interest.

ABBREVIATIONS

AD	=	Alzheimer's disease
Aβ	=	amyloid-β
APP	=	amyloid precursor protein
AHI	=	Apnea-hypopnea index
BACE	=	β-site amyloid precursor protein cleavage enzyme
BFB	=	basal forebrain
BMI	=	body mass index
BOLD	=	blood-oxygenation-level-dependent
BP	=	blood pressure
CAA	=	amyloid angiopathy

CBF	=	cerebral blood flow
CIC	=	circulating Inflammatory Cytokines
CIH	=	chronic intermittent hypoxia
$CMRO_2$	=	cerebral metabolic rate of oxygen
COX-2	=	cycloxygenase-2
CRP	=	C-reactive protein
CPAP	=	continuous Positive Airway Pressure
DM	=	type 2 diabetes mellitus
DMV	=	dorsal motor nucleus V cranial nerve
DMNV	=	dorsal motor nucleus of the vagus
EDS	=	excessive daytime sleepiness
ERC	=	entorhinal cortex
HEP	=	high energy phosphate (ATP)
GMV	=	gray matter volume
ICAM-1 and -2	=	intercellular adhesion molecule-1 and -2
IGF-1R	=	IGF-1 receptor
IH	=	intermittent hypoxia
IR	=	insulin receptor
IL-1β	=	interleukin-1β
LDL	=	low-density lipoprotein

MMP	=	matrix metalloproteinase 2
MTL	=	medial temporal lobe
NMDARs	=	N-methyl-D-aspartate receptors
MCA	=	middle cerebral artery
MCI	=	Mild cognitive impairment
NA	=	nucleus ambiguus
NFTs	=	neurofibrillary tangles
NO	=	nitric oxide
NREM	=	non-rapid eye movement sleep
NF-κB	=	nuclear factor kappa B
NTS	=	nucleus of the solitary tract
OSA	=	obstructive sleep apnea
PCC	=	posterior cingulate cortex
PECAM	=	platelet endothelial cell adhesion molecule-1
PFC	=	prefrontal cortex
RAGE	=	AGE receptor
RAS	=	renin-angiotensin system
ROS	=	reactive oxygen species
SDB	=	sleep disordered breathing
SPs	=	amyloid senile plaques

TNF-α = tumor necrosis factor-α

UA = upper airway

VBM = Voxel-based morphometry

VEGF = vascular endothelial growth factor

REFERENCES

[1] Alzheimer's Association. 2010 Alzheimer's disease facts and figures. Alzheimers Dement 2010; 6: 158-94.

[2] Thal DR, Griffin WS, Braak H. Parenchymal and vascular Aβ-deposition and its effects on the degeneration of neurons and cognition in Alzheimer's disease. J Cell Mol Med 2008; 12: 1848-62.

[3] Braak H, Braak E, Bohl J. Staging of Alzheimer-related cortical destruction. Eur Neurol. 1993; 33: 403-8.

[4] Braak H, Braak E. Neuropil threads occur in dendrites of tangle-bearing nerve cells. Neuropathol Appl Neurobiol 1988;14: 39–44.

[5] Braak H, Braak E. Neuropathological stAging of Alzheimer-related changes. Acta Neuropathol 1991; 82: 239–59.

[6] Braak E, Braak H, Mandelkow EM. A sequence of cytoskeleton changes related to the formation of neurofibrillary tangles and neuropil threads. Acta Neuropathol 1994; 87: 554–67.

[7] Selkoe DJ: Cell biology of protein misfolding: the examples of Alzheimer's and Parkinson's diseases. Nat Cell Biol 2004; 6:1054-61.

[8] Lee SJ, Lim HS, Masliah E, Lee HJ: Protein aggregate spreading in neurodegenerative diseases: problems and perspectives. Neurosci Res 2011; 70: 339-48.

[9] Arriagada PV, Growdon JH, Hedley-Whyte ET, Hyman BT. Neurofibrillary tangles but not senile plaques parallel duration and severity of Alzheimer's disease. Neurology 1992; 42: 631-39

[10] Holmes C, Boche D, Wilkinson D, *et al.* Long-term effects of Abeta42 immunisation in Alzheimer's disease: follow-up of a randomised, placebo-controlled phase trial. Lancet 2008; 372(9634):216-23

[11] Koyama A, Okereke OI, Yang T *et al.* Plasma Amyloid-β as a Predictor of Dementia and Cognitive Decline:A Systematic Review and Meta-analysis. Arch Neurol 2012; 69: 824-31.

[12] Kang J, Lemaire HG, Unterbeck A, *et al.* The precursor of Alzheimer's disease amyloid A4 protein resembles a cell-surface receptor. Nature 1987; 325: 733–36.

[13] Allinson TM, Parkin ET, Turner AJ, Hooper NM. ADAMs family members as amyloid precursor protein alpha-secretases. J Neurosci Res 2003; 74: 342–52.

[14] Sinha S, Anderson JP, Barbour R, *et al.* Purification and cloning of amyloid precursor protein beta-secretase from human brain. Nature 1999; 402: 537–40.

[15] Vassar R, Bennett BD, Babu-Khan S, *et al*. Beta-secretase cleavage of Alzheimer's amyloid precursor protein by the transmembrane aspartic protease BACE. Science 1999; 286: 735–41.

[16] Hussain I, Powell D, Howlett DR, *et al*. Identification of a novel aspartic protease (Asp 2) as beta-secretase. Mol Cell Neurosci 1999; 14: 419–27.

[17] Wolfe MS, Xia W, Ostaszewski BL, Diehl TS, Kimberly WT, Selkoe DJ. Two transmembrane aspar-tates in presenilin-1 required for presenilin endoprote-olysis and gamma-secretase activity. Nature 1999; 398: 513–17.

[18] Steiner H, Winkler E, Edbauer D, *et al*. PEN-2 is an integral component of the gamma-secretase complex required for coordinated expression of presenilin and nicastrin. J Biol Chem 2002; 277: 39062–65.

[19] Francis R, McGrath G, Zhang J, *et al*. aph-1 and pen-2 are required for Notch pathway signaling, gamma-secretase cleavage of betaAPP, and prese-nilin protein accumulation. Dev Cell 2002; 3: 85–97.

[20] Yu G, Nishimura M, Arawaka S, *et al*. Nicastrin modulates presenilin-mediated notch/glp-1 signal transduction and betaAPP processing. Nature 2000; 407: 48–54.

[21] Jarrett JT, Berger EP, Lansbury PT, Jr. The carboxy terminus of the beta amyloid protein is critical for the seeding of amyloid formation: implications for the pathogenesis of Alzheimer's disease. Biochemistry 1993; 32: 4693–7.

[22] Younkin SG. The role of A beta 42 in Alzheimer's disease. J Physiol 1998; 92: 289–92.

[23] Cleary JP, Walsh DM, Hofmeister JJ, Shankar GM, Kuskowski MA, Selkoe DJ, Ashe KH. Natural oligomers of the amyloid-beta protein specifically disrupt cognitive function. Nat Neurosci 2005; 8: 79–84.

[24] Lesne S, Koh MT, Kotilinek L, *et al*. A specific amyloid-beta protein assembly in the brain impairs memory. Nature 2006; 440: 352–7.

[25] Walsh DM, Klyubin I, Fadeeva JV, *et al*. Naturally secreted oligomers of amyloid beta protein potently inhibit hippocampal long-term potentiation *in vivo*. Nature 2002; 416: 535–9.

[26] Ballatore C, Lee VM, Trojanowski JQ. Tau-mediated neurodegeneration in Alzheimer's disease and related disorders. Nat Rev Neurosci 2007; 8: 663–72.

[27] Fulga TA, Elson-Schwab I, Khurana V, *et al*. Abnormal bundling and accumulation of F-actin mediates tau-induced neuronal degeneration *in vivo*. Nat Cell Biol 2007; 9: 139–48.

[28] Lee G, Neve RL, Kosik KS. The microtubule binding domain of tau protein. Neuron 1989; 2: 1615–24.

[29] Goedert M, Wischik CM, Crowther RA, Walker JE, Klug A. Cloning and sequencing of the cDNA encoding a core protein of the paired helical filament of Alzheimer disease: identification as the microtubule-associated protein tau. Proc Natl Acad Sci USA. 1988; 85: 4051–5.

[30] Goedert M, Spillantini MG, Jakes R, Rutherford D, Crowther RA. Multiple isoforms of human micro-tubule-associated protein tau: sequences and localization in neurofibrillary tangles of Alzheimer's disease. Neuron 1989; 3: 519–26.

[31] Neve RL, Harris P, Kosik KS, Kurnit DM, Donlon TA. Identification of cDNA clones for the human microtubule-associated protein tau and chromoso-mal localization of the genes for tau and microtubule-associated protein 2. Brain Res. 1986; 387: 271–80.

[32] Gong CX, Liu F, Grundke-Iqbal I, Iqbal K. Post-translational modifications of tau protein in Alzheimer's disease. J Neural Transm 2005; 112: 813–38.

[33] Kuret J, Congdon EE, Li G, Yin H, Yu X, Zhong Q. Evaluating triggers and enhancers of tau fibrillization. Microsc Res Tech 2005; 67: 141–55.

[34] Mazanetz MP, Fischer PM. Untangling tau hyper-phosphorylation in drug design for neurodegenera-tive diseases. Nat Rev Drug Discov 2007; 6: 464–79.

[35] Iqbal K, Alonso Adel C, Chen S, *et al.* Tau pathology in Alzheimer disease and other tauopathies. Biochim Biophys Acta 2005; 1739: 198–210.

[36] Stoothoff WH, Johnson GV. Tau phosphorylation: physiological and pathological consequences. Biochim Biophys Acta 2005; 1739: 280–97.

[37] Hasegawa M. Biochemistry and molecular biology of tauopathies. Neuropathology 2006; 26: 484–90.

[38] McKee AC, Kosik KS, Kowall NW. Neuritic pathology and dementia in Alzheimer's disease. Ann Neurol 1991; 30:156–65.

[39] Oddo S, Vasilevko V, Caccamo A, Kitazawa M, Cribbs DH, LaFerla FM. Reduction of soluble Abeta and tau, but not soluble Abeta alone, ameliorates cognitive decline in transgenic mice with plaques and tangles. J Biol Chem 2006; 281: 39413–23.

[40] Santacruz K, Lewis J, Spires T, Paulson J, *et al.* Tau suppression in a neurode-generative mouse model improves memory function. Science 2005; 309: 476–81.

[41] Wittmann CW, Wszolek MF, Shulman JM, *et al.* Tauopathy in Drosophila: neurodegeneration without neurofibrillary tangles. Science 2001; 293: 711–4.

[42] Mandelkow EM, Stamer K, Vogel R, Thies E, Mandelkow E. Clogging of axons by tau, inhibition of axonal traffic and starvation of synapses. Neurobiol Aging 2003; 24: 1079–85.

[43] Sherwood CC, Gordon AD, Allen JS *et al.* Aging of the cerebral cortex differs between humans and chimpanzees. PNAS 2011; 108:13029-34.

[44] Smith DE, Roberts J, Gage FH, Tuszynski MH. Age-associated neuronal atrophy occurs in the primate brain and is reversible by growth factor gene therapy. Proc Natl Acad Sci USA 1999; 96:10893-8.

[45] Woodruff-Pak DS, Foy MR, Akopian GG *et al.* Differential effects and rates of normal aging in cerebellum and hippocampus. PNAS 2010; 107:1624-29.

[46] Horner RL, Innes JA, Murphy K, Guz A. Evidence for reflex upper airway dilator muscle activation by sudden negative airway pressure in man. J Physiology 1991; 436: 31–44.

[47] Horner RL, Innes JA, Morrell MJ, Shea SA, Guz A. The effect of sleep on reflex genioglossus muscle activation by stimuli of negative airway pressure in humans. J Physiol 1994; 476: 141-51.

[48] Fogel RB, Trinder J, White DP, Malhotra A. Within-breath control of genioglossus muscle activation in humans: Effect of sleep-wake state. J Physiol 2003; 550: 899-10.

[49] Wheatley JR, Mezzanotte WS, Tangel DJ, *et al.* Influence of sleep on genioglossus muscle activation by negative pressure in normal men. Amer Rev Respir Dis 1993; 148: 597-05.

[50] Mezzanotte WS, Tangel DJ, White DP. Waking genioglossal EMG in sleep apnea patients *vs.* normal controls (a neuromuscular compensatory mechanism). J Clin Invest 1992; 89: 1571–79.

[51] Mezzanotte WS, Tangel DJ, White DP. Influence of sleep onset on upper airway muscle activity in apnea patients vs. normal controls. Amer J Resp Critical Care Med 1996; 153:1880–87.

[52] Remmers JE, DeGroot WJ, Sauerland EK Anch AM. Pathogenesis of upper airway occlusion during sleep. J Appl Physiol 1978; 44: 931-38.

[53] Kay A, Trinder J, Bowes G, Kim Y. Changes in Airway resistance during sleep onset. J Appl Physiol 1994; 76: 1600-07.

[54] Brooks D, Horner RL, Kozar LF, Render-Teixeira CL, Phillipson EA. Obstructive sleep apnoea as a cause of systemic hypertension. Evidence from a canine model. J Clin Invest 1997; 99: 106-09.

[55] Peppard PE, Young T, Palta M, Skatrud J. Prospective study of the association between sleep-disordered breathing and hypertension. N Engl J Med 2000; 342: 1378-84.

[56] Pepperell JC, Ramdassingh-Dow S, Crosthwaite N *et al.* Ambulatory blood pressure after therapeutic and subtherapeutic nasal continuous positive airway pressure for obstructive sleep apnea. a randomized parallel trial. Lancet 2002; 359: 204-10.

[57] Doherty LS, Kiely JL, Swan V, McNichols WT. Long term effects of nasal continuous positive airway pressure therapy on on cardiovascular outcome in sleep apnoea syndrome. Chest 2005; 127: 2076-84.

[58] Marin JM, Carrizo SJ, Vicente E, Agusti E. Long-term cardiovascular outcomes in men with obstructive sleep apnea-hypopnea with or without treatment with continuous positive airway pressure; an observational study. Lancet 2005; 365:1046-53.

[59] Kim HC, Young T, Mathews CG *et al.* Sleep-disordered breathing and neuropsychological deficits. A population-based study. Am J Respir Crit Care Med 1997; 156: 1813-19.

[60] Arzt T, Young T, Finn L, Skatrud JB, Bradley TD. Association of Sleep-disordered breathing and the occurrence of stroke. Am J Respir Crit Care Med 2005; 172: 1447-51.

[61] Yaggi J, Concato J, Kernan WN, Lichtman JH, Brass LM, Mohsenin V. Obstructive sleep apnea as a risk factor for stroke and death. N Engl J Med 2005; 353: 2034-41.

[62] Daulatzai MA. Early stages of pathogenesis in memory impairment during normal senescence and alzheimer's disease. J Alzheimers Dis 2011; 20: 355–67.

[63] Daulatzai MA.Conversion of elderly to Alzheimer's dementia: role of confluence of hypothermia and senescent stigmata—the plausible pathway. J Alzheimers Dis 2011; 21:1039–63.

[64] Daulatzai MA. Role of Sensory Stimulation in Amelioration of Obstructive Sleep Apnea. Sleep Disorders, 2011; vol. 2011; Article ID. 596879, 12 pages, doi:10.1155/2011/596879.

[65] Daulatzai MA. Dysfunctional nucleus tractus solitarius: its crucial role in promoting neuropathogentic cascade of Alzheimer's dementia-a novel hypothesis. Neurochem Res 2012; 37: 846-68.

[66] Daulatzai MA. Pathogenesis of cognitive dysfunction in patients with Obstructive Sleep Apnea: A Hypothesis with Emphasis on the Nucleus Tractus Solitarius. Sleep Disorders 2012; 2012: Article ID 251096, 18 pages, doi:10.1155/2012/251096.

[67] Lis S, Kreiger S, Henning D *et al.* Executive functions and cognitive subprocesses in patients with obstructive sleep apnea. J Sleep Res 2008; 17: 271-80.

[68] Wang P, Rothwell NJ, Pinteaux E, Brough D. Neuronal injury induces the release of pro-interleukin-1β from activated microglia *in vitro*. Brain Research 2008; 1236:1–7.

[69] Drager LF, Jun JC, Polotsky VY. Metabolic consequences of intermittent hypoxia: relevance to obstructive sleep apnea. Best Practice and Research: Clinical Endocrinology and Metabolism 2010; 24: 843–51.

[70] Daulatzai MA. Quintessential risk factors: Their role in promoting cognitive dysfunction and Alzheimer's disease. Neurochem Res 2012; (In Press).

[71] Tan ZS, Seshadri S. Inflammation in the Alzheimer's disease cascade: culprit or innocent bystander? Alzheimers Res Ther 2010; 2:6.

[72] Godbout JP, Chen J, Abraham J, *et al.* Exaggerated neuroinflammation and sickness behavior in aged mice following activation of the peripheral innate immune system. FASEB J 2005; 19:1329–31.

[73] Barrientos RM, Higgins EA, Biedenkapp JC, *et al.* Peripheral infection and aging interact to impair hippocampal memory consolidation. Neurobiol Aging 2006, 27:723–732.

[74] Kohman RA, Crowell B, Kusnecov AW. Differential sensitivity to endotoxin exposure in young and middle-age mice. Brain Behav Immun 2010; 24:486-92.

[75] Choi DY, Zhang J, Bing G. Aging enhances the neuroinflammatory response and alpha-synuclein nitration in rats. Neurobiol Aging 2010; 31:1649-53.

[76] Meydani SN, Wu D. Age-associated inflammatory changes: role of nutritional intervention. Nutr Rev 2007 65:S213-16.

[77] Konsman JP, Vigues S, Mackerlova L, Bristow A, Blomqvist A. Rat brain vascular distribution of interleukin-1 type-1 receptor immunoreactivity: relationship to patterns of inducible cyclooxygenase expression by peripheral inflammatory stimuli. J Comp Neurol 2004; 472:113-129.

[78] Deng XH, Bertini G, Xu YZ, Yan Z, Bentivoglio M. Cytokine-induced activation of glial cells in the mouse brain is enhanced at an advanced age. Neuroscience, 2006; 141: 645–61.

[79] Kalehua AN, Taub DD, Baskar PV, *et al.* Aged mice exhibit greater mortality concomitant to increased brain and plasma TNF-alpha levels following intracerebroventricular injection of lipopolysaccharide. Gerontology, 2000, 46: 115–28.

[80] Zecca L, Youdim MB, Riederer P, Connor JR, Crichton RR. Iron, brain Aging and neurodegenerative disorders. Nat Rev Neurosci 2004; 5: 863–73.

[81] McGeer PL, Itagaki S, Boyes BE, McGeer EG. Reactive microglia are positive for HLA-DR in the substantia nigra of Parkinson's and Alzheimer's disease brains. Neurology 1988; 38:1285–91.

[82] Tansey MG, McCoy MK, Frank-Cannon TC. Neuroinflammatory mechanisms in Parkinson's disease: potential environmental triggers, pathways, and targets for early therapeutic intervention. Exp Neurol 2007; 208:1–25.

[83] Sawada H, Hishida R, Hirata Y, *et al.* Activated microglia affect the nigro-striatal dopamine neurons differently in neonatal and aged mice treated with 1-methyl-4-phenyl-1,2,3,6-tetrahydropyridine. J Neurosci Res 2007; 85:1752–61.

[84] Gaykema RP, Balachandran MK, Godbout JP, Johnson RW, Goehler LE. Enhanced neuronal activation in central autonomic network nuclei in aged mice following acute peripheral immune challenge. Auton Neurosci 2007; 131:137-42.

[85] Wu D, Mura C, Beharka AA, *et al.* Age-associated increase in PGE2 synthesis and COX activity in murine macrophages is reversed by vitamin E. Am J Physiol 1998; 275: C661-68.

[86] Bluthé RM, Layé S, Michaud B, Combe C, Dantzer R, Parnet P. Role of interleukin-1beta and tumour necrosis factor-alpha in lipopolysaccharide-induced sickness behaviour: a study with interleukin-1 type I receptor-deficient mice. Eur J Neurosci 2000; 12: 4447-56.

[87] Rozovsky I, Finch CE, Morgan TE. Age-related activation of microglia and astrocytes: *in vitro* studies show persistent phenotypes of aging, increased proliferation, and resistance to down-regulation. Neurobiol Aging 1998; 19: 97-103.

[88] Bodles AM, Barger SW. Cytokines and the aging brain -what we don't know might help us. Trends Neurosci 2004; 27:621-26.

[89] Hayek MG, Mura C, Wu D, *et al.* Enhanced expression of inducible cyclooxygenase with age in murine macrophages. J Immunol 1997; 159: 2445-51.

[90] Chung HY, Cesari M, Anton S, *et al*. Molecular inflammation: underpinnings of aging and age-related diseases. Aging Res Rev 2009; 8: 18-30.

[91] Zou Y, Jung KJ, Kim JW, Yu BP, Chung HY. Alteration of soluble adhesion molecules during aging and their modulation by calorie restriction. FASEB J 2004; 18: 320-22.

[92] Makarov SS. NF-kappaB as a therapeutic target in chronic inflammation: recent advances. Mol Med Today 2000; 6: 441-48.

[93] Böhrer H, Qiu F, Zimmermann T, *et al*. Role of NFkappaB in the mortality of sepsis. J Clin Invest 1997; 100: 972-85.

[94] Wilkinson J, Radkowski M, Eschbacher JM, Laskus T. Activation of brain macrophages/microglia cells in hepatitis C infection. Gut 2010; 59:1394-400.

[95] Geisler MW, Morgan CD, Covington JW, Murphy C. Neuropsychological performance and cognitive olfactory event-related brain potentials in young and elderly adults. J Clin Exp Neuropsychol 1999; 21:108–26.

[96] Lynch G. 1986. Synapses, Circuits and the Beginnings of Memory. MIT Press, Cambridge, MA, USA.

[97] Eichenbaum H. Using olfaction to study memory. Ann NY Acad Sci 1998; 855: 657–69.

[98] Larsson, H. Semb, B. Winblad, K. Amberla, L.O. Wahlund, L. Backman. Odor identification in normal aging and early Alzheimer's disease: effects of retrieval support. Neuropsychology 1999; 131: 47–53.

[99] Mesholam RI, Moberg PJ, Mahr RN, Doty RL. Olfaction in neurodegenerative disease: a meta-analysis of olfactory functioning in Alzheimer's and Parkinson's diseases. Arch Neurol 1998; 551:84–90.

[100] Murphy C, Gilmore MM, Seery CS, Salmon DP, Lasker BR. Olfactory thresholds are associated with degree of dementia in Alzheimer's disease. Neurobiol Aging 1990; 114: 465–69.

[101] Nordin, O. Almkvist, B. Berglund, L.O. Wahlund. Olfactory dysfunction for pyridine and dementia progression in Alzheimer disease. Arch. Neurol.1997: 548: 993–98

[102] Thomann PA, Dos Santos V, Toro P, Schönknecht P, Essig M, Schröder J. Reduced olfactory bulb and tract volume in early Alzheimer's disease--a MRI study. Neurobiol Aging 2009; 30: 838-41.

[103] Attems J, Jellinger KA. Olfactory tau pathology in Alzheimer disease and mild cognitive impairment. Clin Neuropathol 2006; 256: 265–71.

[104] Kovacs T, Cairns NJ, Lantos PL. Beta-amyloid deposition and neurofibrillary tangle formation in the olfactory bulb in Aging and Alzheimer's disease. Neuropathol Appl Neurobiol 1999: 256:481–91.

[105] Tsuboi Y, Wszolek ZK, Graff-Radford NR, Cookson N, Dickson DW. Tau pathology in the olfactory bulb correlates with Braak stage, Lewy body pathology and apolipoprotein epsilon4, Neuropathol Appl Neurobiol 2003; 295: 503–10.

[106] Ogden CL, Carroll MD, Curtin LR, McDowell MA, Tabak CJ, Flegal KM. Prevalence of overweight and obesity in the United States, 1999-2004. JAMA 2006; 295:1549-55.

[107] Kopelman PG. Obesity as a medical problem. Nature 2000; 404: 635-643.

[108] Rosengren A, Skoog I, Gustafson D, Wilhelmsen L. Body mass index, other cardiovascular risk factors, and hospitalization for dementia. Arch Intern Med 2005; 165: 321-26.

[109] Gustafson D, Lissner L, Bengtsson C, Björkelund C, Skoog I. A 24-year follow-up of body mass index and cerebral atrophy. Neurology 2004; 63:1876-81.

[110] Kivipelto M, Ngandu T, Fratiglioni L, *et al.* Obesity and vascular risk factors at midlife and the risk of dementia and Alzheimer disease. Arch Neurol 2005; 62: 1556-60.

[111] Whitmer RA, Gunderson EP, Barrett-Connor E. Obesity in middle age and future risk of dementia: A 27 year longitudinal population based study. BMJ 2005; 330:1360.

[112] Iannone F, Lapadula G. Obesity and inflammation--targets for OA therapy. Curr Drug Targets 2010; 11: 586-98.

[113] Depalo A, Carpagnano GE, Spanevello A, Sabato R, Cagnazzo MG, Gramiccioni C, Foschino-Barbaro MP. Exhaled NO and iNOS expression in sputum cells of healthy, obese and OSA subjects. J Intern Med 2008; 263: 70-8.

[114] Carpagnano GE, Spanevello A, Sabato R, *et al.* Systemic and airway inflammation in sleep apnea and obesity: the role of ICAM-1 and IL-8. Transl Res 2010;155: 35-43.

[115] Ho L, Qin W, Pompl PN, *et al.* Diet-induced insulin resistance promotes amyloidosis in a transgenic mouse model of Alzheimer's disease. FASEB J 2004; 18: 902–04.

[116] Julien C, Tremblay C, Phivilay A, *et al.* High-fat diet aggravates amyloid-beta and tau pathologies in the 3xTg-AD mouse model. Neurobiol Aging 2010; 31: 1516-31.

[117] Erol A. An integrated and unifying hypothesis for the metabolic basis of sporadic Alzheimer's disease. J Alzheimers Dis 2008; 13: 241–53.

[118] Luchsinger JA, Mayeux R. Adiposity and Alzheimer's disease. Curr Alzheimer Res 2007; 4:127–34.

[119] Wyss-Coray T. Inflammation in Alzheimer disease: driving force, bystander or beneficial response? Nat Med 2006; 12: 1005–15.

[120] Calon F, Cole G. Neuroprotective action of omega-3 polyunsaturated fatty acids against neurodegenerative diseases: evidence from animal studies. Prostaglandins Leukot Essent Fatty Acids 2007; 77: 287–93.

[121] Frautschy SA, Yang F, Irrizarry M, *et al.* Microglial response to amyloid plaques in APPsw transgenic mice. Am J Pathol 1998; 152: 307–17.

[122] Jacobsen JS, Wu CC, Redwine JM, *et al.* Early-onset behavioral and synaptic deficits in a mouse model of Alzheimer's disease. Proc Natl Acad Sci USA 2006; 103: 5161–66.

[123] Lim GP, Yang F, Chu T, *et al.* Ibuprofen suppresses plaque pathology and inflammation in a mouse model for Alzheimer's disease. J Neurosci 2000; 20: 5709–14.

[124] Wirths O, Breyhan H, Marcello A, Cotel MC, Brück W, Bayer TA. Inflammatory changes are tightly associated with neurodegeneration in the brain and spinal cord of the APP/PS1KI mouse model of Alzheimer's disease. Neurobiol Aging 2010; 31:747-57.

[125] Hsuchou H, He Y, Kastin AJ, *et al.* Obesity induces functional astrocytic leptin receptors in hypothalamus. Brain 2009;132: 889-902.

[126] Kitazawa M, Oddo S, Yamasaki TR, Green KN, LaFerla FM. Lipopolysaccharide-induced inflammation exacerbates tau pathology by a cyclin-dependent kinase 5-mediated pathway in a transgenic model of Alzheimer's disease. J Neurosci 2005; 25: 8843–53.

[127] Tan J, Town T, Crawford F, *et al.* Role of CD40 ligand in amyloidosis in transgenic Alzheimer's mice. Nat Neurosci 2002; 5: 1288–93.

[128] Rader DJ. Effect of insulin resistance, dyslipidemia, and intra-abdominal adiposity on the development of cardiovascular disease and diabetes mellitus. Am J Med 2007;120: S12–8.

[129] Mokdad AH, Ford ES, Bowman BA, *et al.* Prevalence of obesity, diabetes, and obesity-related health risk factors -2001. JAMA 2003; 289: 76–9.

[130] Van Gaal LF, Mertens IL, De Block CE. Mechanisms linking obesity with cardiovascular disease. Nature 2006; 444: 875–80.

[131] McFarlane SI, Banerji M, Sowers JR. Insulin resistance and cardiovascular disease. J Clin Endocrinol Metab 2001; 86: 713–8.

[132] Haentjens P, Van Meerhaeghe A, Moscariello A, *et al.* The impact of continuous positive airway pressure on blood pressure in patients with obstructive sleep apnea syndrome: evidence from a meta-analysis of placebo-controlled randomized trials. Arch Intern Med 2007; 167: 757–64.

[133] Eikelis N, Esler M. The neurobiology of human obesity. Exp Physiol 2005; 90: 673–82.

[134] Vincent HK, Taylor AG. Biomarkers and potential mechanisms of obesity-induced oxidant stress in humans. Int J Obes (Lond) 2006; 30: 400–18.

[135] Vincent HK, Innes KE, Vincent KR. Oxidative stress and potential interventions to reduce oxidative stress in overweight and obesity. Diabetes Obes Metab 2007; 9: 813–39.

[136] Clement K, Langin D. Regulation of inflammation-related genes in human adipose tissue. J Intern Med 2007; 262: 422–30.

[137] Kim SK, Kim HJ, Hur KY, *et al.* Visceral fat thickness measured by ultrasonography can estimate not only visceral obesity but also risks of cardiovascular and metabolic diseases. Am J Clin Nutr 2004;79: 593–9.

[138] Fox CS, Massaro JM, Hoffmann U, *et al.* Abdominal visceral and subcutaneous adipose tissue compartments: association with metabolic risk factors in the Framingham Heart Study. Circulation 2007; 116: 39–48.

[139] Rosito GA, Massaro JM, Hoffmann U, *et al.* Pericardial fat, visceral abdominal fat, cardiovascular disease risk factors, and vascular calcification in a community-based sample: the Framingham Heart Study. Circulation 2008; 117: 605–13.

[140] Onat A, Avci GS, Barlan MM, Uyarel H, Uzunlar B, Sansoy V. Measures of abdominal obesity assessed for visceral adiposity and relation to coronary risk. Int J Obes Relat Metab Disord 2004; 28:1018–25.

[141] Pascot A, Despres JP, Lemieux I, *et al.* Contribution of visceral obesity to the deterioration of the metabolic risk profile in men with impaired glucose tolerance. Diabetologia 2000; 43:1126–35.

[142] Vgontzas AN, Papanicolaou DA, Bixler EO, *et al.* Sleep apnea and daytime sleepiness and fatigue: relation to visceral obesity, insulin resistance, and hypercytokinemia. J Clin Endocrinol Metab 2000;85:1151–8.

[143] Coppack SW, Jensen MD, Miles JM. *In vivo* regulation of lipolysis in humans. J Lipid Res 1994; 35: 177–93.

[144] Wisse BE. The inflammatory syndrome: the role of adipose tissue cytokines in metabolic disorders linked to obesity. J Am Soc Nephrol 2004; 15: 2792–800.

[145] Gozal D, Serpero LD, Sans Capdevila O, Kheirandish-Gozal L. Systemic inflammation in non-obese children with obstructive sleep apnea. Sleep Med 2008; 62: 489–94.

[146] Barcelo A, Barbe F, Llompart E, *et al.* Effects of obesity on C-reactive protein level and metabolic disturbances in male patients with obstructive sleep apnea. Am J Med 2004; 117:118–21.

[147] Guilleminault C, Kirisoglu C, Ohayon MM. C-reactive protein and sleep-disordered breathing. Sleep 2004; 27:1507–11.

[148] Sharma SK, Mishra HK, Sharma H, *et al.* Obesity, and not obstructive sleep apnea, is responsible for increased serum hs-CRP levels in patients with sleep-disordered breathing in Delhi. Sleep Med 2008; 9: 149–56.

[149] Mehra R, Storfer-Isser A, Kirchner HL, *et al.* Soluble interleukin 6 receptor: A novel marker of moderate to severe sleep-related breathing disorder. Arch Intern Med 2006; 166:1725–31.

[150] Ryan S, Taylor CT, McNicholas WT. Predictors of elevated nuclear factor-kappaB-dependent genes in obstructive sleep apnea syndrome. Am J Respir Crit Care Med 2006; 174: 824–30.

[151] Hayashi M, Fujimoto K, Urushibata K, Takamizawa A, Kinoshita O, Kubo K. Hypoxia-sensitive molecules may modulate the development of atherosclerosis in sleep apnoea syndrome. Respirology 2006; 11: 24–31.

[152] Paulsen FP, Steven P, Tsokos M, *et al.* Upper airway epithelial structural changes in obstructive sleep-disordered breathing. Am J Respir Crit Care Med 2002; 166: 501–9.

[153] Boyd JH, Petrof BJ, Hamid Q, Fraser R, Kimoff RJ. Upper airway muscle inflammation and denervation changes in obstructive sleep apnea. Am J Respir Crit Care Med 2004; 170: 541–6.

[154] Zanni MV, Stanley TL, Makimura H, Chen CY, Grinspoon SK. Effects of TNF-alpha antagonism on E-selectin in obese subjects with metabolic dysregulation. Clin Endocrinol (Oxf) 2010; 73:48-54.

[155] Raji CA, Ho AJ, Parikshak NN, Becker JT, Lopez OL, Kuller LH, Hua X, Leow AD, Toga AW, Thompson PM. Brain structure and obesity. Hum Brain Mapp 2010; 31: 353-64.

[156] Agostinho P, Cunha RA, Oliveira C. Neuroinflammation, oxidative stress and the pathogenesis of Alzheimer's disease. Curr Pharm Des. 2010; 16: 2766-78.

[157] Walther K, Birdsill AC, Glisky EL, Ryan L. Structural brain differences and cognitive functioning related to body mass index in older females. Hum Brain Mapp 2010; 31: 1052-64.

[158] Davies RJO, Crosby J, Vardy-Visy K *et al.* Non-invasive beat to beat arterial blood pressure during non-REM sleep in obstructive sleep apnea and snoring. Thorax 1994; 49: 335-9.

[159] Lofaso F, Coste A, Gilain L, Harf A, Guilleminault C, Goldenberg F. Sleep fragmentation as a risk factor for hypertension in middle-aged nonapneic snorers. Chest 1996; 109: 896-900.

[160] Young T, Palata, M, Dempsey, J. *et al.* The occurrence of sleep-disordered breathing among middle-aged adults. N Engl J Med 1993; 328: 1230-35.

[161] Hila KM, Young TB, Bidwell T. *et al.* Sleep apnea and hypertension: a population-based study. Ann Intern Med 1994; 120: 382-8.

[162] Zahorka M, Hess U, Himmelmann H. *et al.* Snoring, sleep apnea and hypertension in a field study. In: Sleep-relared Disorders and Internal Medicine. Eds: Peters JH, Podszus T, von Wichert P. Berlin, Springer Verlag 1987; pp. 219-23.

[163] Lindberg E, Janson C, Gislason T, Svärdsudd K, Hetta J, Boman G. Snoring and hypertension: a 10 year follow-up. Eur Respir J 1998; 11: 884-9.

[164] Young T, Finn L, Hla KM, Morgan B, Palta M. Snoring as part of a dose-response relationship between sleep-disordered breathing and blood pressure. Sleep 1996; 19: S202-5.

[165] Bixler EO, Vgontzas AN, Lin HM, Ten Have T, Leiby BE, Vela-Bueno A, Kales A. Association of hypertension and sleep-disordered breathing. Arch Intern Med 2000; 160: 2289-95.

[166] Hu FB, Willett WC, Colditz GA, Ascherio A, Speizer FE, Rosner B, Hennekens CH, Stampfer MJ. Prospective study of snoring and risk of hypertension in women. Am J Epidemiol 1999; 150: 806-16.

[167] Silverberg DS, Oksenberg A. Essential and secondary hypertension and sleep-disordered breathing: a unifying hypothesis. J Human Hypertens 1996;10: 353-63.

[168] Reimann R, Volk R, Müller A, Herzog M. The influence of nocturnal alcohol ingestion on snoring. Eur Arch Otorhinolaryngol 2010; 267:1147-56.

[169] Herzog M, Reimann R. Alcohol ingestion influences the nocturnal cardio-respiratory activity in snoring and non-snoring males. Eur Arch Otorhinolaryngol 2004; 261: 459-62.

[170] Sekosan M, Zakkar M, Wenig BL, Olopade CO, Rubinstein I. Inflammation in the uvula mucosa of patients with obstructive sleep apnea. Laryngoscope 1996; 106: 1018-20.

[171] Rubinstein I. Nasal inflammation in patients with obstructive sleep apnea. Laryngoscope 1995; 105: 175-7.

[172] Phillips BG, Kato M, Narkiewicz K, Choe I, Somers VK. Increases in leptin levels, sympathetic drive, and weight gain in obstructive sleep apnea. Am J Physiol Heart Circ Physiol 2000; 279: H234-37.

[173] Shamsuzzaman ASM, Gersh BJ, Somers VK. Obstructive Sleep Apnea: Implications for Cardiac and Vascular Disease. J Amer Med Assoc 2003; 290: 1906-14.

[174] Caro JF, Sinha MK, Kolaczynski JW, Zhang PL, Considine RV. Leptin: the tale of an obesity gene. Diabetes 1996; 45:1455-62.

[175] Considine RV, Sinha MK, Heiman ML, *et al.* Serum immunoreactive-leptin concentrations in normal-weight and obese humans. N Engl J Med 1996; 334: 292-95.

[176] Wallace AM, McMahon AD, Packard CJ, *et al.* Plasma leptin and the risk of cardiovascular disease in the west of Scotland coronary prevention study (WOSCOPS). Circulation 2001; 104: 3052-56.

[177] Davi G, Guagnano MT, Ciabattoni G, *et al.* Platelet activation in obese women: role of inflammation and oxidant stress. JAMA 2002; 288: 2008-14.

[178] Vgontzas AN. Does obesity play a major role in the pathogenesis of sleep apnoea and its associated manifestations *via* inflammation, visceral adiposity, and insulin resistance? Arch Physiol Biochem 2008; 114: 211-23.

[179] Elmasry A, Lindberg E, Berne C, *et al.* Sleep-disordered breathing and glucose metabolism in hypertensive men: a population-based study. J Intern Med 2001;249:153-161.

[180] Ip MS, Lam B, Ng MM, *et al.* Obstructive sleep apnea is independently associated with insulin resistance. Am J Respir Crit Care Med 2002; 165: 670-76.

[181] Punjabi NM, Sorkin JD, Katzel LI, *et al.* Sleep-disordered breathing and insulin resistance in middle-aged and overweight men. Am J Respir Crit Care Med 2002; 165: 677-82.

[182] Chin K, Shimizu K, Nakamura T, *et al.* Changes in intra-abdominal visceral fat and serum leptin levels in patients with obstructive sleep apnea syndrome following nasal continuous positive airway pressure therapy. Circulation 1999; 100: 706-12.

[183] Punjabi NM, Ahmed MM, Polotsky VY, Beamer BA, O'Donnell CP. Sleep-disordered breathing, glucose intolerance, and insulin resistance. Respir Physiol Neurobiol 2003;136:167-78.

[184] Spiegel K, Leproult R, Van Cauter E. Impact of sleep debt on metabolic and endocrine function. Lancet 1999; 354: 1435-39.

[185] von Känel R, Natarajan L, Ancoli-Israel S, Mills PJ, Loredo JS, Dimsdale JE. Day/Night rhythm of hemostatic factors in obstructive sleep apnea. Sleep 2010; 33: 371-7.

[186] Sanner BM, Konermann M, Tepel M, *et al.* Platelet function in patients with obstructive sleep apnoea syndrome. Eur Respir J 2000;16:648-652.

[187] Eisensehr I, Ehrenberg BL, Noachtar S, *et al.* Platelet activation, epinephrine, and blood pressure in obstructive sleep apnea syndrome. Neurology 1998; 51: 188-95.

[188] Nobili L, Schiavi G, Bozano E, *et al.* Morning increase of whole blood viscosity in obstructive sleep apnea syndrome. Clin Hemorheol Microcirc 2000; 22: 21-27.

[189] Hoffstein V, Herridge M, Mateika S, Redline S, Strohl KP. Hematocrit levels in sleep apnea. Chest 1994; 106: 787-791.

[190] Bokinsky G, Miller M, Ault K, Husband P, Mitchell J. Spontaneous platelet activation and aggregation during obstructive sleep apnea and its response to therapy with nasal continuous positive airway pressure: a preliminary investigation. Chest 1995; 108: 625-630.

[191] Chin K, Ohi M, Kita H, *et al.* Effects of NCPAP therapy on fibrinogen levels in obstructive sleep apnea syndrome. Am J Respir Crit Care Med 1996;153: 1972-76.

[192] Wessendorf TE, Thilmann AF, Wang YM, *et al.* Fibrinogen levels and obstructive sleep apnea in ischemic stroke. Am J Respir Crit Care Med 2000; 162: 2039-42.

[193] Chin K, Kita H, Noguchi T, *et al.* Improvement of factor VII clotting activity following long-term NCPAP treatment in obstructive sleep apnoea syndrome. QJM 1998; 91: 627-33.

[194] Halliwell B. The role of oxygen radicals in human disease, with particular reference to the vascular system. Haemostasis 1993; 23:118-26.

[195] Schulz R, Mahmoudi S, Hattar K, *et al.* Enhanced release of superoxide from polymorphonuclear neutrophils in obstructive sleep apnea: impact of continuous positive airway pressure therapy. Am J Respir Crit Care Med 2000; 162: 566-70.

[196] Prabhakar NR. Sleep apneas: an oxidative stress? Am J Respir Crit Care Med 2002; 165: 859-60.

[197] Teculescu D. Can snoring induce or worsen obstructive sleep apnea? Med Hypotheses1998; 50:125-9.

[198] Bloom JW, Kaltenborn WT, Quan SF. Risk factors in a general population for snoring. Importance of cigarette smoking and obesity. Chest 1988; 93: 678-83.

[199] Honsberg AE, Dodge RR, Cline MG, Quan SF. Incidence and remission of habitual snoring over a 5-to 6-year period. Chest 1995; 108: 604-9.

[200 Namysłowski G, Scierski W, Mrówka-Kata K, Kawecka I, Kawecki D, Czecior E. Sleep study in patients with overweight and obesity. J Physiol Pharmacol 2005; 56: 59-65.

[201] Woodson BT, Garancis JC, Toohill RJ. Histopathologic changes in snoring and obstructive sleep apnea syndrome. Laryngoscope 1991; 101: 1318-22.

[202] Daulatzai MA. The pharyngeal landscape: its length and breadth. J Sleep Res 2009; 18: 483-4.

[203] Bassiouny A, Nasr S, Mashaly M, Ayad E, Qotb M, Atef A. Electron microscopy study of peripheral nerves in the uvulae of snorers and obstructive sleep apnoea patients. JLaryngol Otol 2009; 123:203-7.

[204] Namysłowski G, Scierski W, Zembala-Nozyńska E, Nozyński J, Misiołek M. Histopathologic changes of the soft palate in snoring and obstructive sleep apnea syndrome patients. Otolaryngol Pol 2005; 59: 13-9.

[205] Friberg D, Ansved T, Borg K, Carlsson-Nordlander B, Larsson H, Svanborg E. Histological indications of a progressive snorers disease in an upper airway muscle. Am J Respir Crit Care Med 1998; 157: 586-93.

[206] Takeuchi T, Futatsuka M, Imanishi H, Yamada S. Pathological changes observed in the finger biopsy of patients with vibration-induced white finger. Scand J Work Environ Health 1986; 12: 280-3.

[207] Hamans EP, Van Marck EA, De Backer WA, Creten W, Van de Heyning PH. Morphometric analysis of the uvula in patients with sleep-related breathing disorders. Eur Arch Otorhinolaryngol 2000; 257: 232-6.

[208] Loubaki L, Jacques E, Semlali A, Biardel S, Chakir J, Sériès F. Tumor necrosis factor-alpha expression in uvular tissues differs between snorers and apneic patients. Chest 2008; 134: 911-8.

[209] Almendros I, Acerbi I, Puig F, Montserrat JM, Navajas D, Farré R. Upper-airway inflammation triggered by vibration in a rat model of snoring. Sleep 2007; 30:225-7.

[210] Puig F, Rico F, Almendros I, Montserrat JM, Navajas D, Farre R. Vibration enhances interleukin-8 release in a cell model of snoring-induced airway inflammation. Sleep 2005; 28: 1312-6.

[211] Vgontzas AN, Papanicolaou DA, Bixler EO, *et al.* Elevation of plasma cytokines in disorders of excessive daytime sleepiness: role of sleep disturbance and obesity. J Clin Endocrinol Metab 1997; 82: 1313-16.

[212] Shamsuzzaman AS, Winnicki M, Lanfranchi P, *et al.* Elevated C-reactive protein in patients with obstructive sleep apnea. Circulation 2002; 105: 2462-64.

[213] Hartmann G, Tschop M, Fischer R, *et al.* High altitude increases circulating interleukin-6, interleukin-1 receptor antagonist and C-reactive protein. Cytokine 2000; 12: 246-52.

[214] Venugopal SK, Devaraj S, Yuhanna I, Shaul P, Jialal I. Demonstration that C-reactive protein decreases eNOS expression and bioactivity in human aortic endothelial cells. Circulation 2002; 106: 1439-41.

[215] Woollard KJ, Phillips DC, Griffiths HR. Direct modulatory effect of C-reactive protein on primary human monocyte adhesion to human endothelial cells. Clin Exp Immunol 2002; 130: 256-62.

[216] Ohga E, Nagase T, Tomita T, *et al.* Increased levels of circulating ICAM-1, VCAM-1, and L-selectin in obstructive sleep apnea syndrome. J Appl Physiol 1999; 87: 10-14.

[217] Chin K, Nakamura T, Shimizu K, *et al.* Effects of nasal continuous positive airway pressure on soluble cell adhesion molecules in patients with obstructive sleep apnea syndrome. Am J Med 2000; 109: 562-7.

[218] El-Solh AA, Mador MJ, Sikka P, *et al.* Adhesion molecules in patients with coronary artery disease and moderate-to-severe obstructive sleep apnea. Chest 2002; 121: 1541-47.

[219] Fogel RB, Malhotra A, White DP. Sleep. 2: pathophysiology of obstructive sleep apnoea/hypopnoea syndrome. Thorax 2004; 59: 159-63.

[220] Fogel RB, White DP, Pierce RJ, Malhotra A, Edwards JK, Dunai J, Kleverlaan D, Trinder J. Control of upper airway muscle activity in younger vs. older men during sleep onset. J Physiol 2003; 553: 533-44.

[221] Eckert DJ, McEvoy RD, George KE, Thomson KJ, Catcheside PG. Genioglossus reflex inhibition to upper-airway negative-pressure stimuli during wakefulness and sleep in healthy males. J Physiol 2007; 581: 1193-205.

[222] Doonan RJ, Scheffler P, Lalli M, Kimoff RJ, Petridou ET, Daskalopoulos ME, Daskalopoulou SS. Increased arterial stiffness in obstructive sleep apnea: a systematic review. Hypertens Res. 2011; 34: 23-32.

[223] Peppard, PE; Young, T; Palta, M; Dempsey, J; Skatrud, J. Longitudinal study of moderate weight change and sleep-disordered breathing. JAMA 2000; 284: 3015–21.
[224] Iwasaki YK, Shi Y, Benito B, *et al*. Determinants of atrial fibrillation in an animal model of obesity and acute obstructive sleep apnea. Heart Rhythm 2012 Mar 20. [Epub ahead of print]
[225] Dilaveris PE, Gialofos EF, Sideris SK, *et al*. Simple electrocardiographic markers for the prediction of paroxysmal idiopathic atrial fibrillation. Am Heart J 1998; 135: 733–8.
[226] Sari I, Davutoglu V, Ozbala B, *et al*. Acute sleep deprivation is associated with increased electrocardiographic p-wave dispersion in healthy young men and women. Pacing Clin Electrophysiol 2008; 31: 438–42.
[227] Gami AS, Rader S, Svatikova A, *et al*. Familial premature coronary artery disease mortality and obstructive sleep apnea. Chest 2007; 131: 118–21.
[228] Can I, Aytemir K, Demir AU, *et al*. P-wave duration and dispersion in patients with obstructive sleep apnea. Int J Cardiol 2009; 133: e85-9.
[229] Zhong X, Hilton HJ, Gates GJ, *et al*. Increased sympathetic and decreased parasympathetic cardiovascular modulation in normal humans with acute sleep deprivation. J Appl Physiol 2005; 98: 2024–32.
[230] Depalo A, Carpagnano GE, Spanevello A, *et al*. Exhaled NO and iNOS expression in sputum cells of healthy, obese and OSA subjects. J Intern Med 2008; 263: 70-8.
[231] Aihara K, Oga T, Chihara Y, *et al*. Analysis of systemic and airway inflammation in obstructive sleep apnea. Sleep Breath. 2012 Jun 7. [Epub ahead of print].
[232] Maesako M, Uemura K, Kubota M, *et al*. Exercise Is More Effective than Diet Control in Preventing High Fat Diet-induced β-Amyloid Deposition and Memory Deficit in Amyloid Precursor Protein Transgenic Mice. J Biol Chem. 2012; 287: 23024-33.
[233] Erol A. Unraveling the molecular mechanisms behind the metabolic basis of sporadic Alzheimer's disease. J Alzheimers Dis. 200917: 267-76.
[234] Luchsinger JA. Adiposity, hyperinsulinemia, diabetes and Alzheimer's disease: an epidemiological perspective. Eur J Pharmacol 2008; 585: 119-29.
[235] McGeer PL, Rogers J, McGeer EG. Neuroimmune mechanisms in Alzheimer disease pathogenesis. Alzheimer Dis Assoc Disord 1994; 8: 149-58.
[236] Cole GM, Ma QL, Frautschy SA. Omega-3 fatty acids and dementia. Prostaglandins Leukot Essent Fatty Acids. 2009; 81: 213-21.
[237] Lue LF, Walker DG, Rogers J. Modeling microglial activation in Alzheimer's disease with human postmortem microglial cultures. Neurobiol Aging. 2001; 22: 945-56.
[238] Liu L, Orozco IJ, Planel E, *et al*. A transgenic rat that develops Alzheimer's disease-like amyloid pathology, deficits in synaptic plasticity and cognitive impairment. Neurobiol Dis 2008; 31: 46-57.
[239] Dokmeci D. Ibuprofen and Alzheimer's disease. Folia Med (Plovdiv). 2004; 46: 5-10.
[240] Bayer TA, Wirths O. Review on the APP/PS1KI mouse model: intraneuronal Abeta accumulation triggers axonopathy, neuron loss and working memory impairment. Genes Brain Behav 2008; 7 Suppl 1: 6-11.
[241] Pan W, Hsuchou H, Xu C, Wu X, Bouret SG, Kastin AJ. Astrocytes modulate distribution and neuronal signaling of leptin in the hypothalamus of obese A vy mice. J Mol Neurosci 2011; 43: 478-84.
[242] Lee DC, Rizer J, Selenica ML, *et al*. LPS-induced inflammation exacerbates phospho-tau pathology in rTg4510 mice. J Neuroinflammation 2010; 7: 56.

[243] Dhawan G, Floden AM, Combs CK. Amyloid-β oligomers stimulate microglia through a tyrosine kinase dependent mechanism. Neurobiol Aging 2011 Nov 29. [Epub ahead of print]

[244] Stokin GB, Lillo C, Falzone TL *et al*. Axonopathy and transport deficits early in the pathogenesis of Alzheimer's disease. Science 2005; 307:1282–88.

[245] Lourenssen S, Miller KG, Blennerhassett MG. Discrete responses of myenteric neurons to structural and functional damage by neurotoxins *in vitro*. Am J Physiol Gastrointest Liver Physiol 2009; 297: G228–39.

[246] Elchuri S, Oberley TD, Qi W *et al*. Cu ZnSOD deficiency leads to persistent and widespread oxidative damage and hepatocarcinogenesis later in life. Oncogene 2005; 24: 367–80.

[247] McFadden SL, Ding D, Reaume AG, Flood DG, Salvi RJ. Age-related cochlear hair cell loss is enhanced in mice lacking copper/zinc superoxide dismutase. Neurobiol Aging 1999; 20: 1–8.

[248] Hashizume K, Hirasawa M, Imamura Y *et al*. Retinal dysfunction and progressive retinal cell death in SOD1-deficient mice. Am J Pathol 2008; 172: 1325–31.

[249] Muller F, Song W, Liu Y *et al*. Absence of CuZn superoxide dismutase leads to elevated oxidative stress and acceleration of age-dependent skeletal muscle atrophy. Free Radic Biol Med 2006; 40: 1993–2004.

[250] Reddy VN, Kasahara E, Hiraoka M, Lin LR, Ho YS. Effects of variation in superoxide dismutases (SOD) on oxidative stress and apoptosis in lens epithelium. Exp Eye Res 2004; 79: 859–68.

[251] Rothstein JD, Bristol LA, Hosler B, Brown RH, Kuncl RW. Chronic inhibition of superoxide dismutase produces apoptotic death of spinal neurons. Proc Natl Acad Sci USA 1994; 91: 4155–59.

[252] Troy CM, Shelanski ML. Down-regulation of copper/zinc superoxide dismutase causes apoptotic death in PC12 neuronal cells. Proc Natl Acad Sci USA 1994; 91: 6384–87.

[253] Fischer LR and Glass JD. Oxidative stress induced by loss of Cu, Zn-superoxide dismutase (SOD1) or superoxide-generating herbicides causes axonal degeneration in mouse DRG cultures. Acta Neuropath 2010; 119: 249-59.

[254] Hirooka Y. Role of reactive oxygen species in brainstem in neural mechanisms of hypertension. Auton Neurosci 2008;142: 20-4.

[255] Nozoe M, Hirooka Y, Koga Y, *et al*. Inhibition of Rac1-derived reactive oxygen species in nucleus tractus solitarius decreases blood pressure and heart rate in stroke-prone spontaneously hypertensive rats. Hypertension 2007; 50: 62-8.

[256] Kaur C, Viswanathan S, Ling EA. Hypoxia-induced cellular and vascular changes in the nucleus tractus solitarius and ventrolateral medulla. J Neuropathol Exp Neurol 2011; 70: 201-17.

[257] Porzionato A, Macchi V, Guidolin D, Sarasin G, Parenti A, De Caro R. Anatomic distribution of apoptosis in medulla oblongata of infants and adults. J Anat 2008; 212:106-13

[258] Stecco C, Porzionato A, Macchi V, Sarasin G, Calcagno A, Parenti A, De Caro R. Detection of apoptosis in human brainstem by TUNEL assay. Ital J Anat Embryol 2005; 110: 255-60.

[259] Waki H, Gouraud SS, Maeda M, Paton JF. Specific inflammatory condition in nucleus tractus solitarii of the SHR: novel insight for neurogenic hypertension? Auton Neurosci 2008;142: 25-31.

[260] Waki H, Gouraud SS, Maeda M, Paton JF. Gene expression profiles of major cytokines in the nucleus tractus solitarii of the spontaneously hypertensive rat. Auton Neurosci 2008;142: 40-44.

[261] Zhang JH, Fung SJ, Xi M, Sampogna S, Chase MH. Apnea produces neuronal degeneration in the pons and medulla of guinea pigs. Neurobiol Dis 2010; 40: 251-64.

[262] de la Torre JC. Pathophysiology of neuronal energy crisis in Alzheimer's disease. Neurodegener Dis 2008; 5:126-32.

[263] Drzezga A, Lautenschlager N, Siebner H, *et al.* Cerebral metabolic changes accompanying conversion of mild cognitive impairment into Alzheimer's disease: a PET follow-up study. Eur J Nucl Med Mol Imaging 2003; 30:1104-13.

[264] Perneczky R, Hartmann J, Grimmer T, Drzezga A, Kurz A. Cerebral metabolic correlates of the clinical dementia rating scale in mild cognitive impairment. J Geriatr Psychiatry Neurol 2007; 20: 84-8.

[265] Herholz K. Cerebral glucose metabolism in preclinical and prodromal Alzheimer's disease. Expert Rev Neurother 2010; 10: 1667-73.

[266] Phillips BG, Narkiewicz K, Pesek CA, Haynes WG, Dyken ME, Somers VK. Effects of obstructive sleep apnea on endothelin-1 and blood pressure. J Hypertens 1999; 17: 61-66.

[267] Schulz R, Schmidt D, Blum A, *et al.* Decreased plasma levels of nitric oxide derivatives in obstructive sleep apnoea: response to CPAP therapy. Thorax 2000; 55: 1046-51.

[268] Kato M, Roberts-Thomson P, Phillips BG, *et al.* Impairment of endothelium-dependent vasodilation of resistance vessels in patients with obstructive sleep apnea. Circulation 2000; 102: 2607-10.

[269] Pedrinelli R, Giampietro O, Carmassi F, *et al.* Microalbuminuria and endothelial dysfunction in essential hypertension. Lancet 1994; 344: 14-18.

[270] Lee IK, Kim HS, Bae JH. Endothelial dysfunction: its relationship with acute hyperglycaemia and hyperlipidemia. Int J Clin Pract Suppl 2002;129: 59-64.

[271] Ciarla MV, Bocciarelli A, Di Gregorio S, *et al.* Autoantibodies and endothelial dysfunction in well-controlled, uncomplicated insulin-dependent diabetes mellitus patients. Atherosclerosis 2001; 158: 241-46.

[272] Rizzoni D, Porteri E, Guelfi D, *et al.* Endothelial dysfunction in small resistance arteries of patients with non-insulin-dependent diabetes mellitus. J Hypertens 2001; 19: 913-19.

[273] Hink U, Li H, Mollnau H, *et al.* Mechanisms underlying endothelial dysfunction in diabetes mellitus. Circ Res 2001; 88: E14-22.

[274] Wascher TC, Graier WF, Bahadori B, Toplak H. Time course of endothelial dysfunction in diabetes mellitus. Circulation 1994; 90: 1109-10.

[275] Milionis HJ, Jeremy JY, Mikhailidis DP. Regarding "smoking is associated with dose-related increase of intima-media thickness and endothelial dysfunction." Angiology 1999; 50: 959-61.

[276] Raitakari OT, Adams MR, McCredie RJ, Griffiths KA, Celermajer DS. Arterial endothelial dysfunction related to passive smoking is potentially reversible in healthy young adults. Ann Intern Med 1999; 130: 578-81.

[277] Poredos P, Orehek M, Tratnik E. Smoking is associated with dose-related increase of intima-media thickness and endothelial dysfunction. Angiology 1999; 50: 201-08.

[278] Macey PM, Henderson LA, Macey KE, Alger JR, Frysinger RC, Woo MA, Harper RK, Yan-Go FL, and Harper RM. Brain morphology associated with obstructive sleep apnea. Am J Respir Crit Care Med 2002; 166: 1382–87.

[279] Macey PM, Macey KE, Henderson LA, *et al.* Functional magnetic resonance imaging responses to expiratory loading in obstructive sleep apnea. Respiratory Physiology and Neurobiology 2003; 138: 275–90.

[280] Wang JY, Shum AY, Ho YJ, Wang JY. Oxidative neurotoxicity in rat cerebral cortex neurons: Synergistic effects of h2o2 and no on apoptosis involving activation of p38 mitogen- activated protein kinase and caspase-3. J Neurosci Res 2003; 72: 508–19.

[281] Wang,J.Y, Shum, AY, Wang, J.Y. Hypoxia/reoxygenation induces cell injury *via* different mechanisms in cultured rat cortical neurons and glial cells. Neurosci Lett 2002; 322: 187–91.

[282] Xu W, Chi L, Row BW, *et al.* Increased oxidative stress is associated with chronic intermittent hypoxia-mediated brain cortical neuronal cell apoptosis in a mouse model of sleep apnea. Neuroscience 126: 313–23, 2004.

[283] Xie A, Skatrud JB, Puleo DS, Morgan BJ. Exposure to hypoxia produces long-lasting sympathetic activation in humans. J Appl Physiol 2001; 91:1555-62.

[284] Hayakawa T, Terashima M, Kayukawa Y, Ohta T, Okada T. Changes in cerebral oxygenation and hemodynamics during obstructive sleep apneas. Chest 1996;109: 916-21.

[285] Lavie L, Vishnevsky A, Lavie P. Evidence for lipid peroxidation in obstructive sleep apnea. Sleep 2004; 27:123–28.

[286] Peled N, Abinader EG, Pillar G,Sharif D, Lavie P. Nocturnal ischemic events in patients with obstructive sleep apnea syndrome and ischemic heart disease: effects of continuous positive air pressure treatment. J Am Coll of Cardiol 1999; 34: 1744-49,

[287] Lavie L. Obstructive sleep apnoea syndrome: an oxidative stress disorder. Sleep Med Rev 2003; 7: 35–51.

[288] Valipour A, McGown AD, Makker H, O'Sullivan C, Spiro SG. Some factors affecting cerebral tissue saturation during obstructive sleep apnoea. Eur Respir J 2002; 20: 444-50.

[289] Pizza F, Biallas M, Wolf M, Werth E, Bassetti CL. Nocturnal cerebral hemodynamics in snorers and in patients with obstructive sleep apnea: a near-infrared spectroscopy study. Sleep 2010; 33: 205-10.

[290] Bålfors EM, Franklin KA. Impairment of cerebral perfusion during obstructive sleep apneas. Am J Respir Crit Care Med 1994;150:1587-91.

[291] Beebe DW, Gozal D. Obstructive sleep apnea and the prefrontal cortex: towards a comprehensive model linking nocturnal upper airway obstruction to daytime cognitive and behavioral deficits. J Sleep Res 2002; 11: 1–16.

[292] Kim W, De Vellis S. Microglia in Health and Disease. J Neurosci Res. 2005; 81: 302–13.

[293] Beyer ML, Sherer TB, MacKenzie G, Garcia-Osuna M, Panov AV, Greenamyre JT. Phagocytosis of neuronal or glial debris by microglial cells: upregulation of MHC Class Ï expression and multinuclear giant cell formation *in vitro*. Glia 2000; 31: 262–66.

[294] Barron KD. The microglial cell. A historical review. J Neurol Sci 1995; 134: 57–68.

[295] Milligan CE, Cunningham TJ, Levitt P. Differential immunochemical markers reveal the normal distribution of brain macrophages and microglia in the developing rat brain. J Comp Neurol 1991; 314: 125–35.

[296] Wojtera M, Sikorska B, Sobov T, Liberski PP. Microglial cells in neurodegenerative disorders. Folia Neuropathol 2005; 43: 311–21.

[297] Orr CF, Rowe DB, Halliday GM. An inflammatory review of Parkinson's disease. Prog Neurobiol 2002; 68: 325–40.

[298] Kreutzberg GW. Microglia: a sensor for pathological events in the CNS. Trends Neurosci 1996;19: 312–18.

[299] Aloisi F. Immune function of microglia. Glia 2001; 36:165–79.

[300] Aloisi F, Ria F, Adorini L. Regulation of T cell responses by CNS antigen-presenting cells: different roles for microglia and astrocytes. Immnol Today 2000; 21:141–47.

[301] Raivich G, Botatchek M, Kloss CU, Werner A, Jones LL, Kreutzberg GW. Neuroglial activation repertoire in the injured brain: graded response, molecular mechanisms and cues to physiological function. Brain Res Rev 1999; 30: 77–105.

[302] Marlin SD, Springer TA. Purified intercellular adhesion molecule-1 (ICAM-1) is a ligand for lymphocyte function-associated antigen 1 (LFA-1) Cell 1987; 51: 813–19.

[303] Greenwood J, Amos CL, Walters CE, *et al.* Intracellular domain of brain endothelial intercellular adhesion molecule-1 is essential for T lymphocyte-mediated signalling and migration. J immunol 2003; 171: 2099–108.

[304] Dheen ST, Kaur C, Ling EA. Microglial activation and its implications in the brain diseases. Curr Med Chem. 2007; 14: 1189-97.

[305] Bronstein DM, Perez-Otano I, Sun V, Mullis-Sawin SB, Chan J, Wu GC, *et al.* Glia-dependent neurotoxicity and neuroprotection in mesecephalic cultures. Brain Res 1995; 704: 112–16.

[306] Banati RB, Daniel SE, Blunt SB. Glial pathology but absence of apoptotic nigral neurons in long-standing Parkinson's disease. Movement Dis 1998;13:221–27.

[307] McGeer PL. Rate of cell death in Parkinsonism indicates active neuropathological process. Ann Neurol 1998; 24: 574–76.

[308] Ellis RJ, Olichney JM, Thal LJ, *et al.* Cerebral amyloid angiopathy in the brains of patients with Alzheimer's disease: the CERAD experience, Part XV. Neurology 1996; 46: 1592–96.

[309] Johnson KA, Gregas M, Becker JA, *et al.* Imaging of amyloid burden and distribution in cerebral amyloid angiopathy. Ann Neurol 2007; 62: 229–34.

[310] Ly JV, Donnan GA, Villemagne VL, *et al.* 11C-PIB binding is increased in patients with cerebral amyloid angiopathy-related hemorrhage. Neurolog 2010; 74: 487–93.

[311] Bacskai BJ, Frosch MP, *et al.* Molecular imaging with Pittsburgh Compound B confirmed at autopsy: a case report. Arch Neurol 2007; 64: 431–4.

[312] Lockhart A, Lamb JR, Osredkar T, *et al.* PIB is a non-specific imaging marker of amyloid-beta (Abeta) peptide-related cerebral amyloidosis. Brain. 2007; 130: 2607–15.

[313] Ikonomovic MD, Klunk WE, Abrahamson EE, *et al.* Post-mortem correlates of *in vivo* PiB-PET amyloid imaging in a typical case of Alzheimer's disease. Brain 2008; 131: 1630–45.

[314] Rabinovici GD, Furst AJ, Alkalay A, Racine CA, O'Neil JP, Janabi M, Baker SL, Agarwal N, Bonasera SJ, Mormino EC, Weiner MW, Gorno-Tempini ML, Rosen HJ, Miller BL, Jagust WJ. Increased metabolic vulnerability in early-ons*et Al*zheimer's disease is not related to amyloid burden. Brain 2010; 133: 512-28.

[315] Hiltunen M, van Groen T, Jolkkonen J. Functional roles of amyloid-beta protein precursor and amyloid-beta peptides: evidence from experimental studies. J Alzheimers Dis 2009; 18: 401-12.

[316] Streit WJ, Braak H, Xue QS, Bechmann I. Dystrophic (senescent) rather than activated microglial cells are associated with tau pathology and likely precede neurodegeneration in Alzheimer's disease. Acta Neuropathol 2009; 118: 475-85.

[317] Streit WJ. Microglia and neuroprotection: implications for Alzheimer's disease. Brain Res Brain Res Rev 2005; 48: 234-39.

[318] Goldgaber D, Harris HW, Hla T, *et al.* Interleukin 1 regulates synthesis of amyloid beta-protein precursor mRNA in human endothelial cells. Proc Natl Acad Sci U S A. 1989; 86: 7606-10.

[319] Kadiu I, Glanzer JG, Kipnis J, Gendelman HE, Thomas MP. Mononuclear phagocytes in the pathogenesis of neurodegenerative diseases. Neurotox Res. 2005; 8: 25-50.

[320] McGeer PL, McGeer EG. The inflammatory response system of brain: implications for therapy of Alzheimer and other neurodegenerative diseases. Brain Res Brain Res Rev 1995; 21: 195-218.

[321] Schmechel DE, Goldgaber D, Burkhart DS, Gilbert JR, Gajdusek DC, Roses AD. Cellular localization of messenger RNA encoding amyloid-beta-protein in normal tissue and in Alzheimer disease. Alzheimer Dis Assoc Disord. 1988; 2: 96-111.

[322] Kountouras J, Boziki M, Gavalas E, *et al.* Increased cerebrospinal fluid Helicobacter pylori antibody in Alzheimer's disease. Int J Neurosci 2009; 119: 765-77.

[323] Soscia SJ, Kirby JE, Washicosky KJ, *et al.* The Alzheimer's disease-associated amyloid beta-protein is an antimicrobial peptide. PLoS One 2010; 5: e9505.

[324] Wen Y, Yang SH, Liu R, *et al.* Cdk5 is involved in NFT-like tauopathy induced by transient cerebral ischemia in female rats. Biochim Biophys Acta. 2007; 1772: 473-83.

[325] Dickey C, Kraft C, Jinwal U, *et al.* Aging analysis reveals slowed tau turnover and enhanced stress response in a mouse model of tauopathy. Am J Pathol 2009; 174: 228–38.

[326] Quinn J, Montine T, Morrow J, Woodward WR, Kulhanek D, Eckenstein F. Inflammation and cerebral amyloidosis are disconnected in an animal model of Alzheimer's disease. J Neuroimmunol 2003; 137: 32–41.

[327] Lau LF, Ahlijanian MK. Role of cdk5 in the pathogenesis of Alzheimer's disease. Neurosignals. 2003; 12: 209-14.

[328] Li Y, Liu L, Barger SW, Griffin WS. Interleukin-1 mediates pathological effects of microglia on tau phosphorylation and on synaptophysin synthesis in cortical neurons through a p38-MAPK pathway. J Neurosci 2003; 23:1605-11.

[329] Quintanilla RA, Orellana DI, González-Billault C, Maccioni RB. Interleukin-6 induces Alzheimer-type phosphorylation of tau protein by deregulating the cdk5/p35 pathway. Exp Cell Res 2004; 295: 245-57.

[330] Gorlovoy P, Larionov S, Pham TT, Neumann H. Accumulation of tau induced in neurites by microglial proinflammatory mediators. FASEB J 2009; 23: 2502-13.

[331] Lukiw WJ, Bazan NG. Neuroinflammatory signaling upregulation in Alzheimer's disease. Neurochem Res 2000; 25: 1173-84.

[332] Rojo LE, Fernández JA, Maccioni AA, Jimenez JM, Maccioni RB. Neuroinflammation: implications for the pathogenesis and molecular diagnosis of Alzheimer's disease. Arch Med Res 2008; 39: 1-16.

[333] Vgontzas AN, Bixler EO, Chrousos GP. Sleep apnea is a manifestation of the metabolic syndrome. Sleep Med Rev 2005; 9: 211-24.

[334] Carneiro G, Togeiro SM, Ribeiro-Filho FF, *et al.* Continuous positive airway pressure therapy improves hypoadiponectinemia in severe obese men with obstructive sleep apnea without changes in insulin resistance. Metab Syndr Relat Disord 2009; 7: 537-42.

[335] Mokhlesi B, Scoccia B, Mazzone T, Sam S Risk of obstructive sleep apnea in obese and nonobese women with polycystic ovary syndrome and healthy reproductively normal women. Fertil Steril 2012; 97: 786-91.

[336] Nitsche K, Ehrmann DA. Obstructive sleep apnea and metabolic dysfunction in polycystic ovary syndrome. Best Pract Res Clin Endocrinol Metab 2010; 24: 717-30.

[337] Lin O, Zhang X, Chen G, Huang D, Din H and Tang A. Obstructive sleep apnea syndrome is associated with some components of metabolic syndrome in nonobese adults. Sleep and Breath 2012; 16: 571-78.

[338] Tasali E, Van Cauter E, Hoffman L, Ehrmann DA. Impact of obstructive sleep apnea on insulin resistance and glucose tolerance in women with polycystic ovary syndrome. J Clin Endocrinol Metab 2008; 93: 3878-84.

[339] Liu A, Kushida CA, Reaven GM. Risk for obstructive sleep apnea in obese, nondiabetic adults varies with insulin resistance status. Sleep Breath 2012; Apr 7. [Epub ahead of print].

[340] Moroz N, Tong M, Longato L, Xu H, de la Monte SM. Limited Alzheimer-type neurodegeneration in experimental obesity and type 2 diabetes mellitus. J Alzheimers Dis 2008; 15: 29-44.

[341] Musicco M, Palmer K, Salamone G, *et al.* Predictors of progression of cognitive decline in Alzheimer's disease: the role of vascular and sociodemographic factors. J Neurol 2009; 256: 1288-95.

[342] Dong YF, Kataoka K, Toyama K, *et al.* Attenuation of brain damage and cognitive impairment by direct renin inhibition in mice with chronic cerebral hypoperfusion. Hypertension 2011; 58: 635-42.

[343] Inaba S, Iwai M, Furuno M, *et al.* Continuous activation of renin-angiotensin system impairs cognitive function in renin/angiotensinogen transgenic mice. Hypertension 2009; 53: 356-62.

[344] Inaba S, Iwai M, Tomono Y, *et al.* Exaggeration of focal cerebral ischemia in transgenic mice carrying human Renin and human angiotensinogen genes. Stroke 2009; 40: 597-603

[345] Díaz-Ruiz C, Wang J, Ksiezak-Reding H, *et al.* Role of Hypertension in Aggravating Abeta Neuropathology of AD Type and Tau-Mediated Motor Impairment. Cardiovasc Psychiatry Neurol 2009; 2009: 107286.

[346] Sun X, He G, Qing H, *et al.* Hypoxia facilitates Alzheimer's disease pathogenesis by up-regulating BACE1 gene expression. Proc Natl Acad Sci U S A 2006; 103: 18727–32.

[347] Zhang X, Zhou K, Wang R, *et al.* Hypoxia-inducible factor 1alpha (HIF-1alpha)-mediated hypoxia increases BACE1 expression and beta-amyloid generation. J Biol Chem 2007; 282:10873–80.

Send Orders of Reprints at reprints@benthamscience.net
Frontiers in Clinical Drug Research – Alzheimer Disorders, Vol. 1, 2013, 315-366

CHAPTER 9

Pharmacotherapies for Alzheimer's Disease: From Natural Compounds to Target Synthetic Drugs

Chuang Wang[1,$], Rui Wang[2,$], James M. O'Donnell[3] and Ying Xu[3,*]

[1]Ningbo University School of Medicine, 818 Fenghua Road, Ningbo, Zhejiang 315211, PR China; [2]Program of Pharmacology, Weill Graduate School of Medical Science, Cornell University, New York, NY 10021 and [3]Department of Behavioral Medicine and Psychiatry, West Virginia University, Morgantown, WV 26508, USA

Abstract: Alzheimer's disease (AD) is a most prevalent neurodegenerative disorder and most common cause of dementia, which affects approximately 36 million people worldwide as of 2011 and threatens to be the scourge of the 21st century. AD is characterized by progressive cognitive decline that usually starts with memory deficit and progresses to cause a more generalized cognitive dysfunction, behavioral dysregulation, and neuropsychiatric symptoms, which result in complete loss of life skills and death. The high morbidity and mortality are always the case attributed to lack of effective therapies and prevention strategies because many aspects of its pathogenesis remain unclear. This review discusses the effectiveness and problems with current pharmaceutical treatment. It exemplifies natural compounds and target synthetic drugs as appropriate therapies and formulates rational guidelines for prevention of AD.

Keywords: Alzheimer's disease (AD), Phytochemicals, Resveratrol, Curcumin, EGCG, Ginkgo biloba, Ginseng, Huperzine A/B, Acetylcholinesterase inhibitors, NMDA receptor antagonists, Amyloid β peptide, BACE1, γ-secretase, Tau protein, GSK3, PDEs, Opioid System.

INTRODUCTION

Alzheimer's disease (AD) is a most prevalent neurodegenerative disorder mainly manifested by memory loss, personality changes including psychiatric symptoms, and cognitive dysfunction. It is the most frequent form of dementia found in the elderly. With the increasing number of elderly and growing of life expansion, more people are at the risk of AD. It is estimated that more than 50% of senior citizens aged of 85 are more likely to develop dementia and AD, which requires

***Address correspondence to Ying Xu:** Department of Behavioral Medicine & Psychiatry, West Virginia University, Morgantown, WV, 26505, USA; Tel: 304-293-9062; Fax: 304-293-1634; E-mail: ygxu@hsc.wvu.edu
[$]These authors contribute equally to the manuscript.

expensive care. Although the cause of AD remains unknown, evidence is accumulating that the cell apoptosis or cell death resulting from both genetic and environmental reasons is shown to involve the pathology of AD [1,2]. The amyloid cascade hypothesis, which was first introduced in the early 1990s, states that the accumulation of β-amyloid plaques in the CNS is the primary cause of cell death in AD, which then leads to a series of complex pathogenic consequences [3,4]. Amyloid β (Aβ) is a 38- to 42-amino acid peptide generated from Amyloid Precursor Protein (APP), which is a trans-membrane protein made by neuronal cells, by the sequential cleavage mediated by two proteases: β-site APP amyloid cleaving enzyme (BACE) and γ secretase. Insufficient clearance of Aβ from abnormal cleavage of APP will result in its accumulation.In addition, mutations in amyloid precursor protein (APP) and presenilin 1 &2, as well as the APP gene duplication all belong to the genetic reasons inducing the early onset of AD [5-7]. Therefore, a number of anti-AD compounds are designed to inhibit and reverse the accumulation of amyloid β peptide (APP). Many of the current pharmaceutical therapies are also based on this hypothesis [5,8].

Although Aβ hypothesis is the initiator of AD pathogenesis, it has become increasing clear that AD pathology is related to both extra-cellular accumulation of β-amyloid plaques and intracellular accumulation of tau protein [9-11]. Tau protein deposition happens not only in AD, but also in a number of neurodegenerative diseases such as Parkinson's disease [12,13]. The normal Tau function involves the assembly and stabilization of microtubules which play important roles in cellular functions [11]. The abnormal Tau will aggregate as pairs of filaments and the resultant deposits will cause tubule disintegration and block the neurotransmitters, and thus cell death [10,11,14].

The uncertain therapeutic effects from clinical practice suggest that the amyloid and Tau hypothesizes are feasible but not complete. The link between Aβ accumulation and the development of the neurofibrillary tangles is still not clarified [15]. Therefore, the free radical hypothesis is caught attention which points out that the age-related accumulation of reactive oxygen species (ROS) could also induce neuronal death by damaging major cellular components, such as cell membranes, mitochondria, lipids and proteins, which is also implicated in many other neurological disorders [16,17]. Moreover, excitatory amino acids (EAAs), such as glutamate, are important neurotransmitters involved in learning and memory. The dysregulation of EAA pathways in AD might be also involved in the etiology of AD [18]. The toxic EAA can induce excitotoxicity and cell death [19]. The glutamate receptors such as N-methyl-D-aspartate(NMDA) receptors, when over-activated, will allow high level of Ca^{2+} enter cells and lead to cell death. The EAAs mechanism might be also linked to Aβ aggregation and Tau tangles [20]. In addition, inflammation, which could be due to the damage from chemicals, trauma and toxins, might also contribute to the progression of AD [21-23]. It has been shown that oxidative stress will lead to neuronal inflammation and might be associated with the early onset of AD [24-26]. Moreover, some pathological mechanisms that also are considered to be associated with major or minor symptoms have been proposed. For instance, the cholinergic system degeneration leads to the cell loss in basal forebrain nuclei projecting to cortex and axonal projecting to cortex and thalamus [18,27]; or the cell loss in dopaminergic midbrain and serotonergic raphe nuclei [27,28]. The cholinergic deficient hypothesis indicates that the symptoms of AD is caused by decreased acetylcholine in the presynapse, thus increasing acetylcholine by inhibiting its degradation should improve the memory and cognition impairment in AD [13].

Despite the fact that tireless research is being conducted, the lack of effective therapies and prevention strategies because many aspects of the pathogenesis of AD still remain unclear. This review discusses the effectiveness and problems with current pharmaceutical treatments and exemplifies the natural compounds and target synthetic drugs as appropriate therapies and formulates rational guidelines for prevention of AD.

Phytochemicals as Treatment for AD

Numerous natural compounds have been analyzed for their potentials in the modulation of the pathophysiological events happening in AD [29]. The prevention and treatment strategies of the natural compounds include reducing the aggregation of Aβ and abnormal Tau protein, alleviating the neurotransmitter abnormalities, anti-inflammatory/anti-apoptotic/antioxidant treatment, and improving neuroplasticity [2, 8, 13, 15, 19]. Although the positive effects only are observed in a small group of patients with mild AD, some of them are still regarded as promising strategies because of their greater compliance and milder side effects. The structures of some phytochemicals with anti-amyloid activity are shown in Table. **1**.

Resveratrol

Polygonumcuspidatum is a plant used historically in Asia, known for its medicinal properties and traditionally used in the treatment of neuropsychiatric disorders, such as psychosocial stress, dementia and Parkinson's disease [30,31]. *Trans*-resveratrol is the major active component of *P. cuspidatum*, and it is also enriched in the grapes, red wine and some other dietary products [32]. The aggressive analysis of this phyto-compound has revealed that it has various pharmacological activities, including antioxidant, anti-inflammatory, neuroprotective properties and amelioration of learning and memory impairment [30,31,33,34]. There are several lines of evidence indicating the neuroprotective effect of resveratrol in many neurodegenerative disorders, such as Huntington's disease and Parkinson's disease (PD) [35,36]. Previous *in vitro* studies suggested that *trans*-resveratrol inhibits the activity of monoamine oxidase (MAO) in glial cells involved in the pathology of many neurological disorders [37]. Recently, the neuroprotective effect of *trans*-resveratrol has also come to the attention of scientists [38].

The neuroprotective features of resveratrol have been shown in both *in vitro* and *in vivo* models of AD. It was reported that resveratrol reverses the Aβ-induced toxicity in the PC12 cells [39]. Furthermore, in a transgenic mouse model of Alzheimer's disease (AD), cognitive improvement was observed after treatment with resveratrol [40]. Evidence is also increasing that resveratrol has the anti-amyloidogenic activities. The resveratrol administration can inhibit the β-amyloid peptide polymerization, and this compound can improve the proteasomal clearance of Aβ aggregates [41,42,43]. Recently, the effect of resveratrol is linked to the disruption of β-sheet of amyloid, which is supported by the observation of the conformational changes in fibril or oligomers [42,44].

Another important biological activity that has been attributed to resveratrol is its antioxidant potential. It is known that reactive oxygen species (ROS) plays an important role in mediating the development of oxidative stress, and the ROS accumulation might induce DNA and protein damage, as well as other types of mutations [45,46]. Zini and his co-workers have suggested that resveratrol can decrease the oxidative chain complex III which is the site of ROS generation [47]. A number of studies also demonstrated that resveratrol can scavenge both $O_2^{\cdot-}$ and OH radicals. Moreover, resveratrol has the capacity to enhance the activities of various antioxidant enzymes as a free radical scavenger [48-50]. For instance, the levels of SOD and CAT, which are enzymes protecting cells from ROS damage, will be upregulated after intrapertioneally administration of resveratrol [51]. Chronic treatment of resveratrol will reduce the MDA and nitrite levels, and restore the depleted GSH [33]. Leonard *et al.* have reported that resveratrol demonstrated protective effect against lipid peroxidation and DNA damage induced by ROS generation [49]. In fact, more and more researchers pointed out that the cytoprotective effect of resveratrol is mediated *via* scavenging oxidants and exerting anti-inflammatory activities.

Resveratrol has been shown to regulate a number of signaling pathways and the associated signaling molecules. For instance, it exerts the anti-inflammatory effect against Aβ-induced microglial activation *via* NF-κB/STAT signaling [52]. Resveratrol treatment inhibits the IKB phosphorylation, activates STAT1 and STAT3, and finally NF-κB and IL-6 release [52]. It has been shown that the treatment of resveratrol can dose-dependently reverses the β peptide-induced toxicity to the cells [53]. Han and his co-workers also demonstrated that PKC is involved in this protection effect [53]. Moreover, as mentioned above, the cell death induced by ROS and Aβ can be attenuated by resveratrol, since the compound can inhibit the cleavage of poly (ADP-ribose) polymerase (PARP), which is a substrate of active caspase-3, decrease the Bcl-x to Bax ratio, and activate JNK [39]. In SH-SY5Y cells, resveratrol also activates MAPK/ERK signaling [54]. Since resveratrol does not affect β- and γ-secretases directly, and it also does not work on metalloendopeptidase to promote the amyloid degradation, therefore, people start to think that this compound might activate proteasome which is implicated in the degradation of amyloid peptide [41,55].

The link between resveratrol and sirtuins has also attracted people's attention recently. It is known that the neuroprotective effect of resveratrol against amyloid peptide by regulating NF-κB/SIRT1 pathway. Both the translocation of p65 and degradation of the IkBα induced by the amyloid peptide could be inhibited by resveratrol administration. SIRT1 is a class III histone deacetyltransferase (HDAC) that inhibits RelA/p65 subunit of NF-κB by deacetylating the lysine 310 residue of RelA/p65 because the acetylation of lysine310 is necessary for the transcription activity of RelA/p65. In this way, resveratrol can inhibit NF-κB activity *via* activating SIRT1 [56, 57]. Moreover, people have recently demonstrated that resveratrol can compete with cAMP for the PDE binding sites, which results in the inhibition of the PDEs, enhancing the cellular cAMP levels and finally leading to the activation of SIRT1 [58].

Curcumin

Curcuma longa has been used medically for thousands of years in China. As a main component in several traditional Chinese medicines, such as Xiaoyao-san, it has been found to be effective in treating neuropsychiatric disorders since ancient time in China [59]. As the major component of *C. longa*, curcumin has been found to have a variety of pharmacological activities, including anti-inflammatory, antioxidant, anti-proliferative, and neuroprotective effects [60-62].

AD pathology involves chronic inflammation and oxidative damage in neuronsas discussed above. Curcumin is known to have strong anti-inflammatory and antioxidant activities, which might serve as a promising pharmacotherapeutic compound. Epidemiological studies have already pointed out that Indian population has lower prevalence of AD than US (~4 fold), which might be due to their eating of curried food that contains high level of curcumin [63]. Recently, several lines of investigations have shown that in cultured neurons and neurally related cells, curcumin protects against apoptosis induced by a variety of insults, including N-methyl-d-aspartate (NMDA) treatment [64], 1-methyl-4-phenylpridinium ion (MPP^+) administration [65] and H_2O_2 exposure [66]. The inflammation and oxidative damage symptoms in the transgenic APPsw mice with age-related Aβ accumulation and neuropil abnormalities are dramatically reduced after treatment with curcumin [67]. Some proinflammatory cytokines, such as IL-1β, as well as the amyloid β peptide levels are also dramatically reduced by curcumin administration. Moreover, at high dosage, curcumin can attenuate the deficits in memory in animal model of AD [68].

BDNF exerts its neuronal protective functions through binding to its high-affinity receptor, such as TrkB, which couples to a number of signaling pathways [69]. The binding of BDNF will lead to the phosphorylation of TrkB that activates the downstream molecules, and thus, initiates the signaling event. Certain studies have demonstrated that MAPK/ERK pathway may play an important role in BDNF functioning and activate the downstream molecules required for neuronal survival in depressive disorder [70,71]. The MAPK cascade is mediated by Rsk-mediated suppression of the pro-apoptotic protein BAD as well as activation of the transcription factor cAMP response element-binding protein (CREB) [72]. When CREB is phosphorylated, it activates the downstream targeted gene, Bcl-2. As an anti-apoptotic protein, Bcl-2 plays a vital role in the protection of neural cell death induced by hypoxia and chronic stressor exposure [73,74]. Furthermore, several lines of evidence indicate that CREB activation is also able to stimulate the expression of BDNF [75].

PI-3K/Akt and MAPK/Erk pathways promote neuronal growth and survival; represent two of the key signaling cascades regulated by BDNF/TrkB activation [76,77]. BDNF binding to TrkB is followed by receptor dimerization, kinase activation as well as tyrosine residues phosphorylation [69]. Phosphorylation of the tyrosine 490 leads to the recruitment of the adaptor protein shc, which subsequently activates the PI-3K signaling pathway [78]. Moreover, shc can also activate Ras, which mediates the activation of MAPK/ERK pathway. Akt, a multi-isoform serine/threonine kinase, can be activated by phosphorylation at either Ser473 or Thr308 residues [79]. LY294002, which binds and inhibits PI-3K, can significantly inhibit the phosphorylation of Akt by curcumin [80].

The activation of MAPK/ERK pathway inhibits apoptosis through inactivating the pro-apoptotic protein, Bax, and increasing the content of Bcl-2 [72]. The PI-3K/AKT can also increase the expression of Bcl-2 *via* delaying the onset of p53-mediated apoptosis [81,82]. Previously, we followed these two primary signaling cascades and showed that curcumin activates both MAPK and PI-3K pathways time-dependently. Furthermore, the selective inhibitors of PI-3K and MAPK, LY294002 and U0126, block the activation of the signaling molecules induced by curcumin, respectively. These results suggest that both PI-3K and MAPK pathways play a vital role in the signaling transduction induced by curcumin [80].

CREB is also required for neurotrophins to promote the neuronal survival [78]. The survival-promoting effect of CREB might be attributed to its induction of Bcl-2 transcription [75]. Recent studies have demonstrated that CREB is a common target for different classes of antidepressants. Various kinds of antidepressants significantly increase the phospho-CREB level as well as CREB binding activity in rat hippocampus [83,84]. Moreover, the activation of CREB will in turn promote the transcription and synthesis of more BDNF [85]. We previously found that curcumin application enhances the CREB level dose- and time-dependently. Particularly, LY294002 and U0126, the PI-3K and MAPK inhibitors, dramatically decrease the phosphorylation of CREB caused by curcumin [80]. This result is consistent with our *in vivo* study, which indicates that chronic curcumin treatment reverses the down-regulated expression of phosphorylated-CREB in the stressed rat brain [86]. Taking all these data into consideration, we suspect that curcumin stimulates activation of BDNF/TrkB signaling may result in stimulation of CREB *via* PI-3K and MAPK pathways. The enhanced phosphorylation of CREB will lead to the increased transcription of BDNF, which will finally promote the neuronal survival.

Generally, curcumin is an antioxidant polyphenol that might have strong neuroprotective effect in AD visa preventing Aβ aggregation and activating multiple signaling pathways. Increasing studies confirm this effect that curcumin binds to the redox-active metal ions and protects against the oxidative damages *via* regulating the metal-induced activation of NF-κB [55, 87].

EGCG from Green Tea

Green tea extract received a lot of attention recently due to its multi-fold activities, such as the anti-cancer, anti-inflammation, and neuroprotective effects [88, 89]. Evidence is accumulating that consumption of tea could contribute to reducing the risk of some neurological disorders, such as Parkinson's disease [90]. It has been proved that pretreatment of mice with either green tea extract or (-)-Epigallocatechin-3-gallate (EGCG) can attenuate the damage induced by neurotoxin N-methyl-4-phenyl-1,2,3,6-tetrahydropyridine (MPTP) in the dopaminergic neurons [91]. Green tea has numerous bioactive chemical components such as flavonols, caffeine, theanine, phenolic acids, particularly enriched in flavonoids that include catechins and their derivatives [92]. The analytical results indicate that EGCG is the main polyphenolic ingredient, accounting for more than 10% of the extract [93].

There is increasing evidence that green tea extract, especially EGCG, can prevent neuronal cells against Aβ-induced damage. The protective effect of EGCG against Aβ-induced neurotoxicity might be attributed to 1) change performed fibrils into inactive and non-toxic form of fibrils [94]; 2) bind to and prevent amyloidogenic proteins into forming the amorphous aggregates that is related to fibril formation [95]; 3) keep amyloid-like protein, transthyretin, in a non-aggregated and soluble form [96]; 4) covalently modifying amyloidogenic proteins so as to block the fibrillogenesis directly [97]. In PC12 cells, the anti-β sheet activity of EGCG has been found for both amyloid and α-synuclein [98]. In addition, the anti-β sheet activity of EGCG was also reflected on other amyloidogenic proteins, including plasmodium merozoite surface protein, prion scrapie protein, semen-derived enhancer of virus infection amyloid, and tranthyretin [97,99,100]. Besides amyloid-like proteins, EGCG can also inhibit the formation of other disease-related fibrils, such as huntingtin [101]. Interestingly, in various types of neuronal cells, EGCG can reduce the Aβ generation dose-dependently, while in the presence of some other components in green tea, such as (-)-epigallocatechin (EGC) and (-)-epicatechin (EC), the ability to block the generation of β peptide is inhibited, which might be the reason that the anti-amyloid effect of purified EGCG is greater than the whole green tea extract [102].

It has been shown that EGCG can modulate the proteolytic processing of APP both *in vitro* and *in vivo*. After injecting EGCG directly into the ventricules, the α-secretase activity is increased [91]. Also, α-secretase activity was enhanced in TgAPPsw mice after intraperitoneally injecting 20mg/kg of EGCG daily for 60 days, which indicates that EGCG can regulate α-secretase in the brain. Moreover, amyloid β deposit can be reduced by administration of EGCG, which might be mediated through inhibition of β-secretase [103]. EGCG also has very strong anti-oxidant activity, which might contribute to its anti-amyloid effect. EGCG also serves as an apoptosis inhibitor and activator for neuronal differentiation [104-106].

Recently studies reported that EGCG, *via* activating protein kinase C-dependent pathway, inhibits the microglial activation and neuroinflammation [91,107]. PKC family is known to be involved in multiple cellular events including cell survival, programmed cell death and neuronal differentiation, particularly to have the anti-apoptotic effect [108], which was supported by the observations that PKC activity is reduced after brain damage, such as ischemia [109], and the loss of PKC activator in the substantianigra of PD patients [110]. The involvement of PKC signaling in the neuroprotective effect of EGCG against amyloid β peptide or 6-OHDA has also been discovered [111]. Activation of PKC isoforms by EGCG was also found in the hippocampal region [91]. Moreover, some apoptotic genes are also implicated in the neuroprotective activity of EGCG. For instance, EGCG treatment (1 μM) decreases the expression of Bax, cell cycle inhibitor Gadd45, and Fas ligand; while Bcl-2 and Bcl-x levels are not affected, which might indicate that the neuroprotective effect of EGCG is medicated *via* inhibiting the pro-apoptotic molecules. This is also related to the anti-β peptide activity of EGCG [111].

Nuclear factor kappa B (NF-κB), which was originally discovered as a bacterial lipopolysaccharid-induced nuclear transcription factor, is widely distributed in all cell types and regulates multiple genes crucial to cell survival [55]. There are five NF-κB family members being identified in mammalian cells so far, such as NF-κB1 (p105/p50), NF-κB2 (p100/p52), RELA (p65), cREL, and RELB. When IκB is bound to subunit P50/P65, NF-κB is inactive in the cytoplasma, however, when NF-κB is activated, IκB will be degraded, and then P50/P65 will further translocate into nucleus to activate the transcription of target genes. It has been demonstrated that amyloid β peptide can activate NF-κB pathway while green tea extract can inhibit it [55, 112].

Many other important signaling pathways also play an important role in the neuroprotective effect of EGCG. MAPK and PI-3K pathways, which are crucial in the neuroprotective abilities against a number of insults of many compounds, have been shown to participate in the neuroprotective mechanisms of EGCG [113]. EGCG is demonstrated potent of MAPKs activation (extracellular signal-regulated kinase, c-Jun N-terminal kinase and p38 MAPK) in the *in vitro* studies [114]. For example, the 6-OHDA-induced decrease in ERK1/2 could also be prevented by EGCG [105,111]. In PC12 cells, EGCG also protects against the oxidative radical stress-induced apoptosis *via* PI-3K signaling.

Ginkgo Biloba

Ginkgo biloba has long been used in Asian countries as traditional medicine to improve the vitality in human [115]. The pharmacological actions of Ginkgo biolba are multi-folded, such as anti-apoptosis, anti-inflammatory, neuroprotective, anxiolytic, MAO-inhibitory and antioxidant effects [116,117]. One of the most extensively studied activities is the anti-dementia effect, and particularly for EGb 761 (a standardized extract from Ginkgo biloba with well-defined mixture of ~ 24% flavonoids and ~ 6% terpenes), which might be potential treatment for many neurological disorders, including AD [55]. Ginkgo biloba extract can improve learning and memory in various animal models, and even reverse the scopolamine-induced damage [118]. Clinical investigation indicates that compared to placebo, Ginkgo biloba generates significant improvement on the selective attention and executive function in the patients with dementia [119]. Moreover, for the behavioral and psychological symptoms in dementia (BPSD), Ginkgo biloba is as effective as donepezil, a prescription medicine used to treat mild to severe AD, but with fewer side effects in the patients [120]. The usage of EGb761 to treat dementia has been approved in Germany, and the anti-dementia effect of EGb761 is also found in EPIDOS study with 1465 French women over 75 years [121]. However, it should be noted that Ginkgo biloba has one ingredient, ginkgotoxin with convulsant effects [122]. It has been shown that Ginkgo biloba, when co-treated with anti-platelet or anti-coagulant medicines, might increase the hemorrhage risk [115]. Therefore, further studies on the pharmacokinetics, pharmacodynamics, and adverse effects of Ginkgo biloba should be investigated.

There is increasing studies on the neuroprotective actions of EGb761 and its potential pharmacotherapeutical applications to the neurological disorders. The major constituent of EGb761, flavonoids, might contribute to the neuroprotective effects. For instance, EGb761 can prevent the dopaminergic neurons against the MPTP-induced damage [123]. EGb761 is now in two clinical trials: Ginkgo Evaluation of Memory Study in the United States, and GuidAge study in France [55]. In primary neuronal cells and PC12 cells, EGb761 can prevent cells against Aβ-induced damage [124,125]. The kaempferol and quercetin in EGb761 have been found to protect against the Aβ peptides expressed in *C. elegans*, as well as the SwAPP695/PS1 mutations in neuroblastoma cell lines [126]. In the animal model of AD, the animals would carry a mutant of human APP and demonstrate an age-related cognitive deficit. The treatment of EGb761 can improve the spatial learning and memory compared to the control group with native APP [127]. It is known that high level of cholesterol could affect the APP processing, and EGb761 can lower the levels of free cholesterol so as to interfere with the brain APP and Aβ peptide production [128].

The neuroprotective effects of EGb761 could be mainly due to several reasons: 1) blocking the formation of fibrils. Amyloid peptides, when left in solution for a few days, could form increasing amount of β fibrils, which can enhance the neurotoxicity. The administration of EGb761 prevents this process [129-131]. 2) Anti-metal ions. Transition metal ions, such as Zn^{2+} and Fe(II), increase the β peptide aggregation. Therefore, the chelating property of EGb761 also contributes to its inhibition of amyloid β peptide formation [55,125]. 3) Anti-oxidant activity. It has been reported that EGb761 could inhibit the oxidative stress induced by H_2O_2-related reactive oxygen species in the neuroblastom cell lines carrying PS1 mutation. However, the mechanisms underlying the effects of EGb761 are different from vitamin E because EGb761 treatment can dramatically reduce the caspase-3 signaling activity while vitamin E is not able to [55,132]. 4) Anti-apoptotic activity. In the PC12 cells, EGb761 increases the level of Bcl-2 [132]. 5) Pro-proteolytic degradation. EGb761 has been demonstrated to reduce the neurotoxicity of amyloid β peptides by increasing proteolytic degradation to facilitate the clearance of the fibrils [55].

Most of the studies were performed on the extracts rather than the ingredients in Ginkgo biloba. Understanding the active phytochemicals as well as the pharmacokinetics of the compound could be very useful for the drug development [115]. Ginkgo biloba, especially EGb761, has significant anti-oxidant, anti-apoptotic and anti-amyloidogenic effects, and thus presents as a promising compound for AD treatment.

Ginseng

Ginseng (panax ginseng) has been used in China for thousands of years as one of the most popular traditional medicines for conferring a number of health benefits, such as enhancing people's well-being, treating diverse disorders, and especially its anti-aging effect [115,133]. Since ginseng is famous for its improvement of mental healthcare, it might thus have some potential effects in AD [134]. Moreover, the major active components, Ginsenosides (steroid-like compounds), have also been reported to have multiple pharmacological activities, such as antioxidant, anti-inflammatory, anti-apoptotic and neuroprotective effects. Among the ginsenosides, Rb1, Rh2, Rg2, Rg3 have been documented to prevent the Aβ-induced toxicity. Rb1 prevents the tau phosphorylation; Rg3 and Rh2 antagonize the NMDA receptors, and Re, Rg1, Rg3 improve the cognition and decrease Aβ in the APP transgenic mice [135-138]. Rg1 is reported to inhibit the Aβ production in PC12 cells, and decrease the Aβ level in Tg2576 mice [139,140]. Rg1 can improve learning and memory function in SAMP8 mice and reduce the Aβ in the brain. Rg1, when treated between 0.25-0.5μM, reduces the Aβ peptide in N2a-APP695 cells transfected with APP [141,142]. Rg3 increases β amyloid degradation [143]. All of these might indicate that the ginsenosides can affect the Aβ generation, which might be related to their improvement of cognitive functions.

The meta-analysis indicated that ginseng can improve the cognitive function very effectively; the clinical study also implied that ginseng (9g/day) could promote the Mini-Mental Status Examination (MMSE) and Alzheimer's Disease Assessment Scale (ADAS) scores [134]. During the control trials with the normal young volunteers, ginseng can also improved the memory in the subjects [144]. However, most of the studies for ginseng are performed in Traditional Chinese Medicine and usually given in a mixture with other medicines, which complicates the interpretation of the effects of ginseng. For instance, the prescription DX-9386, which includes panax ginseng, polygala tenuifolia, acorusgramineus, poriacocos (ratio 1:1:25:50), can significantly reverse the cognitive impairments in multiple *in vitro* and animal models [145,146]. Although ginseng is believed to play a key role in improving the cognitive function, other ingredients, such as polygala tenuifolia, might also contribute to the effect. Moreover, ginseng, when used as a treatment of AD, might have much less adverse effects than the conventional drug treatments [134].

Ginseng might improve the cognitive functions *via* influencing the cholinergic and serotoninergic neurotransmitter systems [147]. The *in vitro* study indicates that Rgl increases the α-secretase activity and decreases extracellular release of Aβ. These effects can be blocked by PKC, MAPK and PI-3K inhibitors, which demonstrate their involvement in the neuroprotective actions of Rgl [148]. In addition, 8-week treatment of Rgl in the ovariectomized rats can ameliorate the Aβ damage and promote the spatial learning and memory [148]. In the N2a-APP695 cells, Rgl can reduce the c peptide levels. In the animal model of AD, 3 month treatment of Rgl can inhibit the γ-secretase activity, reduce the cerebral Aβ accumulation, improve the animal behaviors, and activate protective CREB signaling pathway [142]. Rb1 has also been found to reduce the amyloid peptide-induced toxicity and tau phosphorylation *via* the PI-3K/GSK-3β pathway [149]. Some studies also indicated that the neuroprotective effects of ginseng might be related to its antioxidant and anti-apoptotic effects. Rb1 is also found anti-inflammatory effect in a rat model of AD by improving the behavioral performance and reversing the change in neuroinflammatory markers [150].

Huperzine A and Huperzine B

Huperzine A (HupA) is a Lycopodium alkaloid isolated from the Chinese medicinal herb Huperziaserrata used for memory deficiency. Huperziaserrata is widely used as the herbal remedies in the Chinese medicine. It has attracted much attention because it is a highly selective, reversible, and potent AChE inhibitor. In addition, some studies have shown that huperzine A may shift APP metabolism toward the non-amyloidogenic α-secretase pathway. It is shown to reduce glutamate-induced cytotoxicity by antagonizing cerebral NMDA receptors. Doses up to 400 μg twice daily were tested in healthy older individuals in a phase I study in 2010, but did not find any serious adverse effects [151]. A recent phase II study assessed the safety and efficacy of huperzineA in 210 individuals and found the mild to moderate AD. The individuals were randomized to receive placebo, 200 μg twice/day of HupA, or 400 μg twice/day of HupAfor 16 weeks [151]. Therefore, HupA has a higher oral bioavailability compared to tacrine and donepezil. The improvement is more noticeable on working memory than reference memory as compared to tacrine and donepezil as well. HupA at the lower dose did not influence cognitive measures, but secondary analyses suggested that the higher dose shows a modest improvement in cognition. HupA is also in Phase II clinical trials in Europe besides that it is considered as the treatment of memory disorders in China. Phase II trials have also been conducted in over 30 sites in the USA. Since it is marketed as a nutraceutical supplement in the USA, it is not labeled as a therapyto enhance memory. Further studies are required to elucidate whether indeed efficacy could be established at higher doses of the drug.

Huperzine B (HupB), a new Lycopodium alkaloid isolated from the Chinese herb Huperziaserrata, would be a promising candidate in AD therapy, due to a number of advantages over tacrine, such as highly selective inhibition on brain acetylcholinesterase (AChE), and lower peripheral cholinergic side effects. HupB is less potent and selective than HupA, but it has higher therapeutic index and other positive benefits. A novel series of 16-substituted derivatives were synthesized. 9i has the highest potency *in vitro*. The efficacy of 9i is more potent *in vitro* than HupA, galantamine, and rivastigmine, and it is equivalent to donepezil. 9c and 9i were showed moderate neuroprotection against H_2O_2-induced neurotoxicity [152]. Table **1** summarizes the structures of some phytochemicals with anti-amyloid activity.

Target Synthetic Drugs

Acetylcholinesterase Inhibitors

Currently available drug therapies for AD consist primarily of acetylcholinesterase inhibitors (AChEIs). Clinical investigation suggests that AChEIsare efficient in reducing symptoms in patients with mild-to-moderate AD, which leads to the US Food and Drug Administration's (FDA) approval of four AChEIs, such as tacrine, donepezil, rivastigmine and galantamine, for the treatment of AD. Although tacrine was the first approved cholinesterase inhibitor for the treatment of AD in 1993, its use has been largely abandoned due to a high incidence of hepatic adverse effects. The other three AChEIs, donepezil (Aricept®), rivastigmine (Exelon®), and galantamine (Razadyne®) are being used from mild to severe AD patient by now [153]. The pharmacological characteristics and dosage schedules for each of these drugs are described in Table **2**. The clinical benefits of these AChEIs mainly result from preventing the breakdown of acetylcholine and therefore increasing the cholinergic neurotransmission in forebrain regions, such as frontal cortex, hippocampus and amygdala. In addition to inhibition of cholinesterase activity, rivastigmine and galantamine are found to inhibit butylcholinesterase and stimulate nicotinic acetylcholine receptors, respectively.

Table 1: Structures of the Phytochemicals with Anti-Amyloid Activities.

Structure	Systemic name
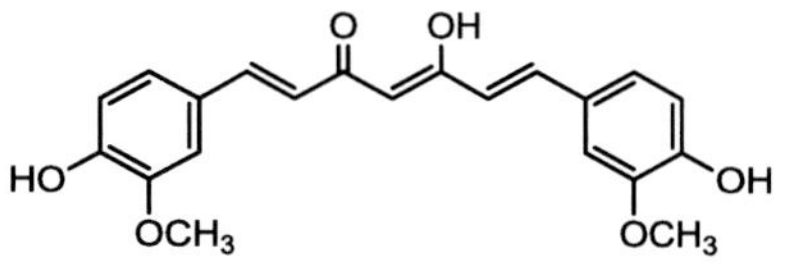	Curcumin
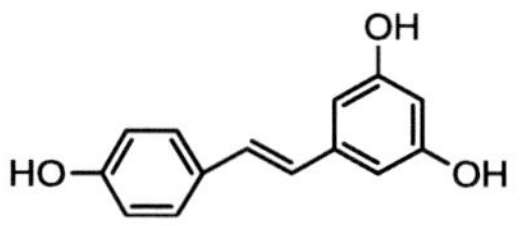	Resveratrol
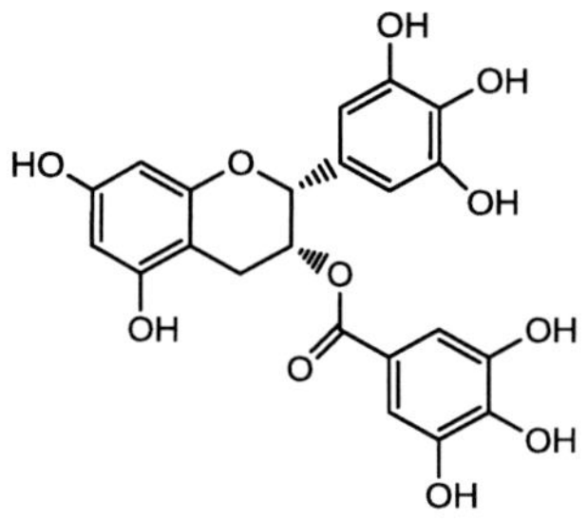	(-)-Epigallocatechin-3-gallate (EGCG)
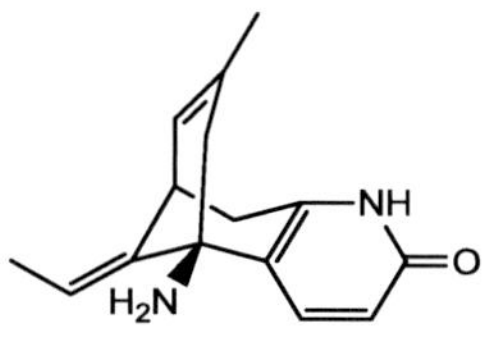	Huperzine A
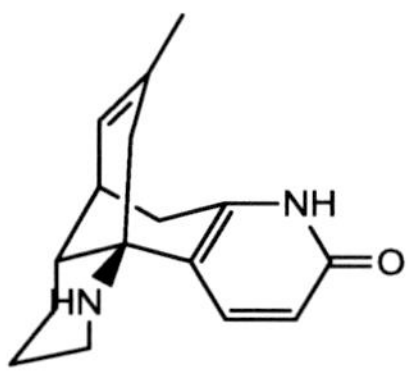	Huperzine B

Table 2: Cholinesterase Inhibitors Profiles

ChEIs	Disposition	Mechanism of Action
Physostigmine (Eserine)	Physostigmine is the first AChEI investigated for the treatment of AD. However, the drug is not approved and is abandoned for AD use due to the side effects such as nausea, vomiting, headaches, diarrhea, and dizziness and short half-life, narrow therapeutic index.	AChEI
Donepezil	Donepezil was approved in 1996 for the treatment of mild-to-moderate AD. Donepezil is well absorbed with a relative oral bioavailability of 100% and reaches peak plasma concentrations in 3 to 4 hours. The elimination half-life of donepezil is about 70 hours and is approximately 96% bound to human plasma proteins.	Reversible, non-competitive AChEI (highly selective)
Tacrine	Tacrine was the first drug approved for treatment of AD in 1993. However, widespread use of tacrine is limited as it is poorly tolerated as it causes a number of side effects including nausea, vomiting, dizziness, diarrhea, seizures, and syncope. In addition, patients who use the drug require periodic blood monitoring due to hepatotoxicity.	Inhibitor of both AChE and BuChE
Rivastigmine	Rivastigmine was approved for the treatment of mild-to-moderate AD and Parkinson's related dementiasin 2000. It is suitable to be converted to a patch delivery system because it is a small molecule (<300 Da in size), and it is both lipophilic and hydrophilic.	Pseudo-irreversible, dualAChEI/BuChEI,G1 isoenzymespecific, selective forhippocampus, neocortex
Galantamine	Galantamine was approved for the treatment of mild-to-moderate AD in 2001. Absorption of galantamine is rapid and complete and shows linear pharmacokinetics. It is well absorbed with absolute oral bioavailability between 80 and 100%. It has a half-life of seven hours.	Reversible, competitive, dual mode: selective AChEI plus allosteric nicotinic receptor modulator
Huperzine A	HupA is natural herb that is marketed as a supplement in the USA. It acts as an AChE inhibitor. Phase II trials in the USA showed a modest but clinical significant effect on cognition in AD.	Highly selective, reversible, and potent AChE inhibitor
Huperzine B	HupB is less potent and selective than HupA, but it has a higher therapeutic index.	Reversible inhibitor of AChE

N-Methyl-D-Aspartate Receptor Antagonists

Several studies have demonstrated that amyloid plaques lead to neuronal loss in AD and are mediated by NMDA receptors. In AD development, tau and amyloid plaques over-activate NMDA receptors and thus can cause the activation of key enzymes, such as calcium/calmodulin-dependent protein kinases and the downstream signaling. Xiong and his coworkers found that the NMDA receptor antagonistsblock inward currents in Xenopus oocytes that express NR1A/NR2B

[154]. The APP-stimulated monocyte-derived macrophages have been suggested to be responsible for the NR1A/NR2B activation due to the direct activation of NMDA receptor subtypes, which is relevant to the pathogenesis of AD, by APPs/Aβ-stimulated monocyte-derived macrophages conditioned media. In addition to amyloid plaques, tau protein also causes NMDA receptor activation and, consequently, NMDA receptor-mediated neurotoxicity in AD.

NMDA receptors in the central nervous system has become increasingly recognized as a key drug target in the neurodegenerative diseases such as AD. Memantine is the first in a novel class of Alzheimer's disease medications acting on the glutamatergic system by blocking NMDA-type glutamate receptors. Memantine is a noncompetitive NMDA receptor antagonist and is thought to protect neurons from excessive glutamate activity that results in excitotoxicity. Other properties of memantine that may also be relevant to AD include its ability to enhance long-term potentiation, and to decrease tau hyper phosphorylation. Memantine was synthesized by Eli Lilly & Co. in 1968 and is approved by the U.S. FDA and the European Medicines Agency for treatment of moderate-to-severe Alzheimer's disease since then [155]. However,it has now received a limited recommendation by the UK's National Institute for Clinical Excellence for patients who fail to other treatment options. For example, it is recommended as an option for managing Alzheimer's disease for people with moderate dementia who are intolerant of or have a contraindication to AChEIs or those with severe Alzheimer's disease. Memantine has been associated with a moderate decrease in clinical deterioration with only a small positive effect on cognition, mood, behavior, and the ability to perform daily activities in moderate to severe Alzheimer's diseases.

Amyloid β Peptide as the Drug Target

One of the hallmarks of AD is the presence of senile plaques (SP) and neurofibrillary tangles (NFT) in the brain, which result from the accumulation and deposition of the β-amyloid peptide (Aβ) and the aggregation of hyperphosphorylated tau protein, respectively. Initially, it was thought that Aβ fibrils, which make up SP, were the root cause of the massive neurodegeneration usually found in the AD brains. Over time, the longstanding emphasis on fibrillar Aβ deposits and neuronal death slowly gave way to a new paradigm involving soluble oligomeric forms of Aβ, which play a prominent role in triggering the cognitive deficits. Understanding how Aβ oligomers target central nervous system provides an important framework for ongoing AD research, which can lead to the development of successful therapeutic strategies designed to alter or perhaps reverse the course of the disease.

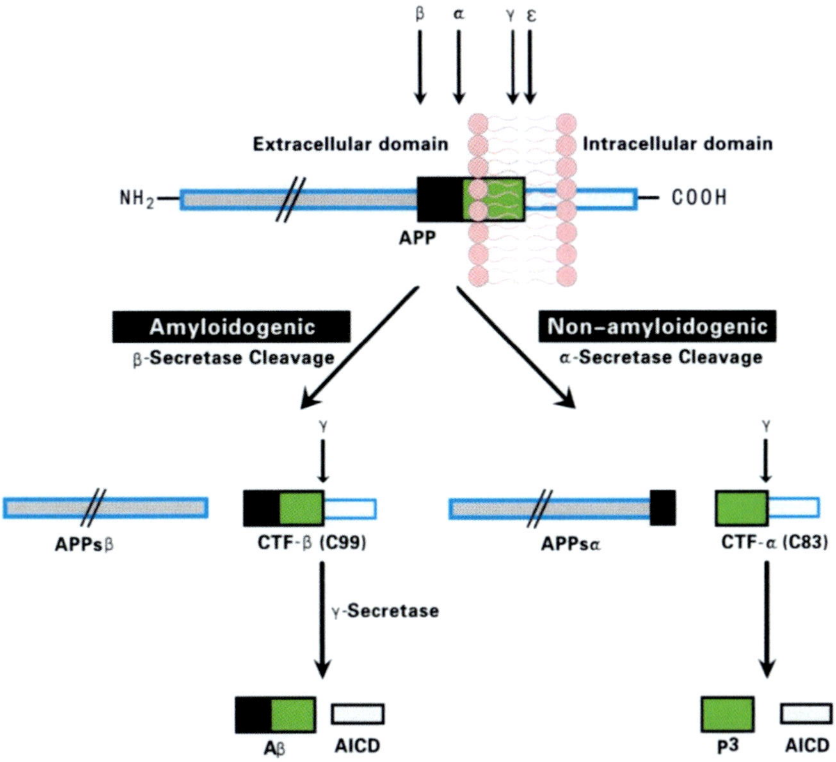

Figure 1: Schematic representation of APP metabolism by the secretase enzymes. APP can be cleaved by either β- or α-secretase, which is then followed by γ-secretase cleavage resulting in the generation of either the p3-fragment (non-amyloidogenic) or an Aβ (amyloigenic pathway).

Amyloid β peptide (Aβ) is cleaved from the amyloid precursor protein (APP) and it usually possesses 36-43 amino acids. Aβ peptides are known to self-assemble into dimer, trimer and higher-order oligomers, which are believed to be the main source of toxicity by causing the death of neurons (Fig. **1**). The mechanism of the toxicity has been studied extensively from both the experimental and theoretical perspectives which are summarized here: (1) activation of inflammatory effects by interacting directly with the membrane; (2) induction of oxidative stress through the formation of metal-Aβ complex; (3) disruption of membrane receptors' function by intimate binding; (4) formation of membrane pore and alteration of ionic homeostasis across the membrane; and (5) modification of the structure of certain DNA by the process of attachment.

Based on the premises of Aβ being the primary causative trigger in the development of AD, a large portion of the many therapeutic approaches currently under development for the treatment of this disease have been directed against the production and accumulation of Aβ. In this category, three sub-classes of medications are tried and have advanced to latter stages of clinical trials permitting meaningful conclusions regarding their efficacy. These include secretase modulators (which decrease Aβ production), anti-aggregants (which prevent Aβ aggregation), and immunotherapies (which increase Aβ removal). Activation of α-secretase is found to decrease levels of Aβ40 and Aβ42 *in vivo*. APP transgenic mice reveal that early administration of either an anti-amyloid vaccine or γ-secretase inhibitor reduces intracellular Aβ accumulation and plaque burden and enhances clearance of tau [156].

β-site APP Cleaving Enzyme I (BACE1) as the Drug Target

Aβ peptides are produced by two consecutive cleavages carried out by β-secretase (BACE) and γ-secretase. The APP β-secretase was first identified by five independent research groups at the end of 1999,which was termed as a new aspartyl protease including BACE 1, Asp 2 or memapsin 2. BACE 1 is ubiquitously expressed with highest levels in the brain and pancreas. BACE 2 also exhibits a α-secretase like activity towards APP, but is only weakly expressed in brain. Therefore, BACE 1 seems to be a highly attractive drug target for AD treatment. Many pharmaceutical groups and academic laboratories have embarked in the search for BACE inhibitors to treat AD. Rational drug design based on substrate analogues and peptidomimetics has provided many compounds that are potent BACE inhibitors *in vitro*. However, some of them have failed in preclinical trials due to insufficient pharmacological properties and to the difficulty in getting high-affinity inhibitors to cross the blood-brain barrier. The BACE inhibitor CTS-21166 from CoMentis/Astellas Pharma has successfully completed Phase I clinical trials and has now moved into Phase II [157].

Compounds capable of interfering with the proteases that generate the Aβ peptide from APP have been actively pursued during the last 5-10 years. Unfortunately, blockage of the most biologically attractive target of these proteases, the β-secretase that performs the first cleavage step of the APP, is found to be particularly problematic, and only a few compounds have reached clinical testing so far. Compounds that stimulate α-secretase that is responsible for the nonamyloidogenic metabolism of APP, are also being developed, and one of them, etazolate (EHT0202, Exonhit; Paris, France), has reached Phase II clinical development. Furthermore, brain penetrate inhibitors of Aβ aggregation have been identified and one such compound, PBT-2 (Prana Biotechnology, Melbourne, Australia), has produced encouraging neuropsychological results in a recently completed Phase II study.

γ-secretase as the Drug Target

γ-secretase inhibition, an attractive and promising target for disease modification, was first shown to decrease *in vivo* Aβ production in 2001. A nonselective γ-secretase inhibitor developed by Eli Lilly& Co., semagacestat (LY450139), is found to have characteristic effects on plasma and cerebrospinal fluid (CSF) levels of Aβ in APP-transgenic mice. It was evaluated in phase I and subsequent phase II trials in healthy human volunteers and patients with mild to moderate AD for safety, tolerability and pharmacodynamic response. The results suggested that semagacestat is generally well tolerated at doses of up to 140 mg/d for 14 weeks, with several findings indicating the need for close clinical monitoring in future studies [158]. However, the clinical development of semagacestat has been halted because preliminary results from two ongoing long-term Phase III studies showed that it is associated with worsening of clinical measures of cognition, clinical global status and functionality. Two multicenter phase III trials, IDENTITY and IDENTITY-2, were designed to assess the efficacy and safety of 140 mg of semagacestat with over 2000 mild to moderate AD patients, using a randomized, double-blind, parallel-group design. The recruitment was terminated when preliminary results at a scheduled interim analysis showed that the treatment group had in fact worse cognitive and functioning measures. Moreover, they had an increased risk of skin cancer compared to the placebo group (ClinicalTrials.gov identifiers: NCT00594568 and NCT00762411). The negative trial results may be attributed to several factors including the stimulation of γ-secretase due to a rebound effect, thus inducing overall stimulation of Aβ production, or the inhibition of γ-secretase resulting in the inhibition of processing of not only APP but also other substrates relevant for cognition.

A group of non-steroidal anti-inflammatory drugs (NSAIDs) were found to lower Aβ42 levels by way of modulating the γ-secretase pathway, which are the first selective Aβ42-lowering agents. Tarenflurbil (MPC-7869, Flurizan) is an R-enantiomer of the NSAID flurbiprofen that inhibits cyclooxygenase activity, and therefore has a benign gastrointestinal profile. Tarenflurbil was shown *in vitro* to modulate γ-secretase activity, shifting amyloid species production from Aβ42 to shorter peptides that are less fibrillogenic. Flurbiprofen was found to reduce amyloid plaque accumulation and to improve memory and behavioral performancein transgenicmice. In a subsequent phase I trial, the drug was well tolerated, but failed to produce significant lowering in Aβ42 CSF or plasma levels. Despite this seemingly low potency, a placebo-controlled, phase II trial was conducted in 210 patients with mild to moderate AD who were randomly assigned to receive R-flurbiprofen (400 mg or 800 mg) or placebo, twice/day for 12 months. It was extended for an additional 12-month period where treated patients were continued to the treatment, while placebo patients were randomized to one of the two doses of R-flurbiprofen. The group with mild AD randomized to the higher dose R-flurbiprofen showed positive effects in daily activities and global functioning, but not in cognitive performance. Those who were treated with this dose for the whole 24 months had lower rates of cognitive decline on all measures than those who were initially given placebo. These findings justified a phase III randomized controlled study of R-flurbiprofen at the 800 mg twice/day dose in 1649 patients with mild AD. Disappointingly, this study revealed no cognitive benefits from treatment. However, this result raised suspicion about the merits of γ-secretase modulation and even in the amyloid cascade hypothesis, tarenflurbil's potency and suboptimal central nervous system levels are obvious targets to blame as potentially important factors contributing to the drug's failure.

In addition, α-secretase pathway stimulation leads to a reduction of the sαAPP substrate that is available for the amyloidogenic pathway, and leads to the formation of a soluble segment (sαAPP), which was suggested to be neuroprotective, and a stimulant for synaptogenesis. Disappointingly, there have been no major compounds modulating these pathways that have emerged from animal studies to reach clinical trials.

Amyloid Antiaggregants and Amyloid Removal Drugs

The next attractive point of intervention is prevention of aggregation of amyloid species. Amyloid species are normally present in healthy brains, and their mere presence does not lead to neurodegeneration. The aggregation of Aβ species will induce oligomers, fibrils and protofibrils, and then deposition in the form of amyloid plaques. Recent studies, however, provided evidence that the earlier soluble oligomers are also neurotoxic. A number of anti-Aβ aggregation agents have been tested in clinical trials.

Tramiprosate (homotaurine, Alzhemed™), 3-amino-1-propanesulfonic acid (3APS), is a sulfonated glycosaminoglycan (GAG) mimetic. GAGs bind to soluble Aβ, promoting fibril formation and deposition in amyloid plaques. GAG mimetics prevent fibril formation and reduce soluble Aβ by competing for GAG-binding sites [159]. In transgenic mice, tramiprosate reduces plaque burden and decreases CSF Aβ levels. A phase I study in healthy subjects demonstrated the drug is well tolerated.A 3-month Phase II study was conducted in 58 patients with mild to moderate AD, who were randomly assigned to tramiprosate 50 mg, 100 mg or 150 mg twice a day, or placebo. Patients who completed the study were continued on for a 21-month open-label extension with 150 mg twice a day. Baseline CSF Aβ42 levels decreased up to 70% after 3 months in patients randomized to 100 mg or 150 mg twice a day. However, no significant differences were observed in cognitive measures between the treatment and placebo groups. A Phase III trial was conducted in 69 centers across the U.S. and Canada, and recruited 1052 patients with mild to moderate AD [160]. Although there was some suggestion of a beneficial effect on post hoc analyses of cognitive sub-scales, and hippocampal volumes on MRI [161], the original primary analysis failed to show any significant difference in cognitive measures between treatment and placebo groups. Its failure shed some doubts on the idea that altering the amyloid cascade could produce meaningful clinical improvement.

Immunotherapy in patients with Alzheimer's disease (AD) is rapidly becoming a hot topic of modern geriatric and clinical gerontology [162]. Anti-Aβ immunotherapies represent the most powerful pharmacological approach in clearing Aβ from the AD brain. Current views extend immunization with Aβ peptide, the amyloidogenic protein found in senile plaque of AD patient's brains, or the infusion of preformed antibody specific for human Aβ, as possible therapeutic approaches to improve the cognitive status in the disease. Animal and human studies have indicated that immunotherapy represents the most powerful pharmacological approach to clear Aβ from the brain. However, vaccine therapy is not always effective in the elderly, since immune defects of variable degree are often present in old persons. An active anti-Aβ vaccine preparation (AN1792) consisting of pre-aggregated Aβ42 and an immune adjuvant (QS-21), has been used in AD patients with questionable clinical efficacy, but it causes meningoencephalitis in approximately 6% of patients and has thus been abandoned.

Passive vaccination may have the advantage of using conformation-specific antibodies designed against specific neurotoxic Aβ species, thus avoiding the removal of 'physiological' monomeric Aβ species from the brain. On the other hand, there is the potential for patients to eventually develop neutralizing antibodies against passive immunotherapy. Anti-Aβ monoclonal antibodies (bapineuzumab and solanezumab) are now being developed. The clinical results of the initial studies with bapineuzumab were equivocal in terms of cognitive benefit. The occurrence of vasogenic edema after bapineuzumab, and more rarely brain micro-hemorrhages (especially in Apo E ε4 carriers), has raised concerns on the safety of these antibodies directed against the N-terminus of the Aβ peptide. Solanezumab, a humanized anti-Aβ monoclonal antibody directed against the mid-region of the Aβ peptide, is shown to neutralize soluble Aβ species. Phase II studies showed a good safety profile of solanezumab, while studies on cerebrospinal and plasma biomarkers documented good signals of pharmacodynamic activity. Bapineuzumab and solanezumab represent the cutting edge of the passive immunotherapy approaches and are presently under extensive clinical testing in large Phase III trials involving more than 6000 AD patients worldwide. These results of the large, ongoing Phase III trials with bapineuzumab and solanezumab will tell us if monoclonal anti-Aβ antibodies may slow down the rate of deterioration of AD.

Tau Protein as the Drug Target

The other hallmark in the AD brain is the NFTs that are formed from insoluble fibrils of tau protein. It is meaningful to discuss the likely involvement of misfolded tau in the neurodegeneration and the memory impairments observed in AD, and present possible therapeutic strategies to ameliorate tau-mediated pathology.

A loss of tau function might be overcome with microtubule-stabilizing agents or inhibitors of tau hyper-phosphorylation and/or acetylation. Inhibitors of tau multimeric assembly might prevent potentially toxic tau oligomers or fibrils. Inhibition of HSP90 and the resulting elevation of the chaperones HSP70/HSP40 may increase proteasomal degradation of hyper-phosphorylated tau. Misfolded tau multimers might be cleared through enhancement of macroautophagy. In addition, misfolded tau species may be released from cells and internalized by nearby neurons, thereby seeding the formation of pathological tau in the recipient cell. This spreading of tau pathology might be inhibited by antibodies that bind misfolded tau in the brain interstitial fluid.

The Glycogen Synthase Kinase 3 (GSK3) as the Drug Target

There are some kinases that modify ser/thr residues that precede a proline residue, including GSK-3 α and β, cyclin-dependent kinase 5 (CDK5), and mitogen-activated kinase 1 (MAPK1). In addition, nonproline directed kinases such as protein kinase A, p38 and microtubule-affinity regulated kinases (MARK1-4) have been implicated in tau phosphorylation. From these various kinases, the existing data arguably support the GSK-3β as being the best-validated tau kinase target.

The GSK3 seems to be linked to all of the major pathological mechanisms that had been identified in Alzheimer's disease ten years ago. Among the early identified links between GSK3 and Alzheimer's disease is the discovery that GSK3 appears to contribute to Aβ-induced tau phosphorylation and neurotoxicity, and promotes the intrinsic apoptotic-signaling pathway that may be partly responsible for neuronal loss in Alzheimer's disease.

GSK3 and Tau

In AD brain, tau phosphorylation is increased three to fourfolds, which is incorporated into neurofibrillary inclusions. Both GSK3β and GSK3α induce the hyper-phosphorylation of tau at both primed and non-primed phosphorylation sites, *in vitro* and in cell culture models of neurodegeneration, implicating GSK3 as an important tau-kinase possibly involved in the formation of NFTs *in vivo*. GSK-3β is one of the main mediators in the process leading to tau hyper-phosphorylation. Increased activity of GSK-3β has been associated with an increased risk of AD, and it is shown to be up-regulated in the frontal cortex of AD patients. Animal studies showed that lithium (GSK3 inhibitor) decreases levels of phosphorylated tau and expression of GSK-3β in the cortex and hippocampus, suggesting that GSK3β is one of the key targets for future disease-modifying therapies for AD. Consistent with this, GSK3β transgenic mice display tau hyper-phosphorylation and neurodegeneration and chronic lithium treatment prevents tau hyper-phosphorylation and NFT formation in double transgenic mice over-expressing GSK3β and tau. Daily administration of the GSK-3 inhibitor LiCl for 30 days to JNPL3 tau Tg mice, which overexpress P301L 4R-tau, results in the inhibition of GSK-3 activity and reduction of both phospho-tau and insoluble tau, as well as lessening of axonal degeneration [163].

GSK3 and Aβ

Increasing evidence has demonstrated that Aβ activates GSK3 by decreasing its inhibitory serine-phosphorylation, which appears to contribute to Aβ-induced increased tau phosphorylation and to Aβ-induced neurotoxicity. These studies showing Aβ-induced activation of GSK3 have used a variety of peptides, including Aβ1-40, Aβ1-42,and the 25-35 peptide fragment, indicating that accumulation of any of these may activate GSK3, although perhaps by utilizing different signaling mechanisms, which remain to be identified. Takashima and colleagues first identified GSK3 (at that time also called tau protein kinase-1) has a neuroprotective effect against Aβ-induced toxicity. They found that Aβ treatment increases GSK3β activity and pretreatment with GSK3β antisense oligonucleotides prevents Aβ-induced cell death and reduces tau phosphorylationin cultured rat hippocampal neurons. These studies indicated that GSK3 is involved in Aβ-induced tau phosphorylation and neurotoxicity. Subsequent reports also confirmed that inhibitors of GSK3, such as lithium and SB216763, reduce Aβ-induced tau phosphorylation and cell death. Fuentealba and colleagues found that treatment with lithium prevents Aβ1-42-induced morphological changes, particularly shrunken soma and affected dendritic and axonal processes, and reduces Aβ-induced decreases in cell viability of primary rat hippocampal neuronal cultures [164]. The subsequent studies suggested that treatment with SB216763 or GSK inhibitor VIII also prevents Aβ-induced caspase-3 activation and tau-phosphorylation *in vivo*, decreases TUNEL positive neurons and cytochrome c release from the mitochondria to the cytosol, which result in learning and memory improvement in the Morris water maze. Administration of the GSK3 inhibitor NP12 is also found to prevent tau phosphorylation and Aβ deposition, as well as improve performance in the Morris water maze in amyloid precursor protein (APP) transgenic mice [165].

GSK3 and Neuroinflammation

Despite being implicated in the core pathogenic events of AD, such as the increased formation of Aβ, tau hyper-phosphorylation and memory impairment, GSK3 might also play a role in other processes thought to impact on the amyloid cascade. There is a growing body of evidence to suggest that inflammation induced by Aβ and tau hyper-phosphorylation plays an important role in AD. Microglia are known to accumulate around Aβ plaques in AD and there is increased expression of inflammatory mediators in AD brain tissue. Moreover, Fc-dependent engagement of microglia is thought to be one mechanism through which Aβ immunotherapy operates. This treatment strategy has proven to be successful in improving cognitive function and reducing Aβ plaque burden in both animal models of AD and in human AD patients. However, phase II clinical trials were aborted as some patients developed symptoms of brain inflammation resembling encephalitis or meningitis following injection with Aβ. In the periphery, the regulation of GSK3 activity is critical for inflammatory cell differentiation, migration and the secretion of pro-inflammatory cytokines [166]. However, little is known about the function of GSK3 in the cerebral inflammatory response, which is predominantly mediated by microglia. If, as we discuss above, GSK3 regulation is aberrant in AD brain, then this might impact upon the cerebral inflammatory response leading to the sustained secretion of neurotoxic inflammatory mediators by microglia, in turn causing by-stander damage to neighboring neurons and contributing to neurodegenerative processes.

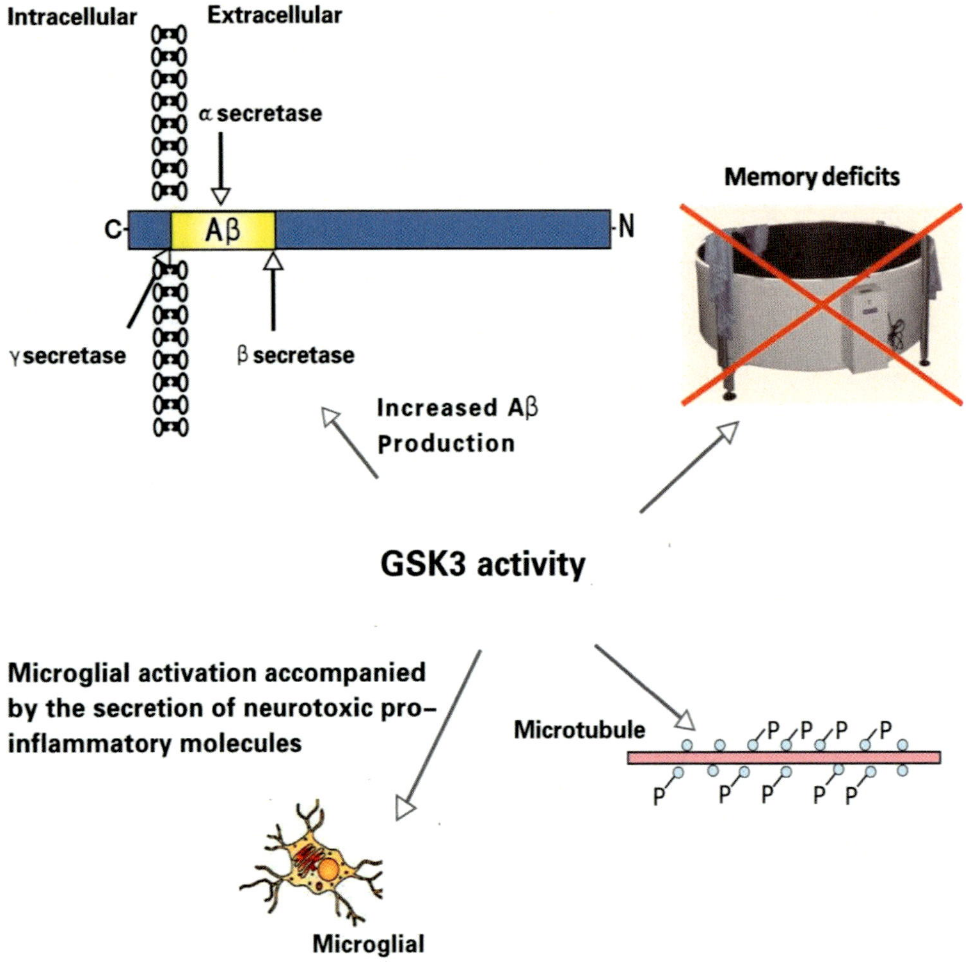

Figure 2: GSK3 and its role in AD. GSK3 is one of the key kinases required for AD-type abnormal hyper-phosphorylation of tau, which is believed to be a critical event in neurofibrillary tangle formation. GSK-3 has also been recently implicated in APP processing/A-beta production, apoptotic cell death, microglial activated and learning and memory deficits.

In various cell cultures, invertebrate and mammalian models of AD increasing GSK3 activity lead to the hyper-phosphorylation of tau, increased Aβ generation and deficits in learning and memory accompanied by neurodegeneration as shown in Fig. **2**. Most importantly, the inhibition of GSK3 activity reverses some of the pathological effects of over-expression of mutated APP and tau in the animal models of AD. This 'GSK3 hypothesis of AD' integrates and extends the well established "amyloid cascade hypothesis of AD", incorporating the known key molecular events and linking these with outcomes such as memory impairment and inflammation. This hypothesis strongly supports the GSK3 inhibitor as a novel treatment strategy for AD.

Phosphodiesterases (PDEs) as the Drug Target

Cyclic nucleotide phosphodiesterases are a super-family of enzymes that are involved in the regulation of the intracellular second messengers the cAMP and cGMP, by controlling their rates of degradation. Cyclic AMP, synthesized from ATP by adenylatecylcase (AC), activates protein kinase A (PKA) that phosphorylates the cAMP response element-binding protein (CREB), and thereby affects transcription of genes related to synaptic plasticity and survival, such as the brain-derived neurotrophic factor (BDNF). On the other hand, cGMP is derived from GTP by guanylylcyclase (GC), which in turn is activated by nitric oxide (NO). The NO/cGMP pathway then activates protein kinase G (PKG), which also induces CREB phosphorylation. Both the cAMP/PKA/ CREB and cGMP/PKG/CREB pathways are implicated in long-term potentiation (LTP), the neurophysiological correlate of memory.

There are 11 different PDE families; each of them typically has different isoforms and splice variants (Fig. **3**). On the basis of substrate specificity, the PDE families can be divided into three groups: (1) the cAMP-specific PDEs that include PDE4, PDE7 and PDE8; (2) the cGMP-selective enzymes that include PDE5, PDE6 and PDE9, (3) and conclude with the dual-substrate PDEs, PDE1, PDE2, PDE3, pde10 and PDE11. Phosphodiesterase inhibitors (PDEIs) are receiving increased attention as putative cognition enhancers. PDEIs prevent the selective breakdown of the cAMP and/or cGMP by inhibiting PDE enzymes. More evidence has been shown that Aβ can have a detrimental effect on LTP, *via* inhibition of both the cAMP/PKA/CREB and cGMP/PKG/CREB pathway. The clinical reports suggest that AC, GC and pCREB levels are reduced in the temporal lobe of AD patients. Given the cognition-enhancing properties of PDEIs in rodents and monkeys and the synaptic dysfunction and cognitive deficits seen in AD patients, PDEIs are currently in development as new therapeutics for the treatment of AD.

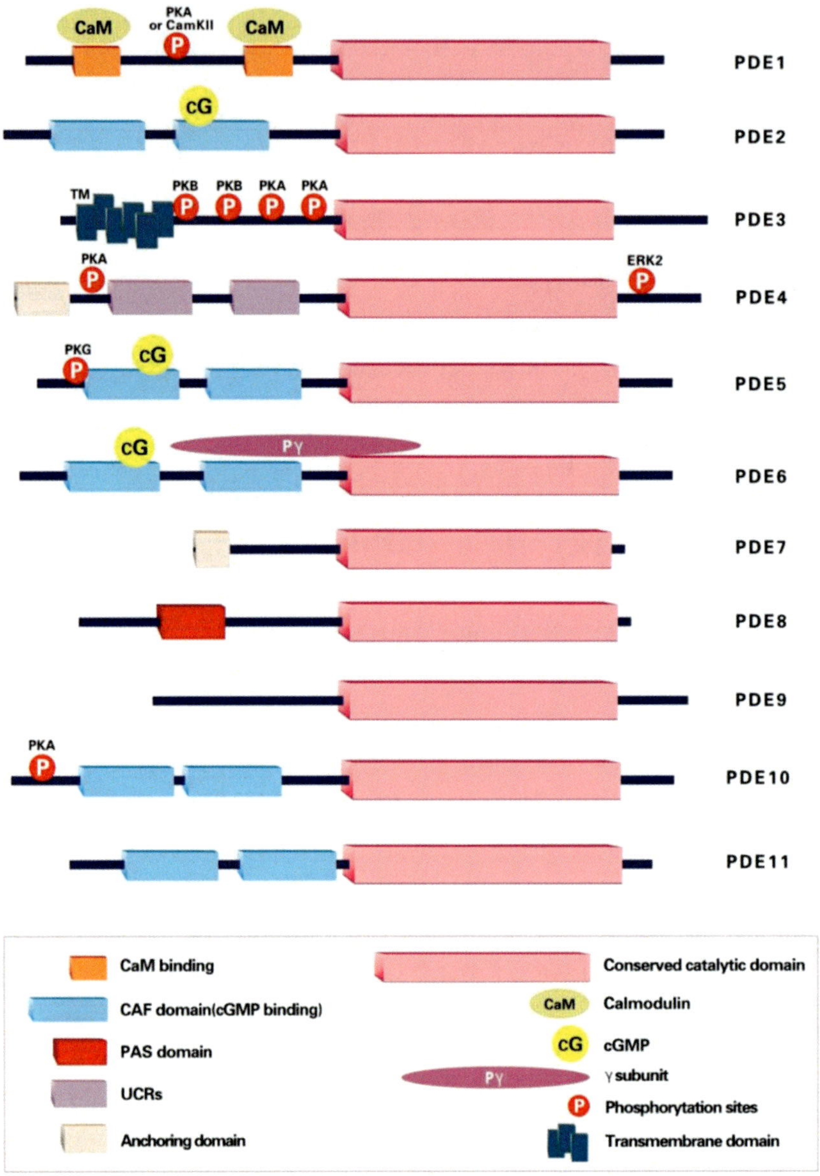

Figure 3: Schematic representation of the structure of the 11 PDE families. The family name is reported on the side of each structure. The conserved catalytic domain is represented as a pink cylinder. Binding proteins are reported in yellow. Phosphorylation sites are reported as a red symbol marked P. Abbreviations: CaM, calmodulin; CamKII, calmodulin-regulated kinase II; cG, cGMP; ERK2, extracellular signal-regulated kinase 2; pγ, PDE γ subunit; GAF, cGMP-activated PDEs, adenylyl cyclase, and Fh1A; PAS, periodaryl-hydrocarbon receptor nuclear translocator (ARNT), and single minded; PKA, protein kinase A; PKB, protein kinase B; TM, transmembrane domain of PDE3; UCRs, upstream conserved regions.

It has been found that specific PDE inhibitors improve different phases of memory consolidation probably by specific modulation of cyclic nucleotide-depending signaling. Currently, PDE inhibitors are considered as a novel treatment for cognitive disorders, and in particular memory improvement beyond the cholinergic-based strategies. Investigations on the involvement of PDEs in learning and memory have mainly focused on PDE1, 2,3, PDE4, PDE5, 9 and PDE10 due to availability of selective inhibitors for these PDEs (Table **3**).

PDE1

PDE1 is also known as calcium-and calmodulin-dependent phosphodiesterase. It has three subtypes, PDE1A, PDE1B and PDE1C. The various isoforms exhibit different affinities for cAMP and cGMP. *In vitro*, PDE1A and PDE1B isozymes hydrolyze cGMP with much higher affinity than cAMP, and PDE1C isozyme hydrolyzes cAMP and cGMP with equally high affinity. *In vivo*, PDE1A and PDE1B primarily regulate cGMP [167]; while PDE1C regulates intracellular cAMP levels in various cell types including neurons [168]. The three PDE1 isoforms are differentially expressed in the brain, leading to different roles in CNS functions. Given that PDE1B is predominantly expressed in striatal neurons expressing dopamine D1 receptor, PDE1B may play a unique role in D1 receptor signaling-mediated cognition. Therefore, PDE1B inhibitors may be a novel class of cognition enhancers for treatment of cognition dysfunction in schizophrenia.

PDE1 inhibition has been reported to cause cerebral artery dilatation by elevations of cGMP levels [169]. The PDE1 inhibitor vinpocetine has been shown to facilitate long-term potentiation [170], enhance the structural dynamics of dendritic spines, and improve aversive-related memory retrieval. A recent study indicates that vinpocetine treatment in streptozotocin-infused rats improves memory consolidation as evidenced by increase time spent in target quadrant in Morris water maze [171]. More recently, Filgueiras and colleagues reported that vinpocetine improves learning and memory deficits in rats exposed to alcohol [172]. However, these beneficial effects of PDE1 inhibition on cognition should be confirmed by the further clinical studies.

PDE2

There are three variants of PDE2 gene, PDE2A1, 2A2 and 2A3, which are found in the brain and other sensory neurons, bulb and tubercle. The biological function of PDE2 may, to some degree, involve interaction between cAMP- and cGMP-mediated signaling mechanisms [173].

The PDE2 inhibitor Bay 60-7550 is proved to enhance long-term potentiation in cultured neurons and hippocampal slices. Bay 60-7550 is also found to fully reverse the working memory deficits induced by the NMDA receptor antagonist MK-801 [174]. It also improves the performance of rats in social and object recognition memory tasks, and reverses MK801-induced deficits in spontaneous alternation in a T-maze [175]. Recently, studies showed that Bay 60-7550 improves short-term object recognition performance after an acute tryptophan depletion-induced memory deficit [176]. The results are in agreement with the report which suggests that Bay 60-7550 can improve acquisition and consolidation in the object recognition task in rats with age-related cognitive impairments [177]. It is known that there is a lower cGMP concentration in the aged brain and an alteration in the activity of cGMP-hydrolyzing PDEs, the mechanism behind PDE2 inhibitor's beneficial effects on cognition defects may be related to stimulating the cGMP-regulated signal cascade and subsequently enhancing neuronal plasticity [175]. In addition, PDE2 inhibition is also able to inhibit oxidative stress-induced anxiety in mice [178,179]. These behavioral and neurochemical findings suggest that PDE2 may be a pharmacological target for treatment of neurodegenerative diseases that involves oxidative stress-related Alzheimer's disease.

PDE3

The PDE3 was purified and described as an enzyme that hydrolyzes both cGMP and cAMP with Km values between 0.1-0.8 μM. The Vmax for cAMP hydrolysis of PDE3 is 4 -10 times higher than Vmax for cGMP hydrolysis.The PDE3 inhibitor cilostazol increases insulin-like growth factor I (IGF-I) and cAMP level in the hippocampus, which may result in improved cognitive function [180]. Some of the PDE3 inhibitors have been developed as one of the strategies for AD treatment, but the arrhythmic effect and the increase in mortality in some applications limit their use.

PDE4

PDE4 is the principal family involved in degradation of cAMP throughout the body. There are four genes that encode different PDE4 enzymes, of which

PDE4A, PDE4B and PDE4D, but not PDE4C, are expressed in the CNS. PDE4 has been implicated in various disorders such as depression, anxiety, and inflammation-related disorders. There is a large body of evidence to suggest that inhibition of PDE4 could have therapeutic benefits for a host of conditions, including several CNS disorders, including Alzheimer's disease.

The increasing studies in mammalian species strongly implicate members of the PDE4 family as the PDEs responsible for regulating pools of cAMP and as being intimately involved in signaling cascades that mediate learning and memory processes. Particularly, augmenting cAMP signaling through PDE4 inhibition is associated with the consolidation and retention of long-term memory. At the molecular level, the prototypical PDE4 inhibitor rolipram facilitates the induction of long-term potentiation (LTP) in the hippocampus, and increases phosphorylation of CREB and expression of Arc. These changes are linked to increased retention of long-term memory in conditioned fear paradigms. Recent studies have shown that repeated treatment with rolipram reverses memory deficits induced by Aβ-25-35 or Aβ-40 in the water maze and passive avoidance tasks [181]. The cognition enhancement by rolipram appears to be attributed to rolipram-induced increases in cAMP/CREB signaling in the prefrontal cortex and hippocampus. These findings are in agreement with rolipram-induced reversal of memory deficits in APP/PS1 transgenic mice [182]. PDE4 inhibitors are also found to reverse the working memory deficits caused by NMDA-receptor antagonist (MK-801), MEK inhibition, electroconvulsive shocks (ECS) and age- or micro-sphere embolism-induced cerebral ischemia in the radial arm maze and the three-panel runway task [183]. Another study shows that two novel PDE4 inhibitors, MEM1018 and MEM1091, both improve the memory processes in radial-arm maze and passive avoidance tests [184]. Consistent with these behavioral data, the PDE4 inhibitors produced parallel and consistent effects on cyclic AMP concentrations in cultured neurons when treatment with the NMDA receptors or MEK [185]. Further studies using selective PDE4 subtype knock out mice demonstrated that PDE4A and PDE4D may be important in the mediation of memory, while PDE4B is more likely involve in dopamine-associated and stress-related processes. Firstly, PDE4D knockout mice exhibit significant changes in associative learning (*i.e.,* fear conditioning). Consistent with this, PDE4D is highly expressed in the hippocampus, a sensitive structure in controlling the progression of depression and memory operations, including the formation of stable declarative (or explicit) memory in humans and spatial (or relational/contextual) memory in rodents. These results provide the evidence that PDE4D may play a crucial role in the mediation of antidepressant activity and memory [186]. Secondly, PDE4A is found to distribute in a similar pattern as PDE4D in the brain, which indicate that PDE4A may play similar roles to those of PDE4D [173]. Evaluation of the behavioral phenotype of PDE4A knockout mice would confirm the functional role for this subtype in the brain.

Despite the abundance of studies indicating a potential role for PDE4A and PDE4D inhibitors in CNS-borne diseases including AD, there are still no ideal PDE4 inhibitors currently available for treatment of CNS illness. One reason may be the incidence and types of adverse effects that occur with existing PDE4 inhibitors. It is well documented that PDE4 inhibitors, including rolipram, can induce a variety of side effects ranging from nausea, vomiting and diarrhea, to more serious conditions such as vasculitis, despite promising preclinical and phase-II study data. Although preliminary adverse effect profiles of the newer PDE4 inhibitors appear to be improved, the mechanisms responsible for these side effects are still not well understood.

PDE5

PDE5 is a phosphoric diester hydrolase enzyme that catalyzes the breakdown of phosphodiester bond by hydrolysis of cyclic nucleotide, particularly cGMP to its corresponding 5' nucleotide monophosphates (5'-GMP). PDE5 has one subtype (PDE5A) that has four isoforms (PDE5A1-4) in human, which is mostly distributed in smooth muscle (visceral, vascular) and skeletal muscles. It can be found in kidneys, prostate, heart, liver, penis, urethra, bladder, uterus, corpus cavernosum, pancreas, clitoris and smooth muscles present in the wall of vagina. PDE5 thus proves to be a novel enzyme target molecule for the discovery of more efficacious drugs for the treatment of sexual and pulmonary diseases. More evidence suggests that PDE5 inhibitors not only constitute an effective treatment for erectile dysfunction; but also play an important role in the CNS disorders because of the fact that cGMP has been recognized as a moderator synaptic plasticity [187]. Moreover, animal studies have shown that sildenafil enhances memory in several models [188] and attenuates memory impairment induced by NO synthase (NOS) inhibition [189]. Another study also shows that sildenafil improves performance dose-dependently in the object retrieval task in cynomolgus macaques [190]. Cognitive dysfunction by blockade of muscarinic cholinergic receptor, diabetes or electroconvulsive shock is also reversed by sildenafil treatment [191,192].

PDE9

PDE9 has the highest affinity of all PDE isoforms for cGMP (Km~170 nM) [193] perhaps indicating its relative importance over other cGMP specific PDEs in regulating signaling at low concentrations of cGMP. Cloning of the mouse and human PDE9A enzymes reveals that it has the highest affinity for cGMP of any of the PDEs identified to date. Furthermore, PDE9A mRNA is widely distributed throughout the rodent brain, mainly expressed in neurons, but occasionally also in astrocytes.

PDE9 inhibition with Bay 73-6691 improved performance in several cognitive tasks including social recognition in rats and mice, novel object recognition in rats and scopolamine-induced impairments in passive avoidance in rats [194]. Consistent with these observations, systemic administration of PF-04447943 (1–30 mg/kg, p.o.) that is a recently described PDE9 inhibitor dose-dependently increases cGMP in the cerebrospinal fluid indicating target engagement in the CNS of rats. PF-04447943 (1-3 mg/kg, p.o.) significantly improves cognitive performance in three rodent cognition assays, *i.e.,* mouse Y maze spatial recognition memory model of natural forgetting, mouse social recognition memory model of natural forgetting and rat novel object recognition with a scopolamine deficit. The mechanism by which PF-04447943 improves cognitive function is currently unclear although modulation of glutamate and/or cholinergic neuron signaling is potentially feasible given that both glutamate (NMDA) and cholinergic receptor activation enhance the formation of cGMP in the brain, and both neural substrates are involved in cognitive function. The further studies are required to define the mechanism by which PDE9 inhibitors improve cognitive function.

PDE10

PDE10A mRNA is highly expressed in the striatum, hippocampus and olfactory tubercle rodents [195]. PDE10A is found to be an important regulator of cGMP in the striatum and is involved in the physiological regulation of motor and cognitive function [196]. Siuciak and his co-workers found that learning and memory in the passive avoidance and water escape tasks are unaffected in PDE10A knockout mice with a DBA1LacJ background [195]. Furthermore, these animals display the same conditioned avoidance response as wild-type mice although they require more training. However, PDE10A knockout mice with a C57BL/6N background are unable to reach the level of performance of the wild-type animals. Hebb and his colleagues demonstrated that chronic treatment with the PDE10 inhibitor papaverine impairs learning and memory in the water escape task in wild-type mice, which could be explained by increased perseveration and impaired locomotor activity [197]. Table **3** summarizes the effects of PDEs inhibitors on mood and related cognitive disorders.

Table 3: The Overview of the Effects of PDE Inhibitors on Mood and Related Cognitive Disorders.

PDE Subtypes	Drugs	Memory Deficits Model	Behavioral Tests	Results	References
PDE1	Vinpocetine	Streptozotocin impaired rat	Morris water maze test	Vinpocetine improves memory consolidation.	[171]
PDE2	Bay60-7550	MK-801 impaired mouse Age-impaired rat.	Object recognition task; T-maze test Object recognition task	BAY60-7550 reverses MK801-induced memory deficits. BAY60-7550 improves acquisition and consolidation.	[175] [198]
PDE4	Rolipram MEM1018 MEM1091	Scopolamine impaired rat MK-801 impaired rat Young rhesus monkey Beta-amyloid impaired rat MK-801 impaired rat MK-801 impaired rat PDE4B knockout mouse.	Radial maze test Three-panel runway task; Radial arm maze test T-maze test Morris water maze test; Passive avoidance test Radial arm maze test; Step through inhibitory avoidance task; Radial arm maze test Step through inhibitory	Rolipram reverses Scopolamine-induced deficits in both working memory and reference memory. Rolipram reverses the working memory deficits caused by MK-801. Rolipram improves working memory in young rhesus monkeys but not in aged ones. Rolipram reverses memory deficits induced by Abeta-25-35. MEM1018 improves the	[199] [184,183] [200] [181] [184] [184] [201]

Table 3: contd....

			avoidance task Hole board test;Light-dark transition test;Open field test	memory processes. MEM1091 improves the memory processes. PDE4B knockout mouse decreases head-dips and time spent in head-dipping in the hole board test, reduces transitions and time on the light side in the light-dark transition test, and decreases initial exploration and rears in the open-field test.	
PDE5	Sildenafil (Viagra)	Muscarinic acetylcholine receptor antagonist impaired rat APP/PS1 transgenic mouse.	T-maze task Fear conditioning task; Morris water maze test	Sildenafil ameliorates memory deficits induced by a muscarinic acetylcholine receptor antagonist. Sildenafil rescues synaptic and memory (fear memory and working memory) deficits in an APP/PS1 transgenic mouse model of amyloid deposition.	[191] [202]
	Vardenafil	Tryptophan deleted rat.	Object recognition task	Vardenafil attenuates the acute tryptophan depletion induced object recognition impairment.	[176]
PDE9	Bay 73-6691 PF-04447943	Scopolamine impaired rat MK-801 impaired rat Scopolamine-induced deficit.	Passive avoidance test; T-maze test Object recognition task; Social recognition task Novel rodent attention task	BAY 73-6691 enhances acquisition, consolidation, and retention of long-term memory in a social recognition task and tends to enhance long-term memory in an object recognition task. PDE9 inhibitor PF-04447943 reversed memory deficit induced byscopolamine similar to donepezil.	[194] [203]
PDE10	PQ-10	PDE10A knockout mouse Scopolamine or MK-801.	Passive avoidance test Object recognition memory	PDE10A knockout mouse exhibits decreased exploratory activity and a significant delay in the acquisition of conditioned avoidance behavior when compared to wild-type mice. PDE10 inhibitor reverses object memory deficits induced by scopolamine or MK-801.	[195] [174]

Opioid System as the Drug Target

Opioid analgesics, commonly used in clinical practice, have been found to prevent neurodegeneration. The opioid receptors have been implicated in several clinical brain disorders, including addiction, epilepsy, Tourette's syndrome and AD including cognitive impairment [204], hyper-phosphorylated tau [205], Aβ production [206] and neuroinflammation [205] (Fig. **4**).

Opioid receptors regulate the neurotransmitters acetylcholine, norepinephrine, gamma aminobutyric acid (GABA), glutamate, and serotonin that have been implicated in the pathogenesis of AD [206]. Sigma-1 receptors, initially considered as subtypes of the opioid receptor family, are unique ligand-regulated molecular chaperones in the endoplasmatic reticulum. Sigma-1 receptors play modulatory roles in the activity of several neurotransmitter systems, particularly the cholinergic and glutamatergic pathways [207]. Sigma-1 agonists show powerful anti-amnesic and neuroprotective effects in a large variety of animal models of cognitive dysfunction affected by AD [208].

Opioid receptors are also linked to AD pathogenesis by affecting Aβ generation under *in vivo* and *in vitro* conditions [209]. Teng and his colleagues found that delta-opioid receptor (DOR) promotes the processing of APP by BACE1 and γ-secretase [210]. Further investigation revealed that DOR forms a complex with BACE1 and γ-secretase, and the activation of DOR mediates the co-endocytic sorting of thesecretases/receptor complex for APP endoproteolysis. Impairment of DOR retards the endocytosis of BACE1 and γ-secretase, and thus the production of Aβ. In another study, opioid-responsive human SH-SY5Y neuroblastoma cells were used to examine the action of morphine sulfate on microtubular tau protein [211]. This study demonstrated that morphine sulfate reduced tau protein in the cytoplasmic fraction of undifferentiated cells, and in the cytoplasmic as well as membrane fractions of differentiated cells. Although opioid system seems to be associated with hyper-phosphorylation of tau, research on this topic is very limited and the corresponding mechanism remains unclear. Some clinical trials have shown association of opioids with AD, but a competitive opiate receptor antagonist, naltrexone, appears not to be useful in AD-type dementia. However, the activation of the opioid system may provide, at least a part, an effective measure to interfere with the development of AD pathogenesis.

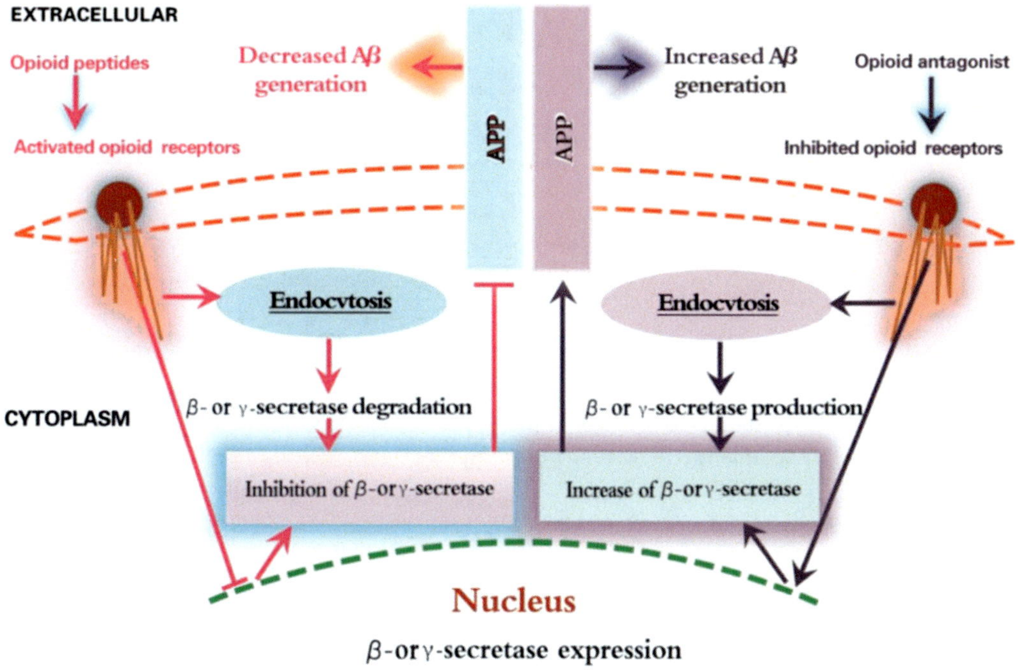

Figure 4: The possible mechanism of opioid system contribution to Aβ generation. Inhibition of opioid receptors retards the endocytosis of BACE1 and γ-secretase and up-regulates BACE1 and c-secretase, and the Aβ generation. Conversely, activation of opioid receptors increases the endocytosis of BACE1 and γ-secretase and down-regulates BACE1 and γ-secretase, limiting the production of Aβ.

CONCLUSIONS

The pathogenesis of Alzheimer's disease is complex with genetic and environmental factors involved. The high morbidity and mortality of AD are always the case attributed to lack of effective therapies and prevention strategies because many aspects of its pathogenesis remain unknown. Current therapies for treatment of AD only provide temporary improvement and reducing the rate of cognitive decline. The researches regarding to the effective components of the natural compounds and the pharmacokinetics of the target synthetic drugs are underway. Given that the wide array of available molecular targets and the rapid progress toward identifying appropriate compounds, it is anticipated that the new generation of drugs that would delay or prevent the progression and pathogenesis of AD will be developed in the near future.

ACKNOWLEDGEMENT

Declared none.

CONFLICT OF INTEREST

The author(s) confirm that this chapter content has no conflict of interest.

REFERENCES

[1] Lahiri DK, Maloney B, Basha MR, Ge YW, Zawia NH. How and when environmental agents and dietary factors affect the course of Alzheimer's disease: the "LEARn" model (latent early-life associated regulation) may explain the triggering of AD. Curr Alzheimer Res 2007; 4(2): 219-228.

[2] Jakob-Roetne R, Jacobsen H. Alzheimer's disease: from pathology to therapeutic approaches. Angew Chem Int Ed Engl 2009; 48(17): 3030-3059.

[3] Hardy J, Selkoe DJ. The amyloid hypothesis of Alzheimer's disease: progress and problems on the road to therapeutics. Science 2002; 297(5580): 353-356.

[4] Hardy JA, Higgins GA. Alzheimer's disease: the amyloid cascade hypothesis. Science 1992; 256(5054): 184-185.

[5] De Felice FG, Ferreira ST. Beta-amyloid production, aggregation, and clearance as targets for therapy in Alzheimer's disease. Cell Mol Neurobiol 2002; 22(5-6): 545-563.

[6] LeVine H, 3rd. The Amyloid Hypothesis and the clearance and degradation of Alzheimer's beta-peptide. J Alzheimers Dis 2004; 6(3): 303-314.

[7] Kurz A, Perneczky R. Amyloid clearance as a treatment target against Alzheimer's disease. J Alzheimers Dis 2011; 24(Suppl 2): 61-73.

[8] van Marum RJ. Current and future therapy in Alzheimer's disease. Fundam Clin Pharmacol 2008; 22(3): 265-274.

[9] Khatoon S, Grundke-Iqbal I, Iqbal K. Brain levels of microtubule-associated protein tau are elevated in Alzheimer's disease: a radioimmuno-slot-blot assay for nanograms of the protein. J Neurochem 1992; 59(2): 750-753.

[10] Lee VM, Trojanowski JQ. The disordered neuronal cytoskeleton in Alzheimer's disease. Curr Opin Neurobiol 1992; 2(5): 653-656.

[11] Wolfe MS. Tau mutations in neurodegenerative diseases. J Biol Chem 2009; 284(10): 6021-6025.

[12] Spillantini MG, Goedert M. Tau protein pathology in neurodegenerative diseases. Trends Neurosci 1998; 21(10): 428-433.

[13] Hong-Qi Y, Zhi-Kun S, Sheng-Di C. Current advances in the treatment of Alzheimer's disease: focused on considerations targeting Abeta and tau. Transl Neurodegener 2012; 1(1): 21.

[14] Pickhardt M, Gazova Z, von Bergen M, *et al.* Anthraquinones inhibit tau aggregation and dissolve Alzheimer's paired helical filaments *in vitro* and in cells. J Biol Chem 2005; 280(5): 3628-3635.

[15] De-Paula VJ, Radanovic M, Diniz BS, Forlenza OV. Alzheimer's disease. Subcell Biochem 2012; 65: 329-352.

[16] Christen Y. Oxidative stress and Alzheimer disease. Am J Clin Nutr 2000; 71(2): 621S-629S.

[17] Tuppo EE, Forman LJ. Free radical oxidative damage and Alzheimer's disease. J Am Osteopath Assoc 2001; 101 (12 Suppl Pt 1): S11-15.

[18] Francis PT, Ramirez MJ, Lai MK. Neurochemical basis for symptomatic treatment of Alzheimer's disease. Neuropharmacology 2010; 59(4-5): 221-229.

[19] Singh S, Kushwah AS, Singh R, Farswan M, Kaur R. Current therapeutic strategy in Alzheimer's disease. Eur Rev Med Pharmacol Sci 2012; 16(12): 1651-1664.

[20] Lipton SA. Pathologically-activated therapeutics for neuroprotection: mechanism of NMDA receptor block by memantine and S-nitrosylation. Curr Drug Targets 2007; 8(5): 621-632.

[21] Akiyama H, Barger S, Barnum S, *et al*. Inflammation and Alzheimer's disease. Neurobiol Aging 2000; 21(3):383-421.

[22] Amor S, Puentes F, Baker D, van der Valk P. Inflammation in neurodegenerative diseases. Immunology 2010; 129(2): 154-169.

[23] Lee YJ, Han SB, Nam SY, Oh KW, Hong JT. Inflammation and Alzheimer's disease. Arch Pharm Res 2010; 33(10): 1539-1556.

[24] Halliday G, Robinson SR, Shepherd C, Kril J. Alzheimer's disease and inflammation: a review of cellular and therapeutic mechanisms. Clin Exp Pharmacol Physiol 2000; 27(1-2): 1-8.

[25] Moreira PI, Santos MS, Oliveira CR, *et al*. Alzheimer disease and the role of free radicals in the pathogenesis of the disease. CNS Neurol Disord Drug Targets 2008; 7(1): 3-10.

[26] Galindo MF, Ikuta I, Zhu X, Casadesus G, Jordan J. Mitochondrial biology in Alzheimer's disease pathogenesis. J Neurochem 2010; 114(4): 933-945.

[27] Court JA, Perry EK. Dementia: the neurochemical basis of putative transmitter orientated therapy. Pharmacol Ther 1991; 52(3): 423-443.

[28] Abdel-Salam OM. Drugs used to treat Parkinson's disease, present status and future directions. CNS Neurol Disord-Dr 2008; 7(4): 321-342.

[29] Thachil AF, Mohan R, Bhugra D. The evidence base of complementary and alternative therapies in depression. J Affect Disord 2007; 97(1-3): 23-35.

[30] Tredici G, Miloso M, Nicolini G, Galbiati S, Cavaletti G, Bertelli A. Resveratrol, map kinases and neuronal cells: might wine be a neuroprotectant? Drugs Exp Clin Res 1999; 25(2-3): 99-103.

[31] Chen LW, Wang YQ, Wei LC, Shi M, Chan YS. Chinese herbs and herbal extracts for neuroprotection of dopaminergic neurons and potential therapeutic treatment of Parkinson's disease. CNS Neurol Disord Drug Targets 2007a; 6(4): 273-281.

[32] Bai Y, Mao QQ, Qin J, *et al*. Resveratrol induces apoptosis and cell cycle arrest of human T24 bladder cancer cells *in vitro* and inhibits tumor growth *in vivo*. Cancer Sci 2010; 101(2): 488-493.

[33] Kumar A, Naidu PS, Seghal N, Padi SS. Neuroprotective effects of resveratrol against intracerebroventricular colchicine-induced cognitive impairment and oxidative stress in rats. Pharmacology 2007; 79(1): 17-26.

[34] Ranney A, Petro MS. Resveratrol protects spatial learning in middle-aged C57BL/6 mice from effects of ethanol. Behav Pharmacol 2009; 20(4): 330-336.

[35] Parker JA, Arango M, Abderrahmane S, *et al*. Resveratrol rescues mutant polyglutamine cytotoxicity in nematode and mammalian neurons. Nat Genet 2005; 37(4): 349-350.

[36] Albani D, Polito L, Batelli S, *et al.* The SIRT1 activator resveratrol protects SK-N-BE cells from oxidative stress and against toxicity caused by alpha-synuclein or amyloid-beta (1-42) peptide. J Neurochem 2009; 110(5): 1445-1456.

[37] Mazzio EA, Harris N, Soliman KF. Food constituents attenuate monoamine oxidase activity and peroxide levels in C6 astrocyte cells. Planta Med 1998; 64(7): 603-606.

[38] Albani D, Polito L, Signorini A, Forloni G. Neuroprotective properties of resveratrol in different neurodegenerative disorders. Biofactors 2010; 36(5): 370-376.

[39] Jang JH, Surh YJ. Protective effect of resveratrol on beta-amyloid-induced oxidative PC12 cell death. Free Radic Biol Med 2003; 34(8): 1100-1110.

[40] Wang J, Ho L, Zhao Z, *et al.* Moderate consumption of Cabernet Sauvignon attenuates Abeta neuropathology in a mouse model of Alzheimer's disease. FASEB J 2006; 20(13): 2313-2320.

[41] Marambaud P, Zhao H, Davies P. Resveratrol promotes clearance of Alzheimer's disease amyloid-beta peptides. J Biol Chem 2005; 280(45): 37377-37382.

[42] Riviere C, Richard T, Quentin L, Krisa S, Merillon JM, Monti JP. Inhibitory activity of stilbenes on Alzheimer's beta-amyloid fibrils *in vitro*. Bioorg Med Chem 2007; 15(2): 1160-1167.

[43] Richard T, Pawlus AD, Iglesias ML, *et al.* Neuroprotective properties of resveratrol and derivatives. Ann N Y Acad Sci 2011; 1215: 103-108.

[44] Ladiwala AR, Lin JC, Bale SS, *et al.* Resveratrol selectively remodels soluble oligomers and fibrils of amyloid Abeta into off-pathway conformers. J Biol Chem 2010; 285(31): 24228-24237.

[45] Caruso F, Tanski J, Villegas-Estrada A, Rossi M. Structural basis for antioxidant activity of trans-resveratrol: ab initio calculations and crystal and molecular structure. J Agric Food Chem 2004; 52(24): 7279-7285.

[46] Baur JA, Sinclair DA. Therapeutic potential of resveratrol: the *in vivo* evidence. Nat Rev Drug Discov 2006; 5(6): 493-506.

[47] Zini R, Morin C, Bertelli A, Bertelli AA, Tillement JP. Effects of resveratrol on the rat brain respiratory chain. Drugs Exp Clin Res 1999; 25(2-3): 87-97.

[48] Martinez J, Moreno JJ. Effect of resveratrol, a natural polyphenolic compound, on reactive oxygen species and prostaglandin production. Biochem Pharmacol 2000; 59(7): 865-870.

[49] Leonard SS, Xia C, Jiang BH, *et al.* Resveratrol scavenges reactive oxygen species and effects radical-induced cellular responses. Biochem Biophys Res Commun 2003; 309(4): 1017-1026.

[50] Losa GA. Resveratrol modulates apoptosis and oxidation in human blood mononuclear cells. Eur J Clin Invest 2003; 33(9): 818-823.

[51] Mokni M, Elkahoui S, Limam F, Amri M, Aouani E. Effect of resveratrol on antioxidant enzyme activities in the brain of healthy rat. Neurochem Res 2007; 32(6): 981-987.

[52] Capiralla H, Vingtdeux V, Zhao H, *et al.* Resveratrol mitigates lipopolysaccharide- and Abeta-mediated microglial inflammation by inhibiting the TLR4/NF-kappaB/STAT signaling cascade. J Neurochem 2012; 120(3): 461-472.

[53] Han YS, Zheng WH, Bastianetto S, Chabot JG, Quirion R. Neuroprotective effects of resveratrol against beta-amyloid-induced neurotoxicity in rat hippocampal neurons: involvement of protein kinase C. Br J Pharmacol 2004; 141(6): 997-1005.

[54] Miloso M, Bertelli AA, Nicolini G, Tredici G,. Resveratrol-induced activation of the mitogen-activated protein kinases, ERK1 and ERK2, in human neuroblastoma SH-SY5Y cells. Neurosci Lett 1999; 264(1-3): 141-144.

[55] Ramassamy C. Emerging role of polyphenolic compounds in the treatment of neurodegenerative diseases: a review of their intracellular targets. Eur J Pharmacol 2006; 545(1): 51-64.

[56] Yeung F, Hoberg JE, Ramsey CS, *et al.* Modulation of NF-kappaB-dependent transcription and cell survival by the SIRT1 deacetylase. EMBO J 2004; 23(12): 2369-2380.

[57] Chen J, Zhou Y, Mueller-Steiner S, *et al.* SIRT1 protects against microglia-dependent amyloid-beta toxicity through inhibiting NF-kappaB signaling. J Biol Chem 2005; 280(48): 40364-40374.

[58] Park SJ, Ahmad F, Philp A, *et al.* Resveratrol ameliorates aging-related metabolic phenotypes by inhibiting cAMP phosphodiesterases. Cell 2012; 148(3): 421-433.

[59] Kong LD, Cheng CHK, Tan RX. Monoamine oxidase inhibitiors from rhizoma of Coptis chinensis. Planta Med 2001; 67(1): 74–6.

[60] Mehta K, Pantazis P, McQueen T, Aggarwal BB. Antiproliferative effect of curcumin (diferuloylmethane) against human breast tumor cell lines. Anticancer Drugs 1997; 8(5): 470-481.

[61] Motterlini R, Foresti R, Bassi R, Green CJ. Curcumin, an antioxidant and anti-inflammatory agent, induces heme oxygenase-1 and protects endothelial cells against oxidative stress. Free Radic Biol Med 2000; 28(8): 1303-1312.

[62] Thiyagarajan M, Sharma SS. Neuroprotective effect of curcumin in middle cerebral artery occlusion induced focal cerebral ischemia in rats. Life Sci 2004; 74(8): 969-985.

[63] Ganguli M, Chandra V, Kamboh MI, *et al.* Apolipoprotein E polymorphism and Alzheimer disease: The Indo-US Cross-National Dementia Study. Arch Neurol 2000; 57(6): 824-830.

[64] Matteucci A, Frank C, Domenici MR, *et al.* Curcumin treatment protects rat retinal neurons against excitotoxicity: effect on N-methyl-D: -aspartate-induced intracellular Ca(2+) increase. Exp Brain Res 2005; 167(4): 641-648.

[65] Chen J, Tang XQ, Zhi JL, *et al.* Curcumin protects PC12 cells against 1-methyl-4-phenylpyridinium ion-induced apoptosis by bcl-2-mitochondria-ROS-iNOS pathway. Apoptosis 2006; 11(6): 943-953.

[66] Vajragupta O, Boonchoong P, Watanabe H, Tohda M, Kummasud N, Sumanont Y. Manganese complexes of curcumin and its derivatives: evaluation for the radical scavenging ability and neuroprotective activity. Free Radic Biol Med 2003; 35(12): 1632-1644.

[67] Lim GP, Chu T, Yang F, Beech W, Frautschy SA, Cole GM. The curry spice curcumin reduces oxidative damage and amyloid pathology in an Alzheimer transgenic mouse. J Neurosci 2001; 21(21): 8370-8377.

[68] Frautschy SA, Hu W, Kim P, *et al.* Phenolic anti-inflammatory antioxidant reversal of Abeta-induced cognitive deficits and neuropathology. Neurobiol Aging 2001; 22(6): 993-1005.

[69] Klein R, Nanduri V, Jing SA, *et al.* The trkB tyrosine protein kinase is a receptor for brain-derived neurotrophic factor and neurotrophin-3. Cell 1991; 66(2): 395-403.

[70] Manji HK, Chen G. PKC, MAP kinases and the bcl-2 family of proteins as long-term targets for mood stabilizers. Mol Psychiatry 2002; 7(Suppl 1): S46-56.

[71] Chen G, Manji HK. The extracellular signal-regulated kinase pathway: an emerging promising target for mood stabilizers. Curr Opin Psychiatry 2006; 19(3): 313-323.

[72] Bonni A, Brunet A, West AE, Datta SR, Takasu MA, Greenberg ME. Cell survival promoted by the Ras-MAPK signaling pathway by transcription-dependent and -independent mechanisms. Science 1999; 286(5443): 1358-1362.

[73] Myers KM, Fiskum G, Liu Y, Simmens SJ, Bredesen DE, Murphy AN. Bcl-2 protects neural cells from cyanide/aglycemia-induced lipid oxidation, mitochondrial injury, and loss of viability. J Neurochem 1995; 65(6): 2432-2440.
[74] Hayley S, Poulter MO, Merali Z, Anisman H. The pathogenesis of clinical depression: stressor- and cytokine-induced alterations of neuroplasticity. Neuroscience 2005; 135(3): 659-678.
[75] Finkbeiner S. Calcium regulation of the brain-derived neurotrophic factor gene. Cell Mol Life Sci 2000; 57(3): 394-401.
[76] Encinas M, Iglesias M, Llecha N, Comella JX. Extracellular-regulated kinases and phosphatidylinositol 3-kinase are involved in brain-derived neurotrophic factor-mediated survival and neuritogenesis of the neuroblastoma cell line SH-SY5Y. J Neurochem 1999; 73(4):1409-1421.
[77] Chen MJ, Nguyen TV, Pike CJ, Russo-Neustadt AA. Norepinephrine induces BDNF and activates the PI-3K and MAPK cascades in embryonic hippocampal neurons. Cell Signal 2007b; 19(1): 114-128.
[78] Leeds P, Leng Y, Chalecka-Franaszek E, Chuang DM. Neurotrophins protect against cytosine arabinoside-induced apoptosis of immature rat cerebellar neurons. Neurochem Int 2005; 46(1): 61-72.
[79] Andjelkovic M, Alessi DR, Meier R, *et al.* Role of translocation in the activation and function of protein kinase B. J Biol Chem 1997; 272(50): 31515-31524.
[80] Wang R, Li YH, Xu Y, *et al.* Curcumin produces neuroprotective effects *via* activating brain-derived neurotrophic factor/TrkB-dependent MAPK and PI-3K cascades in rodent cortical neurons. Prog Neuropsychopharmacol Biol Psychiatry 2010; 34(1): 147-153.
[81] Skorski T, Bellacosa A, Nieborowska-Skorska M, *et al.* Transformation of hematopoietic cells by BCR/ABL requires activation of a PI-3k/Akt-dependent pathway. EMBO J 1997; 16(20): 6151-6161.
[82] Sabbatini P, McCormick F. Phosphoinositide 3-OH kinase (PI3K) and PKB/Akt delay the onset of p53-mediated, transcriptionally dependent apoptosis. J Biol Chem 1999; 274(34): 24263-24269.
[83] Nibuya M, Morinobu S, Duman RS. Regulation of BDNF and trkB mRNA in rat brain by chronic electroconvulsive seizure and antidepressant drug treatments. J Neurosci 1995; 15(11): 7539-7547.
[84] Koch JM, Kell S, Aldenhoff JB. Differential effects of fluoxetine and imipramine on the phosphorylation of the transcription factor CREB and cell-viability. J Psychiatr Res 2003; 37(1): 53-59.
[85] Riccio A, Ahn S, Davenport CM, Blendy JA, Ginty DD. Mediation by a CREB family transcription factor of NGF-dependent survival of sympathetic neurons. Science 1999; 286(5448): 2358-2361.
[86] Xu Y, Ku B, Tie L, *et al.* Curcumin reverses the effects of chronic stress on behavior, the HPA axis, BDNF expression and phosphorylation of CREB. Brain Res 2006; 1122(1): 56-64.
[87] Baum L, Ng A. Curcumin interaction with copper and iron suggests one possible mechanism of action in Alzheimer's disease animal models. J Alzheimers Dis 2004; 6(4):367-377.
[88] Yang F, de Villiers WJ, McClain CJ, Varilek GW. Green tea polyphenols block endotoxin-induced tumor necrosis factor-production and lethality in a murine model. J Nutr 1998; 128(12): 2334-2340.

[89] Hollman PC, Feskens EJ, Katan MB. Tea flavonols in cardiovascular disease and cancer epidemiology. Proc Soc Exp Biol Med 1999; 220(4): 198-202.

[90] Checkoway H, Powers K, Smith-Weller T, *et al*. Parkinson's disease risks associated with cigarette smoking, alcohol consumption, and caffeine intake. Am J Epidemiol 2002; 155(8): 732-738.

[91] Levites Y, Amit T, Mandel S, Youdim MB. Neuroprotection and neurorescue against Abeta toxicity and PKC-dependent release of nonamyloidogenic soluble precursor protein by green tea polyphenol (-)-epigallocatechin-3-gallate. FASEB J 2003; 17(8): 952-954.

[92] Graham HN. Green tea composition, consumption, and polyphenol chemistry. Prev Med 1992; 21(3): 334-350.

[93] Moyers SB, Kumar NB. Green tea polyphenols and cancer chemoprevention: multiple mechanisms and endpoints for phase II trials. Nutr Rev 2004; 62(5): 204-211.

[94] Ehrnhoefer DE, Bieschke J, Boeddrich A, *et al*. EGCG redirects amyloidogenic polypeptides into unstructured, off-pathway oligomers. Nat Struct Mol Biol 2008; 15(6): 558-566.

[95] Hudson SA, Ecroyd H, Dehle FC, Musgrave IF, Carver JA. (-)-epigallocatechin-3-gallate (EGCG) maintains kappa-casein in its pre-fibrillar state without redirecting its aggregation pathway. J Mol Biol 2009; 392(3): 689-700.

[96] Ferreira N, Saraiva MJ, Almeida MR. Natural polyphenols inhibit different steps of the process of transthyretin (TTR) amyloid fibril formation. FEBS Lett 2011; 585(15): 2424-2430.

[97] Chandrashekaran IR, Adda CG, MacRaild CA, Anders RF, Norton RS. Inhibition by flavonoids of amyloid-like fibril formation by Plasmodium falciparum merozoite surface protein 2. Biochemistry 2010; 49(28): 5899-5908.

[98] Bieschke J, Russ J, Friedrich RP, *et al*. EGCG remodels mature alpha-synuclein and amyloid-beta fibrils and reduces cellular toxicity. Proc Natl Acad Sci USA 2010; 107(17): 7710-7715.

[99] Rambold AS, Miesbauer M, Olschewski D, *et al*. Green tea extracts interfere with the stress-protective activity of PrP and the formation of PrP. J Neurochem 2008; 107(1): 218-229.

[100] Hauber I, Hohenberg H, Holstermann B, Hunstein W, Hauber J. The main green tea polyphenol epigallocatechin-3-gallate counteracts semen-mediated enhancement of HIV infection. Proc Natl Acad Sci USA 2009; 106(22): 9033-9038.

[101] Ehrnhoefer DE, Duennwald M, Markovic P, *et al*. Green tea (-)-epigallocatechin-gallate modulates early events in huntingtin misfolding and reduces toxicity in Huntington's disease models. Hum Mol Genet 2006; 15(18): 2743-2751.

[102] Rezai-Zadeh K, Shytle D, Sun N, *et al*. Green tea epigallocatechin-3-gallate (EGCG) modulates amyloid precursor protein cleavage and reduces cerebral amyloidosis in Alzheimer transgenic mice. J Neurosci 2005; 25(38): 8807-8814.

[103] Jeon SY, Bae K, Seong YH, Song KS. Green tea catechins as a BACE1 (beta-secretase) inhibitor. Bioorg Med Chem Lett 2003; 13(22): 3905-3908.

[104] Mandel S, Reznichenko L, Amit T, Youdim MB. Green tea polyphenol (-)-epigallocatechin-3-gallate protects rat PC12 cells from apoptosis induced by serum withdrawal independent of P13-Akt pathway. Neurotox Res 2003; 5(6): 419-424.

[105] Schroeder EK, Kelsey NA, Doyle J, *et al*. Green tea epigallocatechin 3-gallate accumulates in mitochondria and displays a selective antiapoptotic effect against inducers of mitochondrial oxidative stress in neurons. Antioxid Redox Signal 2009; 11(3): 469-480.

[106] Kang KS, Wen Y, Yamabe N, Fukui M, Bishop SC, Zhu BT. Dual beneficial effects of (-)-epigallocatechin-3-gallate on levodopa methylation and hippocampal neurodegeneration: *in vitro* and *in vivo* studies. PLoS One 2010; 5(8): e11951.

[107] Smid SD, Maag JL, Musgrave IF. Dietary polyphenol-derived protection against neurotoxic beta-amyloid protein: from molecular to clinical. Food Funct 2012; 3(12): 1242-1250.

[108] Meinhardt G, Roth J, Totok G. Protein kinase C activation modulates pro- and anti-apoptotic signaling pathways. Eur J Cell Biol 2000; 79(11): 824-833.

[109] Crumrine RC, Dubyak G, LaManna JC. Decreased protein kinase C activity during cerebral ischemia and after reperfusion in the adult rat. J Neurochem 1990; 55(6): 2001-2007.

[110] Nishino N, Kitamura N, Nakai T, Hashimoto T, Tanaka C. Phorbol ester binding sites in human brain: characterization, regional distribution, age-correlation, and alterations in Parkinson's disease. J Mol Neurosci 1989; 1(1): 19-26.

[111] Levites Y, Amit T, Youdim MB, Mandel S. Involvement of protein kinase C activation and cell survival/ cell cycle genes in green tea polyphenol (-)-epigallocatechin 3-gallate neuroprotective action. J Biol Chem 2002; 277(34): 30574-30580.

[112] Lee SY, Lee JW, Lee H, *et al.* Inhibitory effect of green tea extract on beta-amyloid-induced PC12 cell death by inhibition of the activation of NF-kappaB and ERK/p38 MAP kinase pathway through antioxidant mechanisms. Brain Res Mol Brain Res 2005; 140(1-2): 45-54.

[113] Xia Z, Dickens M, Raingeaud J, Davis RJ, Greenberg ME. Opposing effects of ERK and JNK-p38 MAP kinases on apoptosis. Science 1995; 270(5240): 1326-1331.

[114] Owuor ED, Kong AN. Antioxidants and oxidants regulated signal transduction pathways. Biochem Pharmacol 2002; 64(5-6): 765-770.

[115] Howes MJ, Perry E. The role of phytochemicals in the treatment and prevention of dementia. Drugs Aging 2011; 28(6): 439-468.

[116] Gertz HJ, Kiefer M. Review about Ginkgo biloba special extract EGb 761 (Ginkgo). Curr Pharm Des 2004; 10(3): 261-264.

[117] Chan PC, Xia Q, Fu PP. Ginkgo biloba leave extract: biological, medicinal, and toxicological effects. J Environ Sci Health C Environ Carcinog Ecotoxicol Rev 2007; 25(3): 211-244.

[118] Das A, Shanker G, Nath C, Pal R, Singh S, Singh H. A comparative study in rodents of standardized extracts of Bacopa monniera and Ginkgo biloba: anticholinesterase and cognitive enhancing activities. Pharmacol Biochem Behav 2002; 73(4): 893-900.

[119] Kaschel R. Ginkgo biloba: specificity of neuropsychological improvement--a selective review in search of differential effects. Hum Psychopharmacol 2009; 24(5): 345-370.

[120] Yancheva S, Ihl R, Nikolova G, Panayotov P, Schlaefke S, Hoerr R. Ginkgo biloba extract EGb 761(R), donepezil or both combined in the treatment of Alzheimer's disease with neuropsychiatric features: a randomised, double-blind, exploratory trial. Aging Ment Health 2009; 13(2): 183-190.

[121] Andrieu S, Gillette S, Amouyal K, *et al.* Association of Alzheimer's disease onset with ginkgo biloba and other symptomatic cognitive treatments in a population of women aged 75 years and older from the EPIDOS study. J Gerontol A Biol Sci Med Sci 2003; 58(4): 372-377.

[122] Leistner E, Drewke C. Ginkgo biloba and ginkgotoxin. J Nat Prod 2010; 73(1): 86-92.

[123] Ramassamy C, Clostre F, Christen Y, Costentin J. Prevention by a Ginkgo biloba extract (GBE 761) of the dopaminergic neurotoxicity of MPTP. J Pharm Pharmacol 1990; 42(11): 785-789.

[124] Bastianetto S, Ramassamy C, Dore S, Christen Y, Poirier J, Quirion R. The Ginkgo biloba extract (EGb 761) protects hippocampal neurons against cell death induced by beta-amyloid. Eur J Neurosci 2000; 12(6): 1882-1890.
[125] Yao Z, Drieu K, Papadopoulos V. The Ginkgo biloba extract EGb 761 rescues the PC12 neuronal cells from beta-amyloid-induced cell death by inhibiting the formation of beta-amyloid-derived diffusible neurotoxic ligands. Brain Res 2001; 889(1-2): 181-190.
[126] Smith JV, Luo Y. Elevation of oxidative free radicals in Alzheimer's disease models can be attenuated by Ginkgo biloba extract EGb 761. J Alzheimers Dis 2003; 5(4): 287-300.
[127] Colciaghi F, Borroni B, Zimmermann M, *et al.* Amyloid precursor protein metabolism is regulated toward alpha-secretase pathway by Ginkgo biloba extracts. Neurobiol Dis 2004; 16(2): 454-460.
[128] Yao ZX, Han Z, Drieu K, Papadopoulos V. Ginkgo biloba extract (Egb 761) inhibits beta-amyloid production by lowering free cholesterol levels. J Nutr Biochem 2004; 15(12): 749-756.
[129] Pike CJ, Walencewicz AJ, Glabe CG, Cotman CW. *In vitro* aging of beta-amyloid protein causes peptide aggregation and neurotoxicity. Brain Res 1991; 563(1-2): 311-314.
[130] Simmons LK, May PC, Tomaselli KJ, *et al.* Secondary structure of amyloid beta peptide correlates with neurotoxic activity *in vitro*. Mol Pharmacol 1994; 45(3): 373-379.
[131] Soto C, Sigurdsson EM, Morelli L, Kumar RA, Castano EM, Frangione B. Beta-sheet breaker peptides inhibit fibrillogenesis in a rat brain model of amyloidosis: implications for Alzheimer's therapy. Nat Med 1998; 4(7): 822-826.
[132] Luo Y, Smith JV, Paramasivam V, *et al.* Inhibition of amyloid-beta aggregation and caspase-3 activation by the Ginkgo biloba extract EGb761. Proc Natl Acad Sci U S A 2002; 99(19): 12197-12202.
[133] Wollen KA. Alzheimer's disease: the pros and cons of pharmaceutical, nutritional, botanical, and stimulatory therapies, with a discussion of treatment strategies from the perspective of patients and practitioners. Altern Med Rev 2010; 15(3): 223-244.
[134] Lee MS, Yang EJ, Kim JI, Ernst E. Ginseng for cognitive function in Alzheimer's disease: a systematic review. J Alzheimers Dis 2009; 18(2): 339-344.
[135] Radad K, Gille G, Liu L, Rausch WD. Use of ginseng in medicine with emphasis on neurodegenerative disorders. J Pharmacol Sci 2006; 100(3): 175-186.
[136] Nah SY, Kim DH, Rhim H. Ginsenosides: are any of them candidates for drugs acting on the central nervous system? CNS Drug Rev 2007; 13(4): 381-404.
[137] Ho YS, So KF, Chang RC. Anti-aging herbal medicine--how and why can they be used in aging-associated neurodegenerative diseases? Ageing Res Rev 2010; 9(3): 354-362.
[138] Lee B, Park J, Kwon S, *et al.* Effect of wild ginseng on scopolamine-induced acetylcholine depletion in the rat hippocampus. J Pharm Pharmacol 2010; 62(2): 263-271.
[139] Wang YH, Du GH. Ginsenoside Rgl inhibits beta-secretase activity *in vitro* and protects against Abeta-induced cytotoxicity in PC12 cells. J Asian Nat Prod Res 2009; 11(7): 604-612.
[140] Fang F, Chen X, Huang T, Lue LF, Luddy JS, Yan SS. Multi-faced neuroprotective effects of Ginsenoside Rgl in an Alzheimer mouse model. Biochim Biophys Acta 2012; 1822(2): 286-292.
[141] Shi YQ, Huang TW, Chen LM, *et al.* Ginsenoside Rgl attenuates amyloid-beta content, regulates PKA/CREB activity, and improves cognitive performance in SAMP8 mice. J Alzheimers Dis 2010; 19(3): 977-989.

[142] Chen LM, Lin ZY, Zhu YG, *et al.* Ginsenoside Rg1 attenuates beta-amyloid generation *via* suppressing PPARgamma-regulated BACE1 activity in N2a-APP695 cells. Eur J Pharmacol 2012; 675(1-3): 15-21.

[143] Yang L, Hao J, Zhang J, *et al.* Ginsenoside Rg3 promotes beta-amyloid peptide degradation by enhancing gene expression of neprilysin. J Pharm Pharmacol 2009; 61(3): 375-380.

[144] Kennedy DO, Haskell CF, Wesnes KA, Scholey AB. Improved cognitive performance in human volunteers following administration of guarana (Paullinia cupana) extract: comparison and interaction with Panax ginseng. Pharmacol Biochem Behav 2004; 79(3): 401-411.

[145] Howes MJ, Houghton PJ. Plants used in Chinese and Indian traditional medicine for improvement of memory and cognitive function. Pharmacol Biochem Behav 2003; 75(3): 513-527.

[146] Howes MJ, Perry NS, Houghton PJ. Plants with traditional uses and activities, relevant to the management of Alzheimer's disease and other cognitive disorders. Phytother Res 2003; 17(1): 1-18.

[147] Lehmann O, Jeltsch H, Lehnardt O, Pain L, Lazarus C, Cassel JC. Combined lesions of cholinergic and serotonergic neurons in the rat brain using 192 IgG-saporin and 5,7-dihydroxytryptamine: neurochemical and behavioural characterization. Eur J Neurosci 2000; 12(1): 67-79.

[148] Shi C, Zheng DD, Fang L, Wu F, Kwong WH, Xu J. Ginsenoside Rg1 promotes nonamyloidgenic cleavage of APP *via* estrogen receptor signaling to MAPK/ERK and PI3K/Akt. Biochim Biophys Acta 2012; 1820(4): 453-460.

[149] Zhao R, Zhang Z, Song Y, Wang D, Qi J, Wen S. Implication of phosphatidylinositol-3 kinase/Akt/glycogen synthase kinase-3beta pathway in ginsenoside Rb1's attenuation of beta-amyloid-induced neurotoxicity and tau phosphorylation. J Ethnopharmacol 2011; 133(3): 1109-1116.

[150] Wang Y, Liu J, Zhang Z, Bi P, Qi Z, Zhang C. Anti-neuroinflammation effect of ginsenoside Rbl in a rat model of Alzheimer disease. Neurosci Lett 2011; 487(1): 70-72.

[151] Rafii MS, Walsh S, Little JT, *et al.* Alzheimer's Disease Cooperative Study. A phase II trial of huperzine A in mild to moderate Alzheimer disease. Neurology 2011; 76(16): 1389-94.

[152] Pohanka M. Acetylcholinesterase inhibitors: a patent review (2008 - present). Expert Opin Ther Pat 2012; 22(8): 871-86.

[153] Howland RH. Drug therapies for cognitive impairment and dementia. J Psychosoc Nurs Ment Health Serv 2010; 48(4): 11-4.

[154] Xiong H, McCabe L, Costello J, Anderson E, Weber G, Ikezu T. Activation of NR1a/NR2B receptors by soluble factors from APP-stimulated monocyte-derived macrophages: implications for the pathogenesis of Alzheimer's disease. Neurobiol Aging 2004; 25(7): 905-11.

[155] Farrimond LE, Roberts E, McShane R. Memantine and cholinesterase inhibitor combination therapy for Alzheimer's disease: a systematic review. BMJ Open 2012; 2(3): [about 8p]. Available from: http://bmjopen.bmj.com/content/2/3/e000917.short

[156] Golde TE, Petrucelli L, Lewis J. Targeting Abeta and tau in Alzheimer's disease, an early interim report. Exp Neurol 2010; 223(2): 252-66.

[157] Albert JS. Progress in the development of beta-secretase inhibitors for Alzheimer's disease. Prog Med Chem 2009; 48: 133-61.

[158] Fleisher AS, Raman R, Siemers ER, *et al.* Phase 2 safety trial targeting amyloid beta production with a gamma-secretase inhibitor in Alzheimer disease. Arch Neurol 2008; 65(8): 1031-8.

[159] Gervais F, Chalifour R, Garceau D, *et al.* Glycosaminoglycan mimetics: a therapeutic approach to cerebral amyloid angiopathy. Amyloid 2001; 8 (Suppl 1): 28-35.

[160] Aisen PS, Gauthier S, Ferris SH, *et al.* Tramiprosate in mild-to-moderate Alzheimer's disease - a randomized, double-blind, placebo-controlled, multi-centre study (the Alphase Study). Arch Med Sci 2011; 7(1): 102-11.

[161] Gauthier S, Aisen PS, Ferris SH, *et al.* Effect of tramiprosate in patients with mild-to-moderate Alzheimer's disease: exploratory analyses of the MRI sub-group of the Alphase study. J Nutr Health Aging 2009; 13(6): 550-7.

[162] Licastro F, Caruso C. Is immunotherapy an effective treatment for Alzheimer's disease? Immun Ageing 2004; 1(1): 3.

[163] Noble W, Planel E, Zehr C, *et al.* Inhibition of glycogen synthase kinase-3 by lithium correlates with reduced tauopathy and degeneration *in vivo*. Proc Natl Acad Sci U S A 2005; 102(19): 6990-5.

[164] Fuentealba RA, Farias G, Scheu J, Bronfman M, Marzolo MP, Inestrosa NC. Signal transduction during amyloid-beta-peptide neurotoxicity: role in Alzheimer disease. Brain Res Brain Res Rev 2004; 47(1-3): 275-89.

[165] Serenó L, Coma M, Rodríguez M, *et al.* A novel GSK-3beta inhibitor reduces Alzheimer's pathology and rescues neuronal loss *in vivo*. Neurobiol Dis 2009; 35(3): 359-67.

[166] Beurel E. Regulation by glycogen synthase kinase-3 of inflammation and T cells in CNS diseases. Front Mol Neurosci 2011; 4: 18.

[167] Miller CL, Oikawa M, Cai Y, *et al.* Role of Ca2+/calmodulin-stimulated cyclic nucleotide phosphodiesterase 1 in mediating cardiomyocyte hypertrophy. Circ Res 2009; 105(10): 956-64.

[168] Dunkern TR, Hatzelmann A. Characterization of inhibitors of phosphodiesterase 1C on a human cellular system. FEBS J 2007; 274(18): 4812-24.

[169] Kruuse C, Rybalkin SD, Khurana TS, *et al.* The role of cGMP hydrolysing phosphodiesterases 1 and 5 in cerebral artery dilatation. Eur J Pharmacol 2001; 420(1): 55-65.

[170] Molnár P, Gaál L. Effect of different subtypes of cognition enhancers on long-term potentiation in the rat dentate gyrus *in vivo*. Eur J Pharmacol 1992; 215(1): 17-22.

[171] Deshmukh R, Sharma V, Mehan S, Sharma N, Bedi KL. Amelioration of intracerebroventricular streptozotocin induced cognitive dysfunction and oxidative stress by vinpocetine -- a PDE1 inhibitor. Eur J Pharmacol 2009; 620(1-3): 49-56.

[172] Filgueiras CC, Krahe TE, Medina AE. Phosphodiesterase type 1 inhibition improves learning in rats exposed to alcohol during the third trimester equivalent of human gestation. Neurosci Lett 2010; 473(3): 202-7.

[173] Hajjhussein H, Suvarna NU, Gremillion C, Chandler LJ, O'Donnell JM. Changes in NMDA receptor-induced cyclic nucleotide synthesis regulate the age-dependent increase in PDE4A expression in primary cortical cultures. Brain Res 2007; 1149: 58-68.

[174] Reneerkens OA, Rutten K, Bollen E, *et al.* Inhibition of phoshodiesterase type 2 or type 10 reverses object memory deficits induced by scopolamine or MK-801. Behav Brain Res 2013; 236(1): 16-22.

[175] Boess FG, Hendrix M, van der Staay FJ, *et al.* Inhibition of phosphodiesterase 2 increases neuronal cGMP, synaptic plasticity and memory performance. Neuropharmacology 2004; 47(7): 1081-92.

[176] van Donkelaar EL, Rutten K, Blokland A, Akkerman S, Steinbusch HW, Prickaerts J. Phosphodiesterase 2 and 5 inhibition attenuates the object memory deficit induced by acute tryptophan depletion. Eur J Pharmacol 2008; 600(1-3): 98-104.

[177] Domek-Łopacińska K, Strosznajder JB. The effect of selective inhibition of cyclic GMP hydrolyzing phosphodiesterases 2 and 5 on learning and memory processes and nitric oxide synthase activity in brain during aging. Brain Res 2008; 1216: 68-77.

[178] Masood A, Nadeem A, Mustafa SJ, O'Donnell JM. Reversal of oxidative stress-induced anxiety by inhibition of phosphodiesterase-2 in mice. J Pharmacol Exp Ther 2008; 326(2): 369-79.

[179] Masood A, Huang Y, Hajjhussein H, *et al.* Anxiolytic effects of phosphodiesterase-2 inhibitors associated with increased cGMP signaling. J Pharmacol Exp Ther 2009; 331(2): 690-9.

[180] Zhao J, Harada N, Kurihara H, Nakagata N, Okajima K. Cilostazol improves cognitive function in mice by increasing the production of insulin-like growth factor-I in the hippocampus. Neuropharmacology 2010; 58(4-5): 774-83.

[181] Cheng YF, Wang C, Lin HB, *et al.* Inhibition of phosphodiesterase-4 reverses memory deficits produced by Aβ25-35 or Aβ1-40 peptide in rats. Psychopharmacology 2010; 212(2): 181-91.

[182] Gong B, Vitolo OV, Trinchese F, Liu S, Shelanski M, Arancio O. Persistent improvement in synaptic and cognitive functions in an Alzheimer mouse model after rolipram treatment. J Clin Invest 2004; 114(11): 1624-34.

[183] Reneerkens OA, Rutten K, Steinbusch HW, Blokland A, Prickaerts J. Selective phosphodiesterase inhibitors: a promising target for cognition enhancement. Psychopharmacology (Berl) 2009; 202(1-3): 419-43.

[184] Zhang HT, Huang Y, Suvarna NU, *et al.* Effects of the novel PDE4 inhibitors MEM1018 and MEM1091 on memory in the radial-arm maze and inhibitory avoidance tests in rats. Psychopharmacology 2005; 179(3): 613-9.

[185] Suvarna NU, O'Donnell JM. Hydrolysis of N-methyl-D-aspartate receptor-stimulated cAMP and cGMP by PDE4 and PDE2 phosphodiesterases in primary neuronal cultures of rat cerebral cortex and hippocampus. J Pharmacol Exp Ther 2002; 302(1): 249-56.

[186] Rutten K, Misner DL, Works M, *et al.* Enhanced long-term potentiation and impaired learning in phosphodiesterase 4D-knockout (PDE4D) mice. Eur J Neurosci 2008; 28(3): 625-32.

[187] Paul C, Schöberl F, Weinmeister P, *et al.* Signaling through cGMP-dependent protein kinase I in the amygdala is critical for auditory-cued fear memory and long-term potentiation. J Neurosci 2008; 28(52): 14202-12.

[188] Rutten K, Vente JD, Sik A, Ittersum MM, Prickaerts J, Blokland A. The selective PDE5 inhibitor, sildenafil, improves object memory in Swiss mice and increases cGMP levels in hippocampal slices. Behav Brain Res 2005; 164(1): 11-6.

[189] Devan BD, Pistell PJ, Daffin LW Jr, *et al.* Sildenafil citrate attenuates a complex maze impairment induced by intracerebroventricular infusion of the NOS inhibitor Nomega-nitro-L-arginine methyl ester. Eur J Pharmacol 2007; 563(1-3): 134-40.

[190] Rutten K, Basile JL, Prickaerts J, Blokland A, Vivian JA. Selective PDE inhibitors rolipram and sildenafil improve object retrieval performance in adult cynomolgus macaques. Psychopharmacology 2008; 196(4): 643-8.

[191] Devan BD, Sierra-Mercado D Jr, Jimenez M, *et al.* Phosphodiesterase inhibition by sildenafil citrate attenuates the learning impairment induced by blockade of cholinergic muscarinic receptors in rats. Pharmacol Biochem Behav 2004; 79(4): 691-9.

[192] Patil CS, Singh VP, Kulkarni SK. Modulatory effect of sildenafil in diabetes and electroconvulsive shock-induced cognitive dysfunction in rats. Pharmacol Rep 2006; 58(3): 373-80.

[193] Fisher DA, Smith JF, Pillar JS, St Denis SH, Cheng JB. Isolation and characterization of PDE9A, a novel human cGMP-specific phosphodiesterase. J Biol Chem 1998; 273(26): 15559–15564.

[194] van der Staay FJ, Rutten K, Bärfacker L, *et al.* The novel selective PDE9 inhibitor BAY 73-6691 improves learning and memory in rodents. Neuropharmacology 2008; 55(5): 908-18.

[195] Siuciak JA, McCarthy SA, Chapin DS, *et al.* Genetic deletion of the striatum-enriched phosphodiesterase PDE10A: evidence for altered striatal function. Neuropharmacology 2006; 51(2): 374-85.

[196] Lehéricy S, Gerardin E. Normal functional imaging of the basal ganglia. Epileptic Disord 2002; 4 (Suppl 3): 23-30.

[197] Hebb AL, Robertson HA, Denovan-Wright EM. Phosphodiesterase 10A inhibition is associated with locomotor and cognitive deficits and increased anxiety in mice. Eur Neuropsychopharmacol 2008; 18(5): 339-63.

[198] Domek-Łopacińska K, Strosznajder JB. Cyclic GMP and nitric oxide synthase in aging and Alzheimer's disease. Mol Neurobiol 2010; 41(2-3): 129-37.

[199] Zhang HT, O'Donnell JM. Effects of rolipram on scopolamine-induced impairment of working and reference memory in the radial-arm maze tests in rats. Psychopharmacology 2000; 150(3): 311-6.

[200] Ramos BP, Birnbaum SG, Lindenmayer I, Newton SS, Duman RS, Arnsten AF. Dysregulation of protein kinase a signaling in the aged prefrontal cortex: new strategy for treating age-related cognitive decline. Neuron 2003; 40(4): 835-45.

[201] Zhang HT, Huang Y, Masood A, *et al.* Anxiogenic-like behavioral phenotype of mice deficient in phosphodiesterase 4B (PDE4B). Neuropsychopharmacology 2008; 33(7): 1611-23.

[202] Puzzo D, Staniszewski A, Deng SX, *et al.* Phosphodiesterase 5 inhibition improves synaptic function, memory, and amyloid-beta load in an Alzheimer's disease mouse model. J Neurosci 2009; 29(25): 8075-86.

[203] Vardigan JD, Converso A, Hutson PH, Uslaner JM. The selective phosphodiesterase 9 (PDE9) inhibitor PF-04447943 attenuates a scopolamine-induced deficit in a novel rodent attention task. J Neurogenet 2011; 25(4): 120-6.

[204] Yakovleva T, Marinova Z, Kuzmin A, *et al.* Dysregulation of dynorphins in Alzheimer disease. Neurobiol Aging 2007; 28(11): 1700-8.

[205] Anthony IC, Norrby KE, Dingwall T, *et al.* Predisposition to accelerated Alzheimer-related changes in the brains of human immunodeficiency virus negative opiate abusers. Brain 2010; 133(12): 3685-98.

[206] Cai Z, Ratka A. Opioid system and Alzheimer's disease. Neuromolecular Med 2012; 14(2): 91-111.

[207] van Waarde A, Rybczynska AA, Ramakrishnan N, Ishiwata K, Elsinga PH, Dierckx RA. Sigma receptors in oncology: therapeutic and diagnostic applications of sigma ligands. Curr Pharm Des 2010; 16(31): 3519-37.

[208] Singh IN, Goody RJ, Goebel SM, *et al.* Dynorphin A (1-17) induces apoptosis in striatal neurons *in vitro* through alpha-amino-3-hydroxy-5-methylisoxazole-4-propionate/kainate receptor-mediated cytochrome c release and caspase-3 activation. Neuroscience 2003; 122(4): 1013-23.

[209] Sarajärvi T, Tuusa JT, Haapasalo A, *et al.* Cysteine 27 variant of the delta-opioid receptor affects amyloid precursor protein processing through altered endocytic trafficking. Mol Cell Biol 2011; 31(11): 2326-40.

[210] Teng L, Zhao J, Wang F, Ma L, Pei G. A GPCR/secretase complex regulates beta- and gamma-secretase specificity for Abeta production and contributes to AD pathogenesis. Cell Res 2010; 20(2): 138-53.

[211] Lew GM. Changes in microtubular tau protein after morphine in a cultured human neuroblastoma cell line. Gen Pharmacol 1997; 29(5): 869-72.

Send Orders of Reprints at reprints@benthamscience.net
Frontiers in Clinical Drug Research – Alzheimer Disorders, Vol. 1, 2013, 367-410

CHAPTER 10

Challenges in the Morphological Diagnosis of Dementias

Kurt A. Jellinger*

Institute of Clinical Neurobiology, Vienna, Austria

Abstract: Dementia is a major public health problem that threatens to become the scourge of the 21st century. It is caused by dysfunction/loss of synapses and neurons inducing default neuronal networks. Consensus criteria for the diagnosis of the major dementia disorders have recently been updated. Clinical diagnostic accuracy using revised criteria and new biomarkers ranges from 65 to 96% (for Alzheimer disease), with a sensitivity *vs.* other dementias of 71 to 87% and specificity of 44 to 71%. Morphological assessment, using modern molecular biological and genetic methods, and based on homogeneous definitions, harmonized inter-laboratory techniques and assessment standards, can achieve a diagnosis/classification in almost 99%, without, however, clarifying the etiology of most of these disorders. The new National Institute on Aging-Alzheimer Association (ABC) guidelines for the pathological diagnosis of Alzheimer disease combine β-amyloid plaque phases, Braak neurofibrillary and neuritic plaque scores, also considering other pathologies. Major difficulties arise from recent separation of distinct clinico-pathological subtypes (tangle- and plaque intensive, tangle predominant, limbic-predominant and -sparing) from classical Alzheimer disease. Revised research criteria are available for dementia with Lewy bodies, Parkinson disease-dementia, frontotemporal lobe degeneration, vascular cognitive impairment, prion diseases, and other neurodegenerative dementias. However, due to considerable overlap between various neurodegenerative proteinopathies and cases with mixed pathologies, human postmortem studies entail numerous biases that affect both their general applicability and the validity of correlations. Although most neurodegenerative dementias are incurable at present, concerted prospective clinico-pathological studies using validated protocols and data fusion are warranted to overcome the limitations of the current diagnostic framework as a basis for efficient new therapy options.

Keywords: ABC score, α-synucleinopathy, Alzheimer subtypes, Alzheimer's disease, biomarkers, clinico-pathologic correlation, dementia, diagnostic criteria, diagnostic limitations, frontotemporal lobe degeneration, hippocampal sclerosis, Lewy body disorder, mixed pathology, molecular classification, neurodegenerative disorder, neuropathology, NIA-RI criteria, Parkinson's disease-dementia, prion disease, revised guidelines, tangle-only dementia, vascular cognitive impairment/dementia, pathologies, disssorders, prion disorders.

INTRODUCTION

Dementia encompasses deteriorations in several cognitive domains [1]. Although

***Address correspondence to Kurt A. Jellinger:** Institute of Clinical Neurobiology, Vienna, Austria; Tel/Fax: +43-1-5266534; E-mail: kurt.jellinger@univie.ac.at

Atta-ur-Rahman (Ed)

redefined as the manifestation of deteriorating brain functions over time as a part of aging due to cell deaths in the brain caused by neurodegeneration or any other disease [2], it is not only caused by neuronal cell death/loss, but rather by dysfunction and loss of synapses [3] already occurring in early Alzheimer disease (AD) [4] and in α-synucleinopathies, eg, Parkinson disease (PD) and dementia with Lewy bodies (DLB) [5]. In addition, cholinergic neuronal and axonal abnormalities are present in aging and AD [6, 7]. These changes induce disconnection of important nervous circuitries causing default networks [8-11] that have been demonstrated *in vivo* in early AD [12], suggesting that disease progress is transmitted by neuronal pathways [13].

Both the prevalence and incidence of dementia increase exponentially with age and double every five years [14]. According to a report published by the WHO in April 2012 the number of demented persons in 2010 worldwide was estimated at 35.6 millions, rising to 55 millions in 2020 (Europe 14, Asia-Pacific 27 millions) and around 115 millions in 2050 (Europe 16 to 20, Asia-Pacific 35 millions), representing 2.6% of the total and 6.2% of the active population in Europe [15]. About 46% of demented subjects live in Asia, 30% in Europe, and 12% in North America, while their proportion living in developing countries will increase to 70% by 2040 [16]. With the disproportional growth of the elderly population, dementia has become a major public health and socio-economic problem that threatens to become the scourge of our century. The economic burden of dementia was estimated at 141 billion Euro for Europe and 422 billion US$ worldwide, including 34 to 50% for informal care in 2010.

Consensus criteria for the clinical diagnosis of the major dementing disorders have recently been updated, *e.g.*, the revised NICDS-ADRDA and EFNS criteria for AD [1, 17], the National Institute on Aging-Alzheimer's Association (NIA-AA) criteria for AD [18-20], for Parkinson disease-dementia (PDD) [21, 22], DLB [23], frontotemporal lobe degeneration (FTLD) [24, 25], vascular dementia (VaD) or vascular cognitive disorder (SCADDTD) [26, 27]; NINDS-AIREN DSM-IV [28, 29], and other degenerative dementias [30, 31].

Assessment of clinical data with fusion of different biomarkers has already improved the clinical diagnostic accuracy of AD up to 96%, while their sensitivity and specificity *vs.* other dementias is 23-88% [32]. A combination of the best CSF and MRI data will lead to a more precise diagnostic prediction [33, 34], and will be further increased by using multimodal techniques and novel CNS biomarkers [35-43]. The validity of plasma biomarkers for the (preclinical) diagnosis of AD is under discussion [44].

Identification of fibrillar β-amyloid (Aβ) *in vivo* by [(11)C]-PIB PET is a valuable tool for the prospective evaluation of the development and progression of cognitive impairment [45-50]. However, Aβ-associated clinical decline and brain volume loss occur only in the presence of elevated phospho-tau [51, 52]. Evaluation of biopsied or postmortem brain samples indicated that PiB binding *in vivo* correlated with heavy cerebral amyloid deposition [48, 53, 54]. Recent demonstration that SDS-soluble Aβ measured by immunoassay was better than post mortem PiB binding has important implications for imaging-based biomarkers [55]. A selective and specific PET tracer for imaging paired helical filament tau in human brains [56] needs further validation. Hippocampal atrophy in the elderly demonstrated by modern imaging methods is confirmed by postmortem diagnosis of AD [42, 57]. However, histopathology is still considered to add to premortem diagnostic accuracy [58-60]. Clinico-pathological correlations in the most common neurodegenerative dementias were summarized recently [61].

A review of 2861 cases of the National Alzheimer's Coordinating Center Registry (NACCR) showed high diagnostic accuracy for AD (85% sensitivity, 51.1% specificity) and low sensitivity for DLB (32% for pure AD, 12% for AD+DLB) with a specificity over 58% [62]. Evaluation of the accuracy of currently used clinical diagnostic methods for AD in 919 autopsy cases from the database of the NACC (2005-2010) revealed a sensitivity from 70.9 to 87.3% and a specificity of 44.3 to 70.8%. When the clinical diagnosis was not confirmed by minimum levels of AD pathology, the most frequent primary diagnoses were tangle-only dementia, argyrophilic grain disease (AGD), FTLD, CVD, LBD, and hippocampal sclerosis. 39% of these cases met or exceeded minimum threshold levels of AD histopathology. The misdiagnosis rate should be considered when estimating numbers for AD studies [60]. In a recent clinico-pathological study of 200 demented brain donors (mean age 78.7±6.9 years, 26% AD, 15.5% mixed dementia, 28% combined diagnoses), the overall agreement between clinical and postmortem diagnoses was 44% (85% for prion disease, 49% for AD), and there was frequent co-occurrence of multiple pathologies [63].

The aim of this review is to discuss the diagnostic validity of current morphological criteria of neurodegenerative dementias and their limitations as well as to give recommendations for future clinico-pathological research. The diagnosis of young-onset dementia presents challenges that differ from those of older patients who show slow specific pathological presentation. Much of our understanding of the pathogenesis of the degenerative dementias has been driven by the identification of genetic mutations causing early-onset familial disease [64].

THE UPDATED PATHOLOGICAL CRITERIA

At present, several guidelines for the neuropathological diagnosis of most dementing disorders are used, relying on qualitative, semiquantitative, and topographic assessment of morphological and bio/histochemical markers, in particular specific protein inclusions in neurons, glia, and other cells [65-67]. The classification of neurodegenerative disorders, previously based on the anatomical systems involved, has been replaced by molecular diagnosis. A simplified molecular-pathological classification of degenerative dementias is given in Fig. **1**. This will form the basis for their morphological diagnosis in the future. The frequency of the most frequent forms of dementia according to several autopsy series is given in Fig. **2a** and **2b**.

Neuropathological criteria for *Alzheimer disease* (AD), in addition to cut-off quantitative values of senile plaques and neurofibrillary tangles (NFTs) [68, 69], their semiquantitative assessment and age-adjustment in the Consortium to Establish a Registry for Alzheimer's Disease (CERAD) protocol [70], include both the topographical staging of neuritic AD pathology [71], re-evaluated recently [72, 73] and the progress and distribution of Aβ deposition being different [74]. A growing body of evidence supports the idea that plaques and NFTs actually define (but not fully represent) the disease process, which also involves oxidative stress, neuroinflammation, synaptic, neuronal, and axonal dysfunction and loss, *etc.* [69, 75]. All these changes begin before significant cognitive decline occurs. The generally used combination of the CERAD and Braak scores in the National Institute of Aging-Reagan Institute (NIA-RI) criteria relates dementia to AD-typical lesions with high, intermediate and low likelihood [76]. These categories of AD diagnosis apply only to individuals with dementia, and the evaluation of the NIA-RI critera demonstrated their easy and rapid use in AD and non-demented subjects, however, with much less reliability for other dementing disorders. High Braak stages and CERAD criteria identified 54% and 97% of AD cases, respectively, and eliminated between 62 and 100% of non-demented ones with low Braak stages, whereas among non-AD neurodegenerative dementias only between 8 and 42% were identified (Table **1**).

Table 1: Likelihood of dementia (in percent) due to AD according to NIA-R-Institute criteria in various autopsy series. CERAD: Consortium to Establish a Registry for Alzheimer disease; AD: Alzheimer disease; DLB: dementia with Lewy bodies; PD: Parkinson disease; PSP: progressive supranuclear palsy; MMSE: Mini-Mental State Examination score; CDR: Clinical Dementia Rating scale.

Author	**Disorder**	**CERAD / Braak stage**			**Mean age (years)**
		Low A / 0-II	**Interm. B / III-IV**	**High C / V-VI**	
Cochran *et al.* 1998 [77]	Demented (n=17)	47	41	12	?
	Non demented (n=40)	72.5	22.5	5	?
Newell *et al.* 1999 [78]	AD (n=33)	0	3	97	83
	DLB (n=15)	48	26	26	81
	PSP (n=12)	75	17	8	68
	Controls (n=17)	76	24	0	77
Harding *et al.* 2000 [79]	AD (n=31/22-no LB) (CDR 1-3)	26/13	20/27	54/60	77
	DLB, neocortical (n=11)	73	18	9	76
	PD (n=7) (CDR 0-0.5)	83	17	0	79
	Controls (n=18) (CDR 0-0.5)	83	17	0	79
Davis *et al.* 1999 [80]	Controls (n=57, MMSE 27-29)	88	-	12	84
McKee *et al.* 2002 [81]	AD (n=12) (CDR 1-3)0	0	17	83	
	Cognit. normal (n=23) (CDR 0)	62	38	0	83
Jellinger 2006 [82]	AD (n=100) (MMSE 0-17)	0	24	76	85
	DLB (n=36) (MMSE 0-20)	25	33	42	77
	PSP (n=10)	70	20	10	72

Table 1: contd....

	PD dementia (n=20, MMSE 0-20)	25	50	25	83
	PD non demented (n=17, MMSE >20)	70	30	0	72
	Controls (n=20, MMSE 28-30)	100	0	0	81

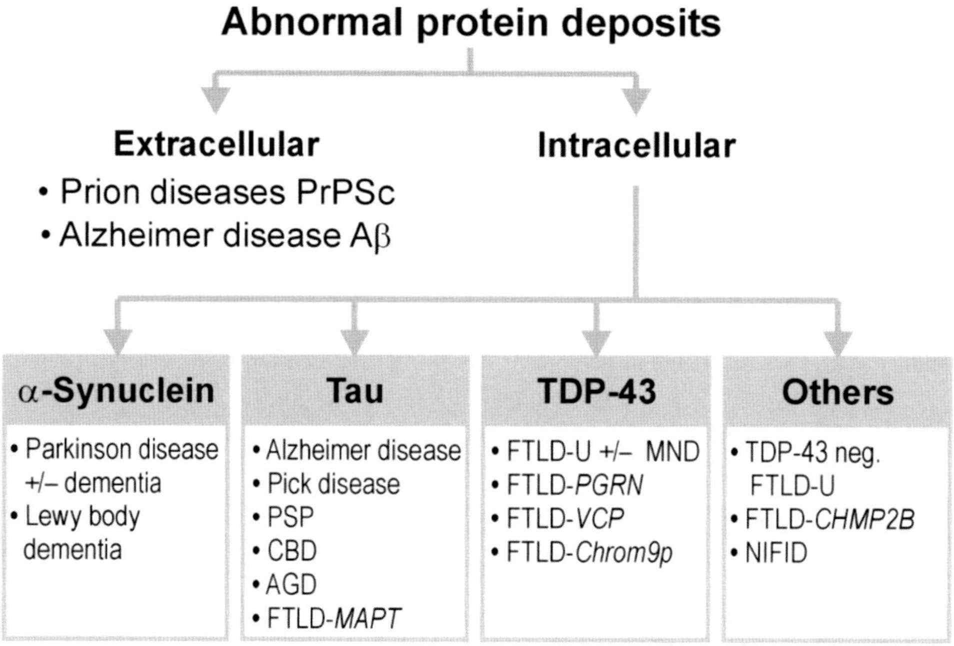

Figure 1: Molecular-pathologic classification of degenerative disorders (abbreviations see Fig. **3**).

A consensus panel recently updated and revised the 1997 guidelines for the neuropathological evaluation of AD and other diseases of brain in the elderly [83, 84]. Major revisions include (1) recognition that AD neuropathology may occur in the apparent absence of cognitive impairment, (2) an "ABC" score for AD neuropathology changes that incorporates histological assessments of Aβ deposits (A), based on the phase assessment of amyloid [74], staging of neurofibrillary tangles /NFT) (B), based on the Braak staging system, and scoring of neuritic plaques (C), based on CERAD criteria [70] (Table **2**), and (3) more detailed approaches for assessing co-morbid conditions, such as Lewy body disease, vascular brain injury, hippocampal sclerosis, and TAR DNA binding protein (TDP-43) immunoreactive lesions [84]. Correlations between neuropathological AD changes with cognitive status were reviewed recently [85].

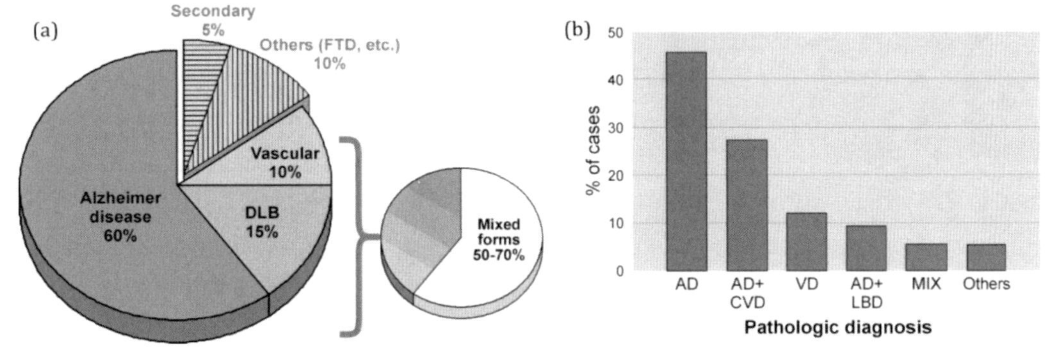

Figure 2: (a) Frequencies of the most common forms of dementia. (b) Frequency of neuropathological diagnosis in an autopsy series of 1700 demented aged persons in Vienna, Austria (from [86]).

Table 2: "ABC" score for AD neuropathological changes (from [84]).

"A"	Thal Phase for Aβ Plaques	"B"	Braak and Braak NFT Stage	"C"	CERAD Neuritic Plaque Score
0	0	0	None	0	None
1	1 or 2	1	I or II	1	Sparse
2	3	2	III or IV	2	Moderate
3	4 or 5	3	V or VI	3	Frequent

AD Neuropathologic Change		B		
A	C	0 or 1	2	3
0	0	Not[a]	Not[a]	Not[a]
1	0 or 1	Low	Low	Low[b]
	2 or 3[d]	Low	Intermediate	Intermediate[e]
2	Any C	Low[c]	Intermediate	Intermediate[e]
3	0 or 1	Low[c]	Intermediate	Intermediate[e]
	2 or 3	Low[c]	Intermediate	High

[a]High levels of neuritic plaques and low Thal Aβ phase are rare and should prompt reconsideration; [b]Widespread NFTs with some Aβ plaques but limited neuritic plaques are infrequent; [c]High levels of Aβ or neuritic plaques with low Braak stage should prompt consideration of co-morbidity, *e.g.*, CVLs, *etc.* [d]Medial temporal lobe NFTs in the absence of significant A-beta or neuritic plaques occur in older people often without or with only mild cognitive impairment, or with dementia from causes other than AD.; [e]Widespread NFTs with some A-beta plaques but limited neuritic plaques are infrequent and may occur in other diseases, particularly tauopathies. Such cases may not fit into a specific neuritic plaque stage.

α-Synucleinopathies, in particular Lewy body disorders (PD and DLB), show a multi-organ distribution of phosporylated α-synuclein (αSyn), with earliest involvement of the gastrointestinal tract and olfactory bulb [72, 87-90]. In addition to assessment criteria [91], staging/classification systems based on the semiquantitative assessment of the distribution and progression pattern of αSyn pathology are used [23, 92] that are suggested to indicate a predictable sequence of lesions [72, 88]. Recent clinico-pathological studies, although partly confirming this system [93], have shown that between 6.3 and 47% of autopsy-proven PD cases did not follow the proposed caudo-rostral spread of Lewy pathology [87, 90, 94-97]. In 7-8.3% of PD cases, the dorsal motor nucleus of vagus was not involved despite definite αSyn inclusions in the higher brainstem or even in cortical regions [96-98]. On the other hand, 30-55% of elderly subjects with widespread/cortical αSyn pathology showed no definite neuropsychiatric symptoms or were not classifiable [98-100]. Therefore, the criteria for classification of Lewy-related pathology have been modified (Table **3**).

Table 3: Classification of Lewy body disease.

None	No Lewy bodies or related changes in IHC for α-synuclein
Brainstem-predominant	Lewy bodies in medulla, pons, or midbrain
Limbic (transitional)	Lewy bodies in cingulate or entorhinal cortices, usually with brainstem involvement
Neocortical (diffuse)	Lewy bodies in frontal, temporal, or parietal cortices usually with involvement of brainstem and limbic sites, which may include amygdala
Amygdala-predominant	Lewy bodies in amygdala with paucity of Lewy bodies in the above regions

It is important to note that this classification is focused on the clinical context of cognitive impairment or dementia. Lewy bodies occur early in olfactory bulb, and are observed outside of the brain in multiple organs [89].

PD-dementia (PDD) and DLB, sharing many clinical and morphological features, are believed to form a continuum within the spectrum of LB diseases [23, 101]. Clinically, an arbitrary cut-off is used: if dementia develops first or within one year of PD diagnosis, then DLB is diagnosed, while PD followed first or within one year of PD diagnosis suggests PDD. Their pathological hallmarks are αSyn/LB pathology or a variable mixture of Lewy and AD pathologies, which may act synergistically [102]. According to revised guidelines, the severity and extent of αSyn are scored semiquantitatively in specific brain regions [23, 92]. DLB differs from PDD by more severe diffuse amyloid load in the striatum [103, 104], and more frequent αSyn affection of the hippocampal CA 2-3 subareas [90].

DLB cases had more severe Aβ load than PDD; but no differences in neuritic and αSyn scores, while others reported higher Aβ load in cortical and subcortical areas [102], the Aβ load being similar to that in AD [105, 106]. Between 10 and 50% of PDD brains had enough AD-related lesions to attain the pathological diagnosis of definite AD [107], but cognitive impairment may also be related to higher Braak Lewy body stages in the absence of significant AD pathology [108]. A recent clinico-pathological study showed that cortical LB/LN pathology is the most significant correlate of dementia in PD, while AD pathology, being abundant in a subset of patients, may modify the clinical phenotype [109]. On the other hand, up to 50% of PD cases exhibit additional Lewy pathology, which is associated with a more aggressive disease course and accelerated cognitive dysfunction [110].

Frontotemporal Lobe Degeneration (FTLD), the third or forth most frequent cause of dementias, includes three major clinical subgroups: behavior variant (bvFTD), semantic dementia (SD), and progressive non-fluent aphasia (PNFA). Their molecular correlates according to the predominant protein aggregates are: microtubulus-associated tau protein (FTLD-tau), TAR-DNA binding protein-43 (FTLD-TP-43), and fusion sarcoma protein (FTLD-FUS), but there are cases with overlapping pathology [25, 111, 112]. Criteria for the neuropathological diagnosis based on biochemical markers are summarized in Fig. **3**. A small group of FTLD patients clinically mimicking DLB has been identified, which morphologically was consistent with TDP-43 proteinopathy [113]. Up to 40% of FTD patients have a familial background, and 30-50% of them (10% of total) show autosomal-dominant traits; furthermore, various gene mutations have been described [111, 114]. FTLD in elderly patients presents features of several phenotypes and morphological subgroups similar to that seen in the presenile group, though patients with MAPT (tau-gene), but no progranulin mutations or FUS pathology are rare or even absent. However, these authors found significantly more frequent hippocampal sclerosis but milder frontal atrophy in the elder group [115]. This was confirmed in two personal cases of FTLD tau at age 85 and 88 years, respectively, but further clinico-pathological studies in this rare subtype of dementia are warranted. TDP-43 positive inclusions also are observed in cases with AD and LB pathologies [116], and FUS-immunoreactive intranuclear inclusions occur in various neurodegenerative diseases [117], but it is not clear whether these changes are a primary, secondary or coincidental event [118].

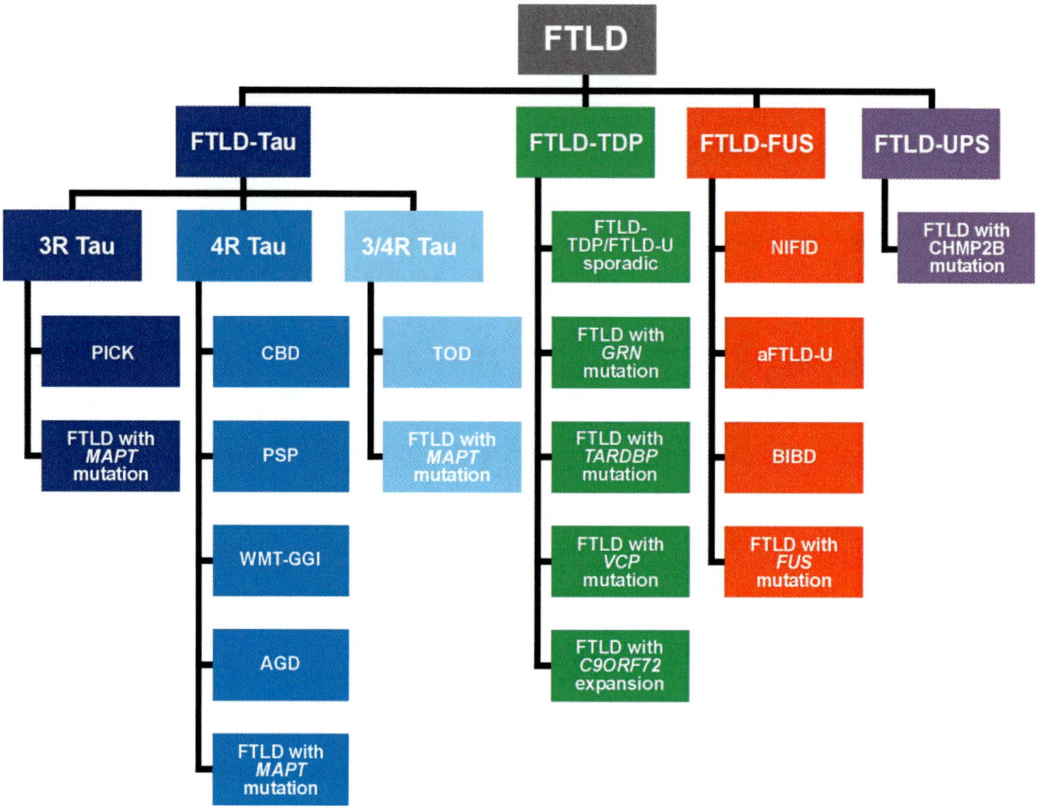

Figure 3: The molecular and genetic classification of FTLD. Three distinct neuropathologic categories may be identified based on the molecular pathology of the misfolded protein within the inclusion: FTLD-Tau, FTLD-TDP, and FTLD-FUS; the molecular pathology of a fourth category, FTLD with epitopes of the ubiquitin–proteasome system (FTLD-UPS), remains indeterminate. 3R, 4R, 3R/4R the predominant tau isoform within the inclusion; PICK, Pick's disease; FTLD with microtubule-associated protein tau (MAPT) mutation with inclusions of 3R, 4R, or 3R and 4R tau protein; CBD, corticobasal degeneration; PSP, progressive supranuclear palsy; AGD, argyrophilic grain disease; TOD, tangle-only dementia; WMT-GGI, white matter tauopathy with globular glial inclusions; FTLD-U, FTLD with ubiquitin inclusions, now called FTLD-TDP; FTLD with progranulin (GRN) mutation; FTLD with TAR DNA-binding protein 43 (TARDBP) gene mutation; FTLD with valosin-containing protein (VCP) mutation; FTLD with C9ORF72 expansion, chromosome 9-linked FTLD with C9ORF72 hexanucleotide repeat expansion; NIFID neuronal intermediate filament inclusion disease; aFTLD-U atypical FTLD with ubiquitin inclusions; BIBD basophilic inclusion body disease; FTLD with fused in sarcoma (FUS) mutation; FTLD with charged multivesicular body protein 2B (CHMP2B) mutation. There may still be unclassified entities within each molecular pathology grouping. After [112].

For **Vascular Dementia** (VaD) (or vascular cognitive impairment/disorder related to cerebrovascular disease), despite various proposals for a categorization of major cerebrovascular lesions [27, 119-121], a harmonization of the criteria and techniques for the assessment of cerebral lesions of presumable/possible vascular origin in cognitively impaired is necessary [122, 123]. The patterns of the vascular brain lesions leading to dementia distinguish three major forms of VaD: 1: multi-infarct dementia, 2: strategic infarct dementia, and 3: subcortical vascular encephalopathy [124]. Due to the high variability of morphological findings and the multifactorial pathogenesis of vascular cognitive impairment, no generally accepted morphological scheme for quantitating cerebrovascular lesions and no validated neuropathological criteria for VaD have been established to date (see [121, 125, 126]. Most authors distinguish familial and sporadic forms of VaD [29, 127]; the most frequent subtype of familial VaD is the cerebral autosomal dominant arteriopathy with subcortical infarcts and leukoencephalopathy (CADASIL), caused by mutations in the Notch3 gene [29, 127]. The revised NIA-AA guidelines recommend reporting all macroscopical and microscopical cerebrovascular lesions, multiple such lesions being associated with increased likelihood of cognitive impairment [84]. Another recent staging strategy proposing semiquantitative assessment of cerebrovascular lesions in 4 brain areas with a score ranging from I to IV/VI (Table **4**) awaits further evaluation.

Table 4: Staging of the cerebrovascular lesions. Modified after [128].

Score	**Staging**
Frontal and Temporal Lobes	
0	Normal appearance of brain, vessels, white matter, and cortex
I	Mild modification of vessel walls, perivascular spaces, or white matter
II	Moderate to severe but isolated modification of the vessel walls (arteriolosclerosis or amyloid angiopathy), usually associated with hemosiderin deposits in the perivascular spaces
III	Moderate to severe perivascular space dilatations either in the deep or the juxtacortical white matter
IV	Moderate to severe myelin loss
V	Presence of cortical microinfarcts
VI	Presence of large infarcts
Hippocampus	
0	Normal appearance
I	Mild modification of vessel walls or perivascular spaces
II	Moderate to severe perivascular space dilatations
III	Presence of microinfarcts (usually in Ammon's horn or the subiculum)

Table 4: contd...

IV	Presence of large infarcts
Basal ganglia	
0	Normal appearance
I	Mild modification of vessel walls or perivascular spaces
II	Moderate to severe perivascular space dilatations
III	Presence of microinfarcts
IV	Presence of large infarcts
Total vascular score	
Frontal lobe + Temporal lobe + Hippocampus + Basal ganglia (/20)	

Mixed Type Dementia, being frequent in elderly subjects, is diagnosed when a combination of several pathologies, *e.g.* AD with cerebrovascular lesions and /or Lewy pathology, is present [127, 129]. Vascular brain injury is frequently encountered in the brains of seniors with and without AD pathology [127, 129, 130], but uniform and reproducible criteria are currently not available.

Prion Diseases (transmissible spongiform encephalopathies/TSE) are rare and fatal neurodegenerative disorders that primarily affect the central nervous system. They are characterized by tissue deposition of a misfolded isoform of the cellular prion protein (PrP), commonly referred to as PrPSc [131]. They are classified by both clinico-pathological syndromes and etiology with subclassification according to molecular criteria [132-134]. Human prion diseases are grouped into three etiologic categories (Table **5**):

(1) sporadic Creutzfeldt-Jakob disease (sCJD) - age at onset 50-78 (mean 60) years, mean duration 5 months -, sporadic familial insomnia (sFI), and the recently described variably protease-sensitive prionopathy (VPSPr) [135, 136];

(2) inherited (genetic or familial) forms such as Gerstmann-Sträussler-Scheinker disease (GSS), fatal familial insomnia (FFI), octapeptide repeat insert (OPRI) [137, 138], rare familial prion disease with AD-like tau pathology [139], genetic CJD associated with the E200K mutation [140], and (3) acquired forms, such as iatrogenic CJD (iCJD), Kuru, and variant CJD (vCJD) (transmission of bovine

spongiform encephalopathy/BSE/ to humans) – age at onset 12-74 (mean 27) years, mean duration 14 months [141]. More than 80% of human prion diseases manifest as sCJD with an incidence of 1-2 cases per million population per year across the world. Around 15% are associated with autosomal dominant pathogenic mutations of the PRNP gene located on the short arm of chromosome 20. To date, 37 pathogenic mutations and 29 non-coding polymorphisms have been described [142]. In contrast, acquired human prion diseases are rare (only 5% of human prion diseases): transmission of sCJD prions occurred through treatment with pituitary hormones pooled from human cadavers, transplantation of dura mater or cornea, the use of contaminated EEG electrodes, and by cannibalism among the Fore linguistic group in Papua New Guinea [143]. Instead, vCJD occurred mainly in the UK and in other countries due to human exposure to BSE prions, and more recently by transmission of vCJD contaminated blood products [144] as well as by aerosols [145].

Table 5: Human prion diseases [146]. CJD: Creutzfeldt–Jakob disease, FFI: fatal familial insomnia, GSS: Gerstmann-Sträussler-Scheinker disease, sFI: sporadic fatal insomnia, VPSPr: variably protease-sensitive prionopathy.

Category	**Etiology**
Idiopathic	
sCJD	Unknown.
sFI	Unknown. Found only in 129 MM patients with PrP^{Sc} type 2.
VPSPr	Unknown.
Hereditary	
Familial CJD	Various PRNP mutations.
GSS	Various PRNP mutations.
FFI	PRNP mutation D178N with M129.
Acquired	
Iatrogenic CJD	Accidental transmission through treatment with prion-contaminated preparations of human growth hormone, dura mater grafts, and, rarely, by contaminated blood products.
Variant CJD (vCJD)	Infection by bovine prions (BSE-contaminated food or other products and aerosols).
Kuru	Infection through ritualistic cannibalism in the Fore population in Papua New Guinea (historical).

Given the biological and phenotypic heterogeneity of prion diseases, the correct identification or "typing" of the prion strain/s associated with each case of human TSE has implications for diagnosis, epidemiological surveillance, and future therapies. Concerning common genetic variations, it is likely that the PRNP locus contains only strong risk factors that act universally across human prion diseases [147]. The origin of prion strains and the mechanisms underlying the strain-specific neuronal targeting remain largely unexplained [148]. Either molecular or histopathological data (or both) may provide reliable surrogate markers for strains typing in humans. The current classification of sporadic human prion diseases recognize six major phenotypic subtypes with distinctive clinico-pathological features, which largely correlate at the molecular level with the genotype at the polymeric codon 129 (methionine, M, or valine, V) in the prion protein gene and with the size of the protease-resistent core of the PrPSc.

To date, six different strains of prions according to the 129 PrP genotype have been distinguished [146]:

(1) Strain A, related to the most common CJD phenotype or myoclonic variant (MM1/MV1); age at onset 48-86 (mean 70) years; about 40% of cases.

(2) Strain B, related to the ataxic and kuru-plaque variants of CJC (VV2 and MV 2K; subtypes) as well as kuru and the plaque-type of dura mater craft-associated CJD; age at onset 40-83, mean 65 years; mean duration 6-16 years; about 23% of cases.

(3) Strain C, related to vCJC; age at onset 12-74, mean 27 years; mean duration 14 months < 1% of cases.

(4) Strain D, related to fatal insomina, either sporadic or familial (MM 2T subtype); age at onset 36-71, mean 52.3 years; mean duration 8-24, mean 15.5 months; around 1% of cases.

(5) Strain D, related to the cortico-striatal variant with synaptic deposits (VV1 subtype); age at onset 24-49, mean 39.3 years, mean duration 15.3 months; around 1% of cases.

(6) Strain F, related to the cortical variant with confluent vacuoles (MM/MV1+ 2C subtype); age at onset.42-89, mean 68.6 years, mean duration 4 months; 28%of cases.

In addition, there are mixed kuru plaques and other mixed morphological cases (MV 2K+2X, and VV2+1 subtypes), about 6 % of cases [146].

Based on this and other, slightly different definitions [136, 149] and on the distribution and severity of histopathological lesions, a general nomenclature for the histopathological diagnosis of sCJD in the absence of molecular data was proposed (Table **6**). According to a recent inter-rater study among surveillance centers in Europa and USA, the use of this consensus classification of human prion disease histotypes allows reliable identification of molecular subtypes with full agreement for the two most common sCJD subtypes and vCJD, and very high concordance for all pure phenotypes and the most common subtypes with mixed phenotypic features [150]. These data support the basis for the current classification of human prion disease and indicate that, besides molecular PrPSc typing, histopathological analysis permits reliable disease classification with high inter-laboratory accuracy.

Table 6: Human prion disease types - terminology [150].

No.	Molecular Types	Histopathological Types (Prion Disease Type with....)
1	MM/MV 1	Diffuse synaptic deposits
2	VV2	Perineuronal and cerebellar plaque-like deposits
3	MV 2K	Kuru plaques
4	MM 2C	Cortical confluent vacuoles
5	MM 2T (=sFI)	Thalamo-olivary atrophy
6	VV1	Cortico-striatal synaptic deposits
7	MM 2V (=vCJD)	Florid plaques
8	MM/MV 1+2C	Mixed diffuse synaptic deposits and cortical confluent vacuoles
9	MV 2K+C	Mixed Kuru plaques and cortical confluent vacuoles
10		Atypical

RAPIDLY PROGRESSIVE AND EARLY ONSET DEMENTIAS

In contrast to most dementing conditions that typically develop over years, rapidly progressive dementia (RPD) being quickly fatal is one of the most challenging problems. The differential diagnosis is often widely ranging, and, in addition to frequent prion diseases includes rapidly progressing neurodegenerative tauopathies and synucleinopathies, autoimmune condition infections, toxic-metabolic and neoplastic diseases [151, 152]. According to the US National Prion Disease Pathology Surveillance Center (NPDPSC) in patients with RPD, treatable disorders are frequently mistaken for CJD. Among 1,106 brain autopsies, 32% were negative for prion disease; most frequent were AD (50%) and VaD (12%), while 23% were potentially treatable diseases, eg., immune-mediated, infectious disorders or tumors [153]. In these and other unclear conditions, both combined CSF markers (14-3-3 protein, extremely high total-tau or phospho-tau and Aβ(42) are sensitive indicators [154]. In a rapidly progressing form of AD that clinically may mimic CJD, the genetic profile (absence of ApoE ε4 homozygenity and biomarkers) differs from classical AD, suggesting that it might represent a distinct subtype of AD [155]. Unfortunately, no detailed morphological data were presented. The roles of prions and inflammation in RPD are unclear, although links between prions and AD are under investigation, and the presence of amyloid plaques in prion diseases like CJD is well documented. Recent studies showed correlations between CSF biomarkers for AD with amyloid plaques and tau pathology in cortical brain biopsies [156], although biopsy findings in demented subjects frequently are non-specific [157], with an overall sensitivity of 65% [158].

Problems may arise in the clinical diagnosis of early onset dementias, but recent studies showed that, despite absence of absolute concordance between clinical phenotype and underlying pathology, they can be distinguished with high accuracy. Among 228 cases with early-onset dementia, 46% were diagnosed with AD and the remaining cases DLB, CJD, VaD, or unclassified dementia. AD was identified with 97% sensitivity and 100% specificity, FTLD with 100% sensitivity and 97% specificity [159]. Non-degenerative non-vascular cases of dementia being more common than expected in patients with a younger onset (30% of younger-onset and 5% of older-onset group aged 70-99 years) include cancer, chronic alcoholism, chronic mental illness, and others [160].

DIAGNOSTIC PROBLEMS FOR ALZHEIMER DISEASE

Histopathological examination of the brain establishes that AD-related lesions are present in sufficient densities to distinguish AD from age-related lesions and allows detection of other dementing disorders [83, 84, 161]. The current algorithms for the morphological diagnosis of AD are based on (semi-) quantitative assessment of plaques and tangles. Despite reasonable inter-rater agreement when using standardized criteria [73, 162-164], no one set of histopathological criteria for AD has been uniformely accepted. The previous algorithms that only considered the classical "plaque and tangle" phenotype of AD did not recognize other subtypes. Analysis of 1677 cases with antemortem diagnosis of dementia from the NACCR showed that 82.4% fell into diagnostic "boxes" that are within the consensus recommendations, while the others were "atypical" cases [165]. An atypical varient of AD, known as posterior cortical atrophy (PCA) morphologically shows greater NFT burden in the occipital and parietal lobes and lower in hippocampus [166]. We recently published a 4-repeat tauopathy clinically presenting as PCA [167] demonstrating the neuropathological variability of this syndrome.

The "tangle predominant dementia" (TPD) with tau pathology often restricted to the limbic system, absence of neuritic plaques, no or very little (diffuse) amyloid plaques and rare amyloid angiopathy accounts for 5-7% of oldest olds frequently presenting with mild to moderate dementia, depression, and low ApoE ε4 frequency. Recent molecular and genetic analysis confirmed an identical tau isoform composition of tangles in TPD and AD [168]. It demonstrated absence of soluble Aβ but elevated soluble amyloid precursor protein α (APPα) in brain tissue, and association with the tau gene MAPT H1 haplotype, classifying it as a specific tauopathy independent of amyloid [169]. The "plaque predominant" type with abundant amyloid plaques, no or very little neuritic pathology restricted to the hippocampus and abnormal phosphorylated tau in neocortical pyramidal cells but lacking overt tangle formation, accounts for 3.5-8% of demented subjects over age 85 [170, 171]. Many of them are associated with cortical Lewy bodies representing a specific type of DLB or LB variant of AD (LBV/AD) [172].

A recent clinico-pathological study of autopsy-proven AD cases separated distinct subtypes:

(a)hippocampal-sparing forms with lower NFT counts in hippocampus but more frequent senile plaques compared to typical AD (11%), (b) limbic-predominant forms (LP-AD) with lower cortical NFT counts and tau burden (14%), and (c) typical AD (75%). Hippocampal-sparing cases were younger at death; while the disease duration in all three types was similar [173]. Additional vascular pathology ranged from 16 to 36%; while Lewy pathology (11-26%) was lowest in hippocampal-sparing form. While MAPT H1H1 genotype was high (~70 %) in TPD and LP-AD and similar to typical AD, tau and Aβ burden in frontal cortex were very low in TPD, differentiating it from AD subtypes, including the LP form. Significant pathological differences between TPD and LP-AD suggest that it may not merely be a variant of AD [174]. Volumetric MRI analysis could reliably track the distribution of NFT pathology and predict pathological subtypes of AD at autopsy [175].

Among 933 autopsy-confirmed AD cases, typical AD accounted for 82.5%, HS and LP form for 8.2 and 8.9%, respectively. Disease duration was lowest in LP cases; age at death highest in typical AD. Additional cerebrovascular pathology was similar in all three types, Lewy pathology most frequent in the HS form [176] (Table 7). These two studies and another one from the NACCR, separating "tangle intensive" and "plaque intensive" cases from "classic" AD [62] emphasize the need for prospective clinico-pathological studies to further elucidate the various phenotypes of AD.

Table 7: Major characteristics of AD subtypes [176].

Characteristics	**Hippocampal-sparing AD (n = 79, 8.2%)**	**Limbic-predom. AD (n = 85, 8.9%)**	**Typical AD (n = 769, 82.5%)**	**P value**
Women	n = 47, 59.5%	n = 55, 64.7%*	n = 520, 67.5%*	* *vs.* HS P < 0.01
Age at death	76.3 ± 8.6	84.9 ± 3.8*	81.3 ± 9.2*	* *vs.* HS P < 0.01
Age at onset	68.0 ± 10.0	73.8 ± 6.4*	79.7 ± 3.8*	P < 0.01; * *vs.* HS <0.001
Disease duration	7.4 ± 3.6**	4.8 ± 2.6***	9.1 ± 4.3*	***P < 0.001; ** *vs.* HS P < 0.001
MMSE final (mean)	10 (n = 20)	11.5 (n = 16)	4.6 (n = 78)*	* *vs.* other forms P < 0.01
Braak NFT stage (mean)	4.5 (2–5)	4.5 (3–5)	5.6 (5–6)	

Table 7: contd....

Cerebrovascular pathology	27 (34.2%)*	35 (41.1%)	286 (36.4%)*	* *vs.*. LP P < 0.01
Lewy body pathology*	19 (24.9%)*	3 (3.5%)	63 (8.2%)	* *vs.*. other forms P < 0.001

Standard metrics for tangles and neuritic plaques are usually semi-quantitative [76, 83, 84, 162] and, according to the BrainNet Europe consortium, good agreement can be reached in the diagnosis only when the lesions are substantial, *e.g.*, when having reached isocortical structures (Braak stage V-VI with absolute agreement 91%), while for mild lesions the agreement was poorer (Braak stage I-II, agreement 50%) [73], thereby limiting the ability to make accurate correlation of antemortem cognitive status and morphological findings [32, 85].

Although the sensitivity and specificity of the NIA-RI criteria is suggested to be 90%, only 30 to 57% of the brains of patients with the clinical diagnosis of probable AD show "pure" AD pathology [32]. Thus, their predictive value may be reduced to 38-44% [177]. Testing the revised NIA-AA guidelines for the neuropathological assessment of AD in 390 autopsy cases (including 199 non-demented individuals) distinguished between pure AD and non-AD dementia as well as non-demented cases with a sensitivity of 71% and a specificity of 99%. The sensitivity increased after neuropathological exclusion of non-AD dementia cases, indicating that cognitive status and assessment according to NIA-AA guidelines appear ideal for distinguishing pure AD from non-AD dementia, preclinical AD (non-demented cases with a low NIA-AA degree of AD pathology), and normal controls (Thal *et al.* submitted).

In a retrospective clinico-pathological study of 1,700 demented elderly persons (66% female; MMSE score <20; mean age at death 84.3±6.0 years; 90% over age 70 years), AD-related lesions were present in 83.2%, but pure AD (ABC 3/3/3) without other pathologies in only 41.0%, AD with other pathologies including mixed dementia in 44.8%, vascular dementia in 10.7%, other disorders in 5.5%, and negative pathology in 0.9% (Jellinger, 2009a, 2011). Although cognitively normal elderly may show variable neocortical AD-related pathology, in general, the number of isocortical tangles correlates best with the severity of cognitive impairment [85, 178-180]. The pattern of gray matter loss associated with tangle pathology is an appropriate *in vivo* surrogate indicator of AD pathology [181]. The predictive value of widespread tau pathology (Braak stages V-VI) for dementia is high [85, 178], while others found that both diffuse and neuritic plaques, rather than tangles in neocortical regions distinguish non-demented and AD subjects with high sensitivity and specificity [182]. A recent study found that the amyloid stage that has progressed to involve the striatum is highly predictive of dementia [183]. However, there is considerable heterogeneity in morphology, extent and topographical distribution of Aβ deposits in brains with fully developed AD. The cortical Aβ burden usually does not correlate with disease duration and the stage of tau pathology [184]. Digital pathology offers a rigorous and reproducible method for quantifying AD pathological changes and may provide additional insights into morphological characteristics that were previously more challenging to assess because of technical limitations [185].

DEMENTIA IN THE OLDEST-OLD

Rates of cognitive decline in AD decrease with age, implying that the phenotypical expression of AD is relatively mild in individuals older than 85 years, and this may affect the possibility to distinguish AD clinically from normal aging in the very old [186]. Neuropathology of AD in oldest-old subjects differs considerably in both intensity and distribution from that in younger age groups [187]. Increased densities of neuritic plaques and NFTs are absent in non-demented patients over age 85-90 [179, 188-192], and there is considerable overlap in the pathologies found in demented and non-demented patients [193]. On the other hand, by age 80-85 years, many cognitively unimpaired subjects may have substantial cortical AD pathology [194], while others found significant positive correlation between the extent of dementia and senile plaque density ($p=0.011$), but not for the NFT density score ($p=0.076$) [195]. Recent studies suggest that dementia in the oldest-old group (90+ years) is only modestly related to AD pathology, while both cardiovascular and cerebrovascular lesions may cause cognitive impairment in most elders with low AD pathology scores [86, 121, 196], and cerebrovascular lesions may contribute to the clinical expression of dementia [197]. However, several clinico-pathological studies clearly showed that Braak NFT staging remains a significant predictor of cognitive status even in oldest-old people [198, 199]. There may be no evidence for a certain number of elderly subjects having dementia without an apparent causative morphologic background [86, 198], although dementia lacking a known pathologic substrate is extremely rare [200].

In a retrospective clinico-pathological study of 1100 demented seniors (66% females, MMSE <20; mean age at death 83.3±5.4 SD years, 90% over age 70 years) AD increased with highest incidence in the 8th and 9th decade, and slightly decreased after age 90, while the relative prevalence of both AD + minor cerebrovascular lesions and mixed dementia significantly increased with age (7.8 to 32.9% and 0 to 7.%, respectively, $p<0.001$). VaD showed a continuing age-related decrease from 15.6 to 9.4% ($p<0.05$), whereas AD + Lewy pathology remained fairly stable (10.3 and 11%) (Fig. **4**). In a prospective study of 180 demented patients (mean age 85±3.4 years), autopsy showed AD in 48%, AD with cerebrovascular pathology in 19%, VaD in 11%, DLB in 9%, and dementia of unknown etiology in 13% (Jellinger and Attems, 2010), confirming the notion that a high percentage of demented persons aged 80+ do not met the pathological criteria of AD or were classified as "dementia of unknown etiology" [201, 202]. Non-AD pathology significantly improved precise differentiation between oldest-old and younger age groups [191].

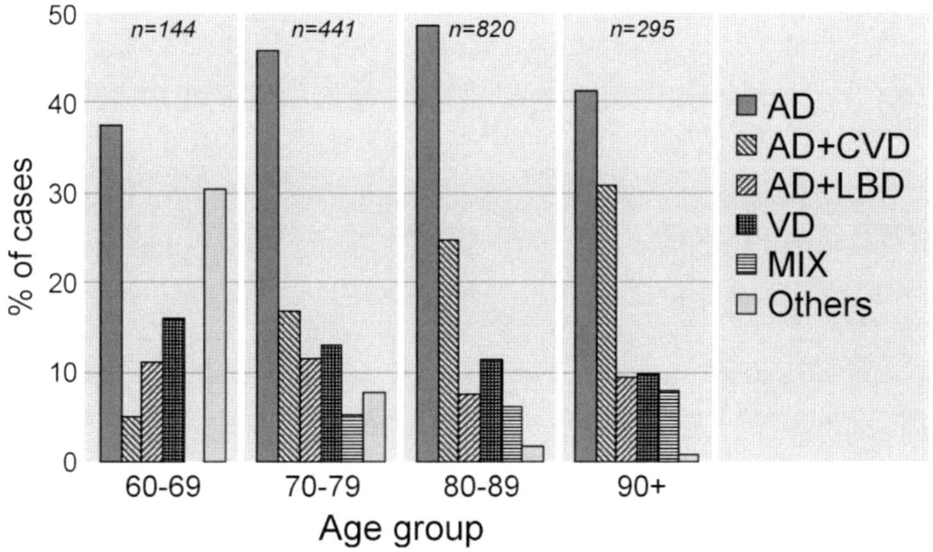

Figure 4: Age-related frequency of neuropathological types of dementing disorders. AD Alzheimer's disease; CVD: cerebrovascular dementia; LBD: Lewy body dementia; VD: vascular dementia; MIX: mixed dementia (AcD + cerebrovascular encephalopathy) From [86].

NEUROPATHOLOGY FINDINGS IN COGNITIVELY UNIMPAIRED AGED SUBJECTS

It is increasingly recognized that the correlation between neuropathological lesions and cognitive impairment is modest and accounts for about a quarter of the variance in cognition of older adults. It has been hypothesized that the concept of resilient aging can be useful to understand mechanisms that underly healthy aging amidst disease-related pathology [203]. Some individuals maintain normal cognitive function despite significant brain pathology, whereas others suffer varying degrees and neurological deterioration. Many aged people do not exhibit cognitive impairment and live "normal" lives despite displaying pathological changes that are characteristic of AD, PD, cerebrovascular disease or other disorders [192, 204-207]. Although the best morphological correlates of cognitive impairment/dementia are (1) the number of NFTs [161, 178, 208], and (2) loss of synapses [209, 210] between 8 and 45% of non-demented, often cognitively stable seniors were found to have AD-related pathologies [80, 194, 205, 208, 211-218]. Many of them correlated with only minimal to mild neuritic changes corresponding to Braak tau stages 0-IV, while 31-88% showed NIA-RI of no likelihood for AD criteria [80, 208, 219]. The frequency of intermediate likelihood of AD criteria ranged from 11.9 to 35.8% [192, 213, 219], and only 1.5 to 3% were scored high likelihood of AD [179, 219], while 18-25% met the CERAD criteria for AD [80, 213, 215, 216].

Review of the data from National Disease Coordinating Center (NDCC) database and the NUN study emphasized that there may be no documented examples of truly endstage neurofibrillary pathology with intact cognition [220]. Although in the Adult Changes in Thought (ACT) and NUN studies, non-demented seniors with severe AD pathology (mean age 89 to 91 years) amounted to 8 and 12%, respectively, most of them showed neuritic Braak stage V and frontal NFT count was slightly lower than in a comparable dementia group [218]. Review of the clinical data from these studies showed that most of the seniors classified as non-demented were indeed significantly memory-impaired [218, 221]. There is obviously no reported case of truly intact cognition despite severe AD-pathology, *i.e.*, widespread neocortical NFTs [85]. A recent study of non-demented elderly demonstrated 62% with low and 28% with high NFT levels [222], whereas clinical AD cases showed much higher cortical neuritic and striatal amyloid plaque scores [183]. In the Nun Study, one-third of the participants fullfilling criteria for the neuropathological diagnosis of AD were not demented at the time of death [223]. A 90+ study revealed significantly less severe Aβ, αSyn and TP'D-43 pathologies and hippocampal lesions in non-demented subjects, while Aβ distribution showed no esssential differences; non-demented individuals had limited hippocampal tau and neocortical Aβ pathology [224]. On the other hand, NFTs in the occipital cortex of 24% of non-demented subjects aged 42-87 years were reported [225]. In another population-based clinico-pathological study in the oldest-old, dementia was present in 61% of cases, but 22% of them did not have sufficient pathology to account for their cognitive loss [226].

The presence of AD lesions in non-demented aged people may represent AD at a stage prior to clinical expression (presymptomatic, preclinical or unrecognized early stage) [194, 211, 227-229]. This is supported by observations that the mechanisms responsible for these changes in non-demented elderly appear similar if not identical to those found in AD [227, 230]. The concept of "preclinical AD"pathology has been solidified by *in vivo* PET scanning, suggesting its high frequency in normal elderly similar to that seen in clinico-pathological cohorts [40, 50, 231-233]. These data further suggest that preclinical stages are not static but progressive over time [47, 54, 234].

Thus, mounting evidence supports the view that AD is a continuous spectrum between between asymptomatic lesions in cognitively unimpaired elderly and dementia, with MCI as a transition between them [235]. Although correlations between cognitive deficits and the severity and extension of plaques and tangles have been found (see [85]), at least in those brains without other pathologies, the distinction between "physiological" (in non-demented subjects) and "pathological" aging may be difficult. Recent biochemical studies found excessive overlap with only subtle quantitative differences between amyloid levels, peptide profiles, solubility, and oligomeric assemblies in pathological aging and AD brains [236], suggesting that pathological aging represents an initial prodromal stage of AD. They further suggest that these individuals would develop clinical symptoms, if they live long enough, or have an inherent individual resistence to the toxic effect of Aβ [237]. A coding mutation (A6773T) in the APP gene that deduces the β-cleavage of APP may protect against AD and also against cognitive in the elderly without AD [238].

A default hypothesis for AD is that it is part of a "normal aging process", such that plaques and tangles are secondary to aging or that the primary effect is on synapses and neurons independent of these morphological AD markers. AD is indeed a disease that accompanies human aging, but it is not an inevitable consequence of it [239, 240]. However, the suggestion that plaques and NFTs, the morphologic markers of AD, may "cause" this disorder is oversimplified or even wrong, since accumulating evidence indicates that AD pathology represents effect rather than cause or at least a host response to injury, equaling adaptative or neuroprotective reaction [241-243]. Larger brain and hippocampal volumes were associated with preserved cognitive function during life despite a high burden of AD pathology [244-246], but the mechanisms that protect the elderly from AD are unknown [247, 248].

The AD resiliant group (pathological AD without cognitive impairment) showed preserved densities of synaptophysin-positive presynaptic terminals and dendritic spines compared with the AD-dementia group, and increased densities of GFAP-positive astrocytes compared with the AD-dementia group and normal controls [246]. Hence, greater amounts of presynaptic proteins and distinct protein-protein interactions may be components of cognitive reserve reducing the risk of dementia with aging [249]. In conclusion, the aging process that results in loss of synapses and neurons may be far more detrimental for those with little cognitive/brain reserve as compared to those with a high one [250, 251].

THE IMPORTANCE OF CONFOUNDING PATHOLOGIES

An important problem is the frequent presence of confounding processes in the aged brain that coexist with AD, as CVD, Lewy body pathology, argyrophilic grain disease, hippocampal sclerosis, and so forth, with about two-thirds of aged human brains containing non-AD-type neuropathology [192, 217, 226, 252, 253], which, however, frequently have been missed clinically and could not be identified without neuropathological examination [63]. The high frequency of mixed pathologies in demented elderly was recently confirmed [254, 255]. Since

50 to 85% of the brains of persons who die aged 80-90+ show appreciable CVLs [256], a specific problem is the impact of vascular brain injury in relation to AD pathology [121, 129, 257-259]. In several autopsy series of very old people, the frequency of AD ranged from 12 to 66%, that of VaD from 9 to 46.8%, that of DLB between 9 and 24%, and that of mixed pathologies between 2 and 86% (!), and was over 40% in a large autopsy series of patients over age 80 [86].

The burden of vascular and AD type pathologies are considered to be independent of each other, and are consistent with an additive or synergistic effect of both types of lesions on cognitive impairment [103, 121, 129, 197, 260-262]. The thresholds for vascular and degenerative lesions in distinguishing "pure" VaD or AD from mixed cases have been critically discussed [199, 263, 264]. AD pathology alone more frequently accounts for dementia than both microscopic and macroscopic infarcts [265], and in advanced or full-blown stages of AD concomitant small vascular lesions do not significantly influence the overall state and progression of cognitive decline, the severity and extent of AD pathology overwhelming the effects of cerebrovascular disease [121, 125, 129, 266, 267]. In a recent autopsy study, global AD pathology significantly correlated with global cognition ($p=0.008$), whereas infarcts and Lewy bodies did not [207]. In the Nun Study, brain infarcts had little effect on cognitive status, but played an important role in increasing the risk of dementia associated with AD pathology [223].

Many studies emphasize multiple confounding pathologies in non-demented elders, in particular cerebrovascular lesions, *e.g.* small or large cerebral infarctions, lacunes, and white matter lesions in up to almost 10% [80, 206, 216, 219, 268]. Evaluation of 336 cognitively normal seniors from four studies revealed moderately to frequent neuritic plaque density in 47%, of these 6% also had Braak stages I to VI, medullary, nigral, and cortical Lewy bodies in 15,8 and 4%, respectively, cerebral microinfarcts in 33%, and high-level microinfarcts in 10%. The burden of brain lesions and comorbidities varied widely within each study but was similar across studies [269]. In an autopsy series of 1000 elderly subjects including 410 with a primary diagnosis of AD, more than 77% of whom had grossly appearing circle of Willis atherosclerosis, which was significantly higher than normal (47%) or other neurodegenerative diseases (43-67%), suggesting that common or reciprocally synergistic mechanisms promote both vascular and AD pathology [270]. In the Medical Research Council Cognitive Function and Ageing Study (MRC CFAS), vascular risk factors were not associated with an increaseed burden of AD pathology in old age, while a positive association between indices of cardiovascular health (treated hypertension and ischemic heart disease) and cerebral microinfarcts emerged, supporting the view that cerebral vessel disease and cardiovascular disease are interrelated [271].

An older study of 382 demented subjects from the State of Florida Brain Bank showed AD as the most frequent pathological finding (77%), followed by DLB (26%), VaD (18%), hippocampal sclerosis (13%), and FTLD (5%). Mixed pathology was common: concomitant AD was present in 66% of LBD, in 77% of VaD, and in 66% of hippocampal sclerosis patients [272]. Among 418 non-demented partizipants of the Religious Order study (mean age 88.5±5.3 years), 35% showed macroscopic infarcts, 8% microinfarcts, 14.8% arteriosclerosis, 5.7% both, only 37.5% being free of cerebrovascular lesions [273]. Up to 75% of cognitively unimpaired seniors had various degrees of cerebral amyloid angiopathy [219], argyrophilic grains in up to 23% [219], Lewy pathologies in up to 18% [192, 206, 213, 216, 217, 268], occasional hippocampal sclerosis [213, 219], and mixed pathologies in 7 to 15% [206, 217, 268]. Among 100 non-demented elderly, mild, moderate and severe intracranial atherosclerosis was present in 31, 17 and 6%, respectively, lacunar state in basal ganglia and/or white matter in 73%, hippocampal sclerosis in 3%, Lewy bodies in 5%, tau pathology in brainstem in 60%, and mixed cerebral pathologies in 6%, whereas only 9% were free of cerebrovascular lesions [192]. A recent British non-demented sample (n=53; mean age 81.5±7.4 years) showed maximum score neuritic plaques in 32-49%, NFTs in hippocampus and neocortex in 81 and 30.8%, respectively, white matter lesions in 55-83%, small vessel disease in 45%, infarcts in 13.7%, lacunes in 6%, and cerebral hemorrhages in 10% [274].

Nevertheless it should be borne in mind that all additional pathologies may interact, although their mutual impact often remains unclear. Therefore, the reliability and clinical relevance of the current diagnostic criteria need better qualification and validation. In addition, considerable phenotypical and morphological differences exist between genetic/familial and sporadic AD [275-278]. A recent autopsy study investigating the association with upstream transcription factor 1 (USF1) genotype at a neuropathological level showed that

USF1 carriers had lower NFT prevalence among 65+ year olds [279]. The synergistic interaction between Aβ, tau, and α-synuclein, accelerating neuropathology and cognitive decline, has been summarized recently [280].

CONCLUSIONS AND FUTURE PERSPECTIVES

There is increasing use of biochemical, genetic, and experimental approaches for refinement of diagnosis and analysis of the relevant contribution of different disease processes to neurodegeneration in AD and other dementias (see [281]). Since the majority of degenerative dementing disorders are associated with intracellular and/or extracellular deposition of misfolded proteins, most of them can be classified and diagnosed by morphological, immunohistochemical and/or molecular (neurochemical) identification of these deposits representing characteristic markers or sign posts of particular disorders. Algorithms for the molecular-pathological classification of sporadic (nongenetic/nonhereditary) forms of neurodegenerative dementias have been proposed [65, 66, 282]. However, due to variable overlap, these changes may fail to distinguish between cognitively intact aged subjects from those with mild cognitive impairment (MCI) or preclinical or mild AD [3, 7, 180, 283, 284]. These latter groups show a wide variety in the intensity and pattern of AD-related lesions and other (vascular) pathologies [206, 274]. Although they often differ from "normal" aging, only a small proportion of cognitively intact aged subjects are free of AD pathology, while up to 50% may show AD-related changes or even definite AD pathology [80, 194, 216, 285, 286]. Additional challenges arise from various subtypes of dementing disorders and frequent coexistence with other pathologies that may have an additive or synergistic effect.

Neuropathology using immunohistochemistry, molecular biological and genetic methods can achieve a diagnosis or classification in up to 99%, using homogeneous and harmonized definitions and standardized inter-laboratory methods, for the assessment of nervous system lesions, and considering exact clinical data. Interdisciplinary projects/initiatives for the standardized assessment of clinical, neuroimaging, biomarker, and neuropathological data are currently under way [33, 40, 43, 287-290]. In the majority of cases except those with known genetic or metabolic backgrounds, however, pathological examination may not be able to clarify the causes or etiology of most dementing disorders [59], while some conservative authors emphasized that autopsy examination of well-studied cases of AD and other dementias still has a critical role to play [291]. Therefore, based on current knowledge and future research, new and more exact criteria for the neuropathological diagnosis of neurodegenerative and dementing disorders are needed in order to find a way out of the current "chaos" regarding histological diagnosis of dementia and their clinical implications [1, 32]. Molecular genetics, biochemistry and animal models, at least in part reproducing the morphology of human AD and related disorders, have produced a large and convincing body of data on the pathogenesis and pathophysiology of these diseases, showing a complex cascade of events leading from preclinical to fully developed neurodegeneration [292-297]. However, both their molecular backgrounds, basic etiologic factors, pathogenic interrelations of various concomitant pathologies and their impact on the manifestation of AD need further validation. Harmonized techniques are required to increase the accuracy and reproducibility of neuropathological diagnosis of dementia disorders as a basis for further successful treatment and neuroprotection; an enormous challenge for modern neurosciences.

ACKNOWLEDGEMENTS

The study was supported in part by the Society for the Promotion of Research in Experimental Neurology, Vienna, Austria. The author thanks Mr. E. Mitter-Ferstl, PhD, for secretarial and computer work.

CONFLICT OF INTEREST

The author(s) confirm that this chapter content has no conflict of interest.

REFERENCES

[1] Dubois B, Feldman HH, Jacova C, *et al.* Revising the definition of Alzheimer's disease: a new lexicon. Lancet Neurol 2010; 9: 1118-27.

[2] Peng FC. Is dementia a disease? Gerontology 2003; 49: 384-91.

[3] Scheff SW, Price DA, Schmitt FA, Scheff MA, Mufson EJ. Synaptic loss in the inferior temporal gyrus in mild cognitive impairment and Alzheimer's disease. J Alzheimers Dis 2011; 24: 547-57.

[4] Reddy PH, Mani G, Park BS, *et al.* Differential loss of synaptic proteins in Alzheimer's disease: implications for synaptic dysfunction. J Alzheimers Dis 2005; 7: 103-17.

[5] Schulz-Schaeffer WJ. The synaptic pathology of alpha-synuclein aggregation in dementia with Lewy bodies, Parkinson's disease and Parkinson's disease dementia. Acta Neuropathol 2010; 120: 131-43.

[6] Geula C, Nagykery N, Nicholas A, Wu CK. Cholinergic neuronal and axonal abnormalities are present early in aging and in Alzheimer disease. J Neuropathol Exp Neurol 2008; 67: 309-18.

[7] Mufson EJ, Binder L, Counts SE, *et al.* Mild cognitive impairment: pathology and mechanisms. Acta Neuropathol 2012; 123: 13-30.

[8] McCaffrey P, Fagan T, Landhuis E. Alzheimer research series on the default network. J Alzheimers Dis 2010; 19: 747-58.

[9] Palop JJ, Mucke L. Amyloid-beta-induced neuronal dysfunction in Alzheimer's disease: from synapses toward neural networks. Nat Neurosci 2010; 13: 812-8.

[10] Pievani M, de Haan W, Wu T, Seeley WW, Frisoni GB. Functional network disruption in the degenerative dementias. Lancet Neurol 2011; 10: 829-43.

[11] Schroeter ML, Vogt B, Frisch S, *et al.* Executive deficits are related to the inferior frontal junction in early dementia. Brain 2012; 135: 201-15.

[12] Wang K, Liang M, Wang L, *et al.* Altered functional connectivity in early Alzheimer's disease: a resting-state fMRI study. Hum Brain Mapp 2007; 28: 967-78.

[13] Raj A, Kuceyeski A, Weiner M. A network diffusion model of disease progression in dementia. Neuron 2012; 73: 1204-15.

[14] Alzheimer's Association. 2012 Alzheimer's disease facts and figures. Alzheimers Dement 2012; 8: 131-68.

[15] WHO, World, Health, Organization. Dementia: a public health priority. Geneva: World Health Organization, http://www.who.int/mental_health/publications/dementia_report_2012/en/index.html, accessed Dec. 2012.

[16] Qiu C, Kivipelto M, von Strauss E. Epidemiology of Alzheimer's disease: occurrence, determinants, and strategies toward intervention. Dialogues Clin Neurosci 2009; 11: 111-28.

[17] Sorbi S, Hort J, Erkinjuntti T, *et al.* EFNS-ENS Guidelines on the diagnosis and management of disorders associated with dementia. Eur J Neurol 2012; 19: 1159-1179.

[18] McKhann GM, Knopman DS, Chertkow H, *et al.* The diagnosis of dementia due to Alzheimer's disease: recommendations from the National Institute on Aging-Alzheimer's Association workgroups on diagnostic guidelines for Alzheimer's disease. Alzheimers Dement 2011; 7: 263-9.

[19] Sperling RA, Aisen PS, Beckett LA, *et al.* Toward defining the preclinical stages of Alzheimer's disease: recommendations from the National Institute on Aging-Alzheimer's Association workgroups on diagnostic guidelines for Alzheimer's disease. Alzheimers Dement 2011; 7: 280-92.

[20] Sarazin M, de Souza LC, Lehericy S, Dubois B. Clinical and research diagnostic criteria for Alzheimer's disease. Neuroimaging Clin N Am 2012; 22: 23-32,viii.

[21] Dubois B, Burn D, Goetz C, *et al.* Diagnostic procedures for Parkinson's disease dementia: recommendations from the Movement Disorder Society task force. Mov Disord 2007; 22: 2314-24.

[22] Emre M, Aarsland D, Brown R, *et al.* Clinical diagnostic criteria for dementia associated with Parkinson's disease. Mov Disord 2007; 22: 1689-707.

[23] McKeith IG, Dickson DW, Lowe J, *et al.* Diagnosis and management of dementia with Lewy bodies: third report of the DLB Consortium. Neurology 2005; 65: 1863-72.

[24] Josephs KA, Hodges JR, Snowden JS, *et al.* Neuropathological background of phenotypical variability in frontotemporal dementia. Acta Neuropathol 2011; 122: 137-53.
[25] Seltman RE, Matthews BR. Frontotemporal lobar degeneration: epidemiology, pathology, diagnosis and management. CNS Drugs 2012; 26: 841-70.
[26] Sachdev P, Kalaria R, O'Brien J, *et al.* A proposal for new criteria for vascular cognitive disorders. Neurology 2012; submitted.
[27] Chui HC, Mack W, Jackson JE, *et al.* Clinical criteria for the diagnosis of vascular dementia: a multicenter study of comparability and interrater reliability. Arch Neurol 2000; 57: 191-6.
[28] Lopez OL, Kuller LH, Becker JT, *et al.* Classification of vascular dementia in the Cardiovascular Health Study Cognition Study. Neurology 2005; 64: 1539-47.
[29] Gorelick PB, Scuteri A, Black SE, *et al.* Vascular contributions to cognitive impairment and dementia: a statement for healthcare professionals from the american heart association/american stroke association. Stroke 2011; 42: 2672-713.
[30] Karantzoulis S, Galvin JE. Distinguishing Alzheimer's disease from other major forms of dementia. Expert Rev Neurother 2011; 11: 1579-91.
[31] Lopez OL, McDade E, Riverol M, Becker JT. Evolution of the diagnostic criteria for degenerative and cognitive disorders. Curr Opin Neurol 2011; 24: 532-41.
[32] Jellinger KA. Criteria for the neuropathological diagnosis of dementing disorders: Routes out of the swamp? In: Braissant O, Wakamatsu H, Liu IKK, Allegaert K, Lenbury Y, Wachholtz A, editors. Recent Researches in Modern Medicine. Proceedings of the WSEAS Intl. Conferences in Cambridge, UK, Feb. 23-25, 2011. Cambridge, UK: WSEAS Press 2011; pp. 71-97.
[33] Vanderstichele H, Bibl M, Engelborghs S, *et al.* Standardization of preanalytical aspects of cerebrospinal fluid biomarker testing for Alzheimer's disease diagnosis: a consensus paper from the Alzheimer's Biomarkers Standardization Initiative. Alzheimers Dement 2012; 8: 65-73.
[34] Zabel M, Schrag M, Mueller C, *et al.* Assessing candidate serum biomarkers for Alzheimer's disease: a longitudinal study. J Alzheimers Dis 2012; 30: 311-21.
[35] Genius J, Klafki H, Benninghoff J, Esselmann H, Wiltfang J. Current application of neurochemical biomarkers in the prediction and differential diagnosis of Alzheimer's disease and other neurodegenerative dementias. Eur Arch Psychiatry Clin Neurosci 2012; 262 Suppl 2: S71-7.
[36] Blennow K, Zetterberg H, Fagan AM. Fluid biomarkers in Alzheimer disease. Cold Spring Harb Perspect Med 2012; 2.
[37] Doecke JD, Laws SM, Faux NG, *et al.* Blood-based protein biomarkers for diagnosis of Alzheimer disease. Arch Neurol 2012; in print: doi: 10.1001/archneurol.2012.1282.
[38] Engelborghs S, Le Bastard N. The impact of cerebrospinal fluid biomarkers on the diagnosis of Alzheimer's disease. Mol Diagn Ther 2012; 16: 135-41.
[39] Fagan AM, Perrin RJ. Upcoming candidate cerebrospinal fluid biomarkers of Alzheimer's disease. Biomark Med 2012; 6: 455-76.
[40] Johnson KA, Fox NC, Sperling RA, Klunk WE. Brain imaging in Alzheimer disease. Cold Spring Harb Perspect Med 2012; 2: a006213.
[41] Price JC. Molecular brain imaging in the multimodality era. J Cereb Blood Flow Metab 2012; 32: 1377-92.

[42] Teipel SJ, Wegrzyn M, Meindl T, *et al.* Anatomical MRI and DTI in the diagnosis of Alzheimer's disease: a European multicenter study. J Alzheimers Dis 2012; 31: S33-47.

[43] Toledo JB, Brettschneider J, Grossman M, *et al.* CSF biomarkers cutoffs: the importance of coincident neuropathological diseases. Acta Neuropathol 2012, 124: 23-35.

[44] Koyama A, Okereke OI, Yang T, Blacker D, Selkoe DJ, Grodstein F. Plasma amyloid-beta as a predictor of dementia and cognitive decline: a systematic review and meta-analysis. Arch Neurol 2012, 69: 824-31.

[45] Kadir A, Marutle A, Gonzalez D, *et al.* Positron emission tomography imaging and clinical progression in relation to molecular pathology in the first Pittsburgh Compound B positron emission tomography patient with Alzheimer's disease. Brain 2011; 134: 301-17.

[46] Villemagne VL, Pike KE, Chetelat G, *et al.* Longitudinal assessment of Abeta and cognition in aging and Alzheimer disease. Ann Neurol 2011; 69: 181-92.

[47] Beckett TL, Webb RL, Niedowicz DM, *et al.* Postmortem Pittsburgh Compound B (PiB) binding increases with Alzheimer's disease progression. J Alzheimers Dis 2012; 32: 127-38.

[48] Clark CM, Pontecorvo MJ, Beach TG, *et al.* Cerebral PET with florbetapir compared with neuropathology at autopsy for detection of neuritic amyloid-beta plaques: a prospective cohort study. Lancet Neurol 2012; 11: 669-78.

[49] Jack CR, Jr., Knopman DS, Weigand SD, *et al.* An operational approach to National Institute on Aging-Alzheimer's Association criteria for preclinical Alzheimer disease. Ann Neurol 2012; 71: 765-75.

[50] Villain N, Chetelat G, Grassiot B, *et al.* Regional dynamics of amyloid-beta deposition in healthy elderly, mild cognitive impairment and Alzheimer's disease: a voxelwise PiB-PET longitudinal study. Brain 2012; 135: 2126-39.

[51] Desikan RS, McEvoy LK, Thompson WK, *et al.* Amyloid-beta associated volume loss occurs only in the presence of phospho-tau. Ann Neurol 2011; 70: 657-61.

[52] Desikan RS, McEvoy LK, Thompson WK, *et al.* Amyloid-beta--associated clinical decline occurs only in the presence of elevated P-tau. Arch Neurol 2012; 69: 709-13.

[53] Svedberg MM, Hall H, Hellstrom-Lindahl E, *et al.* [(11)C]PIB-amyloid binding and levels of Abeta40 and Abeta42 in postmortem brain tissue from Alzheimer patients. Neurochem Int 2009; 54: 347-57.

[54] Sojkova J, Driscoll I, Iacono D, *et al. In vivo* fibrillar beta-amyloid detected using [11C]PiB positron emission tomography and neuropathologic assessment in older adults. Arch Neurol 2011; 68: 232-40.

[55] Niedowicz DM, Beckett TL, Matveev S, *et al.* Pittsburgh compound B and the postmortem diagnosis of Alzheimer's disease. Ann Neurol 2012; 72: 564-70.

[56] Zhang W, Arteaga J, Cashion DK, *et al.* A highly selective and specific PET tracer for imaging of Tau pathologies. J Alzheimers Dis 2012; 31: 601-12.

[57] Dawe RJ, Bennett DA, Schneider JA, Arfanakis K. Neuropathologic correlates of hippocampal atrophy in the elderly: a clinical, pathologic, postmortem MRI study. PLoS One 2011; 6: e26286.

[58] Durand-Martel P, Tremblay D, Brodeur C, Paquet N. Autopsy as gold standard in FDG-PET studies in dementia. Can J Neurol Sci 2010; 37: 336-42.

[59] Jellinger KA. Con: Can neuropathology really confirm the exact diagnosis? Alzheimers Res Ther 2010; 2: 11.

[60] Beach TG, Monsell SE, Phillips LE, Kukull W. Accuracy of the clinical diagnosis of Alzheimer disease at National Institute on Aging Alzheimer Disease Centers, 2005-2010. J Neuropathol Exp Neurol 2012; 71: 266-73.
[61] Taipa R, Pinho J, Melo-Pires M. Clinico-pathological correlations of the most common neurodegenerative dementias. Front Neurol 2012; 3: 68.
[62] Nelson PT, Jicha GA, Kryscio RJ, *et al.* Low sensitivity in clinical diagnoses of dementia with Lewy bodies. J Neurol 2010; 257: 359-66.
[63] Echavarri C, Burgmans S, Caballero MC, Garcia-Bragado F, Verhey FR, Uylings HB. Co-occurrence of different pathologies in dementia: implications for dementia diagnosis. J Alzheimers Dis 2012; 30: 909-17.
[64] Rossor MN, Fox NC, Mummery CJ, Schott JM, Warren JD. The diagnosis of young-onset dementia. Lancet Neurol 2010; 9: 793-806.
[65] Jellinger KA. Basic mechanisms of neurodegeneration: a critical update. J Cell Mol Med 2010; 14: 457-487.
[66] Kovacs GG, Botond G, Budka H. Protein coding of neurodegenerative dementias: the neuropathological basis of biomarker diagnostics. Acta Neuropathol 2010; 119: 389-408.
[67] Kovacs GG, Budka H. Current concepts of neuropathological diagnostics in practice: neurodegenerative diseases. Clin Neuropathol 2010; 29: 271-88.
[68] Duyckaerts C, Dickson DW. Neuropathology of Alzheimer's disease and its variants. In: Dickson DW, Weller RO, editors. Neurodegeneration. The Molecular Pathology of Dementia and Movement Disorders, 2nd Edition: Wiley-Blackwell 2011; pp. 62-91.
[69] Serrano-Pozo A, Frosch MP, Masliah E, Hyman BT. Neuropathological alterations in Alzheimer disease. Cold Spring Harb Perspect Med 2011; 1: a006189.
[70] Mirra SS, Heyman A, McKeel D, *et al.* The Consortium to Establish a Registry for Alzheimer's Disease (CERAD). Part II. Standardization of the neuropathologic assessment of Alzheimer's disease. Neurology 1991; 41: 479-86.
[71] Braak H, Braak E. Neuropathological stageing of Alzheimer-related changes. Acta Neuropathol 1991; 82: 239-59.
[72] Braak H, Bohl JR, Muller CM, Rub U, de Vos RA, Del Tredici K. Stanley Fahn Lecture 2005: The staging procedure for the inclusion body pathology associated with sporadic Parkinson's disease reconsidered. Mov Disord 2006; 21: 2042-51.
[73] Alafuzoff I, Arzberger T, Al-Sarraj S, *et al.* Staging of neurofibrillary pathology in Alzheimer's disease: a study of the BrainNet Europe Consortium. Brain Pathol 2008; 18: 484-96.
[74] Thal DR, Rub U, Orantes M, Braak H. Phases of A beta-deposition in the human brain and its relevance for the development of AD. Neurology 2002; 58: 1791-800.
[75] Krstic D, Knuesel I. Deciphering the mechanism underlying late-onset Alzheimer disease. Nat Rev Neurol 2012; 9: 25-34.
[76] Hyman BT, Trojanowski JQ. Consensus recommendations for the postmortem diagnosis of Alzheimer disease from the National Institute on Aging and the Reagan Institute Working Group on diagnostic criteria for the neuropathological assessment of Alzheimer disease. J Neuropathol Exp Neurol 1997; 56: 1095-7.
[77] Cochran EJ, Schneider JA, Bennett DA, Mufson EJ. Application of NIA/Reagan Institute Working Group Criteria for diagnosis of Alzheimer's disease to members of the Religious Orders Study (abstr.). J Neuropathol Exp Neurol 1998; 57: 508.

[78] Newell KL, Hyman BT, Growdon JH, Hedley-Whyte ET. Application of the National Institute on Aging (NIA)-Reagan Institute criteria for the neuropathological diagnosis of Alzheimer disease. J Neuropathol Exp Neurol 1999; 58: 1147-55.

[79] Harding AJ, Kril JJ, Halliday GM. Practical measures to simplify the Braak tangle staging method for routine pathological screening. Acta Neuropathol 2000; 99: 199-208.

[80] Davis DG, Schmitt FA, Wekstein DR, Markesbery WR. Alzheimer neuropathologic alterations in aged cognitively normal subjects. J Neuropathol Exp Neurol 1999; 58: 376-88.

[81] McKee AC, Kowall NW, Au R. Topography of neurofibrillary tangles distinguishes aging from Alzheimer disease (abstr.). J Neuropathol Exp Neurol 2002; 61: 488.

[82] Jellinger KA. A view on early diagnosis of dementias from neuropathology. In: Herholz K, Morris C, Perani D, editors. The Dementias: Early Diagnosis and Evaluation. New York: Taylor & Francis 2006; pp. 311-428.

[83] Hyman BT, Phelps CH, Beach TG, *et al.* National Institute on Aging-Alzheimer's Association guidelines for the neuropathologic assessment of Alzheimer's disease. Alzheimers Dement 2012; 8: 1-13.

[84] Montine TJ, Phelps CH, Beach TG, *et al.* National Institute on Aging-Alzheimer's Association guidelines for the neuropathologic assessment of Alzheimer's disease: a practical approach. Acta Neuropathol 2012; 123: 1-11.

[85] Nelson PT, Alafuzoff I, Bigio EH, *et al.* Correlation of Alzheimer disease neuropathologic changes with cognitive status: a review of the literature. J Neuropathol Exp Neurol 2012; 71: 362-81.

[86] Jellinger KA, Attems J. Prevalence and pathology of vascular dementia in the oldest-old. J Alzheimers Dis 2010; 21: 1283-1298.

[87] Braak H, Del Tredici K, Rub U, de Vos RA, Jansen Steur EN, Braak E. Staging of brain pathology related to sporadic Parkinson's disease. Neurobiol Aging 2003; 24: 197-211.

[88] Braak H, Alafuzoff I, Arzberger T, Kretzschmar H, Del Tredici K. Staging of Alzheimer disease-associated neurofibrillary pathology using paraffin sections and immunocytochemistry. Acta Neuropathol 2006; 112: 389-404.

[89] Beach TG, Adler CH, Sue LI, *et al.* Multi-organ distribution of phosphorylated alpha-synuclein histopathology in subjects with Lewy body disorders. Acta Neuropathol 2010; 119: 689-702.

[90] Jellinger KA. Neuropathology of sporadic Parkinson's disease: evaluation and changes of concepts. Mov Disord 2012; 27: 8-30.

[91] Alafuzoff I, Parkkinen L, Al-Sarraj S, *et al.* Assessment of alpha-synuclein pathology: a study of the BrainNet Europe Consortium. J Neuropathol Exp Neurol 2008; 67: 125-143.

[92] Alafuzoff I, Ince PG, Arzberger T, *et al.* Staging/typing of Lewy body related alpha-synuclein pathology: a study of the BrainNet Europe Consortium. Acta Neuropathol 2009; 117: 635-52.

[93] Kingsbury AE, Bandopadhyay R, Silveira-Moriyama L, *et al.* Brain stem pathology in Parkinson's disease: an evaluation of the Braak staging model. Mov Disord 2010; 25: 2508-15.

[94] Burke RE, Dauer WT, Vonsattel JP. A critical evaluation of the Braak staging scheme for Parkinson's disease. Ann Neurol 2008; 64: 485-91.

[95] Jellinger KA. A critical reappraisal of current staging of Lewy-related pathology in human brain. Acta Neuropathol 2008; 116: 1-16.

[96] Parkkinen L, Pirttila T, Alafuzoff I. Applicability of current staging/categorization of alpha-synuclein pathology and their clinical relevance. Acta Neuropathol 2008; 115: 399-407.
[97] Jellinger KA. A critical evaluation of current staging of alpha-synuclein pathology in Lewy body disorders. Biochim Biophys Acta 2009; 1792: 730-40.
[98] Zaccai J, Brayne C, McKeith I, Matthews F, Ince PG. Patterns and stages of alpha-synucleinopathy: Relevance in a population-based cohort. Neurology 2008; 70: 1042-8.
[99] Fujishiro H, Ferman TJ, Boeve BF, *et al.* Validation of the neuropathologic criteria of the Third Consortium for Dementia With Lewy Bodies for Prospectively Diagnosed Cases. J Neuropathol Exp Neurol 2008; 67: 649-656.
[100] Leverenz JB, Hamilton R, Tsuang DW, *et al.* Empiric refinement of the pathologic assessment of Lewy-related pathology in the dementia patient. Brain Pathol 2008; 18: 220-4.
[101] Ballard C, Ziabreva I, Perry R, *et al.* Differences in neuropathologic characteristics across the Lewy body dementia spectrum. Neurology 2006; 67: 1931-4.
[102] Compta Y, Parkkinen L, O'Sullivan SS, *et al.* Lewy- and Alzheimer-type pathologies in Parkinson's disease dementia: which is more important? Brain 2011; 134: 1493-505.
[103] Jellinger KA, Attems J. Does striatal pathology distinguish Parkinson disease with dementia and dementia with Lewy bodies? Acta Neuropathol 2006; 112: 253-60.
[104] Halliday GM, Song YJ, Harding AJ. Striatal beta-amyloid in dementia with Lewy bodies but not Parkinson's disease. J Neural Transm 2011; 118: 713-9.
[105] Dugger BN, Serrano GE, Sue LI, *et al.* Presence of striatal amyloid plaques in Parkinson's disease dementia predicts concomitant Alzheimer's disease: usefulness for amyloid imaging. J Parkinsons Dis 2012; 2: 57-65.
[106] Gomperts SN, Locascio JJ, Marquie M, *et al.* Brain amyloid and cognition in Lewy body diseases. Mov Disord 2012; in press: doi: 10.1002/mds.25048.
[107] Sabbagh MN, Adler CH, Lahti TJ, *et al.* Parkinson disease with dementia: comparing patients with and without Alzheimer pathology. Alzheimer Dis Assoc Disord 2009; 23: 295-7.
[108] Braak H, Rub U, Jansen Steur EN, Del Tredici K, de Vos RA. Cognitive status correlates with neuropathologic stage in Parkinson disease. Neurology 2005; 64: 1404-10.
[109] Irwin DJ, White MT, Toledo JB, *et al.* Neuropathologic substrates of Parkinson disease dementia. Ann Neurol 2012; 72: 587-98.
[110] Clinton LK, Blurton-Jones M, Myczek K, Trojanowski JQ, LaFerla FM. Synergistic Interactions between Abeta, tau, and alpha-synuclein: acceleration of neuropathology and cognitive decline. J Neurosci 2010; 30: 7281-9.
[111] Goedert M, Ghetti B, Spillantini MG. Frontotemporal dementia: implications for understanding Alzheimer disease. Cold Spring Harb Perspect Med 2012; 2: a006254.
[112] Halliday GM, Bigio EH, Cairns NJ, Neumann M, Mackenzie IR, Mann DM. Mechanisms of disease in frontotemporal lobar degeneration: gain of function *vs.* loss of function effects. Acta Neuropathol 2012; 124: 373-82.
[113] Claassen DO, Parisi JE, Giannini C, Boeve BF, Dickson DW, Josephs KA. Frontotemporal dementia mimicking dementia with Lewy bodies. Cogn Behav Neurol 2008; 21: 157-63.
[114] Cerami C, Scarpini E, Cappa SF, Galimberti D. Frontotemporal lobar degeneration: current knowledge and future challenges. J Neurol 2012; 259: 2278-86.
[115] Baborie A, Griffiths TD, Jaros E, *et al.* Frontotemporal dementia in elderly individuals. Arch Neurol 2012; 69: 1052-60.

[116] Arai T, Mackenzie IR, Hasegawa M, *et al.* Phosphorylated TDP-43 in Alzheimer's disease and dementia with Lewy bodies. Acta Neuropathol 2009; 117: 125-36.

[117] Woulfe J, Gray DA, Mackenzie IR. FUS-Immunoreactive Intranuclear Inclusions in Neurodegenerative Disease. Brain Pathol 2010; 20: 586-597.

[118] Wilson AC, Dugger BN, Dickson DW, Wang DS. TDP-43 in aging and Alzheimer's disease - a review. Int J Clin Exp Pathol 2011; 4: 147-55.

[119] Román GC, Tatemichi TK, Erkinjuntti T, *et al.* Vascular dementia: diagnostic criteria for research studies. Report of the NINDS-AIREN International Workshop. Neurology 1993; 43: 250-60.

[120] Kalaria RN, Kenny RA, Ballard CG, Perry R, Ince P, Polvikoski T. Towards defining the neuropathological substrates of vascular dementia. J Neurol Sci 2004; 226: 75-80.

[121] Jellinger KA. The enigma of vascular cognitive disorder and vascular dementia. Acta Neuropathol 2007; 113: 349-88.

[122] Pantoni L, Sarti C, Alafuzoff I, *et al.* Postmortem examination of vascular lesions in cognitive impairment: a survey among neuropathological services. Stroke 2006; 37: 1005-9.

[123] Alafuzoff I, Gelpi E, Al-Sarraj S, *et al.* The need to unify neuropathological assessments of vascular alterations in the ageing brain. Multicentre survey by the BrainNet Europe consortium. Exp Gerontol 2012; 47: 825-33.

[124] Thal DR, Grinberg LT, Attems J. Vascular dementia: different forms of vessel disorders contribute to the development of dementia in the elderly brain. Exp Gerontol 2012; 47: 816-24.

[125] Jellinger KA. The pathology of "vascular dementia": a critical update. J Alzheimers Dis 2008; 14: 107-23.

[126] Román GC. The epidemiology of vascular dementia. In: Duyckaerts C, Litvan I, editors. Dementias - Handbook of Clinical Neurology, vol 89 (3rd series). Edinburgh: Elsevier 2008; pp. 639-658.

[127] Ferrer I. Cognitive impairment of vascular origin: neuropathology of cognitive impairment of vascular origin. J Neurol Sci 2010; 299: 139-49.

[128] Deramecourt V, Slade JY, Oakley AE, *et al.* Staging and natural history of cerebrovascular pathology in dementia. Neurology 2012; 78: 1043-50.

[129] Jellinger KA. The enigma of mixed dementia. Alzheimer's & Dementia 2007; 3: 40-53.

[130] Grinberg LT, Thal DR. Vascular pathology in the aged human brain. Acta Neuropathol 2010; 119: 277-90.

[131] Prusiner SB. Prions. Proc Natl Acad Sci U S A 1998; 95: 13363-83.

[132] Wadsworth JD, Collinge J. Molecular pathology of human prion disease. Acta Neuropathol 2011; 121: 69-77.

[133] Safar JG. Molecular pathogenesis of sporadic prion diseases in man. Prion 2012; 6: 108-15.

[134] Head MW, Ironside JW. The contribution of different prion protein types and host polymorphisms to clinicopathological variations in Creutzfeldt-Jakob disease. Rev Med Virol 2012; 22: 214-29.

[135] Zhao QL, Zhou Y, Wang YL, Dong KH, Wang YJ. A new diagnostic algorithm for vascular cognitive impairment: the proposed criteria and evaluation of its reliability and validity. Chin Med J (Engl) 2010; 123: 311-9.

[136] Gambetti P, Puoti G, Zou WQ. Variably protease-sensitive prionopathy: a novel disease of the prion protein. J Mol Neurosci 2011; 45: 422-4.

[137] Cochran EJ, Bennett DA, Cervenakova L, *et al.* Familial Creutzfeldt-Jakob disease with a five-repeat octapeptide insert mutation. Neurology 1996; 47: 727-33.
[138] Jansen C, Voet W, Head MW, *et al.* A novel seven-octapeptide repeat insertion in the prion protein gene (PRNP) in a Dutch pedigree with Gerstmann-Straussler-Scheinker disease phenotype: comparison with similar cases from the literature. Acta Neuropathol 2011; 121: 59-68.
[139] Jayadev S, Nochlin D, Poorkaj P, *et al.* Familial prion disease with Alzheimer disease-like tau pathology and clinical phenotype. Ann Neurol 2011; 69: 712-20.
[140] Kovacs GG, Seguin J, Quadrio I, *et al.* Genetic Creutzfeldt-Jakob disease associated with the E200K mutation: characterization of a complex proteinopathy. Acta Neuropathol 2011; 121: 39-57.
[141] Heath CA, Cooper SA, Murray K, *et al.* Diagnosing variant Creutzfeldt-Jakob disease: a retrospective analysis of the first 150 cases in the UK. J Neurol Neurosurg Psychiatry 2011; 82: 646-51.
[142] Capellari S, Strammiello R, Saverioni D, Kretzschmar H, Parchi P. Genetic Creutzfeldt-Jakob disease and fatal familial insomnia: insights into phenotypic variability and disease pathogenesis. Acta Neuropathol 2011; 121: 21-37.
[143] Collinge J, Whitfield J, McKintosh E, *et al.* Kuru in the 21st century--an acquired human prion disease with very long incubation periods. Lancet 2006; 367: 2068-74.
[144] Brandner S. Diversity of prion diseases: (no) strains attached? Acta Neuropathol 2011; 121: 1-4.
[145] Haybaeck J, Heikenwalder M, Klevenz B, *et al.* Aerosols transmit prions to immunocompetent and immunodeficient mice. PLoS Pathog 2011; 7: e1001257.
[146] Parchi P, Strammiello R, Giese A, Kretzschmar H. Phenotypic variability of sporadic human prion disease and its molecular basis: past, present, and future. Acta Neuropathol 2011; 121: 91-112.
[147] Mead S, Uphill J, Beck J, *et al.* Genome-wide association study in multiple human prion diseases suggests genetic risk factors additional to PRNP. Hum Mol Genet 2012; 21: 1897-906.
[148] Parchi P, Saverioni D. Molecular pathology, classification, and diagnosis of sporadic human prion disease variants. Folia Neuropathol 2012; 50: 20-45.
[149] Gambetti P, Cali I, Notari S, Kong Q, Zou WQ, Surewicz WK. Molecular biology and pathology of prion strains in sporadic human prion diseases. Acta Neuropathol 2011; 121: 79-90.
[150] Parchi P, de Boni L, Saverioni D, *et al.* Consensus classification of human prion disease histotypes allows reliable identification of molecular subtypes: an inter-rater study among surveillance centres in Europe and USA. Acta Neuropathol 2012; 124: 517-29.
[151] Geschwind MD, Shu H, Haman A, Sejvar JJ, Miller BL. Rapidly progressive dementia. Ann Neurol 2008; 64: 97-108.
[152] Josephs KA, Ahlskog JE, Parisi JE, *et al.* Rapidly progressive neurodegenerative dementias. Arch Neurol 2009; 66: 201-7.
[153] Chitravas N, Jung RS, Kofskey DM, *et al.* Treatable neurological disorders misdiagnosed as Creutzfeldt-Jakob disease. Ann Neurol 2011; 70: 437-44.
[154] Zanusso G, Fiorini M, Ferrari S, *et al.* Cerebrospinal fluid markers in sporadic creutzfeldt-jakob disease. Int J Mol Sci 2011; 12: 6281-92.
[155] Schmidt C, Haik S, Satoh K, *et al.* Rapidly progressive Alzheimer's disease: a multicenter update. J Alzheimers Dis 2012; 30: 751-6.

[156] Seppälä TT, Nerg O, Koivisto AM, *et al.* CSF biomarkers for Alzheimer disease correlate with cortical brain biopsy findings. Neurology 2012; 78: 1568-75.
[157] Schott JM, Reiniger L, Thom M, *et al.* Brain biopsy in dementia: clinical indications and diagnostic approach. Acta Neuropathol 2010; 120: 327-41.
[158] Josephson SA, Papanastassiou AM, Berger MS, *et al.* The diagnostic utility of brain biopsy procedures in patients with rapidly deteriorating neurological conditions or dementia. J Neurosurg 2007; 106: 72-5.
[159] Snowden JS, Thompson JC, Stopford CL, *et al.* The clinical diagnosis of early-onset dementias: diagnostic accuracy and clinicopathological relationships. Brain 2011; 134: 2478-92.
[160] Knopman DS, Petersen RC, Cha RH, Edland SD, Rocca WA. Incidence and causes of nondegenerative nonvascular dementia: a population-based study. Arch Neurol 2006; 63: 218-21.
[161] Duyckaerts C, Delatour B, Potier MC. Classification and basic pathology of Alzheimer disease. Acta Neuropathol 2009; 118: 5-36.
[162] Mirra SS, Gearing M, McKeel DW, Jr., *et al.* Interlaboratory comparison of neuropathology assessments in Alzheimer's disease: a study of the Consortium to Establish a Registry for Alzheimer's Disease (CERAD). J Neuropathol Exp Neurol 1994; 53: 303-15.
[163] Bancher C, Paulus W, Paukner K, Jellinger K. Neuropathologic diagnosis of Alzheimer disease: consensus between practicing neuropathologists? Alzheimer Dis Assoc Disord 1997; 11: 207-19.
[164] Nagy Z, Vatter-Bittner B, Braak H, *et al.* Staging of Alzheimer-type pathology: an interrater-intrarater study. Dement Geriatr Cogn Disord 1997; 8: 248-51.
[165] Nelson PT, Kukull WA, Frosch MP. Thinking outside the box: Alzheimer-type neuropathology that does not map directly onto current consensus recommendations. J Neuropathol Exp Neurol 2010; 69: 449-454.
[166] Levine DN, Lee JM, Fisher CM. The visual variant of Alzheimer's disease: a clinicopathologic case study. Neurology 1993; 43: 305-13.
[167] Jellinger KA, Grazer A, Petrovic K, *et al.* Four-repeat tauopathy clinically presenting as posterior cortical atrophy: atypical corticobasal degeneration? Acta Neuropathol 2011; 121: 267-77.
[168] Jellinger KA, Attems J. Neurofibrillary tangle-predominant dementia: comparison with classical Alzheimer disease. Acta Neuropathol 2007; 113: 107-17.
[169] Santa-Maria I, Haggiagi A, Liu X, *et al.* The MAPT H1 haplotype is associated with tangle-predominant dementia. Acta Neuropathol 2012; 124: 693-704.
[170] Terry RD, Hansen LA, DeTeresa R, Davies P, Tobias H, Katzman R. Senile dementia of the Alzheimer type without neocortical neurofibrillary tangles. J Neuropathol Exp Neurol 1987; 46: 262-8.
[171] Tiraboschi P, Hansen LA, Thal LJ, Corey-Bloom J. The importance of neuritic plaques and tangles to the development and evolution of AD. Neurology 2004; 62: 1984-9.
[172] Hansen L, Salmon D, Galasko D, *et al.* The Lewy body variant of Alzheimer's disease: a clinical and pathologic entity. Neurology 1990; 40: 1-8.
[173] Murray ME, Graff-Radford NR, Ross OA, Petersen RC, Duara R, Dickson DW. Neuropathologically defined subtypes of Alzheimer's disease with distinct clinical characteristics: a retrospective study. Lancet Neurol 2011; 10: 785-96.
[174] Janocko NJ, Brodersen KA, Soto-Ortolaza AI, *et al.* Neuropathologically defined subtypes of Alzheimer's disease differ significantly from neurofibrillary tangle-predominant dementia. Acta Neuropathol 2012.

[175] Whitwell JL, Dickson DW, Murray ME, *et al.* Neuroimaging correlates of pathologically defined subtypes of Alzheimer's disease: a case-control study. Lancet Neurol 2012; 11: 868-77.
[176] Jellinger KA. Neuropathological subtypes of Alzheimer's disease (Correspondence). Acta Neuropathol 2012; 123: 153-4.
[177] Bowler JV, Munoz DG, Merskey H, Hachinski V. Fallacies in the pathological confirmation of the diagnosis of Alzheimer's disease. J Neurol Neurosurg Psychiatry 1998; 64: 18-24.
[178] Nelson PT, Jicha GA, Schmitt FA, *et al.* Clinicopathologic correlations in a large Alzheimer disease center autopsy cohort: neuritic plaques and neurofibrillary tangles "do count" when staging disease severity. J Neuropathol Exp Neurol 2007; 66: 1136-46.
[179] Nelson PT, Braak H, Markesbery WR. Neuropathology and cognitive impairment in Alzheimer disease: a complex but coherent relationship. J Neuropathol Exp Neurol 2009; 68: 1-14.
[180] Markesbery WR. Neuropathologic alterations in mild cognitive impairment: a review. J Alzheimers Dis 2010; 19: 221-8.
[181] Whitwell JL, Josephs KA, Murray ME, *et al.* MRI correlates of neurofibrillary tangle pathology at autopsy: a voxel-based morphometry study. Neurology 2008; 71: 743-9.
[182] McKeel DW, Jr., Price JL, Miller JP, *et al.* Neuropathologic criteria for diagnosing Alzheimer disease in persons with pure dementia of Alzheimer type. J Neuropathol Exp Neurol 2004; 63: 1028-37.
[183] Beach TG, Sue LI, Walker DG, *et al.* Striatal amyloid plaque density predicts Braak neurofibrillary stage and clinicopathological Alzheimer's disease: implications for amyloid imaging. J Alzheimers Dis 2011; 28: 869-76.
[184] Cupidi C, Capobianco R, Goffredo D, *et al.* Neocortical variation of Abeta load in fully expressed, pure Alzheimer's disease. J Alzheimers Dis 2010; 19: 57-68.
[185] Neltner JH, Abner EL, Schmitt FA, *et al.* Digital pathology and image analysis for robust high-throughput quantitative assessment of Alzheimer disease neuropathologic changes. J Neuropathol Exp Neurol 2012; 71: 1075-85.
[186] Holland D, Desikan RS, Dale AM, McEvoy LK. Rates of decline in Alzheimer disease decrease with age. PLoS One 2012; 7: e42325.
[187] von Gunten A, Kovari E, Rivara CB, Bouras C, Hof PR, Giannakopoulos P. Stereologic analysis of hippocampal Alzheimer's disease pathology in the oldest-old: evidence for sparing of the entorhinal cortex and CA1 field. Exp Neurol 2005; 193: 198-206.
[188] Polvikoski T, Sulkava R, Rastas S, *et al.* Incidence of dementia in very elderly individuals: a clinical, neuropathological and molecular genetic study. Neuroepidemiology 2006; 26: 76-82.
[189] Prohovnik I, Perl DP, Davis KL, Libow L, Lesser G, Haroutunian V. Dissociation of neuropathology from severity of dementia in late-onset Alzheimer disease. Neurology 2006; 66: 49-55.
[190] Haroutunian V, Schnaider-Beeri M, Schmeidler J, *et al.* Role of the neuropathology of Alzheimer disease in dementia in the oldest-old. Arch Neurol 2008; 65: 1211-7.
[191] Middleton LE, Grinberg LT, Miller B, Kawas C, Yaffe K. Neuropathologic features associated with Alzheimer disease diagnosis: age matters. Neurology 2011; 77: 1737-44.
[192] Jellinger KA, Attems J. Neuropathology and general autopsy findings in nondemented aged subjects. Clin Neuropathol 2012; 31: 87-98.

[193] Fernando MS, Ince PG. Vascular pathologies and cognition in a population-based cohort of elderly people. J Neurol Sci 2004; 226: 13-7.

[194] Price JL, McKeel DW, Jr., Buckles VD, *et al.* Neuropathology of nondemented aging: presumptive evidence for preclinical Alzheimer disease. Neurobiol Aging 2009; 30: 1026-36.

[195] Purohit D, Batheja N, Haroutunian V, Sano M, Grossman H, Perl DP. A clinicopathological correlation study of cognitive status and senile plaques and neurofibrillary tangles in clinically well-characterized individuals of extreme old age (abstract). J Neuropath Exp Neurol 2008; 67: 494.

[196] Laukka EJ, Fratiglioni L, Backman L. The influence of vascular disease on cognitive performance in the preclinical and early phases of Alzheimer's disease. Dement Geriatr Cogn Disord 2010; 29: 498-503.

[197] Strozyk D, Dickson DW, Lipton RB, *et al.* Contribution of vascular pathology to the clinical expression of dementia. Neurobiol Aging 2010; 31: 1710-20.

[198] Dolan D, Troncoso J, Resnick SM, Crain BJ, Zonderman AB, O'Brien RJ. Age, Alzheimer's disease and dementia in the Baltimore Longitudinal Study of Ageing. Brain 2010; 133: 2225-31.

[199] Sinka L, Kovari E, Gold G, *et al.* Small vascular and Alzheimer disease-related pathologic determinants of dementia in the oldest-old. J Neuropathol Exp Neurol 2010; 69: 1247-55.

[200] Jicha GA, Saligram U, Abner EL, Van Eldik L, Nelson PT. Dementia lacking a known pathologic substrate: results from the University of Kentucky Alzheimer's Disease Center Brain Bank (abstr.). Ann Neurol 2010; 68, Suppl 14: S48.

[201] Crystal HA, Dickson D, Davies P, Masur D, Grober E, Lipton RB. The relative frequency of "dementia of unknown etiology" increases with age and is nearly 50% in nonagenarians. Arch Neurol 2000; 57: 713-9.

[202] Jellinger KA. Frequency of "dementia of unknown origin" increases with age (Letter). Arch Neurol 2001; 58: 1498-99.

[203] Yankner BA, Lu T, Loerch P. The aging brain. Annu Rev Pathol 2008; 3: 41-66.

[204] Neuropathology Group. Pathological correlates of late-onset dementia in a multicentre, community-based population in England and Wales. AUTHORS: Neuropathology Group of the Medical Research Council Cognitive Function and Ageing Study (MRC CFAS). Lancet 2001; 357: 169-75.

[205] O'Brien RJ, Resnick SM, Zonderman AB, *et al.* Neuropathologic studies of the Baltimore Longitudinal Study of Aging (BLSA). J Alzheimers Dis 2009; 18: 665-75.

[206] Schneider JA, Aggarwal NT, Barnes L, Boyle P, Bennett DA. The neuropathology of older persons with and without dementia from community *vs.* clinic cohorts. J Alzheimers Dis 2009; 18: 691-701.

[207] Bennett DA, Wilson RS, Boyle PA, Buchman AS, Schneider JA. Relation of neuropathology to cognition in persons without cognitive impairment. Ann Neurol 2012; in print: DOI 10.1002/ana.23654.

[208] Serrano-Pozo A, Mielke ML, Muzitansky A, *et al.* Stable size distribution of amyloid plaques over the course of Alzheimer disease. J Neuropathol Exp Neurol 2012; 71: 694-701.

[209] Scheff SW, Price DA, Schmitt FA, Mufson EJ. Hippocampal synaptic loss in early Alzheimer's disease and mild cognitive impairment. Neurobiol Aging 2006; 27: 1372-84.

[210] Scheff SW, Price DA, Schmitt FA, DeKosky ST, Mufson EJ. Synaptic alterations in CA1 in mild Alzheimer disease and mild cognitive impairment. Neurology 2007; 68: 1501-8.

[211] Morris JC, Storandt M, McKeel DW, Jr., *et al.* Cerebral amyloid deposition and diffuse plaques in "normal" aging: Evidence for presymptomatic and very mild Alzheimer's disease. Neurology 1996; 46: 707-19.

[212] Green MS, Kaye JA, Ball MJ. The Oregon brain aging study: neuropathology accompanying healthy aging in the oldest old. Neurology 2000; 54: 105-13.

[213] Schmitt FA, Davis DG, Wekstein DR, Smith CD, Ashford JW, Markesbery WR. "Preclinical" AD revisited: neuropathology of cognitively normal older adults. Neurology 2000; 55: 370-6.

[214] Mitchell TW, Mufson EJ, Schneider JA, *et al.* Parahippocampal tau pathology in healthy aging, mild cognitive impairment, and early Alzheimer's disease. Ann Neurol 2002; 51: 182-9.

[215] Knopman DS, Parisi JE, Boeve BF, *et al.* Vascular dementia in a population-based autopsy study. Arch Neurol 2003; 60: 569-75.

[216] Knopman DS, Parisi JE, Salviati A, *et al.* Neuropathology of cognitively normal elderly. J Neuropathol Exp Neurol 2003; 62: 1087-95.

[217] White L. Brain lesions at autopsy in older Japanese-American men as related to cognitive impairment and dementia in the final years of life: a summary report from the Honolulu-Asia Aging Study. J Alzheimers Dis 2009; 18: 713-25.

[218] Santacruz KS, Sonnen JA, Pezhouh MK, Desrosiers MF, Nelson PT, Tyas SL. Alzheimer disease pathology in subjects without dementia in 2 studies of aging: the nun study and the adult changes in thought study. J Neuropathol Exp Neurol 2011; 70: 832-40.

[219] Jentoft M, Parisi J, Dickson D, *et al.* Neuropathologic findings in 32 nondemented elderly subjects (abs.). J Neuropathol Exp Neurol 2011; 70: 531.

[220] Abner EL, Kryscio RJ, Schmitt FA, *et al.* "End-stage" neurofibrillary tangle pathology in preclinical Alzheimer's disease: fact or fiction? J Alzheimers Dis 2011; 25: 445-53.

[221] Barnes LL, Schneider JA, Boyle PA, Bienias JL, Bennett DA. Memory complaints are related to Alzheimer disease pathology in older persons. Neurology 2006; 67: 1581-5.

[222] Kramer PL, Xu H, Woltjer RL, *et al.* Alzheimer disease pathology in cognitively healthy elderly: a genome-wide study. Neurobiol Aging 2011; 32: 2113-22.

[223] Mortimer JA. The Nun Study: risk factors for pathology and clinical-pathologic correlations. Curr Alzheimer Res 2012; 9: 621-7.

[224] Robinson JL, Geser F, Corrada MM, *et al.* Neocortical and hippocampal amyloid-beta and tau measures associate with dementia in the oldest-old. Brain 2011; 134: 3708-15.

[225] Pikkarainen M, Kauppinen T, Alafuzoff I. Hyperphosphorylated tau in the occipital cortex in aged nondemented subjects. J Neuropathol Exp Neurol 2009; 68: 653-60.

[226] Corrada MM, Berlau DJ, Kawas CH. A population-based clinicopathological study in the oldest-old: the 90+ study. Curr Alzheimer Res 2012; 9: 709-17.

[227] Troncoso JC, Cataldo AM, Nixon RA, *et al.* Neuropathology of preclinical and clinical late-onset Alzheimer's disease. Ann Neurol 1998; 43: 673-6.

[228] Price JL, Morris JC. Tangles and plaques in nondemented aging and "preclinical" Alzheimer's disease. Ann Neurol 1999; 45: 358-68.

[229] Galvin JE, Powlishta KK, Wilkins K, *et al.* Predictors of preclinical Alzheimer disease and dementia: a clinicopathologic study. Arch Neurol 2005; 62: 758-65.

[230] Haroutunian V, Perl DP, Purohit DP, *et al.* Regional distribution of neuritic plaques in the nondemented elderly and subjects with very mild Alzheimer disease. Arch Neurol 1998; 55: 1185-91.

[231] Mintun MA, Larossa GN, Sheline YI, *et al.* [11C]PIB in a nondemented population: potential antecedent marker of Alzheimer disease. Neurology 2006; 67: 446-52.

[232] Villemagne VL, Pike KE, Darby D, *et al.* Abeta deposits in older non-demented individuals with cognitive decline are indicative of preclinical Alzheimer's disease. Neuropsychologia 2008; 46: 1688-97.

[233] Hampel H, Shen Y, Walsh DM, *et al.* Biological markers of amyloid beta-related mechanisms in Alzheimer's disease. Exp Neurol 2010; 223: 334-46.

[234] Stomrud E, Hansson O, Zetterberg H, Blennow K, Minthon L, Londos E. Correlation of longitudinal cerebrospinal fluid biomarkers with cognitive decline in healthy older adults. Arch Neurol 2010; 67: 217-23.

[235] Petersen RC, Roberts RO, Knopman DS, *et al.* Mild cognitive impairment: ten years later. Arch Neurol 2009; 66: 1447-55.

[236] Maarouf CL, Daugs ID, Kokjohn TA, *et al.* Alzheimer's disease and non-demented high pathology control nonagenarians: comparing and contrasting the biochemistry of cognitively successful aging. PLoS One 2011; 6: e27291.

[237] Moore BD, Chakrabarty P, Levites Y, *et al.* Overlapping profiles of Abeta peptides in the Alzheimer's disease and pathological aging brains. Alzheimers Res Ther 2012; 4: 18.

[238] Jonsson T, Atwal JK, Steinberg S, *et al.* A mutation in APP protects against Alzheimer's disease and age-related cognitive decline. Nature 2012; 488: 96-9.

[239] Swerdlow RH. Is aging part of Alzheimer's disease, or is Alzheimer's disease part of aging? Neurobiol Aging 2007; 28: 1465-80.

[240] Nelson PT, Head E, Schmitt FA, *et al.* Alzheimer's disease is not "brain aging": neuropathological, genetic, and epidemiological human studies. Acta Neuropathol 2011; 121: 571-87.

[241] Castellani RJ, Lee HG, Zhu X, Perry G, Smith MA. Alzheimer disease pathology as a host response. J Neuropathol Exp Neurol 2008; 67: 523-31.

[242] Castellani RJ, Zhu X, Lee HG, Smith MA, Perry G. Molecular Pathogenesis of Alzheimer's Disease: Reductionist *vs.* Expansionist Approaches. Int J Mol Sci 2009; 10: 1386-406.

[243] Castellani RJ, Lee HG, Siedlak SL, *et al.* 2009a - Reexamining Alzheimer's disease: evidence for a protective role for amyloid-beta protein precursor and amyloid-beta. J Alzheimers Dis 2009; 18: 447-52.

[244] Esiri MM, Chance SA. Cognitive reserve, cortical plasticity and resistance to Alzheimer's disease. Alzheimers Res Ther 2012; 4: 7.

[245] Negash S, Xie S, Davatzikos C, *et al.* Cognitive and functional resilience despite molecular evidence of Alzheimer's disease pathology. Alzheimers Dement 2012; published online: doi: 10.1016/j.jalz.2012.01.009.

[246] Arnold SE, Louneva N, Cao K, *et al.* Cellular, synaptic, and biochemical features of resilient cognition in Alzheimer's disease. Neurobiol Aging 2013; 34: 157-68.

[247] Erten-Lyons D, Woltjer RL, Dodge H, *et al.* Factors associated with resistance to dementia despite high Alzheimer disease pathology. Neurology 2009; 72: 354-60.

[248] Jellinger KA, Attems J. Neuropathological approaches to cerebral aging and neuroplasticity. Dialogues Clin Neurosci 2013; 15: in press.

[249] Honer WG, Barr AM, Sawada K, *et al.* Cognitive reserve, presynaptic proteins and dementia in the elderly. Transl Psychiatry 2012; 2: e114.

[250] Salthouse TA. When does age-related cognitive decline begin? Neurobiol Aging 2009; 30: 507-14.

[251] Tucker AM, Stern Y. Cognitive reserve in aging. Curr Alzheimer Res 2011; 8: 354-60.

[252] Nagy Z, Esiri MM, Jobst KA, *et al.* The effects of additional pathology on the cognitive deficit in Alzheimer Disease. J Neuropathol Exp Neurol 1997; 56: 165-170.
[253] Nelson PT, Abner EL, Schmitt FA, *et al.* Modeling the association between 43 different clinical and pathological variables and the severity of cognitive impairment in a large autopsy cohort of elderly persons. Brain Pathol 2010; 20: 66-79.
[254] Kovacs GG, Alafuzoff I, Al-Sarraj S, *et al.* Mixed brain pathologies in dementia: the Brainnet Europe Consortium experience. Dement Geriatr Cogn Disord 2008; 26: 343-350.
[255] Wilson RS, Segawa E, Boyle PA, Anagnos SE, Hizel LP, Bennett DA. The natural history of cognitive decline in Alzheimer's disease. Psychol Aging 2012; 27: 1008-17.
[256] Petrovitch H, Ross GW, Steinhorn SC, *et al.* AD lesions and infarcts in demented and no-demented Japanese-American men. Ann Neurol 2005; 57: 98-103.
[257] Chui HC, Zarow C, Mack WJ, *et al.* Cognitive impact of subcortical vascular and Alzheimer's disease pathology. Ann Neurol 2006; 60: 677-87.
[258] Giannakopoulos P, Gold G, Kovari E, *et al.* Assessing the cognitive impact of Alzheimer disease pathology and vascular burden in the aging brain: the Geneva experience. Acta Neuropathol 2007; 113: 1-12.
[259] Schneider JA, Arvanitakis Z, Bang W, Bennett DA. Mixed brain pathologies account for most dementia cases in community-dwelling older persons. Neurology 2007; 69: 2197-204.
[260] Schneider JA, Wilson RS, Bienias JL, Evans DA, Bennett DA. Cerebral infarctions and the likelihood of dementia from Alzheimer disease pathology. Neurology 2004; 62: 1148-55.
[261] Duyckaerts C, Potier MC, Delatour B. Alzheimer disease models and human neuropathology: similarities and differences. Acta Neuropathol 2008; 115: 5-38.
[262] Launer LJ, Petrovitch H, Ross GW, Markesbery W, White LR. AD brain pathology: vascular origins? Results from the HAAS autopsy study. Neurobiol Aging 2008; 29: 1587-90.
[263] Zekry D, Duyckaerts C, Belmin J, *et al.* The vascular lesions in vascular and mixed dementia: the weight of functional neuroanatomy. Neurobiol Aging 2003; 24: 213-9.
[264] Gold G, Giannakopoulos P, Herrmann FR, Bouras C, Kovari E. Identification of Alzheimer and vascular lesion thresholds for mixed dementia. Brain 2007; 130: 2830-2836.
[265] Troncoso JC, Zonderman AB, Resnick SM, Crain B, Pletnikova O, O'Brien RJ. Effect of infarcts on dementia in the Baltimore longitudinal study of aging. Ann Neurol 2008; 64: 168-76.
[266] Lee JH, Olichney JM, Hansen LA, Hofstetter CR, Thal LJ. Small concomitant vascular lesions do not influence rates of cognitive decline in patients with Alzheimer disease. Arch Neurol 2000; 57: 1474-1479.
[267] Bennett DA, Schneider JA, Bienias JL, Evans DA, Wilson RS. Mild cognitive impairment is related to Alzheimer disease pathology and cerebral infarctions. Neurology 2005; 64: 834-41.
[268] Schneider JA, Arvanitakis Z, Leurgans SE, Bennett DA. The neuropathology of probable Alzheimer disease and mild cognitive impairment. Ann Neurol 2009; 66: 200-8.
[269] Sonnen JA, Santa Cruz K, Hemmy LS, *et al.* Ecology of the aging human brain. Arch Neurol 2011; 68: 1049-56.
[270] Yarchoan M, Xie SX, Kling MA, *et al.* Cerebrovascular atherosclerosis correlates with Alzheimer pathology in neurodegenerative dementias. Brain 2012; 135: 3749-56.
[271] Richardson K, Stephan BC, Ince PG, Brayne C, Matthews FE, Esiri MM. The neuropathology of vascular disease in the Medical Research Council Cognitive Function and Ageing Study (MRC CFAS). Curr Alzheimer Res 2012; 9: 687-96.

[272] Barker WW, Luis CA, Kashuba A, *et al.* Relative frequencies of Alzheimer disease, Lewy body, vascular and frontotemporal dementia, and hippocampal sclerosis in the State of Florida Brain Bank. Alzheimer Dis Assoc Disord 2002; 16: 203-12.
[273] Buchman AS, Leurgans SE, Nag S, Bennett DA, Schneider JA. Cerebrovascular disease pathology and parkinsonian signs in old age. Stroke 2011; 42: 3183-9.
[274] Stephan BC, Matthews FE, Ma B, *et al.* Alzheimer and vascular neuropathological changes associated with different cognitive States in a non-demented sample. J Alzheimers Dis 2012; 29: 309-18.
[275] Maarouf CL, Daugs ID, Spina S, *et al.* Histopathological and molecular heterogeneity among individuals with dementia associated with Presenilin mutations. Mol Neurodegener 2008; 3: 20, doi:10.1186/1750-1326-3-20.
[276] Shepherd C, McCann H, Halliday GM. Variations in the neuropathology of familial Alzheimer's disease. Acta Neuropathol 2009; 118: 37-52.
[277] Woodhouse A, Shepherd CE, Sokolova A, *et al.* Cytoskeletal alterations differentiate presenilin-1 and sporadic Alzheimer's disease. Acta Neuropathol 2009; 117: 19-29.
[278] Martikainen P, Pikkarainen M, Pontynen K, *et al.* Brain pathology in three subjects from the same pedigree with PSEN1 P264L mutation. Neuropathol Appl Neurobiol 2010; 36: 41-54.
[279] Isotalo K, Kok EH, Luoto TM, *et al.* Upstream transcription factor 1 (USF1) polymorphisms associate with Alzheimer's disease-related neuropathological lesions: Tampere autopsy study. Brain Pathol 2012; 22: 765-75.
[280] Jellinger KA. Interaction between pathogenic proteins in neurodegenerative disorders. J Cell Mol Med 2012; 16: 1166-83.
[281] Marchesi VT. Alzheimer's disease 2012: the great amyloid gamble. Am J Pathol 2012; 180: 1762-7.
[282] Dickson DW. Neuropathology of non-Alzheimer degenerative disorders. Int J Clin Exp Pathol 2010; 3: 1-22.
[283] Petersen RC, Parisi JE, Dickson DW, *et al.* Neuropathologic features of amnestic mild cognitive impairment. Arch Neurol 2006; 63: 665-72.
[284] Jicha GA, Abner EL, Schmitt FA, *et al.* Preclinical AD Workgroup staging: pathological correlates and potential challenges. Neurobiol Aging 2012; 33: 622 e1-622 e16.
[285] Arriagada PV, Marzloff K, Hyman BT. Distribution of Alzheimer-type pathologic changes in nondemented elderly individuals matches the pattern in Alzheimer's disease. Neurology 1992; 42: 1681-8.
[286] Bennett DA, Schneider JA, Arvanitakis Z, *et al.* Neuropathology of older persons without cognitive impairment from two community-based studies. Neurology 2006; 66: 1837-44.
[287] Cairns NJ, Taylor-Reinwald L, Morris JC. Autopsy consent, brain collection, and standardized neuropathologic assessment of ADNI participants: the essential role of the neuropathology core. Alzheimers Dement 2010; 6: 274-9.
[288] Frisoni GB. Alzheimer's disease neuroimaging initiative in Europe. Alzheimers Dement 2010; 6: 280-5.
[289] Trojanowski JQ, Vandeerstichele H, Korecka M, *et al.* Update on the biomarker core of the Alzheimer's Disease Neuroimaging Initiative subjects. Alzheimers Dement 2010; 6: 230-8.
[290] Jack CR, Jr. Alzheimer disease: new concepts on its neurobiology and the clinical role imaging will play. Radiology 2012; 263: 344-61.

[291] Esiri MM. Pro: Can neuropathology really confirm the exact diagnosis? Alzheimers Res Ther 2010; 2: 10.
[292] Gotz J, Gotz NN. Animal models for Alzheimer's disease and frontotemporal dementia: a perspective. ASN Neuro 2009; 1.
[293] Balducci C, Forloni G. APP transgenic mice: their use and limitations. Neuromolecular Med 2011; 13: 117-37.
[294] Hunter JM, Bowers WJ, Maarouf CL, *et al.* Biochemical and morphological characterization of the AbetaPP/PS/tau triple transgenic mouse model and its relevance to sporadic alzheimer's Disease. J Alzheimers Dis 2011; 27: 361-76.
[295] Petiet A, Delatour B, Dhenain M. Models of neurodegenerative disease - Alzheimer's anatomical and amyloid plaque imaging. Methods Mol Biol 2011; 771: 293-308.
[296] de Calignon A, Polydoro M, Suarez-Calvet M, *et al.* Propagation of tau pathology in a model of early Alzheimer's disease. Neuron 2012; 73: 685-97.
[297] De Deyn PP, Van Dam D, editors. Animal Models of Dementia. New York, Dordrecht, Heidelberg, London: Humana Press/Springer 2011.

Send Orders of Reprints at reprints@benthamscience.net
Frontiers in Clinical Drug Research – Alzheimer Disorders, Vol. 1, 2013, 411-431

CHAPTER 11

Metal Ions in the Pathogenesis of Alzheimer's Disease: An Open Field

Guido Crisponi[1,*], Valeria M. Nurchi[1], Daniela Fanni[2], Clara Gerosa[2], Sonia Nemolato[2], Miriam Crespo-Alonso[1], Leonardo Toso[1], Joanna I. Lachowicz[1] and Gavino Faa[2]

[1]Department of Chemistry and Geology, University Campus, 09042 Montserrat Cagliari, Italy and [2]Department of Surgical Sciences, Sec., Pathology, Via Ospedale 46, 09124 Cagliari, Italy

Abstract: Despite the large availability of data regarding the mechanisms at the basis of β-amyloid production in Alzheimer's disease (AD), the factors responsible for β-amyloid accumulation in neurons of AD patients until now have not been entirely clarified. Several metal ions have been indicated as possible triggers of the main conformational modifications of β-amyloid protein, and metal homeostasis disarrangement has been proposed as a relevant pathological cofactor of neurodegeneration. Epidemiological evidences suggest a possible association between exposure to increased amounts of multiple metals and onset and progression of AD. According to this hypothesis, metals could induce oxidative stress in neurons, developing neurodegenerative disease. In this chapter, the role of various trace metals as cofactors of a poor cognitive function and of an increased risk of dementia leading to AD will be discussed. In particular, this study is aimed to critically analyze the role played by copper, iron, zinc and aluminium in neurodegeneration. Based on the current evidences concerning the relationship between these metal ions and onset and progression of AD, no definitive recommendations are, at the best of our knowledge, possible. However, foods rich in metals such as iron and copper should be discouraged, as well as caution should be suggested in exposure to aluminium sources. Further studies on the relationship between metal overexposure and AD development hopefully will open new opportunities for the prevention and management of neurodegenerative diseases.

Keywords: Alzheimer's disease, neurodegenerative diseases, β-amyloid protein, oxidative stress, copper, iron, zinc, aluminium.

INTRODUCTION

Alzheimer's disease (AD) may be defined as a disarrangement of the cellular

***Address correspondence to Guido Crisponi:** Department of Chemistry and Geology, Citadel University, Montserrat 09042 Cagliari, Italy; Italy; E-mail crisponi@unica.it

Atta-ur-Rahman (Ed)

control system responsible for the correct folding of proteins during neuronal cell life. Misfolded proteins show a tendency to aggregate, leading to a series of severe pathological consequences and ultimately to AD, suggesting the inclusion of this disease among conformational diseases [1]. Two proteins are characterized by misfolding in AD: β-amyloid protein (Aβ), a proteolytic cleavage product of the Aβ precursor protein (APP), and tau protein, a microtubule-associated protein, prone to aggregation after hyperphosphorylation [2]. Despite a large amount of data available in the recent literature regarding the mechanisms at the basis of Aβ production and of its interactions with other cellular components, the factors responsible for Aβ accumulation in neurons are nowadays still not entirely clarified. Mounting evidences support the hypothesis that both endogenous and exogenous metal ions could be involved, as factors or cofactors, in the pathogenesis of AD [3]. A number of metal ions have been indicated as possible triggers of the main conformational modifications of Aβ, and metal dyshomeostasis has been suggested as a relevant pathological cofactor of neurodegeneration [4]. Gathered evidence suggests that Aβ precipitation could be caused by abnormal interactions with metal ions, including Zn^{II}, Cu^{II} and Fe^{III}. Moreover, in a mouse model of AD, compounds that interdict metal ion binding to Aβ have been reported to inhibit Aβ precipitation and deposition, opening a new field of research: the metallobiology of AD [5]. These data suggest a major role for trace metal ions in the generation of toxic Aβ oligomers in neurons, that are likely to be responsible for the synaptic dysfunction associated with AD. On this basis the so-called "metal protein attenuating compound" approach has been recently proposed as a novel therapy for AD [6].

Whereas the previous data indicate that metal ions - including iron, copper, zinc and aluminium – might increase the neurotoxicity of Aβ, other reports have demonstrated that Aβ could help to limit the toxicity on brain cells exerted by redox-active metal ions, thereby participating to the antioxidant defense of neurons against reactive oxygen species (ROS). These findings were at the basis of the hypothesis of the amyloid paradox, amyloid-beta-metal complexes putatively being neurotoxic and, under some circumstances, neuroprotective [7].

This review will try to report the main data from the recent literature on the role played by metal ions in the onset and development of AD, focusing on the role played by iron, copper, zinc and aluminium in neuronal toxicity and dementia.

METAL-INDUCED OXIDATIVE STRESS AND NEURO-DEGENERATIVE DISEASE

The brain is a highly oxidative organ, accounting for about 2% of total body weight but consuming 20% of the oxygen used by the whole body. Moreover, with normal ageing, human brain accumulates transition metal ions, including iron and copper, which may promote the generation of ROS *via* Fenton chemistry* [9]. Binding of Aβ to redox-reactive metals, which in turn generate free radicals, has been proposed among the several mechanisms recognized to be involved in the molecular pathways to neurodegeneration and, in particular, in the onset and development of AD [10]. Compelling evidences - based on results of studies on *in vitro* systems, experimental animals and humans - indicate that the etiology of AD involves oxidative stress. This implicates that redox active metal ions, such as copper and iron, and redox-inert zinc ions, play a central role in this process [11]. According to this hypothesis, trace metal dyshomeostasis and the consequent abnormal accumulation of transition metals, including iron and copper, in the ageing brain cells might elicit oxidative stress and macromolecular damage that could eventually impede neuronal function leading to neurodegeneration and to AD [12]. It has been claimed that the increased oxidative stress in brain cells of AD patients could be reflected by increased protein and DNA oxidation, enhanced lipid peroxidation, decreased levels of cytochrome c oxidase and increased brain content of transition metal ions iron and copper, both capable of stimulating free radical formation *via* Fenton reaction [13]. Redox active Cu^{II} and Fe^{III} ions have been shown to be reduced in the presence of Aβ protein, with concomitant production of ROS, hydrogen peroxide (H_2O_2) and hydroxyl radical ($OH^{\cdot}$), leading directly to the widespread oxidation damages typical of AD brains [14]. The discovery of a functional iron-response element (IRE) in the 5' untranslated region of APP, consistent with the biochemical evidence that it is a redox-active metalloprotein, is at the basis of the hypothesis that the redox interactions between Aβ protein and metal ions may be at the heart of a pathological positive feedback system, wherein Aβ amyloidosis and oxidative stress promote each other [14].

---

* According to Alexander ".... a potential harmful chemical property of iron and its compounds is the ability to catalytically generate potent hydroxyl radicals that are oxidatively toxic to cells. This catalysis (Fenton reactions) occurs with activated oxygen species (superoxide and hydrogen peroxide) ...." [8].

Immunohistochemical analysis in the brain of subjects affected by AD revealed an increase in neuronal content of ceruloplasmin, a copper protein functioning as a ferroxidase that oxidizes toxic ferrous iron to its non-toxic ferric form, providing the majority of anti-oxidant capacity in human serum [15]. In this study, neuronal ceruloplasmin immunoreactivity resulted particularly strong in the hippocampus. The increased neuronal expression of ceruloplasmin in AD might be related to its antioxidant activity, ceruloplasmin playing an important role in maintaining iron homeostasis and in protecting neurons from iron-mediated free radical injury [16]. On the basis of these data, all together indicating the increased oxidative stress as one of the main underlying factors in the etiology, onset and development of AD, the attention of many researchers has been focused on the possible protective role of antioxidants, including glutathione, vitamin C, vitamin E, lipoic acid, flavonoids and curcumin [13].

Recently, a meta-analysis compiled has critically assessed the literature data on iron and copper storage in the brain of AD patients, leading to very heterogeneous and conflicting results. As for as iron overload is concerned, while data from one laboratory reported a large increase in iron content in AD neocortex compared to age-match controls, several other laboratories failed to reproduce these findings, reporting no significant differences in iron brain content between groups. Regarding zinc, the critical analysis of published data did not show any significant change in zinc content in the neocortex of AD patients, whereas copper appeared to be significantly depleted in AD brain when compared to age-matched controls [17]. Based on these findings, the authors suggested re-evaluate the hypothesis that transition metal overload might account for oxidative injury in AD.

THE ROLE OF COPPER

Copper as a Causative Agent

The risk of copper toxicity, putatively associated with the use of copper plumbing and with copper intake in multi-mineral supplements in developed countries, has been proposed in recent years as the major cause of cognitive decline in the ageing population leading to AD [18]. Excessive copper, causing neuronal toxicity, has been reported as one of the main bio-factors causing neuronal toxicity: it has been hypothesized that the excess of inorganic copper in humans in developed countries in the last 50 years derives from drinking water ingestion, leached from the copper plumbing. In fact the use of such water is related with AD epidemic [19]. Ingestion of inorganic copper has been recently defined as a strong causal factor of cognitive loss and AD on the basis of AD animal models, in which copper in the drinking water markedly worsened AD progression, and of human studies showing a higher mortality in elderly women ingesting mineral supplements containing inorganic copper [20]. Studies on the coordination chemistry of copper ions to Aβ protein showed that the conditional dissociation constant of Cu-Aβ complexes lies in the range of 10pM to 100 nM, with two most likely coordination modes at physiological pH, both being 3N1O (copper coordinated by 3 nitrogen and 1 oxygen binding groups) distorted square planar [21]. Submicromolar concentrations of Cu^{II} have been shown to favor precipitation of Aβ protein in *in vitro* systems [22]. Trace levels of copper, associated with trace levels of iron and zinc, have been identified as the actual initiators of the classic nucleation-dependent Aβ aggregation (seeding) process and of the amyloid fibril-free formation of soluble oligomers (oligomerization), suggesting that there might be an obligatory role for these three metal ions in initiating Aβ amyloidosis [23]. Metal ions have been shown to influence in a different way the aggregation and deposition of Aβ protein: Cu^{II} ions accelerate Aβ deposition, so producing amorphous aggregates, whereas Fe^{III} ions induce the deposition of fibrillar amyloid plaques. Moreover, the binding of Cu^{II} ions to Aβ deposits has been associated with their reduction during their interaction with Aβ [24]. Copper ions might be involved in AD due to their ability to induce aggregation of amyloidogenic peptides and to produce ROS, whose generation by the Cu-Aβ species follows different pathways; in the more relevant one, reduction of dioxygen has been demonstrated to occur by a two-electron process involving two vicinal Cu^{I} ions [25]. The molecular details of the Cu^{II}-Aβ interactions and the definitive picture of the pH-dependent Cu^{II} coordination modes of Aβ protein have been recently clarified by electron paramagnetic resonance [26]. On the basis of these experimental data, many scientists have recommended that people should remove old copper pipes from their homes, or mount special filters, to prevent that copper intake causes serious health problems. According to this hypothesis, tiny traces of copper from pipes solved in water probably lead, over a long period of time, to copper storage in the body which could favor the onset of AD. Studies on redox chemistry of copper-amyloid-beta aggregates have shown that they produce more oxygen species in comparison with copper complexes with amino acids [27]. Studies on redox activity of Cu^{II} ions bound to Aβ in AD plaques gave evidence that the redox activity of copper ions is not silenced in Aβ-Cu^{II} complexes, showing that hydroxyl radicals produced when Cu^{II} reacts with Aβ is quenched by Aβ [28]. These findings might lead to a modified hypothesis on the pathogenesis of AD: the upregulation and the increased neuronal deposition of Aβ should be interpreted as an antioxidant response to oxidative stress, putatively induced by redox-active metal ions, including Cu^{II}. On the other hand, Aβ protein might concentrate the redox-active copper ions at the neuronal membrane, causing lipid peroxidation and neuronal cell death, eventually acting as an "indirect" pro-oxidant. Copper-containing Aβ complexes and aggregates have been shown to facilitate the oxygen reduction with production of H_2O_2 as a major product; this finding involves the ability of Cu-Aβ aggregates to cause oxidative stress through the production of ROS [29].

According to data indicating copper as a causative agent of AD, copper chelation has been proposed as a potential therapy for patients affected by AD [30]. Recently, a new possible therapy has been proposed, aimed at inhibiting the aberrant aggregation of Aβ with copper ions, generally considered as the main responsibilities for Aβ toxicity and for AD onset. This new therapeutic strategy is based on the discovery that human serum albumin, physiologically involved in copper transport both in serum and in cerebrospinal fluid, is able to rapidly and stoichiometrically remove Cu^{II} from Aβ, decreasing Cu-induced aggregation of Aβ and suppressing ROS production. Halting radical production and decreasing Aβ toxicity in neurons could halt, according to the proponents, progression of AD [31]. The reported efficacy of D-penicillamine - a drug widely used in clinical practice to reduce copper systemic overload favoring copper excretion through urine - in reducing oxidative stress in AD patients might confirm the negative role played by brain copper deposits in AD [32].

Copper as a Protective Agent

Copper plays an indispensable role in the development and in physiology of the human central nervous system (CNS), being a cofactor of numerous enzymes, including dopamine-β-hydroxylase, peptidyl-α-monooxygenase, superoxide dismutase (SOD). These enzymes play fundamental roles in catecholamine biosynthesis, activation of neuropeptides and hormones, protection against ROS and in other processes essential for normal CNS function [33].

A central tenet of the Amyloid cascade hypothesis of AD is the aberrant deposition in the brain of Aβ42 in β-sheets in neuritic or senile plaques. The group of Exley [34] has shown that copper Cu^{II} bound to monomeric Aβ prevents it by adopting a β-sheet secondary structure, blocking the deposition of Aβ42 in β-sheets and, consequently, the deposition of senile plaques in the brain. Moreover, the same studies on co-deposition of metals and Aβ clearly showed that Cu^{II} abolishes the β-sheet structure of preformed amyloid fibrils of Aβ, suggesting that copper ions could play a protective role in AD development and that copper-Aβ binding could have implications for AD therapy [35]. A similar finding was made by the group of Jiang [36] for the form Aβ40 of beta amyloid, and taken together these observations strongly suggest that copper prevents both the formation and the accumulation of plaques in the brain. Coincident with the copper-catalysed dissolution of β-sheets of Aβ42, Exley's group made the first observation of the *in vitro* formation of spherulites of this peptide. These spherical globules of amyloid have been previously observed *in vitro* only for the other amyloid-forming proteins insulin and β-lactoglobulin. Copper appeared to have a role in the formation of spherulites of Aβ42 and this will be investigated in future research. The role of metals in the formation, deposition and metabolism of Aβ in AD is greatly debated and these new findings highlight a potential protective role for copper in AD. Moreover, treatment with Cu^{II} associated with the metal ligand clioquinol resulted in approximately 85% reduction of secreted Aβ in Chinese hamster ovary cells overexpressing APP [37]. In this *in vitro* system, Aβ was rapidly degraded through copper-mediated upregulation of matrix metalloproteinase-2 (MMP2) and matrix proteinase-3 (MMP3), suggesting a potential role for copper in reducing Aβ deposition in the brain of AD patients.

The unusually high Cu^{II} ion concentration found in senile plaques in AD has been object of different interpretations during the years. In particular, the relationship between the Aβ found within plaques and redox-active Cu^{II} ions has been analysed by multiple studies, aimed at monitoring the generation of ROS, some indicating Aβ as a prooxidant, others suggesting that Aβ might act as an antioxidant. A recent study, utilizing a variety of fluorescence and absorption assays to monitor the production of ROS, clearly showed that Aβ-Cu^{II} complexes do not generate any more ROS and do not generate H_2O_2 catalytically. In the same experiments, Aβ did not silence the redox activity of Cu^{II} *via* chelation, but it was able to quench hydroxyl radicals produced as the result of Fenton reactions of Cu^{II} and ascorbate by specific oxidation sites at histidine and methionine [28].

Accumulating evidence clearly indicates that the relationship between copper and AD is more complex than previously reported. The problem is not related to the increase (or decrease) of copper levels, but to its abnormal distribution; AD brain is in fact characterized by elevated extracellular and lowered intracellular copper stores, leading to an altered copper homeostasis that might play a key role in the evolution of AD. According to this hypothesis, metal ligands, such as clioquinol, have been shown to increase intracellular copper levels in *in vitro* systems, resulting in up-regulating MMP2 and MMP3, and in enhanced degradation of Aβ [38]. Moreover, clioquinol-Cu^{II} complexes have been shown to induce potent epidermal growth factor receptor (EGFR) phosphorylation; EGFR activation is followed by a significant reduction of the levels of Aβ deposition in cell cultures [39]. In a study on the ability of different metal ions to differently drive Aβ aggregation, both Cu^{II} and Zn^{II} ions completely prevented the formation of soluble fibrillary aggregates [40]. Several investigations indicate that Cu^{II} ions can only cause aggregation of the Aβ peptide giving rise to amorphous aggregates [41], which are thought to be non-toxic to neurons and should have no role in neurodegeneration [42]. Copper ions have also been reported to inhibit fibrillation, blocking the deposition of amyloid fibres, the only amyloid aggregates which are toxic to neurons and actively participate to the process of neurodegeneration [43]. A recent quantitative meta-analysis on the relationship between trace metal ions content in the brain and AD onset gave evidence that, in the majority of studies on this topic copper was significantly depleted in AD [17]. The complexity of the relationship between copper and AD has been recently confirmed by a systemic review of the literature regarding this topic [44]. In that review, eleven trials that tried to either supplement or deplete copper were analysed: none of them provided clear evidence of a beneficial effect on cognitive performance in subjects affected by AD.

THE ROLE OF IRON

Iron as a Causative Agent

One of the earliest arguments suggesting a role for iron overload in AD onset was presented in 1953 in a detailed histochemical study on a series of autoptic AD cases, in which haemosiderin deposits described in some of these patients lead to the hypothesis that iron overload might underlie late-onset cognitive loss and AD insurgence [45]. Iron, like other redox active metals, undergoes redox cycling reactions and possesses the ability to produce reactive radicals such as superoxide anion radical and nitric oxide in biological systems [46]. The interaction of Aβ and redox-active metals like Fe^{II} and its potential role in the etiology of neurodegeneration, in particular of AD, have been debated for decades. The origin of the pathological iron storage in AD brain remains controversial. Recently, amyloid angiopathy has been indicated as a possible cause of iron depositing: according to this hypothesis, amyloid depositing in the arterial wall, constantly associated with AD, might cause vascular lesions and produce micro-bleeds and extracellular iron depositing [47]. It has been suggested that complexes Fe^{II}-Aβ could enhance the aggregation and deposition of Aβ or, alternatively, might facilitate the generation of ROS leading to iron-induced oxidative stress in cortical neurons [48]. Recent studies on the redox potential of the Fe^{II}-Aβ complexes have shown that the binding of iron ions by Aβ modulates the redox potential to a level at which its redox cycling occurs. As a consequence, not only ROS is continuously produced but, since complexes also catalyze the reduction of oxygen to produce H_2O_2, oxygen and biological reducing agents can also be depleted leading to oxidative stress in brain cells [49]. Moreover, in the same study it was found that the affinity of Aβ for Fe^{II} is 8 orders of magnitude stronger than that for transferrin, suggesting that Aβ might compete for iron against transferrin, seriously interfering with iron homeostasis in cortical neurons. When all these data regarding the "metal hypothesis" of AD were taken together a consensus that iron is elevated in the AD neocortex emerged in the literature at the end of the last century [50], iron overload gradually became widely accepted as a typical feature of AD [51], and involved in its pathogenesis through the generation of ROS [52]. Iron accumulation in AD patients was hypothesized to represent an important source of redox-generated ROS, leading to neuronal cell death [53]. The discovery that Aβ-associated iron ions increase plaque toxicity, leading to neurotoxicity in the elderly [54], became the basis for the introduction in clinical practice of the therapeutic use of iron chelators in neurodegenerative conditions and in particular in AD patients [30, 55]. Recently, a role for ceruloplasmin defective activity, associated with iron metabolism disorders, has been hypothesized in the pathogenesis of AD. Ceruloplasmin is a multi-copper oxidase that oxidizes Fe^{II} to Fe^{III}, keeping low noxious Fe^{II} and allowing Fe^{III} to be uptaken and transported out of brain cells by transferrin. Since serum ceruloplasmin levels have been reported to be significantly lowered, it has been hypothesized that defective ceruloplasmin synthesis could be at the basis of iron disorders, ending with storage of the Fe^{II} noxious ions in cortical neurons, followed by ROS hyper production, neuronal cell death and AD onset [56]. Ceruloplasmin deficient activity might be related to the low levels of copper reported in AD brain, which presumably result in the deficient activity of many copper-enzymes in AD, in addition to ceruloplasmin [57].

Taking into account that the presence of an IRE in the 5' untranslated region of APP transcript has been shown to regulate the APP translation, a chelation therapy has been proposed in recent years aimed at disrupting the aberrant metal-peptide interactions that could be at the basis of the insurgence of AD. The iron and copper chelators desferrioxamine, clioquinol, tetrathiolmolybdate and dimercaptopropanol are some novel potential chelating agents, highly promising for a rational therapeutic strategy against AD [58].

Conflicting Results on the Role of Iron

All living cells need a supply of iron, an essential metal ion for life, for oxygen reduction (respiration), for carbon dioxide reduction (photosynthesis), for reduction of dinitrogen and for other fundamental biological processes [59]. A recent meta-analysis has given clear evidence that there is a widespread misconception in the scientific literature regarding the levels of several transition metals, most prominently iron [17]. The author conclusions of this meta-analysis indicate that neocortical iron levels are not altered in AD brain, the putamen being the only cerebral region consistently found to contain increased iron levels in AD [60]. The complexity of the relationship between iron and AD has been recently underlined by a review which analyzed eleven interventional trials that tried to either supplement or deplete iron: no one of these trials provided clear evidence of a beneficial effect on cognitive performance in patients with AD, indicating that further clinical investigations are needed in order to clarify if diets excessive in iron should be avoided in the elderly [44].

THE ROLE OF ZINC

Zinc probably plays a key role in AD, taking a central role in neurodegeneration and in neuroprotection. From one hand, zinc probably facilitates copper binding to Aβ, putatively accelerating the disease process. On the other hand, zinc has been reported to exert protective effects in neurons protecting sulfhydryl groups from peroxidative damage caused by the pro-oxidant activity mediated by transition metal ions, including iron and copper. The role of zinc ions in the origin and evolution of AD has been hypothesized to be paradoxical, and accordingly the theory of the "zinc paradox" has been proposed [11].

Zinc as a Causative Agent

There is evidence that binding of metal ions like Zn^{II} to Aβ may contribute to the pathogenesis of AD, interactions between Zn^{II} and Aβ peptides being reported in brain areas affected by AD. Binding of Zn^{II} to the full-length Alzheimer's Aβ peptide has been demonstrated to induce transition in the peptide structure, that might contribute to amyloid plaque formation and to AD onset and evolution [61].

Conflicting Results on the Role of Zinc

The redox inert metal zinc occupies a special position among trace metals in the human body. Zinc is an essential component of more than 300 proteins, involved in cell growth and development, in different metabolic processes, in the immune function, and in skin and bone metabolism. Zinc also possesses neuroprotective properties, participating actively in neuro-psychiatric performance. Moreover, zinc plays a relevant role in the defense against cellular oxidative stress: this zinc activity is probably, at least in part, mediated by its ability to bind to sulfhydryl groups of structural and metabolically active cell components, protecting them from ROS produced by pro-oxidant factors, including transition metal ions such as iron and copper [62]. Zinc ions are basic structural components and are critical for the function of a large number of metalloproteins, including SOD. As a consequence, zinc burden is fundamental in the antioxidant activity to protect human cells, including neurons, against oxidative stress [46, 63].

In recent years, zinc deficiency has been reported in the brain of patients affected by AD, when compared with age-matched controls. Zinc deficiency might be at the basis of neuronal death, and this hypothesis has induced several researchers to introduce zinc supplementation in the therapy of AD that resulted in a significant cognitive benefit [19].

THE ROLE OF ALUMINIUM

Aluminium as a Causative Agent

The potential role of aluminium as a possible etiopathogenetic factor in AD and in particular in the precipitation, abnormal aggregation and cytotoxicity of beta-amyloid protein in the brain, has been particularly controversial [64]. Aluminium overload is generally considered to be potentially neurotoxic, as demonstrated in multiple experimental animals, but the controversy on the role played by aluminium ions in AD has not been resolved [65]. The link between aluminium and dialysis-dementia is widely accepted in clinical practice by the scientific community. The numerous literature reports regarding cognition/neuro-behavioural deficits associated with increased aluminium body burden have stimulated the research interest [66]. Aluminium enters the brain through multiple routes: from blood, either through choroid plexus or across the blood brain barrier (BBB), and from the nasal cavity into olfactory nerves followed by direct distribution into the brain [67]. Data on tissue aluminium content in patients with chronic renal failure undergoing dialysis for long periods show high levels of aluminium in the brain, clearly demonstrating the ability of this metal to penetrate the BBB [68, 69]. About 90% of aluminium in brain extracellular fluid is predicted to be Al-citrate and only 4% Al-transferrin [70]. BBB permeation of aluminium ions cannot solely be attributed to diffusion and it is probably carrier-mediated as aluminium citrate [67].

Aluminium has been implicated in AD, although the existence of a direct causal dependence has not been demonstrated. The exact mechanism of aluminium toxicity on brain cells is not known, but several lines of evidence show that this simple trivalent cation, incapable of redox changes, might worsen oxidative events and activate ROS generation, linking aluminium storage to the pathogenesis of AD [71]. The hypothesis that aluminium might exacerbate events associated with neurodegenerative diseases, and in particular with AD progression, is clearly emerging. The role of aluminium could be related to its ability to potentiate iron-induced oxidative events, increasing the formation of ROS. Moreover, collected evidence suggests that aluminium potentiates inflammatory events by inducing the synthesis and/or release of interleukins and other inflammatory cytokines [72]. The controversy on the role played by aluminium in the pathogenesis of AD is going on from many decades, but the questions whether aluminium represents a health hazard and whether it plays a relevant pathogenetic role in the development of AD is still the subject of debate. Two key questions that must be answered before a causal relationship between aluminium exposure and AD can be stated have been addressed [65]:

i) Is aluminium content elevated in the brain of AD patients?

ii) Is it possible to demonstrate a strict relationship between aluminium exposure and an increased risk of AD?

Regarding the intimate relationship between aluminium overexposure and AD progression, although aluminium exists only in the trivalent form and is redox-inert, this metal ion has been shown to interact with transition metals such as copper ions in the promotion of oxidative events. This synergistic interaction increases Cu-induced ROS production, oxidative cell stress and cell death, adding a new piece to the puzzle of the multifactorial etiology of AD [73]. Moreover, an important role in the neurotoxicity played by aluminium ions has been recently related to their ability to interact with copper ions at subcellular level. According to this hypothesis, the interaction between aluminium and copper, both present in trace amounts in drinking water, could initiate and propagate an inflammatory response in the ageing brain, participating to the neuroinflammation reported in the AD brain, eventually inducing or potentiating neurodegeneration and boosting progression of dementia [74]. Evidences for participation of aluminium in the formation of neurofibrillary tangles, a histological marker of degenerating neuronal cells in patients affected by AD, have been recently obtained in hippocampal pyramidal cells from brains of aged humans with AD. In this experiment a portion of neuron containing short filaments was found, suggestive for early neurofibrillary tangles, which turned out to be aggregates of an aluminium/hyperphosphorilated tau complex. These facts support a role for aluminium in the formation and growth of neurofibrillary tangles in AD [75]. Even though many AD animal models have been utilized for better understanding the role of aluminium in the origin and progression of the disease, the neurotoxicity and the role of this metal in AD is still considered a problem [76]. Aluminium salts were able to increase levels of glial activation and inflammatory cytokines, favoring the appearance of excess levels of brain inflammation in animal models treated with aluminium at low doses, reflecting those found in some water supplies. This suggests the hypothesis that aluminium could act as a subtle promoter, by accelerating some characteristic indices of brain ageing [77].

The following factors in favor of a strict correlation between aluminium exposure and AD progression have been proposed: the abundance of the neurotoxic aluminium on earth and its bioavailability to humans; limited amounts of aluminium are needed to cause neurotoxicity; the ability of aluminium ions to cross the BBB; the experimental evidences of neuronal death caused by aluminium intoxication [78]. In a recent review on the effects of chronic exposure of human brain to aluminium ions, the ability of aluminium ions to substitute essential metals such as Mg^{II}, Fe^{II} and Ca^{II} ions in intraneuronal proteins was underlined. On the basis of these data, the author hypothesized that aluminium might disrupt calcium homeostasis in human neurons with remarkable signal transduction changes, as frequently observed in AD patients [69]. The intimate mechanism of aluminium-induced neurotoxicity has been recently linked to the accumulation of iron and to the production of ROS, suggesting that aluminium-induced neurodegeneration could be halted by therapeutic agents able to chelate excess iron [79]. In a recent study on the interaction of different biologically relevant metal ions with Aβ, aimed at verifying whether different metal ions might differently drive Aβ aggregation and toxicity, Aβ-aluminium complexes turned out to be the only species capable of triggering amyloid precursor overproduction; this consequently induces peculiar aggregation properties to Aβ that eventually lead to the formation of fibrillar oligomers [40]. The concern on a possible role of aluminium overload in the brain in the onset of AD is at the basis of numerous studies on the use of metal chelators, able to scavenge the aluminium deposited in brain, as a possible new therapy in AD patients [80, 81]. The half-life of brain aluminium in rats, estimated in about 150 days, decreases to 55 days in animals receiving desferrioxamine injections [82]. In a recent paper the good coordinating properties of iron chelators toward the hard cation Al^{III} was remarked [81]. These data point out that aluminium in brain can be successfully mobilized by an iron chelator such as desferrioxamine, and validates the use of chelating agents as a possible therapy for AD.

Conflicting Results on the Role of Aluminium

Aluminium, a biologically reactive metal ion ubiquitous in our environment, and the most abundant metal in nature representing about 8% of the total components of the earth crust is not an essential element in biology. In fact, biochemical evolution has proceeded in the absence of biological reactive forms of aluminium, with no essential role for aluminium in any biochemical system in any living organism, including humans [83].

Humans are experiencing, in recent years, a new massive exposure to biologically reactive aluminium ions, with possible serious consequences on human health and disease [80]. This potentially dangerous exposure is related in a great extent to the atmospheric acidification due to acid rains, which is causing a progressive acidification of the soil, followed by a massive export of aluminium from the earth crust to surface waters, putting vegetables, animals and humans in contact with absorbable cationic aluminium species, probably for the first time in their history [84]. Owing to acid rains, numerous metal ions including aluminium are escaping from mineral deposits where they had been stored for billions of years as hydroxy-aluminosilicates [85, 86], increasing the biological availability of aluminium to living organisms. According to this hypothesis, acid rain is acting as a key to spring the lock for aluminium release, causing its appearance in polluted waters. Given the absence of aluminium in biology, no protective role in halting AD onset and evolution may be, at the best of our knowledge, hypothesized.

Regarding the hypothesis of a certain role for aluminium in AD onset, conflicting results have been published. The oral intake of aluminium in mice did not accelerate progression of brain pathological changes typical of AD, contrasting the notion that excessive oral intake of aluminium is a risk factor for AD [87]. Moreover, oral exposure to aluminium in mice did not alter the recognition memory and did not accelerate the deposition of beta-amyloid plaques [88]. At present, there is no certain evidence to substantiate the hypothesis that the pathogenesis of AD is caused by aluminium found in food and drinking water as monomeric inorganic aluminium and as hydrated aluminium silicates [89].

CONCLUDING REMARKS

Despite the large amount of data presented by the recent literature regarding the mechanisms at the basis of Aβ production and its interactions with other cellular components, including endogenous and exogenous metal ions, the factors responsible for Aβ accumulation in neurons are not so far entirely clarified. Whereas multiple studies indicate that metal ions - including iron, copper, zinc and aluminium – might increase the neurotoxicity of Aβ, other reports have demonstrated that Aβ could help to limit the toxicity on brain cells exerted by redox-active metal ions, contributing to the antioxidant defense of neurons against ROS. These conflicting data are at the basis of the hypothesis known as "amyloid paradox", amyloid-beta-metal complexes putatively being neurotoxic and, under some circumstances, neuroprotective [7]. After the analysis of the literature regarding the relationship between trace metal ions and AD, we might also introduce the concept of the "metal paradox", endogenous trace metal ions putatively being neurotoxic and, under many circumstances, neuroprotective, as clearly reported in this review.

Based on the current evidences concerning the role of metal ions on AD onset and progression, no definitive recommendations are, at the best of our knowledge, possible. However, foods rich in metals such as iron and copper should be discouraged, as well as caution should be suggested in exposure to aluminium sources. Further studies on the relationship between metals overexposure and AD development hopefully will open new opportunities for the prevention and management of neurodegenerative diseases.

ACKNOWLEDGEMENTS

GC and JIL acknowledge Regione Sardegna for the financial support CRP-27564 to the project "Integrated approach in the design of chelators for the treatment of metal overload diseases". MCA is grateful to RAS for the program Master and Back – Percorsi di rientro, PRR-MAB-A2011-19107.

CONFLICT OF INTEREST

The author(s) confirm that this chapter content has no conflict of interest.

REFERENCES

[1] Kopito RR, Ron D. Conformational disease. Nat Cell Biol 2000; 2: E207-E209.

[2] Querfurth HW, LaFerla FM. Alzheimer's disease. N Engl J Med 2010; 362: 329-344.

[3] Zatta P, Drago D, Bolognin S, Sensi SL. Alzheimer's disease, metal ions and metal homeostatic therapy. Trends Pharmacol Sci 2009; 30: 346-355.

[4] Granzotto A, Zatta P. Metal ions and beta amyloid: conformational modifications and biological aspects. Metal Ions in Neurological Systems. Linert W. Kozlowski H Eds. Springer-Verlag,Wien; 2012. p. 77-83.

[5] Bush AI. The metallobiology of Alzheimer's disease. Trends Neurosci 2003; 26: 207-214.

[6] Kenche VB, Barnham KJ. Alzheimer's disease & metals: Therapeutic opportunities. Br J Pharmacol 2011; 163: 211-219.

[7] Bishop GM, Robinson SR. The amyloid paradox: Amyloid-β-metal complexes be neurotoxic and neuroprotective. Brain Pathol 2004; 14: 448-452.

[8] Alexander NM. In: Seiler HG, Sigel A, Sigel H, eds. Handbook on Metals in Clinical and Analytical Chemistry, Marcel Dekker Inc 1994.

[9] Smith DG, Cappai R, Barnham KJ. The redox chemistry of the Alzheimer's disease amyloid β peptide. Biochimica et Biophysica Acta - Biomembranes 2007; 1768: 1976-1990.

[10] Bossy-Wetzel E, Schwarzenbacher R, Lipton SA. Molecular pathways to neurodegeneration. Nat Med 2004; 10: S2-S9.

[11] Cuajungco MP, Fagét KY. Zinc takes the center stage: Its paradoxical role in Alzheimer's disease. Brain Res Rev 2003; 41: 44-56.

[12] Bonda DJ, Lee HG, Blair JA, Zhu X, Perry G, Smith MA. Role of metal dyshomeostasis in Alzheimer's disease. Metallomics 2011; 3: 267-270.

[13] Jomova K, Vondrakova D, Lawson M, Valko M. Metals, oxidative stress and neurodegenerative disorders. Mol Cell Biochem 2010; 345: 91-104.

[14] Huang X, Moir RD, Tanzi RE, Bush AI, Rogers JT. Redox-active metals, oxidative stress, and Alzheimer's disease pathology. Ann N Y Acad Sci 2004; 1012: 153-163.

[15] Loeffler DA, Sima AAF, LeWitt PA. Ceruloplasmin immunoreactivity in neurodegenerative disorders. Free Radic Res 2001; 35: 111-118.

[16] Patel BN, Dunn RJ, Jeong SY, Zhu Q, Julien JP, David S. Ceruloplasmin regulates iron levels in the CNS and prevents free radical injury. J Neurosci 2002; 22: 6578-6586.
[17] Schrag M, Mueller C, Oyoyo U, Smith MA, Kirsch WM. Iron, zinc and copper in the Alzheimer's disease brain: A quantitative meta-analysis. Some insight on the influence of citation bias on scientific opinion. Prog Neurobiol 2011; 94: 296-306.
[18] Brewer GJ. The risks of copper toxicity contributing to cognitive decline in the aging population and to Alzheimer's disease. J Am Coll Nutr 2009; 28: 238-242.
[19] Brewer GJ. Copper excess, zinc deficiency, and cognition loss in Alzheimer's disease. Biofactors 2012; 38: 107-113.
[20] Brewer GJ. Copper toxicity in Alzheimer's disease: Cognitive loss from ingestion of inorganic copper. J Trace Elem Med Biol 2012; 26: 89-92.
[21] Faller P, Hureau C. Bioinorganic chemistry of copper and zinc ions coordinated to amyloid-β peptide. Dalton Transactions 2009: 1080-1094.
[22] Atwood CS, Moir RD, Huang X, Scarpa RC, Bacarra NME, Romano DM, *et al.* Dramatic aggregation of Alzheimer Aβ by Cu(II) is induced by conditions representing physiological acidosis. J Biol Chem 1998; 273: 12817-12826.
[23] Huang X, Atwood CS, Moir RD, Hartshorn MA, Tanzi RE, Bush AI. Trace metal contamination initiates the apparent auto-aggregation, amyloidosis, and oligomerization of Alzheimer's Aβ peptides. J Biol Inorg Chem 2004; 9: 954-960.
[24] Ha C, Ryu J, Chan BP. Metal ions differentially influence the aggregation and deposition of Alzheimer's β-amyloid on a solid template. Biochemistry (Mosc) 2007; 46: 6118-6125.
[25] Hureau C, Faller P. Aβ-mediated ROS production by Cu ions: Structural insights, mechanisms and relevance to Alzheimer's disease. Biochimie 2009; 91: 1212-1217.
[26] Drew SC, Barnham KJ. The heterogeneous nature of Cu2+ interactions with Alzheimer's amyloid-β peptide. Acc Chem Res 2011; 44: 1146-1155.
[27] Guilloreau L, Combalbert S, Sournia-Saquet M, Mazarguil H, Faller P. Redox chemistry of copper-amyloid-β: The generation of hydroxyl radical in the presence of ascorbate is linked to redox-potentials and aggregation state. ChemBioChem 2007; 8: 1317-1325.
[28] Nadal RC, Rigby SEJ, Viles JH. Amyloid β-Cu2+ complexes in both monomeric and fibrillar forms do not generate H2O2 catalytically but quench hydroxyl radicals. Biochemistry (Mosc) 2008; 47: 11653-11664.
[29] Jiang D, Li X, Liu L, Yagnik GB, Zhou F. Reaction rates and mechanism of the ascorbic acid oxidation by molecular oxygen facilitated by Cu(II)-containing amyloid-β complexes and aggregates. J Phys Chem B 2010; 114: 4896-4903.
[30] Cuajungco MP, Fagét KY, Huang X, Tanzi RE, Bush AI. Metal chelation as a potential therapy for Alzheimer's disease. Ann N Y Acad Sci. 2000; 920: 292-304.
[31] Perrone L, Mothes E, Vignes M, Mockel A, Figueroa C, Miquel MC, *et al.* Copper transfer from Cu-Aβ to human serum albumin inhibits aggregation, radical production and reduces Aβ toxicity. ChemBioChem 2010; 11: 110-118.
[32] Squitti R, Rossini PM, Cassetta E, Moffa F, Pasqualetti P, Cortesi M, *et al.* D-penicillamine reduces serum oxidative stress in Alzheimer's disease patients. Eur J Clin Invest 2002; 32: 51-59.
[33] Lutsenko S, Bhattacharjee A, Hubbard AL. Copper handling machinery of the brain. Metallomics 2010; 2: 596-608.
[34] Exley C, House E, Collingwood JF, Davidson MR, Cannon D, Donald AM. Spherulites of amyloid-β{42} *in vitro* and in alzheimer's disease. J Alzheimer's Dis 2010; 20: 1159-1165.

[35] House E, Mold M, Collingwood J, Baldwin A, Goodwin S, Exley C. Copper abolishes the β-sheet secondary structure of preformed amyloid fibrils of amyloid-β42. J Alzheimer's Dis 2009; 18: 811-817.
[36] Huang HC, Jiang ZF. Accumulated amyloid-β peptide and hyperphosphorylated tau protein: Relationship and links in Alzheimer's disease. J Alzheimer's Dis 2009; 16: 15-27.
[37] White AR, Du T, Laughton KM, Volitakis I, Sharples RA, Xilinas ME, *et al.* Degradation of the Alzheimer disease amyloid β-peptide by metal-dependent up-regulation of metalloprotease activity. J Biol Chem 2006; 281: 17670-17680.
[38] Filiz G, Price KA, Caragounis A, Du T, Crouch PJ, White AR. The role of metals in modulating metalloprotease activity in the AD brain. Eur Biophys J 2008; 37: 315-321.
[39] Price KA, Filiz G, Caragounis A, Du T, Laughton KM, Masters CL, *et al.* Activation of epidermal growth factor receptor by metal-ligand complexes decreases levels of extracellular amyloid beta peptide. Int J Biochem Cell Biol 2008; 40: 1901-1917.
[40] Bolognin S, Messori L, Drago D, Gabbiani C, Cendron L, Zatta P. Aluminum, copper, iron and zinc differentially alter amyloid-Aβ 1-42 aggregation and toxicity. Int J Biochem Cell Biol 2011; 43: 877-885.
[41] Chen WT, Liao YH, Yu HM, Cheng IH, Chen YR. Distinct effects of Zn2+, Cu2+, Fe3+, and Al3+ on amyloid-β stability, oligomerization, and aggregation: Amyloid-β destabilization promotes annular protofibril formation. J Biol Chem 2011; 286: 9646-9656.
[42] Tõugu V, Karafin A, Zovo K, Chung RS, Howells C, West AK, *et al.* Zn(II)- and Cu(II)-induced non-fibrillar aggregates of amyloid-β (1-42) peptide are transformed to amyloid fibrils, both spontaneously and under the influence of metal chelators. J Neurochem 2009; 110: 1784-1795.
[43] Yoshiike Y, Tanemura K, Murayama O, Akagi T, Murayama M, Sato S, *et al.* New Insights on How Metals Disrupt Amyloid β-Aggregation and Their Effects on Amyloid-β Cytotoxicity. J Biol Chem 2001; 276: 32293-32299.
[44] Loef M, Walach H. Copper and iron in Alzheimer's disease: A systematic review and its dietary implications. Br J Nutr 2012; 107: 7-19.
[45] Goodman L. Alzheimer's disease: a clinic-pathologic analysis of 23 cases with a theory on pathogenesis. J Nerv Ment Dis 1953; 118: 97-130.
[46] Jomova K, Valko M. Advances in metal-induced oxidative stress and human disease. Toxicology 2011; 283: 65-87.
[47] Schrag M, Crofton A, Zabel M, Jiffry A, Kirsch D, Dickson A, *et al.* Effect of cerebral amyloid angiopathy on brain iron, copper, and zinc in alzheimer's disease. J Alzheimer's Dis 2011; 24: 137-149.
[48] Brewer GJ. Iron and copper toxicity in diseases of aging, particularly atherosclerosis and Alzheimer's disease. Exp Biol Med 2007; 232: 323-335.
[49] Jiang D, Li X, Williams R, Patel S, Men L, Wang Y, *et al.* Ternary complexes of iron, amyloid-β, and nitrilotriacetic acid: Binding affinities, redox properties, and relevance to iron-induced oxidative stress in Alzheimer's disease. Biochemistry (Mosc) 2009; 48: 7939-7947.
[50] Bush AI. Metals and neuroscience. Curr Opin Chem Biol 2000; 4: 184-191.
[51] Gerlach M, Ben-Shachar D, Riederer P, Youdim MBH. Altered brain metabolism of iron as a cause of neurodegenerative diseases? J Neurochem 1994; 63: 793-807.
[52] Benzi G, Moretti A. Are reactive oxygen species involved in Alzheimer's disease? Neurobiol Aging 1995; 16: 661-674.

[53] Smith MA, Harris PLR, Sayre LM, Perry G. Iron accumulation in Alzheimer disease is a source of redox-generated free radicals. Proc Natl Acad Sci U S A 1997; 94: 9866-9868.
[54] Stankiewicz JM, Brass SD. Role of iron in neurotoxicity: A cause for concern in the elderly? Current Opinion in Clinical Nutrition and Metabolic Care 2009; 12: 22-29.
[55] Liu G, Men P, Perry G, Smith MA. Nanoparticle and iron chelators as a potential novel Alzheimer therapy. Methods in molecular biology (Clifton, NJ) 2010; 610: 123-144.
[56] Johannesson T, Kristinsson J, Torsdottir G, Snaedal J. Ceruloplasmin (Cp) and iron in connection with Parkinson's disease (PD) and Alzheimer's disease (AD). Laeknabladid, 2012; 98: 531-537.
[57] Kristinsson J, Snaedal J, Tórsdóttir G, Jóhannesson T. Ceruloplasmin and iron in Alzheimer's disease and Parkinson's disease: A synopsis of recent studies. Neuropsychiatric Disease and Treatment 2012; 8: 515-521.
[58] Bandyopadhyay S, Huang X, Lahiri DK, Rogers JT. Novel drug targets based on metallobiology of Alzheimer's disease. Expert Opinion on Therapeutic Targets 2010; 14: 1177-1197.
[59] Faa G, Crisponi G. Iron chelating agents in clinical practice. Coord Chem Rev 1999; 184: 291-310.
[60] Ding B, Chen K, Ling H, Sun F, Li X, Wan T, *et al.* Correlation of iron in the hippocampus with MMSE in patients with Alzheimer's disease. J Magn Res Imaging 2009; 29: 793-798.
[61] Tõugu V, Karafin A, Palumaa P. Binding of zinc(II) and copper(II) to the full-length Alzheimer's amyloid-β peptide. J Neurochem 2008; 104: 1249-1259.
[62] Faa G, Nurchi VM, Ravarino A, Fanni D, Nemolato S, Gerosa C, *et al.* Zinc in gastrointestinal and liver disease. Coord Chem Rev 2008; 252: 1257-1269.
[63] Powell SR. The antioxidant properties of zinc. J Nutr 2000; 130: 1447S-1454S.
[64] Domingo JL. Aluminum and other metals in Alzheimer's disease: A review of potential therapy with chelating agents. J Alzheimer's Dis 2006; 10: 331-341.
[65] Savory J, Exley C, Forbes WF, Huang Y, Joshi JG, Kruck T, *et al.* Can the controversy of the role of aluminum in Alzheimer's disease be resolved? What are the suggested approaches to this controversy and methodological issues to be considered? Journal of Toxicology and Environmental Health - Part A 1996; 48: 615-635.
[66] Bishop NJ, Morley R, Day JP, Lucas A. Aluminum neurotoxicity in preterm infants receiving intravenous-feeding solutions. N Engl J Med, 1997; 336: 1557-1561.
[67] Yokel RA. Brain uptake, retention, and efflux of aluminum and manganese. Environ Health Perspect 2002; 110: 699-704.
[68] Flendrig JA, Kruis H, Das HA. Aluminium and dialysis dementia. Lancet 1976; 1: 1235.
[69] Walton JR. Aluminum disruption of calcium homeostasis and signal transduction resembles change that occurs in aging and alzheimer's disease. J Alzheimer's Dis 2012; 29: 255-273.
[70] Yokel RA. Aluminium toxicokinetics: An updated minireview. Pharmacol Toxicol 2001; 88: 159-167.
[71] Campbell A, Bondy SC. Aluminum induced oxidative events and its relation to inflammation: a role for the metal in Alzheimer's disease. Cell Mol Biol 2000; 46: 721-730.
[72] Campbell A. The potential role of aluminium in Alzheimer's disease. Nephrol Dial Transplant 2002; 17: 17-20.
[73] Becaria A, Bondy SC, Campbell A. Aluminum and copper interact in the promotion of oxidative but not inflammatory events: Implications for Alzheimer's disease. J Alzheimer's Dis 2003; 5: 31-38.

[74] Campbell A. The role of aluminum and copper on neuroinflammation and Alzheimer's disease. J Alzheimer's Dis 2006; 10: 165-172.
[75] Walton JR. Evidence for participation of aluminum in neurofibrillary tangle formation and growth in Alzheimer's disease. J Alzheimer's Dis 2010; 22: 65-72.
[76] Obulesu M, Rao DM. Animal models of alzheimer's disease: An understanding of pathology and therapeutic avenues. Int J Neurosci 2010; 120: 531-537.
[77] Bondy SC. The neurotoxicity of environmental aluminum is still an issue. Neurotoxicology 2010; 31: 575-581.
[78] Tomljenovic L. Aluminum and Alzheimer's disease: After a century of controversy, is there a plausible link? J Alzheimer's Dis 2011; 23: 567-598.
[79] Wu Z, Du Y, Xue H, Wu Y, Zhou B. Aluminum induces neurodegeneration and its toxicity arises from increased iron accumulation and reactive oxygen species (ROS) production. Neurobiol Aging 2012; 33: 199.e191-199.e112.
[80] Crisponi G, Nurchi VM, Bertolasi V, Remelli M, Faa G. Chelating agents for human diseases related to aluminium overload. Coord Chem Rev, 2012; 256: 89-104.
[81] Crisponi G, Nurchi VM. Thermodynamic remarks on chelating ligands for aluminium related diseases. J Inorg Biochem 2011; 105: 1518-1522.
[82] Yokel RA, Rhineheimer SS, Sharma P, Elmore D, McNamara PJ. Entry, half-life, and desferrioxamine-accelerated clearance of brain aluminum after a single 26Al exposure. Toxicol Sci 2001; 64: 77-82.
[83] Exley C. A biogeochemical cycle for aluminium? J Inorg Biochem 2003; 97: 1-7.
[84] Kopacek J, Hejzlar J, Kana J, Norton SA, Porcal P, Turek J. Trends in aluminium export from a mountainous area to surface waters, from deglaciation to the recent: effects of vegetation and soil development, atmospheric acidification, and nitrogen-saturation. J Inorg Biochem, 2009; 103: 1439-1448.
[85] Exley C, Pinnegar JK, Taylor H. Hydroxyaluminosilicates and acute aluminium toxicity in fish. J Theor Biol 1997; 189: 133-139.
[86] Exley C, Schneider C, Doucet FJ. The reaction of aluminium with silicic acid in acidic solution: an important mechanism in controlling the biological availability of aluminium? Coord Chem Rev 2002; 228: 127-135.
[87] Akiyama H, Hosokawa M, Kametani F, Kondo H, Chiba M, Fukushima M, *et al.* Long-term oral intake of aluminium or zinc does not accelerate Alzheimer pathology in AβPP and AβPP/tau transgenic mice. Neuropathology 2012; 32: 390-397.
[88] Ribes D, Torrente M, Vicens P, Colomina MT, Gómez M, Domingo JL. Recognition memory and β-amyloid plaques in adult Tg2576 mice are not modified after oral exposure to aluminum. Alzheimer Dis Assoc Disord 2012; 26: 179-185.
[89] Willhite CC, Ball GL, McLellan CJ. Total allowable concentrations of monomeric inorganic aluminum and hydrated aluminum silicates in drinking water. Crit Rev Toxicol 2012; 42: 358-442.

Index

C

D

G

H

I

N

O

P

R

S

T

U

V

Z

Made in United States
Orlando, FL
04 November 2026

90205134R00278